Seong Uk
Min

Symmetries and Curvature Structure in General Relativity

World Scientific Lecture Notes in Physics

Published

Vol. 49: Field Theory, Disorder and Simulations
G Parisi

Vol. 50: The Quark Structure of Matter
M Jacob

Vol. 51: Selected Topics on the General Properties of Quantum Field Theory
F Strocchi

Vol. 52: Field Theory: A Path Integral Approach
A Das

Vol. 53: Introduction to Nonlinear Dynamics for Physicists
H D I Abarbanel, et al.

Vol. 54: Introduction to the Theory of Spin Glasses and Neural Networks
V Dotsenko

Vol. 55: Lectures in Particle Physics
D Green

Vol. 56: Chaos and Gauge Field Theory
T S Biro, et al.

Vol. 57: Foundations of Quantum Chromodynamics (2nd ed.)
T Muta

Vol. 59: Lattice Gauge Theories: An Introduction (2nd ed.)
H J Rothe

Vol. 60: Massive Neutrinos in Physics and Astrophysics
R N Mohapatra and P B Pal

Vol. 61: Modern Differential Geometry for Physicists (2nd ed.)
C J Isham

Vol. 62: ITEP Lectures on Particle Physics and Field Theory (In 2 Volumes)
M A Shifman

Vol. 64: Fluctuations and Localization in Mesoscopic Electron Systems
M Janssen

Vol. 65: Universal Fluctuations: The Phenomenology of Hadronic Matter
R Botet and M Ploszajczak

Vol. 66: Microcanonical Thermodynamics: Phase Transitions in "Small" Systems
D H E Gross

Vol. 67: Quantum Scaling in Many-Body Systems
M A Continentino

Vol. 69: Deparametrization and Path Integral Quantization of Cosmological Models
C Simeone

Vol. 70: Noise Sustained Patterns
Markus Loecher

Vol. 73: The Elmentary Process of Bremsstrahlung
W. Nakel and E. Haug

World Scientific Lecture Notes in Physics — Vol. 46

Symmetries and Curvature Structure in General Relativity

G S Hall
University of Aberdeen, UK

World Scientific

NEW JERSEY • LONDON • SINGAPORE • BEIJING • SHANGHAI • HONG KONG • TAIPEI • CHENNAI

Published by

World Scientific Publishing Co. Pte. Ltd.
5 Toh Tuck Link, Singapore 596224
USA office: Suite 202, 1060 Main Street, River Edge, NJ 07661
UK office: 57 Shelton Street, Covent Garden, London WC2H 9HE

British Library Cataloguing-in-Publication Data
A catalogue record for this book is available from the British Library.

SYMMETRIES AND CURVATURE STRUCTURE IN GENERAL RELATIVITY

ISBN 981-02-1051-5

Printed in Singapore.

Preface

A significant amount of relativistic literature has, in the past few decades, been devoted to the study of symmetries in general relativity. This book attempts to collect together the theoretical aspects of this work into something like a coherent whole. However, it should be stressed that it is not a textbook on those exact solutions of Einstein's equations which exhibit such symmetries; an excellent text covering these topics already exists in the literature. It is essentially a study of certain aspects of 4-dimensional Lorentzian differential geometry, but directed towards the special requirements of Einstein's general theory of relativity. Its main objective is to present a mathematical approach to symmetries and to the related topic of the connection and curvature structure of space-time and to pay attention to certain problems which arise and which worry mathematicians more than physicists. It is also hoped that certain of the chapters, for example those on manifold theory and holonomy groups, may serve as more or less self contained introductions to these important subjects.

The book arose out of many years work with my research students in the University of Aberdeen and I thus wish to record my gratitude to M.S. Capocci, W.J. Cormack, J. da Costa, R.F. Crade, B.M. Haddow, Hassan Alam, C.G. Hewitt, A.D. Hossack, S. Khan, M. Krauss, W. Kay, M. Lampe, D.P. Lonie, D.J. Low, L.E.K. MacNay, N.M. Palfreyman, M.T. Patel, M.P. Ramos, A.D. Rendall, M. Robertson, I.M. Roy, G. Shabbir, J.D. Steele, J. Sun, J.D. Whiteley and the late D. Negm. I would also like to thank the many friends and colleagues, both in the University of Aberdeen and elsewhere, who contributed in one way or another to the preparation of this book and, in particular, D. Alexeevski, A. Barnes, P. Bueken, J. Carot, M.C. Crabb, S. Hildebrandt, C.B.G. McIntosh, M. Sharif, Ibohal Singh, I.H. Thomson, E.H. Hall, S. Munro and the late C. Gilbert.

Finally I wish to record my sincere thanks to John Pulham, whose conversations on mathematics and physics and help in preparing this book have left me with a debt I can never repay, to Louise Thomson for her excellent typing of most of the manuscript, to the World Scientific Publishing Company for their infinite patience and to Aileen Sylvester for her unselfish understanding and support.

Contents

Chapter 1

Introduction

1.1 Geometry and Physics

The use of geometrical techniques in modelling physical problems goes back over two thousand years to the Greek civilisation. But for these people physics was a prisoner of geometry (not to mention of egocentricity) and, in particular, of the circle. So, for example, whilst the Ptolemaic geocentric model of the solar system was a powerful model for its time, the geometrical input was not only restricted by the knowledge and prejudices of the period, but provided only an arena in which physics took place. In this sense geometry was not part of the physics but merely a mathematical convenience for its description. In the scientific renaissance of the 16th century onwards the Greek dominance and prejudices were gradually eroded but even Copernicus still felt the need of the beloved circle and his heliocentric theory still needed epicycles. Not until Kepler came upon the scene was the circle finally dethroned when, by the introduction of eliptical orbits, he was able to explain all, and more, than Copernicus could, without the need of epicycles.

The Keplerian system, precisely stated, turns out to be equivalent to a model of the solar system of a Newtonian type based on the inverse square law attraction of a central force. This was one of the most outstanding achievements of Newtonian physics and the Kepler-Newton system has been the standard theory of the solar system ever since (the modifications arising from general relativity theory, although important theoretically, for example in the solving of the Mercury orbital problem, are minor in practice). But even in Newtonian mechanics the geometry of Euclid was still dominant and provided the unquestioned background in which all else took place.

The geometrical description of physics alluded to above has, however, a formal beauty in that it is perhaps to be considered remarkable that the Greek description of, for example, planetary events using one of the most important constructions in Euclidean geometry, the circle, gives a reasonably accurate picture of the solar system. The other important object in Euclid's geometry, the straight line, can also be thought of as having a physical context in the following simple way. Consider the Euclidean plane $\mathbb{R}^2$ with its usual system of "straight" lines etc. Now let f be a bijective continuous map $f : \mathbb{R}^2 \to \mathbb{R}^2$ (although continuity hardly matters here) and consider the set $\mathbb{R}^2$ now with the straight line structure understood to be the images of the original straight lines under f. Thus the straight lines in the new system are the subsets of $\mathbb{R}^2$ of the form $f(L)$ for all straight lines L in the original system. This set with this new linear structure formed by taking across from the original in an obvious way the concepts of length, angle etc under f is mathematically indistinguishable from the original and constitutes a perfectly good model of Euclidean geometry. However, one can distinguish them physically by, for example, regarding the copy of $\mathbb{R}^2$ as a horizontal surface and appealing to Galileo's law of inertia for free particles traversing this surface or to the paths of light beams crossing it (or, quite simply, to the lines of shortest distance measured by taut pieces of string held on the surface) to determine "straight lines". Each of these would presumably single out, at least in a local sense,the original copy of Euclid's geometry.

These examples may be considered early references to the interplay between geometry and physics but their true significance was presumably not fully realised until Hilbert set Euclidean geometry in proper perspective with his axiomatisation of this system (Gauss, Bolyai and Lobachevski having earlier confirmed the existence of alternative geometries). In this sense, these mathematicians played important roles in the understanding of physics.

Another example of the role of geometry in physics comes from the development in the 18th and 19th centuries of analytical mechanics and Lagrangian techniques. The work of D'Alembert, Euler, Lagrange, Hamilton and others cast the formulation of Newtonian mechanics into the geometry of the configuration space (or extended configuration space). From this, the calculus of variations, and the work of Riemann on differential geometry, Newtonian theory was in some sense "geometrised" by reformulating it in a setting of the Riemannian type. As a simple example, consider an n-particle system under a time independent conservative force described in the usual $3n$-dimensional Euclidean (configuration) space. The Euclidean nature (metric) of this space can be thought of as derivable from the kinetic energy of the unconstrained system in a standard way. Suppose now that a holonomic constraint is imposed on the system by restricting the particle to move on some subspace (submanifold) of the original Euclidean space. One could attempt the solve this problem by working in the submanifold of constraint. In doing so, this submanifold inherits a metric from the original Euclidean metric and which, in general, is no longer Euclidean but is rather a "curved metric" of the general type envisaged by Riemann. An elegant finale to the problem was then given by Routh's reduced Lagrangian (Routhian). The problem has in a sense been geometrised (and the constraint removed). Also the theory assumes a "generally covariant" form since the original inertial frame structure has been replaced by "generalised coordinates". A similar geometrisation occurs when one rewrites Maxwell's equations in generally covariant form by changing the usual Maxwell equations, written in an inertial frame in special relativity, by replacing the Minkowski metric that arises there by some general metric and partial derivatives by covariant derivatives.

However, there is a criticism of the claim that this procedure has progressed further with the geometrising of the problem than the earlier examples discussed above. The geometry in these examples really entered the proceedings with the imposition of the original Euclidean structure (and, in the first example, before the constraints were introduced). Thus the geometrical structure was on the arena initially and is again merely a convenience (albeit a useful one) for the solution of the problem. Similar comments apply to the more sophisticated geometrical approaches to mechanics inaugurated by Jacobi and Cartan (the latter albeit after the advent of Einstein's general relativity). In this sense, the geometry was given a-priori and was not subject to any restrictions ("field equations"). Thus it invites the criticism that it can influence the physics without any reciprocal influence on itself and was subject to unfavourable philosophical scrutiny by amongst others, Berkeley [1] and Mach [2]. Also the introduction of "absolute" variables such as the background Euclidean metric in classical mechanics and Maxwell theory (essentially Newton's absolute space) seems to suggest that making a theory "covariant" is a rather trivial matter. One more or less writes down the theory in its original "Euclidean" form and then allows everything to transform as a tensor when moving to some other coordinate system. This problem was apparently first recognised by Kretchmann [3] and elaborated on by Anderson [4] and Trautmann [5]).

The advantage of general relativity is that it contains the space-time metric as a "dynamical" variable (to be determined by solving field equations) and that it contains no such "absolute" variables (except one must, perhaps, concede that the imposition of zero torsion on the Levi-Civita connection derived from the space-time metric amounts to making the torsion an absolute variable in this theory [5]. In this stronger sense general relativity is "generally covariant". On the other hand, Newtonian classical theory has an absolute space-time splitting into absolute space and absolute time together with privileged (inertial) observers and, in addition, a privileged frame in which the ether is at rest if classical electromagnetic theory is to be accomodated within it. Special relativity has an absolute space-time metric and privileged inertial observers whereas, for example, the "bimetric" and "tetrad" variants of Einstein's theory have absolute variables in the form of a flat metric [6] or a privileged tetrad system [7], respectively. Classical theory with its concept of force has to be able to distinguish between "real" forces (*e.g.* the gravitational attraction of one body on another) and the so called "fictitious" (accelerative) forces which arise in non-inertial frames. The ability to distinguish between these types of force is essentially the ability to distinguish between inertial and non-inertial frames. Newton's theory claims each of these abilities (and thus introduces absolute variables) and Einstein's claims neither. Thus there are no a priori distinguished reference frames (coordinate systems) in general relativity (although there are, of course, convenient ones!) and no concept of force (beyond that required in order to describe situations in relativity theory using Newtonian language!) Such remarks as these go under the name of the principal of covariance and its role in the foundations of general relativity is by no means agreed upon. A related (and also contentious) issue is the principle of equivalence. In its strong form it advocates the (local) indistinguishability of a gravitational field and an appropriately chosen acceleration field. In its weaker form it simply reiterates the results of the Eotvos type experiments thus effectively saying that there is a unique symmetric connection in space-time for determining space-time paths (this latter statement leaving room for curvature coupling terms). These topics will not be discussed any further here except to say that they can be used to suggest that a theory of gravitation could be based on a 4-dimensional manifold and whose dynamical variable is a metric of Lorentz signature which is determined by field equations of a tensorial nature. Such a theory is Einstein' general relativity and it will be accepted, henceforth, without question.

Einstein's general theory of relativity is the most successful theory of the gravitational field so far proposed. It describes the gravitational field by a mathematical object called a space-time and is formalised as a pair (M, g) where M is a certain type of 4-dimensional manifold (see chapter 7) and g is a Lorentz metric on M. The metric g together with its connection and curvature "represent" the gravitational field. The restrictions on g are the Einstein field equations (together with boundary and other initial conditions) and are ten second order partial differential equations for g. Since the theory was first published in 1916, it has progressed through early uncertain beginnings, when it played second fiddle to quantum theory and suffered from the "cosmological time problem" (until the latter's correction), to a renaissance in the last fifty years. The problem of the lack of exact solutions to Einstein's field equations has, to some extent, been overcome. In addition, the theory has been put on a much sounder footing both mathematically and physically.

1.2 Preview of Future Chapters

This book is not a text book on general relativity. There are numerous excellent such texts available and thus no point in further duplication. Rather, it concentrates first on the topics of the connection and curvature structure of space-times and then, later, on the various symmetries that are commonly studied in Einstein's theory. In spite of this, an attempt will be made to make the book in some restricted sense, self-contained and so the chapters on certain prerequisite branches of mathematics will sometimes contain a little more than is strictly demanded by the remainder of the text. But the excess over absolute necessity will be restricted to that required for sensible self-containment. The reader will be assumed familiar with basic mathematics although, as stated below, algebra, topology, manifold theory and Lie groups will be treated *ab-initio*. However, although general relativity will be formally introduced, some basic familiarity with it will inevitably have to be assumed. This will mainly take the form of assuming that the reader has some knowledge of the better known exact solutions of Einstein's equations such as the Schwarzschild, Riessner-Nordstrom, Friedmann-Robertson-Walker and plane wave metrics. These metrics can be found treated in detail in many places and references will be given when appropriate. Also, some elementary knowledge of the symmetries (Killing vector fields) they possess will be desirable but not necessary.

Chapter 2 is a review of elementary group theory, linear algebra and the Jordan canonical form. The classification of matrices using Jordan-Segre theory is rather useful in general relativity and is treated in some detail. It is used in several different forms in chapter 7. Chapter 3 gives, along similar lines, a summary of elementary topology. This latter topic is still, unfortunately, used rather sparingly in certain branches of relativity theory. It is used in this text only in a somewhat primitive way but the advantages gained seem, at least to the author, to make it worthwhile. It consists of mainly point set topology together with a brief discussion of the fundamental group and the "rank" theorem. An attempt is made to ensure that these briefest of introductions are sufficient not only for that which is required later but also for some basic understanding of the subject matter.

Chapter 4 is a lengthy chapter on manifold theory. This starts, not surprisingly, with the definition of a manifold and its topology. Then the various mathematical objects required for the study of general relativity are introduced such as the tensor bundles and tensor fields, vector fields and their integral curves, submanifolds, distributions and metrics together with their associated Levi-Civita connections and curvature and Weyl tensors. Although this chapter will introduce the "coordinate free" approach (and this will be used occasionally in the text where convenient) the use of coordinate expressions will be exploited where appropriate. In this respect, the important thing is to recognise when a coordinate-free (or component) expression is unnecessarily clumsy. But this, in many cases, merely reflects personal preference. The calculations in this text are mostly in coordinate notation. In fact, on occasions one will find both used in the same calculation if, for some reason, this led to an economy of expression. This chapter also contains a discussion of some of the troublesome properties of certain types of submanifolds and which is needed in later chapters.

In chapter 5 the concept of a Lie group and its Lie algebra is introduced. The main aims here are firstly to introduce notation for the idea of a (local and global) transformation group and the associated Lie algebras of vector fields on a manifold and secondly to prepare the way for the discussion of the Lorentz group in chapter 6. In chapter 5 a discussion is given of Palais' theorem on the necessary and sufficient conditions for a Lie algebra of vector fields on a manifold to be regarded as arising from a global Lie group action. Such a global action is usually assumed in the literature without justification. In chapter 6 an attempt is made to promote the usefulness of a reasonable knowledge of the Lorentz group beyond that of merely something which occurs in special relativity. This group is investigated both algebraically and also as a Lie group. Its (connected) Lie subgroups are listed and their properties derived.

In chapter 7, general relativity is finally introduced. Here a brief summary of the properties of a space-time are laid down and Einstein's equations are given. The energy-momentum tensors for the "standard" gravitational fields encountered in general relativity are described. There then follows sections on algebraic classification theory on a space-time. Thus the classification of bivectors, the Petrov classification of the Weyl tensor and the classification of second order symmetric tensors (usually the energy-momentum tensor) are described in some detail. Heavy use will be made of them in later chapters. Here the Jordan-Segre theory developed in chapter 2 is justified. The chapter concludes with some comments on the topological decomposition of a space-time with respect to the algebraic types of the Weyl and energy-momentum tensors and the local and global nature of these classification schemes are described.

Chapter 8 is on holonomy theory. Its introduction can be regarded as two-fold. Firstly, the techniques derived from it are useful elsewhere in later chapters and secondly, it provides a classification of space-times which (unlike the ones in chapter 7) is not pointwise but applies to the space-time as a whole. Unfortunately (and not unlike those of chapter 7) it is somewhat too coarse in places. However, it displays the curvature structure clearly and its usefulness in later descriptions of symmetries justifies its inclusion.

Chapter 9 concentrates on the general relations between the metric and its Levi-Civita connection and associated curvature and sectional curvature functions. Thus the problem of the extent to which the prescription of one of these objects determines any other is studied. Holonomy theory is useful here as is a convenient classification of the curvature tensor which is developed in this chapter. In this section the sectional curvature function is discussed and its (generic) equivalence to the metric tensor displayed. This raises the prospect of regarding the sectional curvature function as an alternative field variable in general relativity, at least for vacuum space-times.

The remaining four chapters of the book are on symmetries in general relativity. The idea here is to present some techniques for studying such symmetries and the presentation will be guided by an attempt to achieve a certain reasonable level of rigour and elegance. These symmetries are defined in terms of local transformations and are then described in terms of certain families of vector fields. The emphasis will be placed on technique rather than a multitude of examples, although the salient points will be exemplified. Here, some elementary knowledge of Killing vector theory in the more well known exact solutions is desirable but not necessary. The essential philosophy behind these sections is the theory and the finding of symmetries. The symmetries treated are those described by Killing, homothetic and affine vector fields (chapter 10), conformal vector fields (chapter 11), projective vector fields (chapter 12) together with symmetries of the curvature tensor (curvature collineations) which will be treated in chapter 13. Of some importance in these matters is the study of the zeros of such vector fields (that is the "fixed points" of the associated local transformations) and the description of the consequent orbit structure that they exhibit. Sections on the zeros of Killing, homothetic and conformal vector fields and on the orbit structure of Killing vector fields are given in the relevant chapters. This is applied, in chapter 11, to the study of the conformal reduction of conformal vector fields to Killing and homothetic vector fields and the "linearisation" problem. Holonomy theory is used significantly in the treatment of proper affine and projective symmetry.

The general approach of this book is geometrical, in keeping with the spirit of general relativity. If a geometrical argument could be found then it was used in place of a (usually more) lengthy calculation. Unfortunately the author was sometimes unable to find an elegant argument and thus in these cases a clumsy calculation must suffice. Also, if a certain result depended on a space-time M satisfying a certain algebraic type (*e.g.* its Petrov type), then rather than assume that M had that Petrov type everywhere, M was topologically decomposed into a union of open subsets in each of which the Petrov type was constant (together with a nowhere dense "leftover" subset) and an attempt was then made to prove that result more generally. The author apologises if some of the proofs seem unnecessarily pedantic. This is simply his way of avoiding "cheating at patience" and, in many cases, is probably nothing more than a statement of his ignorance of a better proof. A similar apology is offered for the terse nature of some of the arguments. Conservation of space prevented more than this but it is hoped that enough is given for the details to be traced. This lack of space also prevents a proper discussion of the history of the topics treated. One unfortunate but inevitable consequence of this is the omission of many references and yet another apology must be offered.

The notation used is a fairly standard one with references given numerically in square brackets. Sections, theorems and equations are numbered within each chapter in an obvious way with equation numbers in round brackets. In chapter 2 vectors are introduced in bold face type as is the usual convention. However, later in the text when many other geometrical objects are brought into play, this procedure is dropped. When the Σ symbol is used for summation and the limits are obvious, they are sometimes omitted (and, of course, for tensor notation, the Einstein summation convention is eventually adopted). The end of a proof is denoted $\Box$.

Chapter 2

Algebraic Concepts

2.1 Introduction

This chapter will be devoted to a brief survey of those topics in group theory and linear algebra required for what is to follow. In the interests of completeness and understanding, however, a little more will be covered than is necessary. Because of the large number of excellent texts on these topics [8]–[13] it is necessary only to define those terms used and to state the important theorems with reference to places where the proofs can be found. The main topics dealt with are elementary group theory, vector spaces, linear (and multilinear) algebra and transformations, dual spaces, inner products and the Jordan canonical form. The sets of real and complex numbers are denoted by $\mathbb{R}$ and $\mathbb{C}$ respectively, $\imath^2 = -1$ and a bar over a symbol denotes complex conjugation. The set of all integers is denoted by $\mathbb{Z}$, the set of positive integers by $\mathbb{Z}^+$ and the set of rational numbers by $\mathbb{Q}$. The notation of elementary set theory is assumed. If X and Y are sets, $A \subseteq X$ and f is a map $f : X \to Y$ the restriction of f to A is denoted by $f_{|A}$ and the restriction of the identity map $\imath : X \to X$ to A is then the *inclusion* map $\imath_{|A}$. The empty set is $\emptyset$. If X is a set, a family $\{A_i : i \in I\}$ of disjoint subsets of X, for some indexing set I, satisfying $X = \bigcup_{i \in I} A_i$ is called a *partition* or a *disjoint decomposition* of X.

2.2 Groups

A *group* is a pair $(G, \cdot)$ where G is a non-empty set and $\cdot$ a binary operation on G, that is, a map $G \times G \to G$, $(a, b) \to a \cdot b$, $a, b \in G$, such that

(G1) $a \cdot (b \cdot c) = (a \cdot b) \cdot c$ (the *associative law* for $\cdot$).

(G2) there exists $e \in G$ such that $a \cdot e = e \cdot a = a \quad \forall a \in G$.
(G3) for every $a \in G$ there exists an element $a^{-1} \in G$ such that
$a \cdot a^{-1} = a^{-1} \cdot a = e$.

In any group G the element e is called the *identity* of G and is *unique*. For any $a \in G$ the element a^{-1} in (G3) is also unique and is called the *inverse* of a. Necessarily $(a^{-1})^{-1} = a$ and $(a \cdot b)^{-1} = b^{-1} \cdot a^{-1}$. The group $(G, \cdot)$ (written simply as G if the group operation $\cdot$ is clear) is called *abelian* or *commutative* if $a \cdot b = b \cdot a$, $\forall a, b \in G$. With the group operation (sometimes called the group product) given, $a \cdot b$ will usually be written as ab. If G is an infinite set it is sometimes said to have *infinite order*. If G is a finite set with n members it is said to have *order* n.

If H is a subset of G such that H, together with the (obvious) inherited operation $\cdot$ from G, is a group then H is called a *subgroup* of G and one writes $H < G$. It follows that the identities of H and G coincide and that the inverse of a member $a \in H$ is the same whether taken in the group H or in the group G. It is then clear that a subset H of G is a subgroup of G if and only if, given any $a, b \in H$ then $ab \in H$ and $a^{-1} \in H$ or, alternatively, if and only if given any $a, b \in H$ then $ab^{-1} \in H$. Clearly $\{e\} < G$ and $\{e\}$ is called the *trivial* subgroup.

Some standard examples of groups are:

i) the group of complex numbers $\mathbb{C}$ with binary operation the usual addition operation and with identity 0 (so that the inverse of $z \in \mathbb{C}$ is $-z$), of which the subsets of real numbers $\mathbb{R}$, rational numbers $\mathbb{Q}$ and integers $\mathbb{Z}$ form subgroups,
ii) the group of non-zero complex numbers $\mathbb{C}\backslash\{0\}$ under the usual multiplication operation and with identity 1 of which the non-zero reals $\mathbb{R}\backslash\{0\}$, the positive reals $\mathbb{R}^+$, the non-zero rationals $\mathbb{Q}\backslash\{0\}$ and the complex numbers of unit modulus form subgroups,
iii) the group of all $m \times n$ complex matrices $M_{m\times n}\mathbb{C}$ (written $M_n\mathbb{C}$ if $m = n$) under the usual matrix addition operation and with identity the zero $m \times n$ matrix, of which the set $M_{m\times n}\mathbb{R}$ (written $M_n\mathbb{R}$ if $m = n$) of all real $m \times n$ real matrices forms a subgroup,
iv) the group $GL(n, \mathbb{C})$ of all $n \times n$ complex non-singular matrices under the usual matrix multiplication and with identity the $n \times n$ unit matrix, of which the subset $GL(n, \mathbb{R})$ of $n \times n$ real non-singular matrices and the subset $SL(n, \mathbb{R})$ of $n \times n$ real matrices with unit determinant form subgroups,
v) the set $G = \{e, a, b, c\}$ with binary operation defined by allowing e the

properties of the identity and $aa = bb = cc = e$, $ab = ba = c$, $ac = ca = b$ and $bc = cb = a$ (the Klein 4-group).

If G_1 and G_2 are groups with group operations $\cdot$ and $\times$ respectively, a map $f : G_1 \to G_2$ is called a (group) *homomorphism* of G_1 into G_2 if for each $a, b \in G_1$ $f(a \cdot b) = f(a) \times f(b)$ or, if there is no risk of ambiguity, $f(ab) = f(a)f(b)$. It follows that if e is the identity of G_1 then $f(e)$ is the identity of G_2 and that for each $a \in G_1$, $f(a^{-1})$ is the inverse in G_2 of $f(a)$. If f is *bijective*, *i.e.* if it is both *injective* (*i.e.* one-to-one) *and surjective* (*i.e.* onto) then it follows that f^{-1} is a homomorphism from G_2 to G_1 and under these circumstances f is called a (group) *isomorphism* from G_1 to G_2 (or between G_1 and G_2) and the groups G_1 and G_2 are said to be *isomorphic*. If G is a group and $f : G \to G$ is an isomorphism then f is called a (group) *automorphism* . If G_1 and G_2 are groups and $f : G_1 \to G_2$ is a homomorphism then the set $f(G_1) = \{f(g) : g \in G_1\}$ is a subgroup of G_2 and if e_2 is the identity of G_2 the set $\{g \in G_1 : f(g) = e_2\}$, called the *kernel* of f, is a subgroup of G_1. The kernel of f is identical to the trivial subgroup of G_1 if and only if f is one-to-one.

Let G be a group and let S be a subset of G. The family of all (finite) products of those members of G which are either the identity of G, a member of S or the inverse of a member of S is a subgroup of G containing S called *the subgroup of G generated by S*. This subgroup is just the intersection of all subgroups of G containing S and is thus the smallest subgroup of G containing S, that is, it is a subgroup of any subgroup of G containing S. If S contains a single member $g \in G$ then the subgroup generated by S is called *cyclic*.

Let G be a group and let $H < G$. If $g \in G$, a *right coset of H in G* is the subset of G denoted by Hg and defined by $Hg = \{hg : h \in H\}$. A *left coset gH* of H in G is defined similarly by $gH = \{gh : h \in H\}$. The right cosets (respectively the left cosets) of H in G constitute a partition of G arising from the equivalence relation $\sim$ on G given for $a, b \in G$ by $a \sim b \Leftrightarrow ab^{-1} \in H$ (respectively by $a \sim b \Leftrightarrow a^{-1}b \in H$). Thus a and b are in the same right coset (respectively left coset) of H in G if and only if $ab^{-1} \in H$ (respectively $a^{-1}b \in H$) which is equivalent to $Ha = Hb$ (respectively $aH = bH$). Thus $Ha = H \Leftrightarrow aH = H \Leftrightarrow a \in H$. The map $Hg \to g^{-1}H$ is, for each $g \in G$, easily seen to be a well defined map from right to left cosets of H in G which is bijective and so the families of right and left cosets of H in G are in one-to-one correspondence. The sets of left and right cosets of H in G are denoted, respectively, by $L(G, H)$

and $R(G, H)$. If $g \in G$, the cosets Hg and gH are not necessarily equal. If $Hg = gH, \forall g \in G$, H is called a *normal* (or *invariant*) subgroup of G and one writes $H \triangleleft G$. Thus H is a normal subgroup of G if and only if $g^{-1}Hg \equiv \{g^{-1}hg : h \in H\} = H$. It follows that every subgroup of an abelian group is normal. One can approach the idea of a normal subgroup in the following alternative way. Let A and B be subsets of the group G and define the product of A and B by $A \cdot B = \{ab : a \in A, b \in B\}$. One can then ask, for a subgroup H of G, if the right (or left) cosets of H in G form a group under this product. A necessary condition is that if $a, b \in G$ then $Ha \cdot Hb = Hc$ (or $aH \cdot bH = dH$) for some $c, d \in G$. It then follows from the above remarks that this last condition on the right (or the left) cosets of H in G is equivalent to H being a normal subgroup of G. The rest of the group axioms then follow and show that the right cosets (which are the same as the left cosets) of H in G form a group under this product with identity H and inverse operation $(Ha)^{-1} = Ha^{-1}$ for each $a \in G$. it also follows that $Ha \cdot Hb = Hab, \forall a, b \in G$. The group formed by the left cosets is identical since $H \triangleleft G \Rightarrow Hg = gH$, $\forall g \in G$. The resulting group is called the *quotient* (or *factor*) *group of G by H (or of H in G)* and is denoted by G/H.

If G is a group and H is a normal subgroup of G the map $f : G \to G/H$ given by $g \to Hg$ is a homomorphism (called the natural homomorphism) of G onto G/H and its kernel is easily seen to be H. This fact can be easily reformulated in the following slightly more general way. Let G_1 and G_2 be groups and $f : G_1 \to G_2$ a homomorphism. Then the kernel K of f is a *normal* subgroup of G_1 and G_1/K is isomorphic to $f(G_1)$, the isomorphism being given by the (well-defined) map $Kg \to f(g)$ for $g \in G_1$. If the homomorphism f is onto G_2 then G_1/K and G_2 are isomorphic.

If H is a subgroup of the group G and if $g \in G$ the set $g^{-1}Hg \equiv \{g^{-1}hg : h \in H\}$ is easily seen to be a subgroup of G and is said to be *conjugate to* H. It is clearly isomorphic to H under the isomorphism $h \to g^{-1}hg$. The subgroups of G conjugate to H all coincide with H if and only if H is a *normal* subgroup of G.

Now suppose that $G_1, \cdots G_n$ are groups. The *direct product* $G_1 \otimes \cdots \otimes G_n$ of these groups is the Cartesian product $G_1 \times \cdots \times G_n$ together with the binary operation $(a_1, \ldots, a_n) \cdot (b_1, \ldots, b_n) = (a_1b_1, \ldots, a_nb_n)$ where $a_i, b_i \in G_i$ $(i \leqslant i \leqslant n)$ and is easily found to be a group in an obvious way.

As a general remark one notes that the binary operation for a group G is written $(a, b) \to ab$ $(a, b \in G)$. Suppose one defines a binary operation $(a, b) \to ba$. Clearly this also leads to a group structure for G with the same identity and the same inverse for each member of G. Also the two resulting group structures are equal if and only if the original structure (and hence both structures) are abelian. In general they are not equal but they are isomorphic under the map $f : G \to G$ given by $f(g) = g^{-1}$. This follows since, if the respective operations are denoted by $\cdot$ and $\times$, $f(a \cdot b) = b^{-1} \cdot a^{-1} = f(b) \cdot f(a) = f(a) \times f(b)$.

2.3 Vector Spaces

A *field* is a triple $(F, +, \cdot)$ where F is a non-empty set and $+$ and $\cdot$ are binary operations on F such that

(F1) $(F, +)$ is an abelian group with identity element 0,
(F2) $(F \setminus \{0\}, \cdot)$ is an abelian group with identity element 1,
(F3) the operation $\cdot$ is *distributive* over $+$, that is

$$a \cdot (b + c) = a \cdot b + a \cdot c \qquad \forall a, b, c \in F.$$

The operation $+$ is called *addition* and $\cdot$ *multiplication* and so 0 and 1 are referred to as the *additive and multiplicative identities*, respectively. Then for $a \in F$ the additive inverse is written $-a$ and, if $a \neq 0$, the multiplicative inverse is written a^{-1}. The members 0 and 1 are not allowed to be equal by the axioms and so F contains a least two members. In fact the axioms allow F to be finite or infinite. Examples of fields are the rationals $\mathbb{Q}$, the reals $\mathbb{R}$ and the complex numbers $\mathbb{C}$ with the usual addition and multiplication in each case. In this book only the fields $\mathbb{R}$ and $\mathbb{C}$ will be required. It is usual to omit the $\cdot$ in writing the second of these operations provided this causes no confusion. A subset F' of a field F which is a field under the inherited operations $+$ and $\cdot$ from F is called a *subfield* of F. The identity elements of F' and F coincide.

A *vector space V over a field F* is an abelian group $(V, \oplus)$ and a field $(F, +, \cdot)$ together with an operation $\odot$ of members of F on members of V such that if $a \in F$ and $\mathbf{v} \in V$ then $a \odot \mathbf{v} \in V$ and which satisfy

(V1) $(a + b) \odot \mathbf{v} = a \odot \mathbf{v} \oplus b \odot \mathbf{v}$,
(V2) $a \odot (\mathbf{u} \oplus \mathbf{v}) = a \odot \mathbf{u} \oplus a \odot \mathbf{v}$,
(V3) $(a \cdot b) \odot \mathbf{v} = a \odot (b \odot \mathbf{v})$,
(V4) $1 \odot \mathbf{v} = \mathbf{v}$,

for $a, b \in F$, $\mathbf{u}, \mathbf{v} \in V$ and where 1 is the multiplicative identity of F. This proliferation of symbols can and will be simplified without causing any ambiguity by writing $+$ for both $+$ and $\oplus$ above and by omitting the symbols $\cdot$ and $\odot$. The members of V are called *vectors* , the members of F *scalars* and the operation $\odot$ *scalar multiplication*. Although the theory discussed here will apply to all vector spaces this book will only really be concerned with vector spaces over $\mathbb{R}$ (*real vector spaces*) and over $\mathbb{C}$ (*complex vector spaces*). It can now be deduced from the above axioms that if $\mathbf{0}$ is the identity of V (referred to as the *zero vector* of V) and 0 the additive identity of F then for $a \in F$ and $\mathbf{v} \in V$,

i) $a\mathbf{0} = \mathbf{0}$,
ii) $0\mathbf{v} = \mathbf{0}$,
iii) $(-a)\mathbf{v} = -(a\mathbf{v})$, (where the minus sign denotes additive inverses in both the groups $(F+)$ and V)
iv) if $\mathbf{v} \neq \mathbf{0}$ then $a\mathbf{v} = \mathbf{0} \Rightarrow a = 0$.

The most common examples of vector spaces are obtained when $V = \mathbb{R}^n$ and $F = \mathbb{R}$ or when $V = \mathbb{C}^n$ and $F = \mathbb{C}$ and where in each case addition in V and scalar multiplication by members of F are the standard componentwise operations. These will be referred to as the *usual* (vector space) *structures* structures on $\mathbb{R}^n$ and $\mathbb{C}^n$.

If V is a vector space over F and W is a subset of V such that, together with the operations on W naturally induced from those on V and those on V and F, W is a vector space over F then W is called a *subspace* of V. An equivalent statement is that W is a subspace of V if for each $\mathbf{u}, \mathbf{v} \in W$ and $a, b \in F$, $a\mathbf{u} + b\mathbf{v}$ (which is a well defined member of V) is a member of W. Clearly $\{0\}$ is a subspace of V called the *trivial subspace*. If W is a subspace of V the identity (zero) element and the operation of taking additive inverses are identical in W and V. It should be remarked here in this context that the fact that a subset of V is a subgroup of V is not sufficient to make it a subspace of V. In fact, the relation between a vector space and its associated field of scalars is rather subtle and leads to the theory of extension fields (see, for example, [12]). For example, $\mathbb{R}$ with its usual structure is a real vector space but $\mathbb{Q}$, which is a subgroup of $\mathbb{R}$, is not a subspace of $\mathbb{R}$. However, since $\mathbb{Q}$ is a subset of $\mathbb{R}$ which induces from $\mathbb{R}$ the structure of a field (*i.e.* $\mathbb{Q}$ is a *subfield* of $\mathbb{R}$) $\mathbb{R}$ may be regarded as a vector space *over* $\mathbb{Q}$ and then $\mathbb{Q}$ is a subspace of $\mathbb{R}$. But in this case $\mathbb{R}$ is not finite dimensional (dimension will be defined later).

Let U and V be vector spaces over F. A map $f : U \to V$ is called a *homomorphism* (of vector spaces) or a *linear transformation* (or just a *linear map*) if for each $\mathbf{u}, \mathbf{v} \in V$ and $a \in F$

i) $f(\mathbf{u}+\mathbf{v}) = f(\mathbf{u}) + f(\mathbf{v})$
ii) $f(a\mathbf{u}) = af(\mathbf{u})$

where an obvious abbreviation of notation has been used. These two conditions can be combined into a single condition equivalent to them and which is that for any $\mathbf{u}, \mathbf{v} \in V$ and $a, b \in F$, $f(a\mathbf{u} + b\mathbf{v}) = af(\mathbf{u}) + bf(\mathbf{v})$. If f is bijective then it easily follows that $f^{-1} : V \to U$ is necessarily linear and in this case f is called an *isomorphism* (of vector spaces) and U and V are then *isomorphic* vector spaces. At the other extreme, the map which takes the whole of U to the zero member of V is also linear (and is called the *zero* linear map). For a general linear transformation $f : U \to V$, the subset $\{\mathbf{u} \in U : f(\mathbf{u}) = \mathbf{0}\}$ is a *subspace* of U called the *kernel* of f whilst the subset $\{f(\mathbf{u}) : \mathbf{u} \in U\}$ is a subspace of V called the *range* of f. For any vector spaces U and V over the same field F if f and g are linear maps from U to V and $a \in F$, $\mathbf{u} \in U$ one can define linear maps $f + g$ and af from U to V by $(f + g)(\mathbf{u}) = f(\mathbf{u}) + g(\mathbf{u})$ and $(a \cdot f)(\mathbf{u}) = af(\mathbf{u})$. It then follows easily that the set of all linear maps from U to V is itself a vector space over F denoted by $L(U, V)$. Also if U, V and W are vector spaces over the same field F and if $f : U \to V$ and $g : V \to W$ are linear then $g \circ f : U \to W$ is linear.

If V is a vector space over F and W is a subspace of V then, from what has been said above, W is a normal subgroup of the abelian group V and so one can construct the (abelian) quotient group V/W. The members of this quotient group are written $\mathbf{v} + W$ for $\mathbf{v} \in V$ and this group can be converted into a vector space over F by defining $a(\mathbf{v} + W) = a\mathbf{v} + W$ and checking that this operation is well-defined. The vector space V/W is called the *quotient (vector) space of V by W*. It follows easily that if U and V are vector spaces over F and $f : U \to V$ a surjective homomorphism then V is isomorphic to the quotient space of U by the kernel of f. Also, if W is a subspace of the vector space V there exists a homomorphism of V onto V/W (i.e. the map $\mathbf{v} \to \mathbf{v} + W$, $\mathbf{v} \in V$).

Associated with any vector space are the important concepts of *basis* and *dimension* and a brief discussion of these ideas can now be given. It should be remarked before starting this discussion that one often requires sums of members of a vector space. Clearly, within the algebraic structure defined, one can only make sense of *finite* sums. If V is a vector space over F and if $\mathbf{v}_1, \ldots, \mathbf{v}_n \in V$, $a_1, \ldots, a_n \in F$, $(n \in \mathbb{Z}^+)$, the member $\sum_{i=1}^{n} a_i \mathbf{v}_i$ of V is called a *linear combination* (over F) of $\mathbf{v}_1, \ldots, \mathbf{v}_n$. If S is a non-empty subset of V the set of all linear combinations of (finite) subsets of S is called the (*linear*) *span* of S and denoted by $Sp(S)$. It easily follows that $Sp(S)$ is a subspace of V. Next, a non-empty subset S of V is called *linearly independent* (over F) if given any $\mathbf{v}_1, \ldots, \mathbf{v}_n \in S$ the equation $a_1\mathbf{v}_1 + \cdots + a_n\mathbf{v}_n = \mathbf{0}$ for $a_1, \ldots, a_n \in F$ has only the solution $a_1 = \cdots = a_n = 0$. Otherwise S is called *linearly dependent* (over F). Clearly if S is linearly independent and $\mathbf{v} \in V$ is a linear combination of the members $\mathbf{v}_1, \ldots, \mathbf{v}_n$ of S, $\mathbf{v} = \sum_{i=1}^{n} a_i\mathbf{v}_i$, then the members $a_1, \ldots a_n \in F$ are uniquely determined. It is also clear that any subset of a linearly independent subset of V is itself linearly independent. If T is a subset of V with the property that $V = Sp(T)$ then T is called a *spanning set* for V (or T *spans* V). Of particular interest is the case when T not only spans V but is also linearly independent. It is a very important result in vector space theory that *given any vector space V over any field F there exists a subset S of V such that S is a linearly independent spanning set for V*. Such a subset S is called a *basis* for V and the proof of the existence of S requires Zorn's lemma (*i.e.* the axiom of choice—see *e.g.* [13]). In fact if S_1 is a linearly independent subset of V and S_2 a spanning set for V with $S_1 \subseteq S_2$ then there exists a basis S for V such that $S_1 \subseteq S \subseteq S_2$. It follows that any linearly independent subset of V can be enlarged to a basis for V and that any spanning set for V can be reduced to a basis for V. A vector space V over F which admits a *finite* spanning subset is called *finite dimensional (over F)* and in such a case the existence of a finite basis for V is easily established without appeal to Zorn's lemma. In fact it then follows that the number of members in any basis for V is the same positive integer and this number is called the *dimension* of V over F. For a finite dimensional vector space V over F of dimension n, any spanning set or linearly independent subset of V which consists of n members is necessarily a basis for V. Also the obvious uniqueness of the member $(a_1, \ldots, a_n) \in F^n = F \times \cdots \times F$ associated with each $\mathbf{v} \in V$ when expressed in terms of a fixed basis $\{\mathbf{v}_i\} \equiv \{\mathbf{v}_1, \ldots, \mathbf{v}_n\}$ for V (*i.e.* $\mathbf{v} = \sum_{i=1}^{n} a_i\mathbf{v}_i$) shows that V is isomorphic to the vector space F^n over F with the usual componentwise operations. (The scalars $a_1, \ldots, a_n$ are called the *components* of $\mathbf{v}$ with respect to the basis $\{\mathbf{v}_i\}$.) Thus any two finite-dimensional vector spaces over the same field and of the same dimension are isomorphic. If V is not finite-dimensional (over F) it is called *infinite-dimensional* (*over F*).

It should be remarked at this point that if the same abelian group V can be given vector space structures over different fields F_1 and F_2 the concepts of linear independence, basis, dimension *etc* can be quite different for the vector spaces V over F_1 and V over F_2. If the field F over which V is considered as a vector space is clear and if V is then finite dimensional, the dimension of V is written $\dim V$. In this case any subspace W of V is finite dimensional and $\dim W \leqslant \dim V$ and $\dim V/W = \dim V - \dim W$. Any field F is a 1-dimensional vector space over itself and is isomorphic to any (1-dimensional) subspace of any vector space V over F. If U and V are finite-dimensional vector spaces over F and $f : U \to V$ is linear the kernel and range of f are finite-dimensional and their dimensions are called, respectively, the *nullity* and *rank* of f *and their sum equals* $\dim U$.

Let $W_1, \ldots, W_n$ be subspaces of a vector space V over F. If every $\mathbf{v} \in V$ can be written *in exactly one way* in the form $\mathbf{v} = \mathbf{w}_1 + \cdots + \mathbf{w}_n$ $(\mathbf{w}_i \in W_i)$ then V is called the *internal direct sum* of $W_1, \ldots, W_n$. If $V_1, \ldots, V_n$ are each vector spaces over F then the Cartesian product $V_1 \times \cdots \times V_n$ can be given the structure of a vector space over F by the obvious componentwise addition and scalar multiplication by members of F. This vector space is called the *external direct sum* of $V_1, \ldots, V_n$ and is denoted by $V_1 \oplus \cdots \oplus V_n$. If V is the internal direct sum of $W_1, \ldots, W_n$ then it is clearly isomorphic in an obvious sense to the external direct sum of them. Also if V is the external direct sum of $V_1, \ldots, V_n$ then it is easy to see that V is the internal direct sum of subspaces $W_1, \ldots, W_n$ such that W_i is isomorphic to V_i. Because of this close relation between internal and external direct sums one simply uses the term *direct sum* (and the same notation) for either. If V is the direct sum of $V_i, \ldots, V_n$ then V is finite-dimensional if and only if each V_i is and then $\dim V = \dim V_1 + \cdots + \dim V_n$. A related result arises for arbitrary subspaces U and W of a finite-dimensional vector space V. Define the *sum* of U and W, denoted $U + W$, by $U + W = \{\mathbf{v} \in V : \mathbf{v} = \mathbf{u} + \mathbf{w}, \mathbf{u} \in U, \mathbf{w} \in W\}$. Then $U + W$ is a subspace of V. Also the intersection $U \cap W$ is a subspace of V and $\dim(U + W) + \dim U \cap W = \dim U + \dim W$. The vector space V is the direct sum of U and W if and only if $U + W = V$ and $U \cap W = \emptyset$.

Let U, V be finite-dimensional vector spaces over F with dimensions m and n, respectively, let $\{\mathbf{u}_i\} \equiv \{\mathbf{u}_i, \ldots, \mathbf{u}_m\}$ be a basis for U, $\{\mathbf{v}_j\} \equiv \{\mathbf{v}_1, \ldots, \mathbf{v}_n\}$ a basis for V and $f : U \to V$ a linear transformation. The *matrix (representation) of f with respect to the bases $\{\mathbf{u}_i\}$ and $\{\mathbf{v}_j\}$* is the $m \times n$ matrix $A = (a_{ij})$ $(1 \leqslant i \leqslant m,\ 1 \leqslant j \leqslant n)$ with entries in F determined by the m relations $f(\mathbf{u}_i) = \sum_{j=1}^{n} a_{ij}\mathbf{v}_j$. Under a change of basis $\mathbf{u}_i \to \mathbf{u}_i' = \sum_{k=1}^{m} s_{ik}\mathbf{u}_k$ in U and $\mathbf{v}_j \to \mathbf{v}_j' = \sum_{\ell=1}^{n} t_{j\ell}\mathbf{v}_\ell$ in V where $S = (s_{ik})$ and $T = (t_{je})$ are non-singular (invertible) $m \times m$ and $n \times n$ matrices, respectively, *with entries in F*, the matrix of f with respect to the new bases $\{\mathbf{u}'_i\}$ and $\{\mathbf{v}'_j\}$ is SAT^{-1}. In the case when $U = V$ and $\dim U (= \dim V) = n$ one can choose the same basis $\mathbf{v}_1, \ldots, \mathbf{v}_n$ in U and V and *the $n \times n$ matrix A representing f with respect to $\mathbf{v}_1, \ldots, \mathbf{v}_n$* is determined by the relations $f(\mathbf{v}_i) = \sum_{j=1}^n a_{ij}\mathbf{v}_j$. Under a change of basis $\mathbf{v}_i \to \mathbf{v}'_i = \sum_{j=1}^n P_{ij}\mathbf{v}_j$ for some $n \times n$ non-singular matrix $P = (p_{ij})$ the matrix of f with respect to the new basis $\{\mathbf{v}'_i\}$ is PAP^{-1}. Two $n \times n$ matrices A and B with entries in F are called *similar* (over F) if $B = PAP^{-1}$ for some non-singular $n \times n$ matrix P with entries in F (*i.e.* if A and B represent the same linear map in different bases). Similarity is easily seen to be an equivalence relation on the set M_nF of such matrices and prompts the question of whether there is, in any such equivalence class, a particularly simple (canonical) matrix which characterises the equivalence class. Thus one seeks a basis of V for which the matrix representation of a given linear transformation $f : V \to V$ is "canonical". This procedure will be dealt with in more detail in section 2.6. It is noted here that with U and V as above and if f is a linear transformation $f : U \to V$, the rank of f equals the row and the column rank of any matrix representing f. If $f : V \to V$ is linear then f is called *non-singular* if one (and hence any) of its representative matrices is non-singular and then f is non-singular if and only if f is of rank n which is equivalent to f being an isomorphism of V. The set of all such isomorphisms of V is a group under the usual combination and inverse operations for such maps and is denoted by $GL(V)$ (or $GL(n, \mathbb{R})$ or $GL(n, \mathbb{C})$ if $F = \mathbb{R}$ or $\mathbb{C}$).

In this book only real and complex vector spaces (*i.e.* $F = \mathbb{R}$ or $F = \mathbb{C}$) will be needed and it will prove useful to consider briefly the relationship between these two important structures. First one notes that if V is a vector space over F and F' is a subfield of F then clearly V is a vector space over F'. Since $\mathbb{R}$ is a subfield of $\mathbb{C}$ any complex vector space V can be regarded as a real vector space by simply restricting scalar multiplication from $\mathbb{C}$ to $\mathbb{R}$. However, if $\mathbf{v}$ is a non-zero member of V, the set $\{\mathbf{v}, i\mathbf{v}\}$, whilst linearly *dependent* over $\mathbb{C}$, is linearly *independent* over $\mathbb{R}$. By extending this argument one can show that if the dimension of V over $\mathbb{C}$ is finite and equal to n, then the dimension of V over $\mathbb{R}$ is $2n$. Starting now with a *real* vector space V there are two questions regarding a natural extension to a complex vector space. The first question asks if one can in any way regard V itself as a complex vector space and the second if one can extend V to another abelian group V' such that V' can be regarded as a complex vector space. The answer in general to the first question is no, since it asks if one can reverse the argument earlier in this paragraph about turning complex vector spaces into real ones. A necessary condition is that the dimension of V over $\mathbb{R}$, if finite, be *even*. Another necessary condition is the existence of an operation on V corresponding to "multiplication by i", that is, a linear map $f : V \to V$ such that $f \circ f(\mathbf{v}) = -\mathbf{v}$ for each $v \in V$. Such a map is called a *complex structure* on V. It can now be easily shown that if V admits a complex structure then it *can* be regarded as a vector space over $\mathbb{C}$ by simply extending scalar multiplication from $\mathbb{R}$ to $\mathbb{C}$ according to the definition $(a + ib)\mathbf{v} = a\mathbf{v} + bf(\mathbf{v})$ $(a, b \in \mathbb{R},\ \mathbf{v} \in V)$. The existence of the complex structure would, if the original real vector space V were finite dimensional, force the latter to have even dimension $2n$ and the complex vector space thus constructed would then have dimension n. In fact the condition that the original real vector space has (finite) even dimension is itself sufficient for it to be regarded as a complex vector space in the above sense. As for the second question, one starts from *any real* vector space V and constructs the real vector space $V \oplus V$. The map f from $V \oplus V$ to itself defined by $f : (\mathbf{u}, \mathbf{v}) \to (-\mathbf{v}, \mathbf{u})$ is linear and satisfies the condition that $f \circ f$ is the negative of the identity map on $V \oplus V$. Hence $V \oplus V$ can be regarded as a complex vector space and is the extension of V sought. It is called the *complexification* of V. One may picture such a procedure as starting from the real vector space V and building a vector space over $\mathbb{C}$ from "vectors like $\mathbf{u} + i\mathbf{v}$" $(\mathbf{u}, \mathbf{v} \in V)$. Thus complex scalar multiplication is, for $a, b \in \mathbb{R}$, $\mathbf{u}, \mathbf{v} \in V$,

$$\begin{aligned}(a + ib)(\mathbf{u}, \mathbf{v}) &= a(\mathbf{u}, \mathbf{v}) + bf((\mathbf{u}, \mathbf{v}))\\ &= a(\mathbf{u}, \mathbf{v}) + b(-\mathbf{v}, \mathbf{u})\\ &= (a\mathbf{u} - b\mathbf{v}, a\mathbf{v} + b\mathbf{u}).\end{aligned}$$

If either the original vector space V (over $\mathbb{R}$) or the extended one $V \oplus V$ (over $\mathbb{C}$) is finite dimensional then their dimensions are equal.

It is remarked here for completeness that, concerning the general situation described at the beginning of the last paragraph, if F' is a subfield of F then F is a vector space over F' in an obvious way. Now suppose that V is a vector space over F (and hence over F'). Then if the two vector spaces F (over F') and V (over F) have finite dimension, it is not hard to show that the vector space V (over F') is also finite-dimensional and that $\dim V$ (over F') $= \dim V$ (over F). $\dim F$ (over F'). As examples of the above discussion it is noted that with their usual structures $\mathbb{R}$ is a subfield of $\mathbb{C}$ and that as a *real* vector space $\dim \mathbb{C} = 2$. Also the complexification of the n-dimensional real vector space $\mathbb{R}^n$ is the n-dimensional complex vector space $\mathbb{C}^n$ and, as a *real* vector space, $\dim \mathbb{C}^n = 2n$.

Now let V be an n-dimensional real vector space with a chosen fixed basis and let f, g be non-singular maps $V \to V$ with matrices $A = (a_{ij})$ and $B = (b_{ij})$, respectively, in this basis. Then the map $g \circ f$ has matrix AB in this basis. Thus the group of non-singular maps $V \to V$ and the group $GL(n, \mathbb{R})$ of non-singular $n \times n$ real matrices are isomorphic but, in the above notation, the isomorphism $f \to A$ requires that the "map product" $f \cdot g$ be $g \circ f$ (see the remark at the end of section 2.2).

2.4 Dual Spaces

Let V be a *finite-dimensional* vector space over F of dimension n and consider the vector space (over F) $\overset{*}{V} = L(V, F) \equiv$ the set of all linear maps from V to F (with F regarded as a 1-dimensional vector space over F). The vector space $\overset{*}{V}$ is called the *dual (space) of* V *(over* F*)*. The dimensionality of $\overset{*}{V}$ is easily revealed by first noting that if $\{\mathbf{v}_i\}$ is a basis for V and $a_1, \ldots, a_n \in F$ then (from the linearity of members of $\overset{*}{V}$) there is exactly one $\mathbf{w} \in \overset{*}{V}$ such that $\mathbf{w}(\mathbf{v}_i) = a_i$ for $i = 1, \ldots, n$. A straightforward argument then shows that there is a uniquely determined set $\mathbf{w}_1, \ldots, \mathbf{w}_n \in \overset{*}{V}$ such that $\mathbf{w}_i(\mathbf{v}_j) = \delta_{ij}$ for each i, j, $1 \leqslant i, j \leqslant n$, and where the Kronecker δ is defined by $\delta_{ij} = 1$ $(i = j)$ and $\delta_{ij} = 0$ $(i \neq j)$. This set constitutes a basis for $\overset{*}{V}$ called the *dual basis* of $(\mathbf{v}_1, \ldots, \mathbf{v}_n)$ showing that $\overset{*}{V}$ is a finite-dimensional vector space over F of dimension n. As a consequence, V and $\overset{*}{V}$ are isomorphic since they are each isomorphic to F^n. However, it should be noted that the isomorphism $V \to \overset{*}{V}$ defined uniquely by $\mathbf{v}_i \to \mathbf{w}_i$ (and then by linearity) depends on the original basis chosen for V and is not, in the usual sense of the word, *natural*.

With V as above it now follows that one may construct the dual of $\overset{*}{V}$ and denote it by $\overset{**}{V}$. The previous construction shows that $\overset{**}{V}$ is a vector space over F of dimension n consisting of all linear maps $\overset{*}{V} \to F$ and that V and $\overset{**}{V}$ are thus isomorphic. In this case, however, a *natural* isomorphism $f : V \to \overset{**}{V}$ can be constructed as follows; for each $\mathbf{v} \in V$ define $f(v) \in \overset{**}{V}$ to be the map $\overset{*}{V} \to F$ given by $(f(\mathbf{v}))(\mathbf{w}) = \mathbf{w}(\mathbf{v})$ for each $\mathbf{w} \in \overset{*}{V}$. The linearity and bijective nature of f can then be checked and the result follows. The isomorphism f is called the *natural isomorphism* between V and $\overset{**}{V}$ and sometimes it is convenient to identify V and $\overset{**}{V}$ by means of f. For any given basis in V, the dual basis of its dual basis is the image in $\overset{**}{V}$ under f of the given basis of V.

For completeness it is remarked that a linear map from a vector space V (over F) to F (*i.e.* a member of $\overset{*}{V}$) is sometimes referred to in the literature as a *linear functional (on V)*. It should also be stressed that the brief remarks above on dual spaces depend on the *finite dimensionality* of V.

2.5 Forms and Inner Products

Let U and V be finite-dimensional vector spaces over F of dimension m and n, respectively, and let $W = U \oplus V$ so that W is a vector space of dimension $m+n$ over F. A map $f : W \to F$ is called a *bilinear form (on W) if*

$$f(a_1\mathbf{u}_1 + a_2\mathbf{u}_2, \mathbf{v}) = a_1 f(\mathbf{u}_1, \mathbf{v}) + a_2 f(\mathbf{u}_2, \mathbf{v})$$

and

$$f(\mathbf{u}, a_1\mathbf{v}_1 + a_2\mathbf{v}_2) = a_1 f(\mathbf{u}, \mathbf{v}_1) + a_2 f(\mathbf{u}, \mathbf{v}_2)$$

are true for each $\mathbf{u}, \mathbf{u}_1, \mathbf{u}_2 \in U$, $\mathbf{v}, \mathbf{v}_1, \mathbf{v}_2 \in V$ and $a_1, a_2 \in F$. Thus f is "linear" in each of its arguments. If f_1 and f_2 are bilinear forms and if $a_1, a_2 \in F$ then if one defines $(a_1 f_1 + a_2 f_2)(\mathbf{u}, \mathbf{v}) = a_1 f_1(\mathbf{u}, \mathbf{v}) + a_2 f_2(\mathbf{u}, \mathbf{v})$ it easily follows that $a_1 f_1 + a_2 f_2$ is a bilinear form on W and that the set of all bilinear forms on W is a vector space over F. Also if $\{\mathbf{u}_i\}$ is a basis for U and $\{\mathbf{v}_j\}$ a basis for V and if $A = (a_{ij})$ $(1 \leqslant i \leqslant m,\ 1 \leqslant j \leqslant n)$ is any $m \times n$ matrix of members of F there is exactly one bilinear form f on W such that $f(\mathbf{u}_i, \mathbf{v}_j) = a_{ij}$. It follows that the bilinear forms f_{pq} corresponding to the arrays $a_{ij} = \delta_{ip}\delta_{jq}$ for each p, q with $1 \leqslant p \leqslant m$, $1 \leqslant q \leqslant n$, constitute a basis for the vector space of bilinear forms on W and so this vector space has dimension mn. The matrix A above is called the *matrix of f with respect to the bases $\{\mathbf{u}_i\}$ and $\{\mathbf{v}_i\}$*. Thus if $\mathbf{u} \in U$, $\mathbf{v} \in V$, $\mathbf{u} = \sum_{i=1}^m x_i \mathbf{u}_i$, $\mathbf{v} = \sum_{i=1}^n y_i \mathbf{v}_i$ then $f(\mathbf{u}, \mathbf{v}) = \sum_{i=1}^m \sum_{j=1}^n a_{ij} x_i y_j$. Under a change of basis in both U and V given by $\mathbf{u}_i \to \mathbf{u}'_i = \sum_{r=1}^m S_{ir}\mathbf{u}_r$ and $\mathbf{v}_j \to \mathbf{v}'_j = \sum_{s=1}^n P_{js}\mathbf{v}_s$ for appropriate non-singular matrices S and P the matrix of f changes from A (with respect to $\{\mathbf{u}_i\}$ and $\{\mathbf{v}_j\}$) to SAP^{T}

(with respect to $\{\mathbf{u}'_i\}$ and $\{\mathbf{v}'_j\}$) where P^{T} is the *transpose* of P.

Now suppose that in the above paragraph U and V are equal (and say labelled by V and of dimension n). A bilinear form on $V \oplus V$ is now called simply a *bilinear form on* V. If $\{\mathbf{v}_i\}$ is a basis for V *the matrix of* f *with respect to the basis* $\{\mathbf{v}_i\}$ *is the matrix* $A = (a_{ij})$ *where* $a_{ij} = f(\mathbf{v}_i, \mathbf{v}_j)$. Under a change of basis $\mathbf{v}_i \to \mathbf{v}'_i = \sum_{r=1}^n S_{ir}\mathbf{v}_r$ for some non-singular $n \times n$ matrix S, the matrix of f changes to SAS^{T}. In this case it makes sense to ask if the matrix representing f is symmetric or non-singular since it is a square matrix and since each of these properties, if satisfied with respect to some basis of V, would be satisfied with respect to all bases of V. A bilinear form on V is called *symmetric* if its matrix is symmetric and *non-degenerate* if its matrix is non-singular. Thus a bilinear form f on V is symmetric if and only if $f(\mathbf{v}, \mathbf{v}') = f(\mathbf{v}', \mathbf{v})$ $\forall \mathbf{v}, \mathbf{v}' \in V$ and non-degenerate if whenever $f(\mathbf{v}, \mathbf{v}') = 0$, $\forall \mathbf{v}' \in V$, then $\mathbf{v} = 0$.

Now specialise to the case where V is a finite dimensional vector space over $\mathbb{R}$ (a *real* vector space) of dimension n and let f be a bilinear form on V (a *real bilinear form*). The mapping $V \to \mathbb{R}$ given by $\mathbf{v} \to f(\mathbf{v}, \mathbf{v})$ is called a (*real*) *quadratic form* (*on* V) (more precisely the quadratic form on V associated with f). If $A = (a_{ij})$ is the matrix representing f with respect to the basis $\mathbf{v}_i$ of V then this quadratic form is the map $q : \mathbf{v} \to \sum_{i,j=1}^n a_{ij}x_i x_j$ where $\mathbf{v} = \sum_{i=1}^n x_i \mathbf{v}_i$ $(x_i \in \mathbb{R})$. Thus only the "symmetric part" of f (*i.e.* the symmetric part of A) matters in constructing its associated quadratic form and the correspondence between real *symmetric* bilinear forms on V and real quadratic forms on V is, in fact, one to one. This is because such a quadratic form q uniquely determines the original symmetric bilinear form f according to $f(\mathbf{u}, \mathbf{v}) = \frac{1}{4}[q(\mathbf{u} + \mathbf{v}) - q(\mathbf{u} - \mathbf{v})]$. A quadratic form is called *non-degenerate* if its corresponding symmetric bilinear form is non-degenerate. Given a basis $\{\mathbf{v}_i\}$ in V then to each real quadratic form there corresponds a unique real symmetric matrix A, and conversely. Under a change of basis $\mathbf{v}_i \to \mathbf{v}'_i = \sum_{r=1}^n S_{ir}\mathbf{v}_r$, the matrix representing f changes to SAS^{T}. One calls real symmetric $n \times n$ matrices A and B *congruent* if there exists a non-singular $n \times n$ real matrix S such that $B = SAS^{\mathrm{T}}$. Congruence is an equivalence relation on such matrices and one thus asks if there is a set of conditions which apply to symmetric real matrices and which serves to characterise the particular equivalence classes. This leads to the question of whether there is a particularly simple (canonical) matrix in each equivalence class. The answer to both questions is provided by a very important theorem (*Sylvester's law of inertia*) which states that *for a given real symmetric* $n \times n$ *matrix* A *there exists a non-singular* $n \times n$ *real matrix S such that*

$$SAS^{\mathrm{T}} = \mathrm{diag}(\underbrace{1,\ldots,1}_{r \text{ terms}}, \underbrace{-1,\ldots,-1}_{s \text{ terms}}, \underbrace{0,\ldots,0}_{t \text{ terms}}) \equiv \overset{(n)}{I}{}^{r}_{s} \tag{2.1}$$

The ordered set (r, s, t) of integers $(r + s + t = n)$ characterises the equivalence class of A and is (collectively) called the *signature* of A. The rank of A is $r + s$ and A is non-singular if and only if $t = 0$. The right hand side of (2.1) is called the *Sylvester canonical form* or the *Sylvester matrix* for A. It is understood that the entries ± 1 and 0 are in the order indicated in (2.1). Often one denotes the signature of A by the symbol (r, s, t), or just by (r, s) in the important non-degenerate case $t = 0$, (or even by simply writing out the diagonal entries in the Sylvester matrix.) If $s = t = 0$ one sometimes refers to the corresponding signature $(n, 0)$ as *positive definite* (and *negative definite* if $r = t = 0$). The signature is called *Lorentz* if $t = 0$ and if either $r = n - 1$, $s = 1$ or $r = 1$, $s = n - 1$. If $t = 0$ the Sylvester matrix is denoted by I^r_s.

The final concept to be discussed in this section is that of an *inner product* . Before the definition is given, two remarks are appropriate. Firstly, the most general definition involves a *complex* inner product. Here, however, only the *real* inner product will be introduced. Although the complex field will be used later in a significant way, the real inner product will be sufficient for the purposes of this book. Secondly, the usual definition of an inner product arose out of the (positive-definite, Euclidean) concept of length and angle and is not wide enough to embrace the metric concepts required in general relativity. Here the more general definition, sufficient to cover the requirements of the latter, will be used, thus risking offending those who feel that a term other than inner product should be employed. Let V be a finite dimensional vector space over $\mathbb{R}$ of dimension n. An *inner product* on V is a *symmetric non-degenerate* bilinear form $f : V \times V \to \mathbb{R}$. Such a real vector space V equipped with such an inner product will be called an *inner product space* . [The usual definition differs from this one only by the replacing of the non-degenerate condition by the (stronger) *positive definite* condition that $f(\mathbf{v}, \mathbf{v}) > 0$ whenever $\mathbf{v} \neq 0$. This latter condition clearly implies the non-degenerate condition and gives rise to what will here be called a *positive definite inner product*. A real vector space possessing such a structure is called a *Euclidean* (vector) space.] An inner product (respectively a positive-definite inner product) on V is sometimes called a *metric on V* (respectively a *positive definite metric on*

V). If it is not positive definite it is called *indefinite*.

If V is an inner product space with inner product f, and if $\mathbf{u}, \mathbf{v} \in V$, then the real number $f(\mathbf{u}, \mathbf{v})$ is called the *inner product of* $\mathbf{u}$ *and* $\mathbf{v}$. Clearly $f(\mathbf{v}, \mathbf{0}) = 0$ for any $\mathbf{v} \in V$ and for $\mathbf{v} \in \mathbf{V}$, $\mathbf{v} \neq \mathbf{0}$, $f(\mathbf{v}, \mathbf{v})$ may be positive, negative or zero. Two non-zero vectors $\mathbf{u}, \mathbf{v}$ are called *orthogonal* if $f(\mathbf{u}, \mathbf{v}) = 0$. For any subspace W of V, the associated *subspace* $W^{\perp} = \{\mathbf{v} \in V : f(\mathbf{u}, \mathbf{v}) = 0, \forall \mathbf{u} \in W\}$ is called the *orthogonal complement* of W. Clearly $(W^{\perp})^{\perp} = W$. This notation is a little unfortunate since, except in the positive or negative definite cases, it may be false that $V = W \oplus W^{\perp}$. In the positive (or negative) definite cases, however, it is always true that $V = W \oplus W^{\perp}$. Also, in the positive (negative) definite cases, any subspace W of V inherits a naturally induced, necessarily positive (negative) definite, inner product by restriction from that on V in an obvious way. Again this result may fail for a general inner product space in that, although there is an induced bilinear form, it may fail to be non-degenerate. In any case if a subspace W of V does inherit an inner product from f then $V = W \oplus W^{\perp}$. The moral which emerges is that one should be careful with general inner products in the sense defined here since many results for positive definite inner products are not valid for them. For any inner product f, two subspaces U, W of V are called *orthogonal* if for each $\mathbf{u} \in U$ and $\mathbf{w} \in W$, $f(\mathbf{u}, \mathbf{w}) = 0$. A basis $\{\mathbf{v}_i\}$ for V is called *orthonormal* if $f(\mathbf{v}_i, \mathbf{v}_j) = 0$ for $i \neq j$ and $f(\mathbf{v}_i, \mathbf{v}_i) = \pm 1$. Orthonormal bases always exist. Any $\mathbf{v} \in V$ such that $f(\mathbf{v}, \mathbf{v}) = \pm 1$ is called a *unit vector* . Inner products are classified by their signature as in (2.1) where now the non-degenerate condition means that $t = 0$. It is remarked that any real finite-dimensional vector space V admits a real inner product of any desired signature (r, s) with $r + s = \dim V$.

The above discussion shows that a given inner product f on V takes its canonical Sylvester form (with $t = 0$) with respect to a suitably ordered orthonormal basis of V. If $\{\mathbf{v}_i\}$ is such a basis and I_s^r is the Sylvester matrix corresponding to f, then the set of all such bases for V is in one to one correspondence with the set of all $n \times n$ non-singular real matrices Q satisfying $I_s^r = Q I_s^r Q^{\mathrm{T}}$. This can be described group theoretically by first considering the set $GL(n, \mathbb{R})$ of all $n \times n$ non-singular real matrices. The set $\{Q : I_s^r = Q I_s^r Q^{\mathrm{T}}\}$ is then a *subgroup* of $GL(n, \mathbb{R})$ called *the orthogonal group of the Sylvester matrix* I_s^r and denoted by $O(r, s)$ (or just the *orthogonal group* and denoted by $O(n)$ if $s = 0$). There are many isomorphic copies of this subgroup in $GL(n, \mathbb{R})$ because if A is an $n \times n$ real symmetric matrix whose Sylvester matrix is I_s^r (and so $SAS^{\mathrm{T}} = I_s^r$ for some $S \in GL(n,\mathbb{R})$) the set $\{P : PAP^{\mathrm{T}} = A\}$ is a subgroup of $GL(n\mathbb{R})$ isomorphic to $O(r,s)$ under the isomorphism $P \to SPS^{-1} \in O(r,s)$ (and so these subgroups are actually conjugate). In other words, when considering an inner product f on V it doesn't matter which particular basis one uses to lay down a "canonical" form for f since the changes of basis which preserve it give rise to a subgroup of $GL(n\mathbb{R})$ isomorphic to $O(r,s)$. The choice of a suitably ordered *orthonormal* basis and the consequent *Sylvester* canonical form is usually (but not always) the most convenient. If f is a positive definite inner product for V the corresponding orthogonal group $O(n,0) \equiv \{Q : I_n = QI_nQ^{\mathrm{T}}\} \equiv \{Q : I_n = QQ^{\mathrm{T}}\}$ where $I_n = I_0^n$, is denoted by $O(n)$. Also $Q \in O(r,s) \Rightarrow \det Q = \pm 1$. The subgroup $\{Q \in O(r,s) : \det Q = 1\}$ of $O(r,s)$ is called the *special orthogonal group* of the Sylvester matrix I_s^r and is denoted by $SO(r,s)$.

The material of the preceding paragraph can be viewed in an alternative way. Let V be as above and let f be an inner product for V with associated quadratic form q. If g is a linear transformation on V then g is said to *preserve* f if for all $\mathbf{u},\mathbf{v} \in V$, $f(\mathbf{u},\mathbf{v}) = f(g(\mathbf{u}),g(\mathbf{v}))$, and to preserve q if for each $\mathbf{v} \in V$, $q(\mathbf{v}) = q(g(\mathbf{v}))$. It is then straightforward to show the equivalence of the statements (i) that g preserves f, (ii) that g preserves q and (iii) that the matrix G which represents g with respect to a basis $\{\mathbf{v}_i\}$ of V for which the matrix A of f is $a_{ij} \equiv f(\mathbf{v}_i\mathbf{v}_j)$ satisfies $GAG^{\mathrm{T}} = A$. The third condition here is, for an orthonormal basis, the statement that $G \in O(r,s)$ where r and s characterise the signature of f. A transformation such as g which preserves the "size" $|f(\mathbf{v},\mathbf{v})|^{1/2}$ of each $\mathbf{v} \in V$ with respect to f is called *f-orthogonal* (or just *orthogonal* if the inner product is clear). If $\dim V = 2 \,\mathrm{or}\, 3$ and f is positive definite such transformations are, up to reflections, the usual Euclidean rotations.

Now let $V_1,\ldots,V_m$ each be a finite-dimensional vector space over the field F such that $\dim V_i = n_i$ $(1 \leqslant i \leqslant m)$. A *multilinear map* (or *form*) f on $V = V_1 \oplus \cdots \oplus V_m$ is a map $f : V \to F$ which is linear in each of its arguments (in an obvious way as a generalisation of a bilinear form) and then the set of all such maps on V is with the obvious operations a vector space over F. Further, if $\{\mathbf{e}_i^1\},\ldots,\{\mathbf{e}_k^m\}$ are bases for $V_1,\ldots,V_m$, respectively, with corresponding dual bases $\{\overset{*}{\mathbf{e}}{}_i^1\},\ldots,\{\overset{*}{\mathbf{e}}{}_k^m\}$ the multilinear maps $V \to F$ denoted by $\overset{*}{\mathbf{e}}{}_i^1 \otimes \cdots \otimes \overset{*}{\mathbf{e}}{}_k^m$ where

$$\overset{*}{\mathbf{e}}{}_i^1 \otimes \cdots \otimes \overset{*}{\mathbf{e}}{}_k^m(\mathbf{e}_a^1,\ldots,\mathbf{e}_b^m) = \overset{*}{\mathbf{e}}{}_i^1(\mathbf{e}_a^1)\ldots\overset{*}{\mathbf{e}}{}_k^m(\mathbf{e}_b^m) = \delta_a^i\ldots\delta_b^k$$

(extended to the whole of V by linearity) constitute a basis for the vector space of all multilinear maps $V \to F$. Thus its dimension is $n_1 n_2 \dots n_m$.

2.6 Similarity, Jordan Canonical Forms and Segre Types

Throughout this section V will denote an n-dimensional vector space over the field F, where $F = \mathbb{R}$ or $F = \mathbb{C}$. Occasionally, a particular result will necessitate a distinction between these fields. Now let f be a linear transformation on V (*i.e.* $f : V \to V$). A vector $\mathbf{v} \in V$, $\mathbf{v} \neq \mathbf{0}$, is called an *eigenvector* of f if $f(\mathbf{v}) = \lambda\mathbf{v}$ for some scalar $\lambda \in F$ and λ is called the *eigenvalue* of f associated with the eigenvector $\mathbf{v}$. (Sometimes the alternative terms *characteristic vector* and *characteristic value* are used.) If $\mathbf{v}$ is an eigenvector of f with eigenvalue λ then any non-zero member of the 1-dimensional subspace of V spanned by $\mathbf{v}$ (the *eigendirection* or *characteristic direction* determined by $\mathbf{v}$) is an eigenvector of f with the same eigenvalue λ. One can describe this situation with respect to any basis $\{\mathbf{e}_i\}$ of V in terms of the matrix $A = (a_{ij})$ which represents f in this basis. If $\mathbf{v} = \sum_{i=1}^{n} v_i \mathbf{e}_i$ then the condition that $\mathbf{v}$ be an eigenvector of f with eigenvalue λ is equivalent to the condition $\sum_{i=1}^{n} v_i a_{ij} = \lambda v_j$ (or $\mathbf{v}A = \lambda\mathbf{v}$) on the components of $\mathbf{v}$ in the basis $\{\mathbf{e}_i\}$ (and the components v_i are said to be the *components of an eigenvector*, and λ an *eigenvalue of the matrix* A). It then follows from elementary algebra that a scalar λ is an eigenvalue for f (corresponding to some eigenvector for f) if and only if $x = \lambda$ satisfies the equation $\det(A - xI_n) = 0$. The left hand side of this equation is a polynomial of degree n in x called the *characteristic polynomial* of A and the equation itself is called the *characteristic equation* of A. If $\lambda \in F$ is an eigenvalue of f, the number of times the factor $(x - \lambda)$ occurs in the characteristic equation is called the *multiplicity* of λ (*i.e.* the number of times "properly counted" that λ occurs as a root of the characteristic equation). It should be borne in mind, of course, that the characteristic polynomial (and hence its roots) are determined uniquely by f even though it is defined through the (basis-dependent) representative matrix A. In fact it is a straightforward matter to show that two similar matrices have the same characteristic polynomial. Hence one speaks of the characteristic polynomial of f. If λ is an eigenvalue of f the set of all eigenvectors of f with eigenvalue λ, together with the zero vector form a subspace of V called the λ-*eigenspace* of f. Its dimension may be any integer between 1 and n and is not necessarily equal to the multiplicity of λ as will be seen later. It should be noted that *the kernel of f consists of* *precisely the eigenvectors of f with zero eigenvalue* (*together with the zero vector*), that is, it is the 0-eigenspace of f and so if the dimension of the latter is m then $\operatorname{rank} f = n - m$. A λ-eigenspace of f has the property that its image under f is contained in itself (and equals itself if $\lambda \neq 0$). A generalisation of this concept is sometimes useful. A subspace U of V is called an *invariant subspace* of f if $f(U) \subseteq U$, that is, if f maps each member $\mathbf{u}$ of U into the set U. Thus a λ-eigenspace of f is an invariant subspace of f but a non-zero member of an invariant subspace of f need not be an eigenvector of f—in fact a non-trivial invariant subspace of f may contain *no* eigenvectors of f as will be seen later.

The linear map f on V above is uniquely determined (by linearity) if its action on a basis of V is known. Since the action of f on its eigenvectors (if it has any) is particularly simple it follows that one should perhaps attempt to construct a basis for V containing the maximum possible number of eigenvectors of f. Now a set of eigenvectors of f with pairwise distinct eigenvalues can be shown to be linearly independent. If a basis consisting entirely of eigenvectors of f can be found then the matrix of f in this basis is *diagonal* and its diagonal entries are just the eigenvalues of f. Thus if A was the matrix of f in some original (arbitrary) basis of V and T the non-singular matrix responsible for the change of basis to the basis of eigenvectors of f then TAT^{-1} is the diagonal matrix described above. So in this case some matrix in the similarity equivalence class containing A is a diagonal matrix and A is said to be *diagonalisable* (over F). Conversely if a matrix representing f in some basis of V is diagonalisable (over F) then V admits a basis of eigenvectors of f whose corresponding eigenvalues are the entries in this diagonal matrix. The map f is also called *diagonalisable* over F in this case. From a remark earlier in this paragraph it follows that a sufficient (but by no means necessary) condition for the existence of a basis of eigenvectors of f is that the characteristic equation associated with f admits n *distinct* solutions in the associated field F of V. Suppose now that n distinct eigenvalues cannot be found but that the characteristic polynomial of f still factorises over F into n linear factors. Then diagonalisability will follow if and only if the dimension of the λ–eigenspace equals the multiplicity of λ for each eigenvalue λ. Otherwise (but still retaining the assumption about the complete linear factorisation of the characteristic polynomial) one must accept that no matrix A representing f in any basis for V is diagonal and a different "almost diagonal" canonical form must be sought. These diagonal and "almost diagonal" canonical forms are the *Jordan matrices* to be described below. The situation when the characteristic polynomial of f does not produce n linear factors over F is described in the next paragraph.

When V is a complex vector space, that is when $F = \mathbb{C}$, these *Jordan canonical forms* are the final step in what is essentially a classification of linear maps on V. This is because $\mathbb{C}$ is an *algebraically closed* field, that is, any nth degree polynomial with coefficients in $\mathbb{C}$ factorises into n linear factors over $\mathbb{C}$. Thus the characteristic polynomial of any linear map f on V factorises over $\mathbb{C}$ into n linear factors. If V is a real vector space ($F = \mathbb{R}$) a problem arises in that $\mathbb{R}$ is not algebraically closed and the characteristic polynomial of a linear map f on V may not factorise into n linear factors over $\mathbb{R}$. If it does factorise the Jordan theory still applies. If it does not, then this (real) characteristic polynomial must factorise into a product of linear and (irreducible) quadratic factors over $\mathbb{R}$, there being at least one of the latter in this case. Thus complex (non-real) solutions of the characteristic polynomial arise—and in complex conjugate pairs. These solutions, strictly speaking, are not eigenvalues of f since they do not belong to the appropriate field and since, if λ is such a solution, then for no $\mathbf{v} \in V$, $\mathbf{v} \neq \mathbf{0}$, is $f(\mathbf{v}) = \lambda\mathbf{v}$ for this would imply $\lambda \in \mathbb{R}$. However, if one extends the real vector space V to its complexification $V \oplus V$ which is a complex vector space (as described earlier in this chapter) then f easily extends to a *linear* map $\tilde{f}$ on $V \oplus V$ by defining $\tilde{f}((\mathbf{u}, \mathbf{v})) = (f(\mathbf{u}), f(\mathbf{v}))$ for $\mathbf{u}, \mathbf{v} \in V$. Now if $\lambda \in \mathbb{C}$ is a solution of the original real characteristic polynomial of f it easily follows that if Id denotes the identity map on $V \oplus V$ then the linear map $\tilde{f} - \lambda Id$ on $V \oplus V$ fails to be non-singular and so some non-zero member $(\mathbf{x}, \mathbf{y}) \in V \oplus V$ satisfies $(\tilde{f} - \lambda Id)(\mathbf{x}, \mathbf{y}) = (0, 0)$. (To see this choose a basis $\{\mathbf{e}_i\}$ $(1 \leq i \leq n)$ for V. Then if $(\mathbf{u}, \mathbf{v}) \in V \oplus V$ with $\mathbf{u} = \sum_{i=1}^{n} u_i\mathbf{e}_i, \mathbf{v} = \sum_{i=1}^{n} v_i\mathbf{e}_i \in V$, $u_i, v_i \in \mathbb{R}$, one has according to the rules given earlier for the complex structure and complexification of V

$$\begin{aligned}(\mathbf{u}, \mathbf{v}) &= (u_1\mathbf{e}_1, v_1\mathbf{e}_1) + \ldots + (u_n\mathbf{e}_n, v_n\mathbf{e}_n) \\ &= (u_1 + iv_1)(\mathbf{e}_1, \mathbf{0}) + \ldots + (u_n + iv_n)(\mathbf{e}_n, \mathbf{0}).\end{aligned}$$

Hence $\{(\mathbf{e}_i, \mathbf{0})\}$ is a basis for the complexification of V. The matrices representing f on V and $\tilde{f}$ on $V \oplus V$ with respect to the bases $\{\mathbf{e}_i\}$ and $\{(\mathbf{e}_i, \mathbf{0})\}$ are then equal since $f(\mathbf{e}_i) = \sum_{j=1}^{n} a_{ij}\mathbf{e}_j \Rightarrow \tilde{f}(\mathbf{e}_i, \mathbf{0}) = \sum_{j=1}^{n} a_{ij}(\mathbf{e}_j, \mathbf{0})$ where $A = (a_{ij})$ is this common (real) matrix. The information given in the real vector space V then leads to the equation $det(A - \lambda I_n) = 0$ in the complexification and the result follows.) Thus $(\mathbf{x}, \mathbf{y})$ is an eigenvector of $\tilde{f}$ and, if $\lambda = a+ib$ $(b \neq 0)$, this gives $\tilde{f}(\mathbf{x}, \mathbf{y}) = (a+ib)(\mathbf{x}, \mathbf{y})$ or $f(\mathbf{x}) = a\mathbf{x}-b\mathbf{y}$

and $f(\mathbf{y}) = a\mathbf{y} + b\mathbf{x}$. From this it follows that *the vectors* $\mathbf{x}, \mathbf{y} \in V$ *span an invariant 2-dimensional subspace of* f (which, it is easily shown, contains no eigenvectors of f). In other words if f is a linear map on a real vector space and λ is a complex solution of the (real) characteristic equation of f then the obvious extension of f to the complexification of V has an eigenvector with λ as corresponding eigenvalue and whose "real and imaginary parts" span an invariant 2-dimensional subspace of f in V. (Under such circumstances the terms *complex eigenvector* and *complex eigenvalue* will be used.) Thus whilst an eigenvector of f with a corresponding real eigenvalue spans a "special" 1-dimensional subspace of V, a complex eigenvector of f with corresponding complex (non-real) eigenvalue gives rise to a "special" 2-dimensional subspace of V. This approach will be all that is required here to handle the cases when the characteristic polynomial of a linear map on a real vector space does not decompose completely into linear factors over $\mathbb{R}$ and the technique using the so called *rational canonical form* will not be required.

The matrix version of the above discussion starts by selecting some matrix $A \in M_n(F)$ ($F = \mathbb{R}$ or $\mathbb{C}$). Now consider all those linear maps from an n-dimensional vector space V (over F) to itself which are represented by A in some basis of V. Then A is diagonalisable over F if and only if one (and hence each) of these maps admits n independent (*i.e.* a basis of) eigenvectors.

There is an important result that is conveniently discussed here. Let $A = (a_{ij}) \in M_n(\mathbb{R})$ be a *real symmetric* matrix. Then it is well known from elementary algebra that the roots of the characteristic polynomial of A (*i.e.* the eigenvalues of A) are all real and that A is diagonalisable over $\mathbb{R}$ (*i.e.* PAP^{-1} is diagonal for some real non-singular matrix P). Thus any linear map f from a real (or complex) vector space V to itself which can be represented in some basis of V by a *real symmetric* matrix admits a basis of eigenvectors of V. Moreover, if V is a *real* vector space and $\{\mathbf{e}_i\}$ is a basis of V in which f is represented by A, define a positive definite inner product (metric) h on V by $h(\mathbf{e}_i, \mathbf{e}_j) = \delta_{ij}$. Then any two eigenvectors of f with distinct eigenvalues are *orthogonal* with respect to h [$\mathbf{u} = \sum u_i \mathbf{e}_i$, $\mathbf{v} = \sum_{i=1}^n v_i e_i$, $u_i, v_i \in \mathbb{R}$, $\sum_{i=1}^n u_i a_{ij} = \lambda u_j$ and $\sum_{i=1}^n v_i a_{ij} = \mu v_j$, $\lambda, \mu \in \mathbb{R}$, $\lambda \neq \mu$, $\Rightarrow \sum_{i=1}^n u_i v_i = 0$ $(= h(\mathbf{u}, \mathbf{v}))$] and it is not hard to show that one may always select a basis of eigenvectors of f of unit size and which are mutually *orthogonal* with respect to h. Standard theory then shows that there exists $P \in O(n)$ such that PAP^{-1} is diagonal. But then since $P \in O(n)$, $P^{-1} = P^{\mathrm{T}}$ and so PAP^{T} is diagonal. So A is simultaneously similar to and congruent to a diagonal matrix. This collection of results is usually referred to as *the principal axes theorem* . Whilst it is a pretty and very useful result from the matrix viewpoint it is puzzling from the point of view of linear transformations since it requires the finding of a basis in which the associated matrix of the transformation is symmetric and then alludes to curious metric statements arising from h which were not part of the original problem. These curiosities arise from the fact that, in terms of matrices, the above results are heavily basis dependent. The general situation takes a more natural form when stated in terms of linear transformations in a basis independent way. So let V be a real vector space with $\dim V = n$, let $f : V \to V$ be linear and let h be a positive definite metric on V. Then f is called *self adjoint (with respect to h)* if $h(\mathbf{x}, f(\mathbf{y})) = h(f(\mathbf{x}), \mathbf{y})$ for each $\mathbf{x}, \mathbf{y} \in V$. This condition is defined without appeal to any basis but implies the statement that in any basis $(\mathbf{e}_i)$ for V the matrix AH is symmetric, where A and H are the matrices representing f and h in that basis (so that $H = (h_{ij})$, $h_{ij} = H(\mathbf{e}_i, \mathbf{e}_j)$). Conversely, if in any basis of V AH is symmetric then f is self adjoint with respect to h. Now suppose P is a real non-singular matrix "taking" H to its Sylvester canonical form, $PHP^{\mathrm{T}} = I_n$. In this new "Sylvester" basis the matrix representing f is given by $Q = PAP^{-1}$. Now $(P^{-1})^{\mathrm{T}} = PH$ and so since H is symmetric and $HP^{\mathrm{T}} = P^{-1}$

$$Q^{\mathrm{T}} = PHA^{\mathrm{T}}P^{\mathrm{T}} = P(AH)^{\mathrm{T}}P^{\mathrm{T}}.$$

From this it follows that Q is symmetric if and only if AH is symmetric, that is, if and only if f is self adjoint with respect to h. Thus if f is self adjoint with respect to some positive definite metric h on V there exists an orthonormal (with respect to h) basis of eigenvectors of f. Conversely the existence of a basis of eigenvectors of f implies that f is self adjoint with respect to the positive definite metric for V for which this basis is orthonormal. Hence the self adjoint condition (with respect to some positive definite metric h) for f is equivalent to the condition that the (similarity equivalence) class of matrices representing f contains a diagonal matrix.

The previous paragraph involved a canonical form for the special case of a self adjoint transformation on a real vector space. Here, V will be taken as an n-dimensional *complex* vector space and f *any* linear map from V to itself (the case when V is a real vector space will be considered afterwards). First, a definition is required. A linear transformation $g : V \to V$ is called *nilpotent* of index p if $g \circ \cdots \circ g$ (p times) is the zero map on V and $g \circ \cdots \circ g$ ($p-1$ times) is not the zero map on V. Thus every eigenvalue of a nilpotent map is zero and such a map is diagonalisable if and only if it is the zero map. In the special case when f is diagonalisable let $\lambda_1, \ldots, \lambda_r \in \mathbb{C}$ be the distinct eigenvalues of f with multiplicity $m_1, \ldots, m_r$ (so that $\sum_{i=1}^r m_i = n$). Then one may write $V = V_1 \oplus \cdots \oplus V_r$ where, for each i, V_i is the λ_i-eigenspace of f and is an invariant subspace of f of dimension m_i. Further, the restriction of f to V_i is a linear map $V_i \to V_i$ of the form $\lambda_i I^i$ where I^i is the identity map $V_i \to V_i$. The theory of the Jordan canonical form generalises this simple result to the case when f is an arbitrary linear map $V \to V$. It turns out (see, *e.g.* [8]) that if $\lambda_1, \ldots, \lambda_r \in \mathbb{C}$ are the distinct eigenvalues of f with multiplicity $m_1, \ldots, m_r$ ($\sum_{i=1}^r m_i = n$) then again one may write $V = V_1 \oplus \cdots \oplus V_r$ where each V_i is an invariant subspace of f of dimension m_i. However, V_i is not necessarily the λ_i–eigenspace of f (though it certainly contains it). In fact the restriction of f to V_i is of the form $\lambda_i I^i + N_i$ where the linear map $N_i : V_i \to V_i$ is *nilpotent*. Now it can be shown that one may choose a basis in V_i in which the map N_i is represented by a matrix with some arrangements of zeros and ones at each place on the superdiagonal and zeros elsewhere. Since the matrix of the map $\lambda_i I^i$ on V_i is the same diagonal matrix in every basis of V_i one chooses the basis described for N_i above in each V_i to obtain a basis for V. This basis is called a *Jordan basis for* f and the matrix A representing f in this basis is then of the form

$$A = \begin{pmatrix} A_1 & & & \\ & A_2 & & \\ & & \ddots & \\ & & & A_r \end{pmatrix}$$

where A_i is an $m_i \times m_i$ matrix with λ_i in every diagonal position and some arrangement of zeros and ones in each superdiagonal position and where all other entries in A_i and in A are zero. Further each matrix A_i can be written in the form

$$A_i = \begin{pmatrix} B_{i1} & & & \\ & B_{i2} & & \\ & & \ddots & \\ & & & B_{ik(i)} \end{pmatrix}$$

where B_{ij} is a $p_{ij} \times p_{ij}$ matrix whose diagonal entries are each equal to λ_i and whose superdiagonal elements are each equal to one (and all other entries are zero) and where $p_{i1} \geqslant \cdots \geqslant p_{ik(i)}$ (and is called a *basic Jordan block*). Once an ordering is established for the eigenvalues $\lambda_1, \ldots, \lambda_r$ then, with the above conventions, the canonical form for A above is uniquely determined. In other words, there is associated with any such linear transformation f the eigenvalues $\lambda_1, \ldots, \lambda_r$ (in some order), their associated multiplicities $m_1, \ldots, m_r$ and for each i $(1 \leqslant i \leqslant r)$ the numbers p_{ij} chosen such that $p_{i1} \geqslant \cdots \geqslant p_{ik(i)}$ (and, of course, $m_i = p_{i1} + \cdots + p_{ik(i)}$). These quantities (and their orderings) associated with f uniquely determine the matrix A above and this matrix is called the *Jordan (canonical) form for f* or *the Jordan matrix for f*. (If one ignores the orderings described above then A is a Jordan form or a Jordan matrix for f.) In matrix language any matrix representing f in some basis of V is similar to A and A is then referred to as the Jordan form of this representative matrix. With the above quantities and orderings given, two matrices are similar if and only if they are similar to the same Jordan matrix (*i.e.* if and only if they have the same Jordan form).

The general Jordan structure of f or any of its representative matrices (*i.e.* ignoring the actual values of $\lambda_1, \ldots, \lambda_r$ but retaining the numerical ordering of the p_{ij} for convenience) can be encoded in the symbol

$$\{(p_{11}, \ldots, p_{1k(1)})(p_{21}, \ldots, p_{2k(2)}) \cdots (p_{r1}, \ldots, p_{rk(r)})\} \tag{2.2}$$

called the *Segre type* (or *Segre characteristic* or *Segre symbol*) of f or of a matrix representing f. This rather useful notation gathers together inside each set of round brackets the sizes of the B_{ij} sub-matrices within each matrix A_i which are associated with the same eigenvalue λ_i. In particular if f were diagonalisable over $\mathbb{C}$ with eigenvalues $\lambda_1, \ldots, \lambda_r$ of multiplicity $m_1, \ldots, m_r$, respectively, then in the above notation A_i would be an $m_i \times m_i$ *diagonal* matrix with λ_i in every diagonal position (and so the matrices $B_{i1}, \ldots, B_{ik(i)}$ are each 1×1 matrices with entry λ_i and $k(i) = m_i$). Thus each $p_{ij} = 1$ and the Segre symbol for f would be $\{(1 \ldots 1) \ldots (1 \ldots 1)\}$ where the number of entries inside the ith pair of round brackets equals m_i.

Returning to the general case it follows that the Jordan matrix A representing f can be written as $A = D + N$ where D is an $n \times n$ diagonal matrix whose entries are the eigenvalues $\lambda_1, \ldots, \lambda_r$ and N is a matrix with some arrangement of zeros and ones on the superdiagonal and zeros elsewhere (from which one can easily deduce that N^p is the zero matrix for some positive integer p). Hence, with respect to the Jordan basis, D represents a diagonalisable linear map and N a nilpotent linear map $V \to V$. This decomposition of the map f shows the role played by the nilpotent part. The nilpotent part is absent when f is diagonalisable. In fact it now easily follows that a linear map $f : V \to V$ is nilpotent if and only if all its eigenvalues are zero. This result helps considerably with the geometrical interpretation of the Jordan theory. The effects of the restricted maps N_i on V_i can be seen by noting that whereas every non-zero member of V_i would be an eigenvector of f if f were diagonalisable, there is, in the general case, only one independent eigenvector associated with each B_{ij} block within each A_i. The dimension of the λ_i–eigenspace is sometimes called the *geometric multiplicity* of λ_i (and the term *algebraic multiplicity* reserved for what was earlier called *multiplicity*). *The algebraic multiplicity is always greater than or equal to the geometric multiplicity for each* λ_i *and they are equal for each* λ_i *if and only if* f *is diagonalisable.*

For the general linear map f above or any of its representative matrices the characteristic polynomial is

$$(-1)^n(x-\lambda_1)^{m_1}(x-\lambda_2)^{m_2}\dots(x-\lambda_r)^{m_r}.$$

There is another polynomial associated with f which contains additional information about f. First one can prove directly from the Jordan forms the *Cayley-Hamilton theorem*, namely that every matrix such as A above satisfies its own characteristic equation. It follows that there exists a polynomial of least degree m $(1 \leqslant m \leqslant n)$ which is satisfied by A and, moreover, this polynomial is unique if it is agreed that it should be monic. Such a polynomial is called the *minimal polynomial of* A and can be shown to be the same for all matrices in the same similarity class. Hence it may be called the *minimal polynomial of* f. It can be shown that for the general map f considered here, the minimal polynomial of f is

$$(x-\lambda_1)^{p_{11}}(x-\lambda_2)^{p_{21}}\dots(x-\lambda_r)^{p_{r1}}.$$

The power to which $(x-\lambda_i)$ is raised in the minimal polynomial is the largest integer (*i.e.* p_{i1}) amongst the associated $p_{i1},\dots p_{ik(i)}$ (and so the minimal polynomial divides the characteristic polynomial). The polynomials $(x-\lambda_i)^{p_{ij}}$ are called the *elementary divisors* (associated with the eigenvalue λ_i) of A (or indeed of f or any matrix representing f) since they each divide the characteristic polynomial (but only the one of highest power for each i is in the minimal polynomial). An elementary divisor associated with λ_i and with $p_{ij} = 1$ is called *simple* . Otherwise it is *non-simple* of order p_{ij}.

As an example consider the matrix

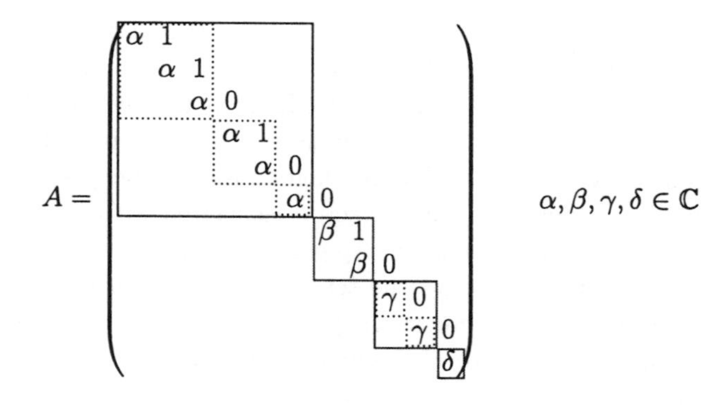

where α, β, γ and δ are distinct and all non-recorded entries are zero. In the above notation the A_i blocks are shown with unbroken lines whilst the B_{ij} blocks within them are indicated with broken lines. The eigenvalues and the algebraic multiplicities are $\alpha(6)$, $\beta(2)$, $\gamma(2)$ and $\delta(1)$. The characteristic polynomial is $(x-\alpha)^6(x-\beta)^2(x-\gamma)^2(x-\delta)$ whilst the minimal polynomial is $(x-\alpha)^3(x-\beta)^2(x-\gamma)(x-\delta)$. The Segre type is $\{(321)(2)(11)(1)\}$ and the elementary divisors are $(x-\alpha)^3, (x-\alpha)^2, (x-\alpha), (x-\beta)^2, (x-\gamma)$ and $(x-\delta)$. There are seven independent eigenvectors which may be chosen as having a one in the third (respectively the fifth, sixth, eighth, ninth, tenth and eleventh) slots, respectively, in $\mathbb{C}^{11}$ and zero elsewhere, the eigenvalues being, respectively, $\alpha, \alpha, \alpha, \beta, \gamma, \gamma$ and δ. The geometric multiplicities of the eigenvalue α is thus 3, of β is 1, of γ is 2 and of δ is 1. In writing out Segre types it is convenient to omit the round brackets around a *single* digit. Thus the above Segre type would be $\{(321)2(11)1\}$.

The above discussion of the Jordan canonical form was applied to a vector space V over the *algebraically closed field* $\mathbb{C}$. However, some of the applications of this theory are to the case of a vector space V over $\mathbb{R}$ and, as discussed earlier, $\mathbb{R}$ is not algebraically closed. Nevertheless if f is a linear transformation from an n-dimensional *real* vector space V to itself *and all the roots of the characteristic equation of f lie in* $\mathbb{R}$ then the Jordan theory discussed here and all the attendant notation still applies. If, on the other hand, a complex conjugate (non-real) pair of solutions of the characteristic polynomial arise they lead to a 2-dimensional invariant subspace of f as described earlier and this remark will be sufficient for the purposes of this text. One could handle this situation by extending f to the complexification of V where the Jordan theory applies. In fact, if one regards this theory as a classification of matrices by similarity, the problem is not affected by such an extension. This follows because two *real* matrices are similar over $\mathbb{R}$ if and only if they are similar over $\mathbb{C}$. (In fact, more can be proved [12].)

A few closing remarks regarding the Jordan canonical form can now be made. Again let V be a vector space over F ($= \mathbb{R}$ or $\mathbb{C}$) of finite dimension n and let f be a linear map $f : V \to V$. An eigenvalue λ of f may be "degenerate" in two different ways; firstly due to the appearance of two elementary divisors of f (simple or non-simple) each associated with the same eigenvalue λ of f and secondly due to the appearance of a *non-simple* elementary divisor associated with λ. The former can, in a sense, be regarded as an "accidental" degeneracy (*i.e.* the accidental equality of the two elementary divisors) and is indicated by the use of round brackets in the Segre characteristic of f. The latter is indicated by a digit greater than one in the Segre characteristic. An eigenvalue can, of course, be degenerate in both ways. The difference between these two types of degeneracy is reflected to some extent in the differences between the geometric and algebraic multiplicity of the eigenvalue. Here, an eigenvalue λ will be called *non-degenerate* (respectively, *degenerate*) if the λ-eigenspace has dimension 1 (respectively, > 1).

Using the notation of the preceding paragraph one might ask how tightly the Jordan basis for f is determined (assuming that, if $F = \mathbb{R}$, all eigenvalues of f are real). If f may be represented by a diagonal matrix (in a Jordan basis of eigenvectors $\{\mathbf{e}_i\}$) then any other Jordan basis $\{\mathbf{e}'_i\}$ for f is related to the original one by linear transformations of the basis members within the eigenspaces of f (and, in particular, $\mathbf{e}'_1 = \mu_1\mathbf{e}_1, \ldots, \mathbf{e}'_n = \mu_n\mathbf{e}_n$, $\mu_1, \ldots, \mu_n \in F$ if the eigenvalues of f are distinct.) Other types of ambiguities in the Jordan basis are introduced (by the non-simple elementary divisors) if f is not diagonalisable. Now let $A \in M_n(F)$ ($F = \mathbb{R}$ or $\mathbb{C}$). It is a well-known result that *A is similar to its transpose*. This result will be useful in a later chapter. It also shows that if (as some authors do) one takes the Jordan canonical matrices as being the transposes of the ones defined here, then nothing is changed. It is remarked here, for later reference, that if two matrices in M_nF have the same characteristic and minimal polynomials they are not necessarily similar. This follows, for example, by considering two matrices of Segre types $\{(32)1\}$ and $\{(311)1\}$, respectively, which each have characteristic polynomial $(x-\alpha)^5(x-\beta)$ and minimal polynomial $(x-\alpha)^3(x-\beta)$ for $\alpha, \beta \in F$.

Let V be a real vector space and $f : V \to V$ a linear map with a non-real conjugate pair of eigenvalues. Suppose that, when the situation is "complexified" (as described earlier) $\tilde{f}$ is diagonalisable. Then one often says that "f is diagonalisable over $\mathbb{C}$ but not over $\mathbb{R}$" to describe this situation and the pair of symbols $z\bar{z}$ will be inserted in the Segre symbol for f. Thus if $\dim V = 4$ and f has, say, two distinct real eigenvalues and a non-real conjugate pair of eigenvalues its Segre symbol will be written $\{z\bar{z}11\}$ (or $\{11z\bar{z}\}$).

2.7 Lie Algebras

Let V be a vector space over a field F ($= \mathbb{R}$ or $\mathbb{C}$). Then V is called an *algebra* if there is a binary operation on V, $(\mathbf{u}, \mathbf{v}) \to \mathbf{uv}$ ($\mathbf{u}, \mathbf{v} \in V$) such that for $a \in F$ and $\mathbf{u}, \mathbf{v}, \mathbf{w} \in V$

(A1) $(\mathbf{u}+\mathbf{v})\mathbf{w} = \mathbf{uw} + \mathbf{vw}$
(A2) $\mathbf{u}(\mathbf{v}+\mathbf{w}) = \mathbf{uv} + \mathbf{uw}$
(A3) $a(\mathbf{uv}) = (a\mathbf{u})\mathbf{v} = \mathbf{u}(a\mathbf{v})$

[An algebra V satisfying the extra condition $(\mathbf{uv})\mathbf{w} = \mathbf{u}(\mathbf{vw})$ is called *associative*.] If further V is a real vector space and satisfies, in addition to (A1), (A2) and (A3) the conditions

(A4) $(\mathbf{uv}) = -(\mathbf{vu})$
(A5) $\mathbf{u}(\mathbf{vw}) + \mathbf{v}(\mathbf{wu}) + \mathbf{w}(\mathbf{uv}) = 0$

then V is called a *Lie algebra*. If V is a finite-dimensional Lie algebra of dimension n and $\{\mathbf{e}_i\}$ is a basis for V then there exist real numbers C^c_{ab} such that $\mathbf{e}_a\mathbf{e}_b = \sum_{c=1}^{n} C^c_{ab}\mathbf{e}_c$. The numbers C^c_{ab} are called the *structure constants* for V in the basis $(\mathbf{e}_a)$ and, from (A4) and (A5) satisfy

$$C^c_{ab} + C^c_{ba} = 0, \qquad \sum (C^e_{ab}C^c_{ed} + C^e_{bd}C^c_{ea} + C^e_{da}C^c_{eb}) = 0.$$

The "associative" condition is not necessarily satisfied. The condition (A5) is sometimes referred to as the *Jacobi identity*. A subspace U of V which is itself a Lie algebra under the binary operation induced on it by V is called a (Lie) *subalgebra* of V. If V, W are Lie algebras and $f : V \to W$ is a vector space homomorphism (respectively, isomorphism) such that (with an obvious simplifying of notation) $f(\mathbf{uv}) = f(\mathbf{u})f(\mathbf{v})$ ($\mathbf{u}, \mathbf{v} \in V$) then f is called a *Lie algebra homomorphism* (respectively, *isomorphism*), and V and W are said to be *Lie algebra homomorphic* (respectively, *isomorphic*).

The set $M_n\mathbb{R}$ is a finite dimensional real vector space which becomes a Lie algebra if one defines the binary operation $(A, B) \to AB - BA$.

Chapter 3

Topology

3.1 Introduction

Topology is one of the most important branches of mathematics. A topology for a set is a certain type of "structure" for that set which allows one to discuss sensibly such important concepts as "convergence" and "limits" of sequences in this set and the idea of a "continuous function" between such sets. It also permits the description of certain subsets of a set as "large" or "small" and of points which are arbitrarily close to certain subsets of the set. Most geometrical "pictures" of mathematical constructions assume some topology on the set in question and hence its importance.

Perhaps a more natural "structure" for a set is a "distance structure" namely the ability to attach some positive real number to each pair of distinct points of the set (subject to certain sensible requirements) and which is interpreted as the "distance" between them. Such a "metric" structure should make sense of those important concepts described above which follow from a topological structure and hence should give rise to a natural topological structure for the set in question. However, the converse is not true in the sense that a definition of a topological structure has evolved during the 20th century which is sufficiently strong to do all that is required of it but which does not imply the existence of a "compatible" metric structure for the set concerned. Thus a topological structure is more general then a metric structure and the latter is a special case of the former. It follows that one can proceed either by starting with the axioms for a topological structure and show how a metric structure is a special case or by starting with the ideas of a metric structure and generalising to those of a topology. The former approach has much aesthetic appeal but its generality in the initial stages tends to obscure the proceedings. The second approach will be adopted here. The multitude of texts on topology render it unnecessary for this chapter to be detailed. Its intention is to briefly sketch out the ideas and give the definitions of those concepts required for a reasonable working knowledge of metric spaces and topology. The proofs of results which require more than a few elementary remarks will be omitted and can be found in the extensive bibliography of topology texts of which [14],[15],[16] are excellent for the general sections 3.2 to 3.9 and [17],[18] are especially useful for section 3.10. The reader should be warned that not all definitions in topology are standardised and that they should be checked carefully, especially in older texts.

3.2 Metric Spaces

Let X be any set. A *metric* (structure) for X is a map d from the Cartesian product $X \times X$ to the non-negative real numbers such that for any $x, y, z \in X$ the following hold.

(M1) $d(x,y) = d(y,x)$
(M2) $d(x,y) = 0 \Leftrightarrow x = y$
(M3) $d(x,y) + d(y,z) \geqslant d(x,z)$

Although these axioms can be recast in many equivalent ways the above set seems the most natural and no member of it is redundant. The pair (X,d) is called a *metric space* . For any set X the function d defined by $d(x,y) = 1$ $(x \neq y)$ and $d(x,x) = 0$ is a metric for X called the *discrete* metric for X. Thus *any* set can be given a metric structure. [It might perhaps be remarked at this point that for most purposes the axiom M2 could be weakened to $d(x,x) = 0$, $\forall x \in X$, without affecting the general ideas involved. Such a map d is then called a *pseudo-metric* for X [14] and (X,d) a *pseudometric space.* For any set X with at least two members the map d given by $d(x,y) = 0$, $\forall x, y \in X$, is a pseudometric (but not a metric) for X. Pseudometrics will not be required in this text.] Each of the maps d, d_1 and d_2 given by

$$d(\mathbf{x},\mathbf{y}) = \left(\sum_{i=1}^{n}(x^i - y^i)^2\right)^{1/2}, \quad d_1(\mathbf{x},\mathbf{y}) = \max\left(|x^i - y^i|, 1 \le i \le n\right)$$

$$d_2(\mathbf{x},\mathbf{y}) = \sum_{i=1}^{n} |x^i - y^i|, \quad \mathbf{x} = (x^1,\ldots,x^n) \quad \text{and} \quad \mathbf{y} = (y^1,\ldots,y^n) \in \mathbb{R}^n$$

is a metric for $\mathbb{R}^n$ and d is called the *standard* (or *usual*) metric for $\mathbb{R}^n$. If (X, d) is a metric space and $A \subseteq X$ the restriction $\bar{d}$ of d to $A \times A$ is a metric for A called the *induced metric* on A. If $(X_1, d_1), \ldots, (X_n, d_n)$ are metric spaces then $d(x, y) = \sum_{i=1}^{n} d_i(x_i, y_i)$ is a metric for $X_1 \times \cdots \times X_n$ where $x = (x_1, \ldots, x_n)$, $y = (y_1, \ldots, y_n)$ and $x_i, y_i \in X_i$.

Let (X, d) and (Y, d') be metric spaces and $f : X \to Y$ a map. It is important to note that the existence of metrics on both X and Y allows a criterion of continuity for f to be stated in that f would be regarded as "continuous" if for each $x_0 \in X$ the point $f(x) \in Y$ can be made arbitrarily close to $f(x_0)$ (with respect to the metric d' on Y) by choosing x arbitrarily close to x_0 (with respect to the metric d on X). Thus f is called *continuous at* x_0 if given $\varepsilon > 0$ there exists $\delta > 0$ such that $d(x, x_0) < \delta \Rightarrow d'(f(x), f(x_0)) < \varepsilon$. The map f is then called *continuous* if it is continuous at all points of X.

Next let (X, d) be a metric space. A *sequence* in X (in fact in any set) is a map $s : \mathbb{Z}^+ \to X$ and is usually "identified" with the ordered set of values of the function s and denoted by $\{s_n\}$ where $s_n = s(n)$. A sequence $\{s_n\}$ in X might, roughly speaking, be regarded as "converging" to a point $s \in X$ if by "proceeding sufficiently far" along the sequence $s_1, s_2, s_3, \ldots$ one can *get* and *stay* arbitrarily close to s. Again the existence of a metric d on X allows this concept to be stated naturally and precisely. The sequence $\{s_n\}$ in X *converges* to $s \in X$ if given any $\varepsilon > 0$ there exists $N \in \mathbb{Z}^+$ such that $n \geqslant N \Rightarrow d(s_n, s) < \varepsilon$. If $\{s_n\}$ converges to s one writes $s_n \to s$ or $\lim_{n\to\infty} s_n = s$ and s is the *limit* of $\{s_n\}$.

The ideas of these last two paragraphs can be combined to give another perhaps more familiar notion of continuity of the above function f. It can be shown that f is continuous at x_0 if and only if for any sequence $\{s_n\}$ such that $s_n \to x_0$ then the sequence $\{f(s_n)\}$ in Y converges to $f(x_0)$. Unfortunately this result relies on a specific property of metric spaces and is no longer true, as stated, for general topological spaces unless one generalises the concept of a sequence to that of a *net* (or a *filter*). However, for the topological spaces dealt with in this text, the result is true and no generalised concepts of convergence over and above the usual (sequential) convergence as defined earlier will be required.

One can also decide whether two metric spaces (X, d) and (Y, d') are essentially the same metric space. Such a pair of spaces will be called *isometric* if there exists a surjective map $f : X \to Y$ which preserves distance, that is, for each $x, y \in X$, $d(x, y) = d'(f(x), f(y))$. The map f is called an *isometry* and it follows that f is necessarily injective, that the resulting inverse map f^{-1} is also an isometry and that f and f^{-1} are necessarily continuous.

Again, if (X, d) is a metric space, $A \subseteq X$ and $x \in X$, $x \notin A$, one can define the statement "x is close to A" as meaning that for any $\varepsilon > 0$ there exists $y \in A$ such that $d(x, y) < \varepsilon$. This concept together with those of continuity and convergence are all available for sets with a metric structure.

However, as will be indicated below, a closer inspection of these concepts reveals that although defined in terms of metric structure they can each be cast into another equivalent form which does not use the actual existence of the metric but rather only some consequences of this existence. Further these consequences do not need the existence of the metric from whence they came (or any other metric) but can stand alone as a basis for the axiomatisation of another more general structure for X within which continuity, convergence, "closeness" and other similar topological concepts make good sense.

To see this let (X, d) be a metric space, let $p \in X$ and let $\varepsilon > 0$. Define $N_\varepsilon(p) = \{x \in X : d(x, p) < \varepsilon\}$ (called *the open ε-ball about* p) and $B_\varepsilon(p) = \{x \in X : d(x, p) \leq \varepsilon\}$ (called the *closed ε-ball about* p). Now call a subset U of X *open* (in (X, d)) if for any $p \in U$ there exists $\varepsilon > 0$ such that $N_\varepsilon(p) \subseteq U$. The family of open subsets of X (which includes all the $N_\varepsilon(p)$ for all choices of p and ε) are all that is required to develop the topological concepts introduced above. In fact it is straightforward to show that

i) if (X, d) and (Y, d') are metric spaces a map $f : X \to Y$ is continuous at $x_0 \in X$ if and only if whenever V is a subset of Y satisfying $f(x_0) \in U \subseteq V$ for some open subset U in Y there exists an open subset W in X such that $x_0 \in W \subseteq f^{-1}(V)$ and that f is continuous (*i.e.* continuous at each $x \in X$) if and only if whenever U is open in (Y, d') then $f^{-1}(U)$ is open in (X, d),

ii) a sequence $\{s_n\}$ in X converges to $s \in X$ if and only if for any open set U in (X, d) containing s there exists $N \in \mathbb{Z}^+$ such that $n \geqslant N \Rightarrow s_n \in U$ and

iii) if $A \subseteq X$ and $x \in X$ then "x is close to A" if and only if whenever U is an open set in (X, d) containing x then $U \cap A \neq \emptyset$.

Of course, an open set was defined in terms of a metric. In order that the idea of an open set can break free from the idea of a metric one needs an axiomatisation which starts with the set X and decides which subsets of X are to be regarded as "open". The rules governing such a decision must, of course, be satisfied by the open subsets defined above for a metric space.

There are several ways of describing this axiomatisation and the one given here is the usual one and perhaps the simplest.

3.3 Topological Spaces

Let X be a set. A *topology* (or *topological structure*) for X is a family $\mathcal{T}$ of subsets of X such that the following conditions hold.

(T1) $\emptyset \in \mathcal{T}$
(T2) $X \in \mathcal{T}$
(T3) The union of an *arbitrary* family of members of $\mathcal{T}$ is itself a member of $\mathcal{T}$.
(T4) The intersection of a *finite* family of members of $\mathcal{T}$ is itself a member of $\mathcal{T}$.

The pair $(X, \mathcal{T})$ is called a *topological space* and the members of $\mathcal{T}$ are called *open* subsets of X. It should be noted that axiom (T3) involves *arbitrary* families whereas axiom (T4) involves *finite* families (and, in fact, (T4) could be replaced by the insistence that the intersection of any *two* members of $\mathcal{T}$ is a member of $\mathcal{T}$). It is easy to verify that the open sets defined above in a metric space (X, d) satisfy these axioms and the resulting topology for X is sometimes denoted $\mathcal{T}_d$ and called the *standard (metric) topology for* X (or for (X, d)). However, as will be established more conveniently a little later, not every topology $\mathcal{T}$ for a set X arises from some metric on X. Thus the concept of a topological space is more general than a metric space. (A review of the work which establishes necessary and sufficient conditions for a topological space $(X, \mathcal{T})$ to be *metrisable*, that is for there to exist a metric d on X such that $\mathcal{T}$ is the metric topology $\mathcal{T}_d$ on X arising from d, is given in [14],[15]). A subset of X is called *closed* if its complement is open. Thus, from the De-Morgan laws in set theory, the above four axioms now become *properties* of closed sets provided that the word "open" is replaced by the word "closed" and that the words "arbitrary" and "finite" in axioms (T3) and (T4), respectively, are interchanged. In fact these four properties of *closed* sets can be used as an alternative axiom system for a topology; one simply postulates the existence of a family of subsets of X with these (closed) set properties and then the complements of the members of this family give a topology for X.

The examples of metric spaces given above serve as examples of (metrisable) topological spaces as indicated earlier and the topology resulting from the standard metric on $\mathbb{R}^n$ is called the standard (or usual) topology on $\mathbb{R}^n$. In fact $\mathbb{R}^n$ will always be assumed to have this topology unless otherwise stated. Some other less standard examples can now be given. Let X be any set. If $\mathcal{T}$ is the family of *all* subsets of X then $\mathcal{T}$ is a topology for X called the *discrete* topology (and is the metric topology for X with the discrete metric). Alternatively one could take $\mathcal{T} = \{\emptyset, X\}$ as a topology for X (the *indiscrete* topology for X). In this case $\mathcal{T}$ is not metrisable if X contains at least two members. Another topology for X is the family of subsets of X whose complement is finite, together with the set $\emptyset$. This topology, called the *co-finite topology* for X, is just the discrete topology for X if X is finite. If in the last two sentences finite is replaced by countable then one obtains the *co-countable* topology for X. Again, one could start with a pseudometric space and construct a topology for it from the pseudometric just as was done above for a metric space. For a metric space, any closed ball is easily shown to be a closed set in the resulting topology and thus open and closed intervals in $\mathbb{R}$ are open and closed sets respectively in the (standard) topology resulting from the standard metric on $\mathbb{R}$.

Let $(X, \mathcal{T})$ be a topological space and let $x \in X$. A subset N of X containing x is called a *neighbourhood of* x if there exists an open set U such that $x \in U \subseteq N$. A neighbourhood N of x need not be open nor need it be a neighbourhood of some point $y \in N$, $y \neq x$. A subset N of X is a neighbourhood of each of its points if and only if N is open in X. Again let $x \in X$ and $A \subseteq X$. Then x is a *limit point* of A if every open set containing x intersects A in some point other than (possibly) x. (For $x \notin A$ this is the idea of "closeness" mentioned earlier.) This definition has the intuitively pleasing consequence that a subset $A \subseteq X$ is closed if and only if A contains all its limit points. These concepts lead to two operations applicable to any subset A of X which, respectively, reduce it to its "largest contained" open subset and increase it to its "smallest containing" closed subset. For the first one define the *interior of* A, $\operatorname{int} A$, to be the set of all points in A of which A is a neighbourhood and for the second one define the *closure* $\bar{A}$ of A to be the union of A and all its limit points. It is then straightforward to show that $\operatorname{int} A$ is open, that $\bar{A}$ is closed, and that if U is an open set and B a closed set such that $U \subseteq A \subseteq B$ then $U \subseteq \operatorname{int} A \subseteq A \subseteq \bar{A} \subseteq B$. It follows for any subset A that A is open if and only if $A = \operatorname{int} A$, that A is closed if and only if $A = \bar{A}$ and that $\overline{X \setminus A} = X \setminus \operatorname{int} A$. A closely related concept is that of boundary. A point $x \in X$ is called a *boundary point of* $A \subseteq X$ if $x \notin \operatorname{int} A$ *and* $x \notin \operatorname{int}(X \setminus A)$. The *boundary* $\delta(A)$ of A is the set of all boundary points of A. Thus $\delta(A) = \bar{A} \cap (\overline{X \setminus A})$ (and is hence always a closed set) and $\delta(A) = \delta(X \setminus A)$. An intuitively pleasing result regarding the boundaries of open and closed sets is the pair of facts that A is closed if and only if $\delta(A) \subseteq A$ and A is open if only only if $A \cap \delta(A) = \emptyset$. For any topological space $(X, \mathcal{T})$ and subset A, X admits a disjoint decomposition $X = \operatorname{int} A \cup \delta(A) \cup \operatorname{int}(X \setminus A)$. A subset A of X is called *dense* (in X) if $\bar{A} = X$ and a subset B of X is called *nowhere dense* if $\operatorname{int} \bar{B} = \emptyset$. These definitions allow a statement of "topological" size. One can think of a subset A of X as being "large" in (the topology of) X if it is *open and dense* and a subset B of X as being "small" in (the topology of) X if it is *closed and nowhere dense*. Also a subset A of X is dense if and only if $\operatorname{int}(X \setminus A) = \emptyset$ and so the complement of an open, dense subset is closed and nowhere dense, and vice versa. It should be understood here that a subset A of X which is merely dense need not be "large" if it is not open; it may simply be "well distributed" as the subset $\mathbb{Q}$ of $\mathbb{R}$ shows since, in the usual topology on $\mathbb{R}$, $\mathbb{Q}$ is dense but not open. It is easily checked that the intersection of *finitely many* open dense subsets of a topological space is open and dense and hence that the union of finitely many closed, nowhere dense sets is closed and nowhere dense.

The concepts of continuity and convergence described earlier in a metric space are conveniently described in any topological space $(X, \mathcal{T})$. A sequence $\{x_n\}$ in X is said to *converge with limit* x if given any neighbourhood A of x there exists $N \in \mathbb{Z}^+$ such that $n \geqslant N \Rightarrow x_n \in A$. Also, if $(X, \mathcal{T})$ and $(Y, \mathcal{T}')$ are topological spaces and $f : X \to Y$ a map then f is called *continuous* at $x \in X$ if $f^{-1}(N)$ is a neighbourhood of x whenever N is a neighbourhood of $f(x)$. Then f is called *continuous* if it is continuous at each $x \in X$. It follows that f is continuous if and only if whenever U is open in Y, $f^{-1}(U)$ is open in X (and that f is continuous if and only if whenever V is closed in Y, $f^{-1}(V)$ is closed in X). Further, if $(X, \mathcal{T})$, $(Y, \mathcal{T}')$ and $(Z, \mathcal{T}'')$ are topological spaces and $f : X \to Y$ and $g : Y \to Z$ are continuous then $g \circ f : X \to Z$ is continuous. If $(X, \mathcal{T})$ and $(Y, \mathcal{T}')$ are topological spaces and $f : X \to Y$ is a bijective map such that f and f^{-1} are continuous then f is called a *homeomorphism* and $(X, \mathcal{T})$ and $(Y, \mathcal{T}')$ are said to be *homeomorphic* (*i.e.* topologically identical). The continuity of both f and f^{-1} is essential here since one of these maps could be continuous and the other not (as for example occurs if $X = Y$ with X containing more than one member, if f is the identity map on X and if $\mathcal{T}$ and $\mathcal{T}'$ are the discrete and the indiscrete topologies for X, respectively). Clearly, the relation of being homeomorphic is an equivalence relation on topological spaces.

The above paragraphs give the definition and basic properties of topological spaces. They apply to metric spaces through the natural topology on such spaces induced by the metric. The *topology* of a metric space does not require all the detail of the metric but only the open sets that it leads to. For example, the three metrics on $\mathbb{R}^n$ described in the previous section are *different* metrics for $\mathbb{R}^n$ but it is easy to show that each gives rise to the *same topology* for $\mathbb{R}^n$ (*i.e.* the standard topology). The topology is thus blind to the differences between these metrics. One calls two metrics d and d' for the same set X *equivalent* if the resulting metric topologies $\mathcal{T}_d$ and $\mathcal{T}_{d'}$ are equal. If such is the case, whilst d and d' can be different metrics for X (*i.e.* the identity map $i_X : X \to X$ is not an isometry), the topological spaces $(X, \mathcal{T}_d)$ and $(X, \mathcal{T}_{d'})$ are identical (*i.e.* i_X is a homeomorphism between them). Of course there may still exist an isometry f between (X, d) and (X, d') with $f \neq i_X$ but there need not. For let $X = \mathbb{R}$ and d be the usual metric for $\mathbb{R}$. The map $f : x \to \tan x$ is a bijection $(-\frac{\pi}{2}, \frac{\pi}{2}) \to \mathbb{R}$ and gives rise to a metric d' on $\mathbb{R}$ by "transferring" the metric d (restricted to $(-\frac{\pi}{2}, \frac{\pi}{2})$) onto $\mathbb{R}$, *i.e.* $d'(x, y) = d\left(f^{-1}(x), f^{-1}(y)\right)$. It clearly follows that there is *no* isometry between $(\mathbb{R}, d)$ and $(\mathbb{R}, d')$ but $(\mathbb{R}, \mathcal{T}_d)$ and $(\mathbb{R}, \mathcal{T}_{d'})$ are homeomorphic (and the identity map $i_{\mathbb{R}}$ is such a homeomorphism). One is thus led to the concept of a *metric property* of a metric space (one which if true for (X, d) is true for all metric spaces isometric to (X, d)) and a *topological property* of (X, d) (one which if true for (X, d) is true for any metric space homeomorphic to (X, d)). Thus topological properties are metric properties but not necessarily vice versa. A metric property (but not a topological property) may be destroyed by changing the metric d for a certain set X to a metric d' equivalent to d. A rather nice example of this feature is the concept of a bounded metric space. A metric space (X, d) is called *bounded* (and d is called a *bounded metric* for X) if there exists $\alpha \in \mathbb{R}$ such that $d(x, y) \leqslant \alpha$ for each $x, y \in X$. A subset A of a metric space (X, d) is called *bounded* if A with its induced metric is a bounded metric space. Now for *any* metric space (X, d) the function d' defined by $d'(x, y) = \min\left(d(x, y), 1\right)$ can be shown to be a bounded metric for X which is equivalent to d. Hence any metric space is homeomorphic to (but not necessarily isometric to) a bounded metric space. Hence "boundedness" is a metric property but not a topological property of metric spaces.

3.4 Bases

If (X, d) is a metric space and U is an open subset of X then for each $x \in U$ there exists $\delta > 0$ such that $x \in N_\delta(x) \subseteq U$. It follows that U is the union of open sets of the form $N_\varepsilon(x)$ where x ranges over all the points of U and the positive real number $\varepsilon = \varepsilon(x)$ is determined by x. Thus the open balls in X constitute a privileged subfamily of $\mathcal{T}_d$ from which *all* open subsets of X can be constructed. This particularly useful construction can be applied to a general topological space $(X, \mathcal{T})$. A subfamily $\mathcal{B}$ of $\mathcal{T}$ is called a *base* (or *basis*) for $\mathcal{T}$ if each member U of $\mathcal{T}$ is a union of members of $\mathcal{B}$. Then $\mathcal{T}$ is said to be *generated* by $\mathcal{B}$. It follows that $\mathcal{B}$ is a base for $\mathcal{T}$ if and only if for each $x \in X$ and each open set U of X containing x there exists $B \in \mathcal{B}$ with $x \in B \subseteq U$. For any topological space $(X, \mathcal{T})$, $\mathcal{T}$ is itself a (rather uninteresting) base for $\mathcal{T}$ whilst the family of (finite) open intervals is a base for $\mathbb{R}$ with its usual topology. A more interesting base for this last topological space is the (countable) family of (finite) open intervals with rational endpoints. As described above, the family of all open balls in a metric space (X, d) is a base for $\mathcal{T}_d$.

The idea of a base can be "reversed" by taking a set X and a family $\mathcal{B}$ of subsets of X and asking if there exists a topology $\mathcal{T}$ for X for which $\mathcal{B}$ is a base. By using the definition of a base above it is not hard to prove that such a topology exists if and only if X equals the union of all members of $\mathcal{B}$ and if whenever $B_1, B_2 \in \mathcal{B}$ and $p \in B_1 \cap B_2$ there exists $B \in \mathcal{B}$ such that $p \in B \subseteq B_1 \cap B_2$. It is clear from the examples given above that a particular topological space may possess many bases. An easily proved criterion restricting this situation says that if X is a set and $\mathcal{B}$ and $\mathcal{B}'$ are families of subsets of X which are bases for the topologies $\mathcal{T}$ and $\mathcal{T}'$ on X then $\mathcal{T} = \mathcal{T}'$ if and only if (i) for each $B \in \mathcal{B}$ and each $x \in B$ there exists $B' \in \mathcal{B}'$ with $x \in B' \subseteq B$ and (ii) for each $B' \in \mathcal{B}'$ and each $x \in B'$ there exists $B \in \mathcal{B}$ with $x \in B \subseteq B'$. If $\mathcal{T} = \mathcal{T}'$ the bases $\mathcal{B}$ and $\mathcal{B}'$ are said to be *equivalent*.

If X is a set and $\mathcal{B}$ a family of subsets of X one sometimes requires a topology $\mathcal{T}$ for X in which the members of $\mathcal{B}$ are open sets (*i.e.* $\mathcal{B} \subseteq \mathcal{T}$). Of course if $\mathcal{B}$ satisfies the conditions for being a base for some topology on X then one could identify this topology with $\mathcal{T}$. In any case one could solve the problem trivially by giving X the discrete topology. However, given that the union of the members of $\mathcal{B}$ equals X, there is a more interesting and usually more useful solution to the problem which, in a precise sense, contains no more sets than is necessary. One simply notes that the family

$\mathcal{B}'$ consisting of all *finite* intersections of members of $\mathcal{B}$ *is* a base for some topology $\mathcal{T}$ on X and that if $\mathcal{T}'$ is any other topology for X for which all members of $\mathcal{B}$ are open, $\mathcal{T} \subseteq \mathcal{T}'$. If this is the case $\mathcal{B}$ is called a *subbase* (or *subbasis*) for the topology $\mathcal{T}$ and is said to *generate* $\mathcal{T}$. For example, the family of subsets of $\mathbb{R}$ which are either of the form $(-\infty, a)$ or (a, ∞) for each $a \in \mathbb{R}$ is a subbase for the usual topology on $\mathbb{R}$.

It is often important to know those open subsets of a topological space $(X, \mathcal{T})$ which contain a particular point $x \in X$. A family $\mathcal{D}$ of open subsets each of which contains x is called a *local base* (or *basis*) *at* x if whenever $U \in \mathcal{T}$ and $x \in U$ there exists $D \in \mathcal{D}$ such that $x \in D \subseteq U$. The connection between a base for $(X, \mathcal{T})$ and the collection of local bases at each $x \in X$ is easily established. In fact if $\mathcal{B}$ is a collection of subsets of X then $\mathcal{B}$ is a base for $(X, \mathcal{T})$ if and only if for each $x \in X$ the family $\mathcal{B}_x = \{B \in \mathcal{B} : x \in B\}$ is a local base at x.

Two rather important definitions for a topological space can now be given. A topological space is called *first countable* if there is a local base at each $x \in X$ with countably many members (*i.e.* a countable local base) and it is called *second countable* if it admits a base with countably many members (a countable base). Clearly, by the result at the end of the last paragraph a second countable topological space is necessarily first countable but the converse result is false as is easily seen by considering an uncountable set X with discrete topology. First countable topological spaces are important in the theory of convergence. In fact there is a sense in which "sequences are adequate" for describing matters of convergence in first countable topological spaces whereas for non-first countable spaces the more general concept of convergence involving nets and filters must be used. A detailed discussion of this can be found in [14]-[16]. Only first countable topological spaces will be required in this text (and it should be pointed out in this context that any metric space is first countable in its metric topology since the family $N_{\frac{1}{n}}(x)$ $(n \in \mathbb{Z}^+)$ is a countable local base at any $x \in X$, but not necessarily second countable). The concept of *second countability* is important in that it leads to several other useful topological properties. One rather interesting one is that a second countable topological space $(X, \mathcal{T})$ is "not too large" in the sense that if $A \subseteq X$ and A is uncountable then A is "cramped" at some point or, more precisely, some point of A is a limit point of A. A metric space need not be second countable but, as shown earlier, the set $\mathbb{R}$ with its usual topology admits a countable base and is thus second (and hence first) countable.

Bases, local bases and subbases can be rather useful. For example, a map $f : X \to Y$ between topological spaces is continuous if and only if the inverse image under f of each member of a base (or, in fact, each member of a subbase) of Y is open in X. Also, a sequence $\{x_n\}$ in a topological space X converges to $x_0 \in X$ if and only if for any member B of a local base at x_0 there exists $N \in \mathbb{Z}^+$ such that $n \geqslant N \Rightarrow x_n \in B$.

3.5 Subspace Topology

Let (X, d) be a metric space, A a subset of X and $\bar{d}$ the induced metric on A. The open ε-balls in $(A, \bar{d})$ are merely the intersections with A of the sets $N_\varepsilon(x)$ for each $x \in A$ and the open subsets of A (in the metric topology on A arising from the metric $\bar{d}$) are then the intersections with A of open subsets in X. Thus one sees how a natural topology might be induced on any subset A of any topological space $(X, \mathcal{T})$. One defines a topology $\bar{\mathcal{T}}$ on A by setting $\bar{\mathcal{T}} = \{U' \subseteq A : U' = A \cap U,\ U \in \mathcal{T}\}$, that is, the open subsets of $(A, \bar{\mathcal{T}})$ are precisely the intersections with A of open subsets of X. That $\bar{\mathcal{T}}$ is a topology for A is easily established and is called the *subspace* or *relative topology on* A (from the topology $\mathcal{T}$ on X). One is then relieved to find that if A and B are subsets of X with $B \subseteq A$ then the subspace topology on B due to its being a subset of X (with the topology $\mathcal{T}$) is the same as its subspace topology as a subset of A when the latter has its subspace topology $\bar{\mathcal{T}}$ from $(X, \mathcal{T})$. This is easily done and the topological space $(A, \bar{\mathcal{T}})$ is called a (topological) *subspace* of $(X, \mathcal{T})$. Subspace topologies are well-behaved. It is easy to check, for example, that the closed subsets in $(A, \bar{\mathcal{T}})$ are precisely the intersections with A of the closed subsets of X and that if C is a subset of A then a point $p \in A$ is a limit point of C in the topology $\bar{\mathcal{T}}$ on A if and only if it is a limit point of C in the original topology $\mathcal{T}$ on X. Closures and interiors in subspace topologies can also be easily handled and if $N \subseteq A$ is a neighbourhood of $x \in A$ in $\mathcal{T}$ then $N \cap A$ is a neighbourhood of x in $\mathcal{T}'$. Also, if $\mathcal{B}$ is a base for X then the intersection of the members of $\mathcal{B}$ with A give a base for the topology $\bar{\mathcal{T}}$ on A. A similar result holds for local bases. Now suppose $(X, \mathcal{T})$ and $(Y, \mathcal{T}')$ are topological spaces, $A \subseteq X$ and $f : X \to Y$ a continuous map. Then the restriction $f_{|A} : A \to Y$ is continuous as a map between the topological spaces $(A, \bar{\mathcal{T}})$ and $(Y, \mathcal{T}')$. A similar result holds for continuity at individual points of A. Further, the continuity of $f : X \to Y$ is equivalent to the continuity of the map $\tilde{f} : X \to f(X)$ defined by $\tilde{f}(x) = f(x)$ when $f(X)$ has subspace topology from $(Y, \mathcal{T}')$.

It also follows that the inclusion map $i_A : A \to X$ is continuous when A has subspace topology. In fact $\mathcal{T}'$ is the "smallest" topology for A (in an obvious sense) which would make i_A continuous and this result is often invoked to give an alternative definition of subspace topology. Finally, if $\{x_n\}$ is a sequence in A and $x_0 \in A$ then $x_n \to x_0$ in the topological space $(X, \mathcal{T})$ if and only if $x_n \to x_0$ in the topological space $(A, \bar{\mathcal{T}})$.

If (X, d) is a metric space and $A \subseteq X$ the topology $\mathcal{T}_{\bar{d}}$ on A arising from the naturally induced metric $\bar{d}$ on A is the same as the subspace topology on A arising from the topology $\mathcal{T}_d$ on X. It is interesting to ask which topological properties of the topological space $(X, \mathcal{T})$ are inherited by its subspaces. Such properties are termed *hereditary* and the topological properties of being first or second countable and of being metrisable are, from the above discussion, each hereditary. Others will be encountered later.

A subset A of a topological space X is called *discrete* (in X) or a *set of isolated points* if given $a \in A$ there is an open subset U of X satisfying $a \in U$ and $U \cap A = \{a\}$. In other words, the induced topology on A from $\mathcal{T}$ is discrete.

3.6 Quotient Spaces

Let $(X, \mathcal{T})$ be a topological space and let $\sim$ be an equivalence relation on X which gives rise to a partition $\mathcal{P}$ of X (so that a member of $\mathcal{P}$ containing $x \in X$ consists of all points y of X such that $x \sim y$). Let p be the map $X \to \mathcal{P}$ which associates with each $x \in X$ the unique member of $\mathcal{P}$ in which it lies. The family $\{U \subseteq \mathcal{P} : p^{-1}(U) \in \mathcal{T}\}$ is a topology for $\mathcal{P}$ called the *quotient topology* for $\mathcal{P}$ and the resulting topological space is called the *quotient space of X by* $\sim$ and will be denoted by $X/\sim$. The map p (called the *projection map*) is now continuous. Sometimes (and for fairly obvious reasons) the quotient topology is called the *identification topology* since one can regard those members of X which lie inside a single member of $\mathcal{P}$ as having been (topologically) "identified". Thus one obtains $\mathcal{P}$ from X by "identification" from the relation $\sim$. As an example let $(X, \mathcal{T})$ be the subset $[0, 1]$ of $\mathbb{R}$ with subspace topology from the usual topology on $\mathbb{R}$ and let $\mathcal{P}$ consist of the subsets $\{x\}$ for each $x \in (0, 1)$ and the subset $\{0, 1\}$. Then the resulting quotient space as described above is homeomorphic to the subset $S^1 \equiv \{(x, y) : x^2 + y^2 = 1\}$ of $\mathbb{R}^2$ with subspace topology from the usual topology on $\mathbb{R}^2$.

3.7 Product Spaces

It was shown in section 3.2 that if $(X_1, d_1), \ldots, (X_n, d_n)$ were metric spaces then a metric could be constructed on the set $X = X_1 \times \cdots \times X_n$. However, the metric given is one of several metrics for this product competing for the title of "natural" (*cf.*the metrics for $\mathbb{R}^n$ given earlier). Fortunately, they are equivalent metrics for this product and the (unique) topology they generate on X has the property that each open set is a union of open sets of the form $U_1 \times \cdots \times U_n$ where U_i is open in (X_i, d_i). This leads to the idea of the (topological) product of topological spaces $(X_1, \mathcal{T}_1), \ldots, (X_n, \mathcal{T}_n)$. Consider the family of subsets of the product set X of the form $U_1 \times \cdots \times U_n$ where $U_i \in \mathcal{T}_i$. This family is readily shown to be a base for a topology $\mathcal{T}$ on X. The topology $\mathcal{T}$ is called the *product topology* for X and $(X, \mathcal{T})$ is called the (topological) product (space) of $(X_1, \mathcal{T}_1), \ldots, (X_n, \mathcal{T}_n)$. The *projection maps* $p_i : X \to X_i$ defined by $p_i(x_1, \ldots, x_n) = x_i$ are then continuous and, in fact, the product topology on X is the smallest topology (in an obvious sense) for which the maps p_i are continuous. The product topology on X can also be regarded as being generated by the subbase of subsets of the form $p_i^{-1}(U) = X_1 \times \cdots \times X_{i-1} \times U \times X_{i+1} \times \cdots \times X_n$ for each $U \in \mathcal{T}_i$ and each i, $1 \leqslant i \leqslant n$. The finite intersections of these subbase members yield the base for $\mathcal{T}$ given above. If $\{x_m\}$ is a sequence in X then $x_m \to x_0 \in X$ if and only if $p_i(x_m) \to p_i(x_0)$ in $(X_i, \mathcal{T}_i)$ for each i. If $A_i \subseteq X_i$ $(1 \leqslant i \leqslant n)$ are subspaces of X_i the product topology on $A_1 \times \cdots \times A_n$ equals its subspace topology from $X_1 \times \cdots \times X_n$. The standard topology on $\mathbb{R}^n$ is the product topology on $\mathbb{R} \times \cdots \times \mathbb{R}$ (n times) when $\mathbb{R}$ has its standard topology. It is often useful to note that a map from a topological space Y into the above product space X is continuous if and only if $p_i \circ f : Y \to X_i$ is continuous for each i. Here only a *finite* product of topological spaces has been considered. Infinite products can be handled but extra difficulties are encountered. They will not be required in this text and are not discussed here. Further details are available in [14]-[16].

3.8 Compactness and Paracompactness

Suppose one considers $\mathbb{R}^n$ with its usual metric and topological structure and asks if one can attach any meaning to the concept of a "small" subset A of $\mathbb{R}^n$. A concept of "smallness" (*i.e.* the closed and nowhere dense concept) was discussed earlier in this chapter. Another such idea can now be introduced. Merely asking that A be a bounded subset is inadequate since such a subset with its subspace topology might be homeomorphic to $\mathbb{R}^n$ itself (*e.g.* the subset $(0, 1)$ of $\mathbb{R}$). More is clearly required. The classical concept of "small" was that A should be a *closed and bounded* subset of $\mathbb{R}^n$. However this definition contains reference to the metric on $\mathbb{R}^n$ and may not be a topological property. Fortunately this definition is, for $\mathbb{R}^n$, equivalent to the *topological* statement that if $\mathcal{B}$ is a family of open subsets of $\mathbb{R}^n$ whose union contains A then the union of some *finite* subfamily of $\mathcal{B}$ will also contain A. The satisfying of this definition by A would thus be unaffected by a change of metric to another *equivalent* metric. Further this equivalent condition can now be taken over to any topological space and leads to the important idea of *compactness*. It was, the author believes, Hermann Weyl who described this concept rather beautifully by defining a compact city as one which could be patrolled by a finite number of arbitrarily short-sighted policemen!

Let $(X, \mathcal{T})$ be a topological space and let $\mathcal{B}$ be a family of subsets of X. If the union of the members of $\mathcal{B}$ equals X, $\mathcal{B}$ is called a *covering* of X (or one says that $\mathcal{B}$ *covers* X) and if each member of $\mathcal{B}$ is open it is an *open covering* of X. If a subfamily $\mathcal{B}'$ of $\mathcal{B}$ also covers X, $\mathcal{B}'$ is called a *subcovering* of $\mathcal{B}$. The topological space $(X, \mathcal{T})$ is called *compact* if each open covering of X contains a *finite* subcovering. A subspace of X is called *compact* if it is compact in its subspace topology. It is easy to see that $\mathbb{R}^n$ is not compact for any $n \in \mathbb{Z}^+$ and less easy to see that the subspace $[a, b]$ of $\mathbb{R}$ ($a, b \in \mathbb{R}$, $a < b$) is a compact subspace of $\mathbb{R}$ as is the sphere $S^{n-1} = \{\mathbf{x} \in \mathbb{R}^n : |\mathbf{x}| = 1\}$ as a subspace of $\mathbb{R}^n$. [Earlier in this chapter it was suggested that second countable topological spaces were in some sense not too "large" because any uncountable subset of such a space necessarily contained a limit point of itself. In fact any second countable space, whilst not in general compact, has the property that any open covering of it contains a *countable* subcovering.] It can be shown that a subspace of a compact topological space need not itself be compact but that a *closed* subspace of a compact space is compact.

There are some rather important results regarding continuous functions on compact topological spaces. Firstly, if X and Y are topological spaces with X compact and if $f : X \to Y$ is continuous then $f(X)$ is a compact subspace of Y (and hence Y is necessarily compact if f is surjective). It follows that any quotient space of a compact space is compact. Secondly, if X is a compact space and $\mathbb{R}$ has its usual topology and if $f : X \to \mathbb{R}$ is a continuous real valued function on X then f is bounded and attains its bounds (that is $f(X)$ is a bounded subset of $\mathbb{R}$ and there exist $x, y \in X$ such that $f(x) = \sup[f(X)]$ and $f(y) = \inf[f(X)]$). It follows that if, in addition, f is a positive function (*i.e.* $f(x) > 0$ for each $x \in X$) then f is "bounded away from zero" (*i.e.* there exists $\varepsilon \in \mathbb{R}$, $\varepsilon > 0$, such that $f(x) \geqslant \varepsilon$ for each $x \in X$). Finally, let $(X, \mathcal{T})$ be the topological product of the spaces $(X_1, \mathcal{T}_1), \ldots, (X_n, \mathcal{T}_n)$. If X is compact then it follows from the above and the continuity and surjectivity of the projection maps $p_i : X \to X_i$ that each of the spaces X_i is compact. The converse of this result is also true, that is, the compactness of each of the X_i implies the compactness of X. This result stated here for finite products can be extended to infinite products and is the theorem of Tychonoff [14], [15].

A topological space can be quite pathological if it is restricted by nothing more than the axioms of a topological space. This generality is, on the one hand, the reason for their wide applicability but, on the other hand, often restricts progress in particular areas of investigation. Thus these axioms are occasionally supplemented by extra restrictions often inspired by the "standard" topological spaces. Extra axioms governing the "topological separation" of points are, in particular, often imposed. These are the so-called separation axioms and only three of these will be discussed here. A topological space X is called a T_1 *space* if given $x, y \in X$, $x \neq y$, there exist open subsets U and V such that $x \in U$, $y \notin U$, $y \in V$, $x \notin V$. This is equivalent to the statement that all subsets of X containing a single point of X are closed. A topological space X is called a T_2 *space* (or more commonly a *Hausdorff space*) if given $x, y \in X$, $x \neq y$, there exist open subsets U and V such that $x \in U$, $y \in V$ and $U \cap V = \emptyset$. Finally, a topological space X is called *normal* if for each disjoint pair of closed subsets A and B of X there exist disjoint open subsets U, V such that $A \subseteq U$, $B \subseteq V$. Every subspace of a Hausdorff space is Hausdorff and every closed subspace of a normal space is normal. The extra "separation" provided in a Hausdorff space removes many pathologies. Clearly every Hausdorff space is a T_1 space and every metric space is Hausdorff and (less obviously) normal. Recalling the discussion above about the definition of compactness it is reassuring to note that any compact subspace of a Hausdorff space is closed. It can now be checked that a continuous bijective map from a compact to a Hausdorff topological space is a homeomorphism. It was stated above that every continuous real valued function on a compact topological space is bounded. For a metric space the converse is true, that is, if every real valued continuous function on a metric space is bounded then it is compact. It is perhaps also relevant to remark that although the definition of compactness used here is now standard in topology there have been several other (non-equivalent) competing definitions.

A covering $\mathcal{B}$ of a topological space X is called *locally finite* if each $x \in X$ is contained in some open set which intersects non-trivially only *finitely many* members of the family $\mathcal{B}$. If $\mathcal{B}'$ is also a covering of X such that for each $B' \in \mathcal{B}'$ there exists $B \in \mathcal{B}$ such that $B' \subseteq B$ then $\mathcal{B}'$ is called a *refinement* of $\mathcal{B}$. A topological space X is called *paracompact* if it is Hausdorff and if every open covering of X admits an open, locally finite refinement. If X is paracompact then it is normal and any closed subspace of X is paracompact. Paracompactness is a rather important property of topological spaces and especially of the topology on a manifold which will be discussed later. It leads to the existence of a certain family of real valued continuous functions on X called a partition of unity [14], [15]. These functions (and especially their differentiable equivalents for a manifold) can be useful in converting certain "local" properties of the topological space or manifold into "global" ones. For example, define a topological space X to be *locally metrisable* if for each $x \in X$ there exists an open set U containing x such that the subspace topology on U is a metric topology (*i.e.* arises from a metric d defined on U). Smirnov has proved that a locally metrisable space is metrisable if and only if it is paracompact (see, *e.g.* [14]). A famous theorem of A.H. Stone shows that every metric space is paracompact (see, *e.g.* [15]).

3.9 Connected Spaces

There are two ways of describing a topological space as being "all in one piece". Unfortunately they are not, in general, equivalent but, thankfully, for many topological spaces, including the ones required here, they are equivalent. A topological space X is called *connected* if X is *not* the union of two disjoint non-empty open sets (and otherwise *disconnected*). This definition is unaffected if "open" is replaced by "closed". A subspace of X is called *connected* if it is connected in the subspace topology (and otherwise *disconnected*). The space $\mathbb{R}$ with its usual topology is connected as are all its interval subspaces. However, $\mathbb{Q}$ is a disconnected subspace of $\mathbb{R}$ since it may be written as $\mathbb{Q} = A \cup B$ with $A = \mathbb{Q} \cap (-\infty, \sqrt{2})$ and $B = \mathbb{Q} \cap (\sqrt{2}, \infty)$. If A is a connected subspace of X then the closure $\bar{A}$ of A is also connected. An important idea in a topological space is that of a component. A subset C of X is called a *component* of X if it is "maximally connected", that is, if it is a connected subspace of X and, if D is a connected subspace of X containing C, then $D = C$. It is straightforward to show that any topological space is the disjoint union of its components and that any component of X must be closed.

To introduce the second criterion for "connectedness" one requires the definition of a *path* (or *curve*) in a topological space as a continuous map c from the closed interval $[a, b] \subseteq \mathbb{R}$ (usual topology) to X. This interval is often chosen as $I \equiv [0, 1]$. The points $c(a)$ and $c(b)$ in X are called the *initial* (or *starting*) and *end* (or *final*) *points* of the path c, respectively. A topological space X is called *path connected* if given any $x, y \in X$ there exists a path c with initial point x and final point y (*i.e. a path from x to y*). A subspace of X is called *path connected* if it is so in its subspace topology. Clearly $\mathbb{R}$ is path connected. *Path components* are defined just as components were above with the obvious modification. A fairly easy argument shows that *a path connected topological space is connected* but the converse statement is false; in fact the subspace $\{(0,0)\} \cup \{(x, \sin(\frac{1}{x})), x \in (0,1)\}$ of $\mathbb{R}^2$ is connected but not path connected since no path connects $(0,0)$ and $(\pi^{-1}, 0)$.

A topological space is called *locally connected* if it has a base consisting of (open) connected subsets of X and *locally path connected* if it has a base of (open) path connected subsets of X. Thus, from a result at the end of the last paragraph, a locally path connected space is locally connected. These two local properties, if satisfied, lead to two satisfying results. The components of a locally connected space X (which are always closed) are *open* subsets of X and *if X is locally path connected then X is connected if and only if it is path connected*. All the topological spaces encountered in this text will be locally connected and locally path connected and hence for such spaces the components and path components coincide and are simultaneously open and closed subsets of X and the conditions of being connected and path connected are equivalent.

If X and Y are topological spaces with X connected (respectively path connected) and $f : X \to Y$ is continuous, then $f(X)$ is a connected (respectively path connected) subspace of Y. Thus a quotient space of a connected (respectively path connected) space is connected (respectively path connected). The topological product of (non-empty) spaces $X_1, \ldots, X_n$ is connected (respectively path connected) if and only if each of the spaces $X_1, \ldots, X_n$ is connected (respectively path connected). Clearly $\mathbb{R}^n$ is connected and path connected.

3.10 Covering Spaces and the Fundamental Group

Let X and Y be topological spaces and f and g continuous maps from X into Y. Then f and g are called *homotopic* (and one writes $f \sim g$) if there exists a continuous map $F : X \times I \to Y$ such that $F(x, 0) = f(x)$ and $F(x, 1) = g(x)$ for each $x \in X$. The function F is called a *homotopy* of f into g. If, for each $t \in I$, $f_t : X \to Y$ is defined by $f_t(x) = F(x, t)$ then f_t is continuous and $f_0 = f$, $f_1 = g$. Thus as t moves from 0 to 1 in I the mappings f_t represent a "continuous deformation" of the map f into the map g through the intermediate stages f_t. A standard trick in elementary homotopy theory can be used to show that $\sim$ is an *equivalence relation* on the set of all continuous maps from X into Y and the resulting equivalence classes are called *homotopy classes*. With only a little more effort one can also show that if X, Y, Z are topological spaces with f_0 and $f_1 : X \to Y$ and g_0 and $g_1 : Y \to Z$ continuous maps with $f_0 \sim f_1$ and $g_0 \sim g_1$ then $g_0 \circ f_0$ and $g_1 \circ f_1$ are homotopic maps $X \to Z$.

Now let X be a topological space and let $\alpha : I \to X$ be a path in X, starting from $\alpha(0)$ and ending at $\alpha(1)$. The *inverse path* α^{-1} to α is the path $\alpha^{-1} : I \to X$ defined by $\alpha^{-1}(t) = \alpha(1 - t)$ and so α^{-1} begins at $\alpha^{-1}(0) = \alpha(1)$ and ends at $\alpha^{-1}(1) = \alpha(0)$. If α and β are paths in X such that the end point of α coincides with the starting point of β then the product path of α and β, denoted by $\alpha \cdot \beta$, can be defined by

$$\alpha \cdot \beta(t) = \begin{cases} \alpha(2t) & 0 \leqslant t \leqslant \frac{1}{2} \\ \beta(2t - 1) & \frac{1}{2} \leqslant t \leqslant 1. \end{cases}$$

A little argument (the so-called *glueing lemma*) is required to show that $\alpha \cdot \beta$ is continuous and clearly $\alpha \cdot \beta(0) = \alpha(0)$, $\alpha \cdot \beta(1) = \beta(1)$. A path α in X satisfying $\alpha(t) = x_0$ for each $t \in I$ and some fixed $x_0 \in X$ is called a *constant* (or *null*) *path at* x_0. A path α in X is called *closed* (at x_0) if $\alpha(0) = \alpha(1)$ $(= x_0)$. For any path α it is clear that $\alpha \cdot \alpha^{-1}$ and $\alpha^{-1} \cdot \alpha$ are closed. Two paths α and β in X with the same beginning and the same endpoint, $\alpha(0) = \beta(0)$, $\alpha(1) = \beta(1)$ are called *homotopic* (and one writes $\alpha \sim \beta$) if there is a continuous map $G : I \times I \to X$ such that $G(t, 0) = \alpha(t)$ and $G(t, 1) = \beta(t)$ for each $t \in I$ *and* if $G(0, t) = \alpha(0)$ $(= \beta(0))$ and $G(1, t) = \alpha(1)$ $(= \beta(1))$ for each $t \in I$. It can now be shown that if α, β, γ and δ are paths in X then

i) $\alpha \sim \beta$ and $\gamma \sim \delta$ and $\alpha \cdot \gamma$ exists $\Rightarrow \beta \cdot \delta$ exists and $\alpha \cdot \gamma \sim \beta \cdot \delta$,
ii) $\alpha \sim \beta \Rightarrow \alpha^{-1} \sim \beta^{-1}$,

iii) if β and γ are constant paths such that $\alpha \cdot \beta$ and $\gamma \cdot \alpha$ exist then $\alpha \cdot \beta \sim \alpha \sim \gamma \cdot \alpha$,
iv) if $\alpha \cdot \beta$ and $\beta \cdot \gamma$ exist then $(\alpha \cdot \beta) \cdot \gamma$ and $\alpha \cdot (\beta \cdot \gamma)$ exist and are homotopic,
v) $\alpha \cdot \alpha^{-1}$ and $\alpha^{-1} \cdot \alpha$ are homotopic to constant paths.

Now for any topological space X with $x_0 \in X$ let $\pi_1(X, x_0)$ denote the family of homotopy equivalence classes of all paths in X which are closed at x_0. If α is a closed path at x_0 denote the equivalence class containing α by $[\alpha]$. Now define a multiplication on $\pi_1(X, x_0)$ by $[\alpha][\beta] = [\alpha \cdot \beta]$ (this being well defined according to (i)). Denote by [1] the member of $\pi_1(X, x_0)$ which contains the constant path at x_0. Then it follows from (iii) that $[1][\alpha] = [\alpha][1] = [\alpha]$ for any $[\alpha] \in \pi_1(X, x_0)$. Further, for any $[\alpha] \in \pi_1(X, x_0)$, $[\alpha^{-1}]$ is well defined by (ii) and $[\alpha][\alpha^{-1}] = [1] = [\alpha^{-1}][\alpha]$ by (v). Finally, if $[\alpha], [\beta], [\gamma] \in \pi_1(X, x_0)$ it follows from (iv) that $([\alpha][\beta])[\gamma] = [\alpha]([\beta][\gamma])$. Thus the set $\pi_1(X, x_0)$ has been given the structure of a group, under the multiplication defined above, called the *fundamental group* (or *the first homotopy group*) *of X at x_0*. The group $\pi_1(X, x_0)$ depends on the point $x_0 \in X$. However, if $x_0, x_1 \in X$ and there exists a path in X joining x_0 to x_1 then $\pi_1(X, x_0)$ and $\pi_1(X, x_1)$ are isomorphic (the isomorphism depending on the path from x_0 to x_1 chosen). In particular, if X is path connected then one can speak of *the fundamental group* of X denoted by $\pi_1(X)$. If X and Y are topological spaces and $f : X \to Y$ is a continuous map then for $x_0 \in X$ one can define a map $f_* : \pi_1(X, x_0) \to \pi_1(Y, y_0)$ where $y_0 = f(x_0)$ by $f_* : [\alpha] \to [f \circ \alpha]$. It is easily checked that f_* is well defined and that it is a group homomorphism of $\pi_1(X, x_0)$ into $\pi_1(Y, y_0)$.

As examples, it can be shown that $\pi_1(\mathbb{R}^n)$ for $n \geqslant 1$ and $\pi_1(S^n)$ for $n \geqslant 2$ consist of the identity element alone whereas $\pi_1(S^1)$ is isomorphic to $\mathbb{Z}$. A path connected topological space whose fundamental group has only one element is called *simply connected* and in such a space every closed path is homotopic to a constant path. A topological space X such that each $x \in X$ lies inside an open subset U of X such that any closed path at x in U is homotopic to the constant path at x (as paths in X) is called *locally simply connected*. Finally it is useful to note that if X and Y are path connected topological spaces then $X \times Y$ is path connected and $\pi_1(X \times Y)$ is isomorphic to the group product $\pi_1(X) \otimes \pi_1(Y)$.

Simply connected topological spaces have several desirable properties and it is thus useful to note that under some fairly mild restrictions a topological space X can always be obtained from a simply connected topological space $\tilde{X}$ "by identification", that is, X is homeomorphic to a quotient space of $\tilde{X}$. This can be made precise by a brief discussion of *covering spaces*. Let X and $\tilde{X}$ be connected, locally path connected (and hence path connected) Hausdorff topological spaces and let $p : \tilde{X} \to X$ be a continuous map. Then $\tilde{X}$ is called a *covering space of X with covering map p* if p is surjective *and* if for each $x \in X$ there exists a (necessarily connected) open set U in X containing x such that each component of $p^{-1}U$ is homeomorphic to U under (the appropriate restriction of) p. Subsets of X such as U are called *admissible* and, clearly, any connected open subset of an admissible set is admissible. As an example take $X = S^1$ where S^1 is regarded as the set $\{z \in \mathbb{C} : |z| = 1\}$ and $\tilde{X} = \mathbb{R}$ and define $p : \mathbb{R} \to S^1$ by $p(x) = e^{2\pi i x}$. Then $\mathbb{R}$ is a covering space of S^1, the details being easily checked, and for each admissible subset $U \subseteq S^1$, $p^{-1}(U)$ consists of infinitely many disjoint open subsets of $\mathbb{R}$ each homeomorphic under p to U. Another example arises if $X = \tilde{X} = S^1$ (with S^1 realised as above) and $p : S^1 \to S^1$ is given by $p(z) = z^2$. In this case, for each admissible subset U in S^1, $p^{-1}(U)$ consists of just two disjoint open subsets of S^1 diffeomorphic under p to U and one calls $\tilde{X}$ a "double covering" of X.

If $\tilde{X}$ is a covering space of X the covering map p is necessarily an *open map* (*i.e.* $p(U)$ is open in X whenever U is open in $\tilde{X}$). It also leads to a homomorphism $p_* : \pi_1(\tilde{X}, x_0) \to \pi_1(X, p(x_0))$ between the fundamental groups of X and $\tilde{X}$. Moreover, the map p_* can be shown to be *injective* and so $p_*[\pi_1(\tilde{X}, x_0)]$ is a subgroup of $\pi_1(X, p(x_0))$ isomorphic to $\pi_1(\tilde{X}, x_0)$ and whose set of (right or left) cosets in $\pi_1(X, p(x_0))$ is in bijective correspondence with the set $p^{-1}(p(x_0))$. With an extra restriction on X there is a powerful converse to this theorem, namely, if X is a connected, locally path connected, locally simply connected Hausdorff topological space and if H is any subgroup of $\pi_1(X, x)$ then there exists a covering space $\tilde{X}$ of X with covering map p such that $p_*[\pi_1(\tilde{X}, x_0)] = H$ where $x_0 \in \tilde{X}$ and $p(x_0) = x$. *By choosing H to consist only of the identity element one can thus find a simply connected covering space of X.* Such a simply connected covering space of X is called a *universal covering space* (or simply a *universal cover*) of X and is unique in the sense that if $\tilde{X}_1$ and $\tilde{X}_2$ are universal covers of X with covering maps p_1 and p_2 then there exists a homeomorphism $f : \tilde{X}_1 \to \tilde{X}_2$ such that $p_2 \circ f = p_1$.

Thus if X is a connected, Hausdorff, locally path connected and locally simply connected topological space let $\tilde{X}$ be a universal cover of X with covering map p. Then let R denote the equivalence relation on $\tilde{X}$ given by $x_1 R x_2 \Leftrightarrow p(x_1) = p(x_2)$ and denote the associated quotient topological space by $\tilde{X}/R$ and projection map by $i : \tilde{X} \to \tilde{X}/R$. There is a natural bijection $f : \tilde{X}/R \to X$ satisfying $p = f \circ i$ which is continuous (since $\tilde{X}/R$ has the quotient topology) and open (since whenever U' is open in $\tilde{X}/R$, $U = i^{-1}(U')$ is open in $\tilde{X}$ and so $p(U) = f(i(U)) = f(U')$ is open in X because p is open). Hence f^{-1} is continuous and f a homeomorphism. It follows that X is homeomorphic to a quotient space of a simply connected space, justifying an earlier remark.

3.11 The Rank Theorems

If X is a topological space and $f : X \to \mathbb{R}$ a continuous map such that for a particular $x_0 \in X$, $f(x_0) \neq 0$ then it is easily proved that there exists an open neighbourhood U of x_0 such that $f(x) \neq 0$ for each $x \in U$. In fact, if $f(x_0) > 0$ (respectively $f(x_0) < 0$) then U may be chosen so that $f(x) > 0$ (respectively $f(x) < 0$) for each $x \in U$. The proof merely consists of choosing an open interval I of $\mathbb{R}$ containing $f(x_0)$ but not 0 and then setting $U = f^{-1}I$. This theorem can be generalised in a way that will be useful in what is to follow. It can be stated and applied in several different ways and is often referred to, somewhat loosely, as the *rank theorem*. Since the generalisation here involves matrices the standard topology on matrix spaces must be discussed first. The set $M_{m\times n}\mathbb{R}$ of real $m \times n$ matrices can be put into a natural bijective correspondence with the set $\mathbb{R}^{mn}$ and thus inherits a natural topology by the insistence that this correspondence be a homeomorphism when $\mathbb{R}^{mn}$ has its standard topology. With this topology, and in the case $m = n$, it is useful to note that the determinant function $\det : M_n\mathbb{R} \to \mathbb{R}$ is continuous and hence, by the remark at the beginning of this section, the subset $GL(n, \mathbb{R})$ is an *open* subset of $M_n\mathbb{R}$. Also, given any $A \in M_n\mathbb{R}$ the map f from $M_n\mathbb{R}$ to itself given by $f : T \to TAT^{\mathrm{T}} - A$ is continuous. If O_n is the zero matrix in $M_n\mathbb{R}$, $\{O_n\}$ is closed in $M_n\mathbb{R}$ and hence $f^{-1}\{O_n\} = \{T : TAT^{\mathrm{T}} = A\}$ is closed in $M_n\mathbb{R}$. By choosing A to be the Sylvester matrix, I_s^r (section 2.5) it follows that $O(r, s)$ is a *closed* subset of $M_n\mathbb{R}$. In the case when $s = 0$ and $r = n$ it is easily shown that the orthogonal group $O(n)$ is also a *bounded* subset of $M_n\mathbb{R}$ with respect to the standard metric on $\mathbb{R}^{n^2}$ (whose associated topology is the standard topology on $\mathbb{R}^{n^2}$) and hence, from section 3.8, $O(n)$ is a *compact* subset of $M_n\mathbb{R}$.

Theorem 3.1 *Let X be a topological space and f a continuous map $X \to$*

$M_{m\times n}(\mathbb{R})$. If $x_0 \in X$ and $f(x_0)$ has rank p then there exists an open subset U of X containing x_0 such that $f(x)$ has rank $\geqslant p$ for each $x \in U$.

Proof. To prove this note that if $f(x_0) = A$ then some $p \times p$ submatrix of A has non-zero determinant. Let this submatrix occupy those entries in A with indices $i_1, \ldots, i_p$ and $j_1, \ldots, j_p$ and consider the (continuous) map $g : M_{m\times n}\mathbb{R} \to M_p\mathbb{R}$ which maps an $m \times n$ matrix on to the submatrix occupying precisely this position. Then $g \circ f$ is a continuous map from X to $M_p(\mathbb{R})$ such that $g \circ f(x_0)$ is non-singular and so the continuous map $h = \det \circ g \circ f : X \to \mathbb{R}$ satisfies $h(x_0) \neq 0$. It follows that there exists an open subset U in X containing x_0 such that $h(x) \neq 0$ for each $x \in U$ and hence that $f(x)$ has a non-singular $p \times p$ submatrix. This implies that the rank of $f(x) \geqslant p$ and the result follows. □

The above theorem is a particularly simple case of the general rank-type theorems. It has several corollaries and restatements of which the following will be noted. Firstly, let X be a topological space and $f : X \to M_n\mathbb{R}$ be continuous. Then if $x_0 \in X$ and $f(x_0)$ is non-singular there exists an open subset U in X containing x_0 such that $f(x)$ is non-singular for each $x \in U$. Secondly, let X be a topological space and $f : X \to \mathbb{R}^n \times \cdots \times \mathbb{R}^n$ (m times with the product topology) be continuous. If $x_0 \in X$ and $f(x_0) = (\mathbf{a}_0^1, \ldots, \mathbf{a}_0^m)$ with $\mathbf{a}_0^1, \ldots, \mathbf{a}_0^m$ spanning a p-dimensional subspace of $\mathbb{R}^n$ then there exists an open subset U of X containing x_0 such that if $x \in U$ and if $f(x) = (\mathbf{a}^1, \ldots, \mathbf{a}^m)$ then $\mathbf{a}^1, \ldots, \mathbf{a}^m$ span a subspace of $\mathbb{R}^n$ of dimension $\geqslant p$ for each $x \in U$.

Other results similar to those mentioned in this section can be established by similar techniques. Such results usually go under the general name of "rank theorems".

Chapter 4

Manifold Theory

4.1 Introduction

In many branches of theoretical physics one is dealing with a description of events in space and time using techniques of the calculus. This is usually accomplished by assuming that occurences in space and time can be described, at least locally, by coordinatising the latter (*i.e.* putting it into a local one-one correspondence) with some open subset of $\mathbb{R}^4$. For this, and other reasons, interest centres on a certain mathematical structure on a set which in some sense makes it locally indistinguishable from $\mathbb{R}^n$. The theory of *manifolds* describes such a structure.

Manifold theory is rather important in general relativity theory. Since the local similarity of a manifold to $\mathbb{R}^n$ is (together with some differentiability requirement on the coordinate changes on the overlap of these localities) essentially the only restriction on the general definition of a manifold, this theory with its wealth of powerful theorems can then be used to study many problems in physics. The aim of this chapter is to review the basic results on manifold theory and to describe the various structures on them. Since a number of results will be needed explicitly such results will be collected together and stated as theorems for ease of reference later. Although all the relevant definitions and constructions will be given the reader will be referred to the bibliography for the proofs of most of the theorems. As in the previous chapters, and even though the contents of this chapter are dictated by the remainder of the text, a certain amount of completeness will, it is hoped, be achieved. Prior to defining a manifold the next section will review some elementary results from the calculus of $\mathbb{R}^n$. This chapter will be guided by the excellent text [19] with much use also being made of the texts [20]-[25].

4.2 Calculus on $\mathbb{R}^n$

In this section attention will be focused on functions $f : A \to \mathbb{R}^m$ where A is a non-empty open subset of $\mathbb{R}^n$. Sets like $\mathbb{R}^n$ have more structure than those which arise from their usual vector space and topological structures. They have a differentiable structure in that it makes sense to ask if a function like f above is differentiable. Standard calculus on $\mathbb{R}^n$ shows how one can define differentiability for such a function f and in this respect much information is stored in the projection functions $f_i \equiv p_i \circ f$ where $p_i : \mathbb{R}^m \to \mathbb{R}$ is the projection function onto the ith coordinate. One then has the following results:

i) f is continuous if and only if each f_i is continuous,
ii) if f is differentiable at $p \in \mathbb{R}^n$ then, at p, f is continuous and the partial derivatives $\frac{\partial f_i}{\partial x_j}$ exist,
iii) if the partial derivatives $\frac{\partial f_i}{\partial x_j}$ exist in some neighbourhood of p and are continuous at p then f is differentiable at p,
iv) if each of the kth partial derivatives $\frac{\partial^k f_i}{\partial x_{j_1},\ldots\partial x_{j_k}}$ exist and are continuous in some neighbourhood of p then the order in which these derivatives are taken is irrelevant.

A function f satisfying the conditions of part (iv) is said to be k times continuously differentiable or to be of *class* C^k at p (or on A if it holds for each $p \in A$). Part (iv) also applies when, with the usual convention, one sets k=∞. Such a function f is then of class C^∞ or *smooth* at p (or on A).

Two rather important theorems from the calculus on $\mathbb{R}^n$ will be required and will be stated here without proof (these being easily available in the standard texts). It is remarked that all results given for C^k functions hold in the case k=∞ unless otherwise stated (so that in this context $k \geq 1$ includes k=∞).

Theorem 4.1 The Inverse Function Theorem for $\mathbb{R}^n$

Let $A \subseteq \mathbb{R}^n$ be non-empty and open and let $f : A \to \mathbb{R}^n$ be a C^k function $(k \geq 1)$. If $\det(\frac{\partial f_i}{\partial x_j})$ $(1 \leq i, j \leq n)$ is non-zero at $\mathbf{x}_0 \in A$ there exists an open subset U of $\mathbb{R}^n$ such that $\mathbf{x}_0 \in U \subseteq A$ and an open subset V in $\mathbb{R}^n$ containing $f(\mathbf{x}_0)$ such that f maps U bijectively onto V and such that the inverse function $f^{-1} : V \to U$ is C^k.

Theorem 4.2 The Implicit Function Theorem for $\mathbb{R}^n$

Let $A \subseteq \mathbb{R}^{n+m}$ $(= \mathbb{R}^n \times \mathbb{R}^m)$ be non-empty and open and let $f : A \to \mathbb{R}^m$ be a C^k function $(k \geq 1)$. Suppose $(\mathbf{x}_0, \mathbf{y}_0) \in \mathbb{R}^{n+m}$ with $\mathbf{x}_0 \in \mathbb{R}^n$,

$\mathbf{y}_0 \in \mathbb{R}^m$ and that $f(\mathbf{x}_0, \mathbf{y}_0) = 0$. If $\det(\frac{\partial f_i(x,y)}{\partial y_j})$ is non-zero at $(\mathbf{x}_0, \mathbf{y}_0)$ $(1 \leq i, j \leq m)$ there exists an open set U in $\mathbb{R}^n$ containing $\mathbf{x}_0$, an open subset V in $\mathbb{R}^m$ containing $\mathbf{y}_0$ (and $U \times V \subseteq A$) and a unique C^k function $g : U \to V$ such that $f(\mathbf{x}, g(\mathbf{x})) = 0$ for each $\mathbf{x} \in U$ (and $g(\mathbf{x}_0) = \mathbf{y}_0$).

4.3 Manifolds

The requirements which are imposed on a set in order to give it a differentiability (*i.e.* a locally $\mathbb{R}^n$) structure can be achieved in (at least) two distinct ways. One could just give the set a topological structure and then impose the differentiability requirements by means of various local homeomorphisms into $\mathbb{R}^n$. Alternatively one could directly impose the differentiability structure initially and deduce the topology from it in a natural way. The second of these approaches will be followed here whilst the first can be found, for example, in [22]. It should be noted that the first approach sometimes imposes topological restrictions that are not necessary for a differentiability structure and which are not assumed here.

Let M be a set. A one-one map x from some subset U of M onto an open subset of $\mathbb{R}^n$ is called an *n-dimensional chart* (of M). If $p_i : \mathbb{R}^n \to \mathbb{R}$ is the projection function onto the ith coordinate one can construct n real valued functions $x^i = p_i \circ x$ on U. The function x^i is called the *ith coordinate function.* The subset U of M is called the *chart domain of* x and if $p \in U$ the n-tuple $(x^1(p), ..., x^n(p))$ is the set of coordinates of p in the chart x. The set U is also called a *coordinate neighbourhood* (of any of its points). A chart is called *global* if its domain is M. These definitions concern the coordinatisation of U. The complete coordinatisation of M consists of postulating the existence of a family of chart domains like U whose union is M. However, some care is required here. Suppose, for example, that f is a real valued function defined on $\mathbb{R}^2$ which is a C^k function of the usual (x, y) coordinates. If the coordinates are changed to (x', y') the function f will not in general be a C^k function of the new coordinates unless the coordinate changes themselves are C^k. Thus without some restriction on the differentiability of the coordinate transformations the concept of a C^k function would make no sense. Returning to the complete coordinatisation of M, a collection A of charts the union of whose domains equals M is called a *C^k atlas of M* $(k \geq 1)$ if whenever x and y are charts in this collection with domains U and V such that $U \cap V \neq \emptyset$ then $x(U \cap V)$ and $y(U \cap V)$ are open subsets of $\mathbb{R}^n$ and the bijective map (coordinate transformation)

$y \circ x^{-1} : x(U \cap V) \to y(U \cap V)$ and its inverse $x \circ y^{-1}$ are C^k.

Thus an atlas gives a local coordinatisation and (as will be seen later) a differentiability structure to M. One may extend the atlas A by adding extra charts provided the above conditions still hold on the extended set. Two C^k atlases are called *equivalent* if their union is also a C^k atlas. This leads to the definition of a *complete* C^k *atlas* for M as a C^k atlas for M which is not contained in any other C^k atlas of M. It is not hard to show that any given C^k atlas of M is contained in a unique complete C^k atlas for M (and so equivalent C^k atlases belong to the same complete C^k atlas). Any C^k atlas on M is called (or said to determine) an *n-dimensional C^k structure on M* and M with such a structure is called an *n-dimensional C^k manifold* and one writes *dim* $M = n$. The identity map on $\mathbb{R}^n$ is clearly a global chart on $\mathbb{R}^n$ and gives rise to what is called the *standard or usual C^∞ manifold structure on $\mathbb{R}^n$* (and this structure will always be assumed on $\mathbb{R}^n$ unless otherwise stated). A C^∞ structure on M is called *smooth* and the resulting manifold a *smooth manifold*. Some other examples of (smooth) manifolds can now be given.

i) Let V be an n-dimensional real vector space and $\{\mathbf{e}_i\}$ a basis for V. The map $x : V \to \mathbb{R}^n$ given by $x(\mathbf{v}) = (a_1, ..., a_n)$ where the a_i are the components of $\mathbf{v}$ in the basis $\{\mathbf{e}_i\}$ is a global chart for V and is thus a C^∞ atlas for V. This gives V the structure of an n-dimensional smooth manifold. The C^∞ atlases arising from different bases are equivalent and the unique smooth manifold structure arising on V is called its *usual or standard manifold structure* and agrees with the one above when $V = \mathbb{R}^n$.

ii) Let V be an n-dimensional complex vector space and $\{\mathbf{e}_i\}$ a basis for V. Let $\mathbf{v} \in V$ with components $a_i = \alpha_i + i\beta_i$ in the basis $\{\mathbf{e}_i\}$. The map $x : V \to \mathbb{R}^{2n}$ given by $\underset{\sim}{x}(\mathbf{v}) = (\alpha_1, ..., \alpha_n, \beta_1, ..., \beta_n)$ is a global chart for V which gives V the structure of a $2n$-dimensional smooth manifold. Again the C^∞ atlases arising from different bases are equivalent and the unique C^∞ manifold structure arising on V is called its *usual or standard manifold structure*.

iii) The set $M_{m\times n}\mathbb{R}$ can be given a smooth mn-dimensional manifold structure, again called its *usual or standard manifold structure*, by noting that the map x given by

$$x : (a_{ij}) \to (a_{11}, ..., a_{1n}, a_{21}, ..., a_{2n}, ..., a_{mn})$$

is an mn-dimensional global chart for $M_{m\times n}\mathbb{R}$.

iv) The sets of symmetric and skew symmetric members of $M_n\mathbb{R}$ can be given smooth $\frac{1}{2}n(n+1)$ and $\frac{1}{2}n(n-1)$–dimensional manifold structures, respectively, by using the global charts

$$x : (a_{ij}) \rightarrow (a_{11}, ..., a_{1n}, a_{22}, ..., a_{2n}, ..., a_{nn})$$

and

$$x : (a_{ij}) \rightarrow (a_{12}, ..., a_{1n}, a_{23}, ..., a_{2n}, ..., a_{(n-1)n})$$

respectively. The former is denoted $S(n, \mathbb{R})$.

v) The set $M_{m\times n}\mathbb{C}$ and the symmetric and skew-symmetric members of $M_n\mathbb{C}$ can be given smooth manifold structures of dimensions $2mn$, $n(n+1)$ and $n(n-1)$, respectively, by adapting examples (ii)–(iv) above.

The examples above all involved the specification of a global chart for M. Other examples, including ones where global charts do not exist can be found in, for example, [19],[22],[24],[25], where the circle S^1 and the sphere S^2 are shown to be smooth manifolds.

vi) If M_1 and M_2 are smooth manifolds of dimensions n_1 and n_2, respectively, the set $M_1 \times M_2$ can be given a smooth $(n_1 + n_2)$–dimensional manifold structure by constructing an atlas of charts on $M_1 \times M_2$ consisting of all maps of the form $x_1 \times x_2$, where x_1 and x_2 are charts belonging to atlases for M_1 and M_2, respectively, with domains $U_1 \subseteq M_1$ and $U_2 \subseteq M_2$, respectively, and where $x_1 \times x_2 : U_1 \times U_2 \rightarrow \mathbb{R}^{n_1+n_2}$ is the map $(p, q) \rightarrow (x_1(p), x_2(q))$. The smoothness of the coordinate transformations for overlapping chart domains is easily checked and the resulting manifold $M_1 \times M_2$ is called the *manifold product of* M_1 *and* M_2. There is no problem in extending this procedure to a finite product of manifolds. In particular the usual manifold structure on $\mathbb{R}^n$ is the manifold product of $\mathbb{R}^{n_1}$ and $\mathbb{R}^{n_2}$ (with their usual manifold structures) provided $n = n_1 + n_2$. The 2-torus $S^1 \times S^1$ is thus a smooth manifold.

vii) Let $M = \mathbb{R}^{m+n}$ $(= \mathbb{R}^m \times \mathbb{R}^n)$ and for each $\mathbf{a} \in \mathbb{R}^n$ define a map $z_a : (\mathbf{x}, \mathbf{a}) \rightarrow \mathbf{x}$ $(\mathbf{x} \in \mathbb{R}^m)$. It is easily checked that the maps z_a are m-dimensional charts for $\mathbb{R}^{m+n}$ with disjoint domains and so they define a smooth m-dimensional manifold structure on $\mathbb{R}^{m+n}$. Thus a given set may be given distinct manifold structures of different dimensions.

viii) The map $f : \mathbb{R} \rightarrow \mathbb{R}$ given by $f(x) = x^3$ is easily seen to be a global chart on M and hence gives rise to a smooth 1-dimensional manifold structure on $\mathbb{R}$. This structure is different from the usual (1-dimensional) manifold structure on $\mathbb{R}$ since the chart given here and the usual global (identity) chart on $\mathbb{R}$ give rise to inequivalent atlases on $\mathbb{R}$. This follows from the fact that the map f^{-1} is not differentiable at $x = 0$. Thus a given set may possess different manifold structures of the same dimension. It will, however, be seen later that these two structures are *essentially the same* (whereas the structure defined in (vii) above is quite different from the usual manifold structure on $\mathbb{R}^{m+n}$).

Most of the manifolds and structures required in this text are smooth. Occasional references to C^k $(i \leq k < \infty)$ structures will be required, however, and so some generality will, in this sense, be permitted. One can also extend these concepts to the C^ω (analytic) case.

4.4 Functions on Manifolds

Let M and M' be an n-dimensional C^k manifold and an n'-dimensional $C^{k'}$ manifold, respectively, and let $f : M \to M'$ be a map. The differentiability structures on M and M' allow a concept of differentiability to be defined for f. Suppose $p \in M$ and let $p' = f(p) \in M'$. If one chooses charts x in M and x' in M' whose domains contain p and p', respectively, the function $F = x' \circ f \circ x^{-1}$ whose domain is an open subset of $\mathbb{R}^n$ is called the *coordinate representative of f with respect to the charts x and x'*. The construction of F is made possible by the manifold structures on M and M' and it makes sense to ask about the differentiability of F. However, one is really seeking a notion of differentiability for f and if this is to be obtained from representatives like F it must be independent of the representative F chosen. Thus if $j \leq \min(k, k')$, f will be called a *C^j function at p* if F is C^j at $x(p)$. This definition is independent of the representative F, that is, of the charts x and x' chosen since if new charts y and y' are selected whose domains contain p and p', respectively, the new representative of F is $G = y' \circ f \circ y^{-1}$. But the map $(y' \circ x'^{-1}) \circ F \circ (x \circ y^{-1})$ is a restriction of G and is C^j at $y(p)$ if F is C^j at $x(p)$. The result follows. A particularly important case of this construction is when $M' = \mathbb{R}$ with its usual manifold structure and so f is a *real valued function* (or, more simply, a *function*) on M. In this case the identity chart on $\mathbb{R}$ will be used to construct its coordinate representative. The original function $f : M \to M'$ is called C^j if it is C^j at each point of M.

In the above discussion the domain of f was taken to be M. However, it still applies when f is only defined on some proper open subset of M. The crucial point is that a map from $\mathbb{R}^n$ to $\mathbb{R}^{n'}$ can, by definition, only be differentiable at $p \in \mathbb{R}^n$ if it is defined in some neighbourhood of p.

It is easily seen that the coordinate functions x^i defined earlier are examples of C^k real valued functions on the corresponding chart domain (or on M if the chart is a global chart). In example (iii) (with $m = n$) the function $A \to det\ A$ is a smooth (*i.e.* C^∞) function on $M_n\mathbb{R}$. If M and M' are as above and M'' is a $C^{k''}$ manifold of dimension n'' and if $f : M \to M'$ and $g : M' \to M''$ are maps, then if $p \leq \min(k, k', k'')$ and f and g are C^p maps then (with an appropriate domain) so is $g \circ f$.

The concept of a differentiable function between manifolds can be used to specify precisely when the two manifolds are essentially the same manifold. Suppose X is a set and M a C^k manifold and that $f : X \to M$ is a bijective map. It is obvious how one can transfer the manifold structure from M to X using f thus making X a C^k manifold of the same dimension as M. The map f and its inverse f^{-1} are then C^k maps. So let M and M' be C^k manifolds and let $f : M \to M'$ be a bijective map such that f *and its inverse* f^{-1} are C^k maps. Then M and M' are said to be C^k*-diffeomorphic* and $f : M \to M'$ and $f^{-1} : M' \to M$ are called C^k*-diffeomorphisms*. In the case $k = \infty$ f and f^{-1} are simply called *diffeomorphisms* and M and M' are said to be *diffeomorphic*. Since the coordinate representatives of such an f give rise to differentiable (and hence continuous) maps between open subsets of $\mathbb{R}^n$ and $\mathbb{R}^{n'}$, where $n = dim\ M$ and $n' = dim\ M'$, it follows that $dim\ M = dim\ M'$ (since $\mathbb{R}^n$ and $\mathbb{R}^{n'}$ are only homeomorphic if $n = n'$). Thus C^k-diffeomorphic (or diffeomorphic) manifolds are necessarily of the same dimension. It should be noted here that it is the continuity of the coordinate representative of f that is used (since it is a map between open subsets of $\mathbb{R}^n$ and $\mathbb{R}^m$ whose topology is known). The concept of a continuous map between manifolds makes no sense yet since no topology has been specified for a manifold. This will be the subject of the next section. It should also be noted that for M and M' to be C^k- diffeomorphic (or diffeomorphic) it is required that both f and f^{-1} be C^k (unlike when defining, for example, isomorphisms of groups or vector spaces where the structure preserving properties of the analogous inverse function follow automatically from those of the function). In fact if M is $\mathbb{R}^{m+n}$ with its usual $(m+n)$-dimensional manifold structure and M' is $\mathbb{R}^{m+n}$ with the m-dimensional manifold structure as in example (vii) above the identity map $i : M' \to M$ is C^∞ but i^{-1} is not differentiable (in fact, with the natural topologies on M and M' defined in the next section, i^{-1} is not even continuous). Thus M

and M' are not diffeomorphic. Also, in example (viii) above two different manifold structures were set up for $\mathbb{R}$. Let M_1 be the manifold $\mathbb{R}$ with its usual structure and M_2 the manifold $\mathbb{R}$ with the global chart $r \to r^3$. Then the map $g : M_2 \to M_1$ given by $x \to x^3$ is smooth with smooth inverse and so M_1 and M_2 are diffeomorphic. If U and U' are open subsets of a smooth manifold M and $f : U \to U'$ is a bijective map with f and f^{-1} smooth then f is called a *local diffeomorphism* of M.

4.5 The Manifold Topology

Let M be an n-dimensional C^k manifold and let B be the collection of all coordinate domains of a complete atlas of M. *Then B is a basis for a topology on M called the manifold topology on M.* The proof follows immediately from the remarks on bases in section 3.4 by noting that if x is a chart on M with domain U and if $V \subseteq U$ is such that $x(V)$ is open in $\mathbb{R}^n$ then the restriction of x to V is also a chart of M with domain V. The topology thus defined on M blends naturally with the differentiable structure on M as the following points below demonstrate.

i) The manifold topology of $\mathbb{R}^n$ (derived from its natural manifold structure) coincides with its usual topology.
ii) If x is a chart of M with domain U then x is a *homeomorphism* from U to an (open) subspace of $\mathbb{R}^n$ where U has subspace topology from the manifold topology on M.
iii) If M and M' are C^k and $C^{k'}$ manifolds of dimension n and n' respectively and if $f : M \to M'$ is C^p ($p \leq \min(k, k')$) then f is continuous with respect to the manifold topologies.
iv) If M_1 and M_2 are as in example (vi) of section 4.3,the manifold topology of the product manifold $M_1 \times M_2$ coincides with the product of the manifold topologies of M_1 and M_2.
v) It should be noted that the manifold topology on $\mathbb{R}^{m+n}$ which arises from the non-standard manifold structure of example (vii) of section 4.3 is *not* the usual topology of $\mathbb{R}^{m+n}$ and the failure of the map i^{-1} described at the end of the last section to be continuous is now clear.

Some general properties of the topology of any manifold M can be deduced directly from the fact that M is, topologically, *locally like* $\mathbb{R}^n$. Some of these are collected together in the following theorem

Theorem 4.3 *Let M be a C^k manifold.*

i) M is locally connected, locally path connected, locally simply connected and locally metric.
ii) M is T_1 but not necessarily Hausdorff.
iii) If M is compact, it does not admit a global chart.
iv) M is connected if and only if it is path connected.
v) M is first countable but not necessarily second countable.
vi) M is second countable if and only if M admits a countable atlas.
vii) If M is compact it admits a finite atlas and is second countable.
viii) Every component of M is open and closed in M.

Proof. The proof of (i) follows directly from the *locally* $\mathbb{R}^n$ nature as does the first parts of (ii) and (v). The second parts of (ii) and (v) will be dealt with in an example below. Part (iii) follows because no non-empty open subset of $\mathbb{R}^n$ is compact and part (iv) follows since M is locally path connected (from (i) and section 3.9). For part (vi) the assumption that M is second countable implies that every open covering of M admits a countable subcovering (section 3.8) and hence a countable atlas exists. Conversely the existence of a countable atlas and the locally $\mathbb{R}^n$ nature of M leads to second countability [19]. Part (vii) is immediate from the definition of compactness and part (vi). Part (viii) follows from part (i) and section 3.9. The second parts of (ii) and (v) can be dealt with by the following example. Let M be the subset of $\mathbb{R}^2$ given by $M = \{(x,0) : x \in \mathbb{R}, x \neq 0\} \cup \{(0,y) : y \in \mathbb{R}\}$ and for each $y \in \mathbb{R}$ define a chart f_y for M with range in $\mathbb{R}$ by

$$f_y : (x,0) \to x \qquad f_y : (0,y) \to 0$$

Clearly the union of the domains of these charts covers M and each pair of distinct chart domains intersect in the set $\{(x,0) : x \in \mathbb{R}, x \neq 0\}$ and give rise to the identity coordinate transformation there. Thus M becomes a smooth manifold of dimension one. It is now straightforward to show that the points $(0, y_1)$ and $(0, y_2)$, $y_1 \neq y_2$, do not lie in disjoint open subsets of M and so M is not Hausdorff. To see that M is not second countable one simply notes that the open covering of M by the domains of the charts f_y contains no countable subcovering and this contradicts second countability (section 3.8). □

All manifolds encountered in this text will be Hausdorff but, for reasons of generality, it has not been incorporated into the definition because some natural constructions with Hausdorff manifolds lead to non-Hausdorff manifolds (see section 4.12). However, from now on *all manifolds will be assumed smooth* (C^∞). This is only a partial restriction because a theorem due to Whitney [26] says that any C^k ($k \geq 1$) manifold contains within its C^k complete atlas a C^∞ atlas. In any case many of the results and constructions needed do not depend on the degree of differentiability (≥ 1) imposed on the manifold.

There is a very important series of results which link certain topological features of a manifold with the existence of a family of rather important functions on the manifold. Let M be a (smooth) manifold. A *partition of unity* on M is a collection $\{f_a\}$ of smooth real valued non-negative functions (with $a \in A$ for some indexing set A) each of whose domain is M and which satisfies the following conditions.

i) The closure (in the manifold topology) of the subset of M on which f_a is not zero is, for each $a \in A$, compact and lies inside a coordinate domain of M. [This set $\overline{\{m \in M : f_a(m) \neq 0\}}$ is called the *support* of f_a and is denoted by C_a.]
ii) The collection $\{C_a : a \in A\}$ is locally finite.
iii) For each $m \in M$, the sum $\sum_{a \in A} f_a(m) = 1$ (and note that, from (ii), only finitely many terms in this sum are non-zero).

It follows that each $f_a : M \to [0, 1]$ and that $\cup_{a \in A} C_a = M$. If M admits a partition of unity it is necessarily Hausdorff.

The main theorem regarding the existence of partitions of unity can now be stated for connected Hausdorff manifolds which will be the ones most relevant here. The theorem is readily extended to non-connected manifolds [19].

Theorem 4.4 *Let M be a smooth connected Hausdorff manifold. Then the following are equivalent.*

i) M admits a partition of unity.
ii) M is second countable.
iii) M is paracompact.

Since it is easily shown that the product of two Hausdorff (respectively connected or second countable) non-empty topological spaces is Hausdorff (respectively connected or second countable) it follows from this theorem that *the product of two (smooth connected Hausdorff) paracompact manifolds is a (smooth connected Hausdorff) paracompact manifold.*

The property of paracompactness is extremely useful for extending certain local constructions on a manifold to global ones, this construction being accomplished by use of the functions in the partition of unity (often affectionately known, for obvious reasons, as *bump functions*).

4.6 The Tangent Space and Tangent Bundle

In this and subsequent sections it will be shown that at each point of an n-dimensional smooth manifold M there is a well defined family of finite dimensional real vector spaces. These are the tensor spaces and they are of fundamental importance. The first of these is the set of all *vectors* at a point in a manifold and the remainder of the tensor spaces will be derived from it.

Although the idea of a differentiable function has been introduced through its coordinate representatives no mention of how to define its derivatives was made. Let g be a smooth function $g : M \to \mathbb{R}$, let x be a chart on M with domain U (and using the usual identity chart on $\mathbb{R}$) and let G be the coordinate representative of g in this chart so that $g = G \circ x$ with G a smooth map from some open subset of $\mathbb{R}^n$ to $\mathbb{R}$. One can now define the functions $g_{,a} : U \to \mathbb{R}$ (sometimes written $\frac{\partial g}{\partial x^a}$) by

$$g_{,a} \equiv \frac{\partial G}{\partial x^a} \circ x \qquad (1 \leq a \leq n)$$

called the *partial derivatives* of g in the chart x.

To give a precise constructive definition of the set of *vectors* at any point $m \in M$ let $F(m)$ denote the family of smooth real valued functions whose domains (which are, by definition, open subsets of M) include m. A *derivation* on M is a map $L : F(m) \to \mathbb{R}$ satisfying

(1) $L(\alpha f + \beta g) = \alpha L(f) + \beta L(g)$
(2) $L(fg) = fL(g) + gL(f)$

for $f,\ g \in F(m)$ and $\alpha,\ \beta \in \mathbb{R}$ where the meaning of $\alpha f \in F(m)$ is clear and where the sum and product of two members of $F(m)$ is defined in the obvious way on the intersection of their domains and, hence, are members of $F(m)$. If x is a chart of M whose domain contains m, the n maps $(\frac{\partial}{\partial x^a})_m : F(m) \to \mathbb{R}$ defined by $g \to g_{,a}(m)$ are derivations on $F(m)$. Their importance stems from the fact that the family of all derivations on $F(m)$ can be shown to be a real vector space for which the $(\frac{\partial}{\partial x^a})_m$ constitute a basis [19]. Thus the vector space of all derivations on $F(m)$ is an n-dimensional real vector space and is called the *tangent space* to M at m and denoted by T_mM. Its members are called *tangent vectors* at m. The geometrical interpretation of the tangent space as constructed here follows from the idea of directional derivatives in the calculus of $\mathbb{R}^n$. If $\mathbf{v} \in T_mM$ then, in the chart x, $\mathbf{v} = \sum_{a=1}^n v^a(\frac{\partial}{\partial x^a})_m$ ($v^a \in \mathbb{R}$) and if $g \in F(m)$, $\mathbf{v}(g) = \sum_{a=1}^n g_{,a}(m)v^a$ is the directional derivative of g in the direction of $\mathbf{v}$.

With the notation of the previous paragraph, if $\mathbf{v} \in T_mM$ then $\mathbf{v} = \sum_{a=1}^n v^a(\frac{\partial}{\partial x^a})_m$. The coordinate functions x^a are members of $F(m)$ and the *components* v^a of $\mathbf{v}$ in the basis $(\frac{\partial}{\partial x^a})_m$ (or in the chart x) are then given by

$$\mathbf{v}(x^a) = \sum_{b=1}^n v^b(\frac{\partial}{\partial x^b})_m(x^a) = \sum_{b=1}^n v^b\delta_b^a = v^a$$

If y is another chart of M with coordinate functions y^a whose domain contains m then $\frac{\partial}{\partial y^a}$ are members of T_mM and so $(\frac{\partial}{\partial y^a})_m = \sum_{b=1}^n A_a^b \left(\frac{\partial}{\partial x^b}\right)_m$ with each $A_a^b \in \mathbb{R}$. But

$$\left(\frac{\partial}{\partial y^a}\right)_m(x^c) = \left(\frac{\partial x^c}{\partial y^a}\right)_m \quad \left(= \sum_{b=1}^n A_a^b\left(\frac{\partial}{\partial x^b}\right)_m(x^c) = A_a^c\right)$$

It follows that if $\mathbf{v} \in T_mM$ and $\mathbf{v} = \sum_{a=1}^n v^a(\frac{\partial}{\partial x^a})_m = \sum_{a=1}^n v'^a(\frac{\partial}{\partial y^a})_m$ then

$$v^a = \mathbf{v}(x^a) = \sum_{b=1}^n \left(\frac{\partial x^a}{\partial y^b}\right)_m v'^b \qquad v'^a = \sum_{b=1}^n \left(\frac{\partial y^a}{\partial x^b}\right)_m v^b \tag{4.1}$$

which gives the relationship (transformation law) between the components of v in the bases $(\frac{\partial}{\partial x^a})_m$ and $(\frac{\partial}{\partial y^a})_m$. A basis for T_mM is sometimes called a *frame* (at m).

It is convenient to have a notation for the set of all tangent vectors at all points of M. This set is called the *tangent bundle* of M, denoted by TM, and defined by

$$TM = \bigcup_{m \in M} T_mM$$

There is an obvious projection map $\pi : TM \to M$ defined by $\pi(\mathbf{v}) = m$ if $\mathbf{v} \in T_mM$. Thus π maps each vector in TM to the point of M to which it is attached and $\pi^{-1}\{m\} = T_mM$. The set TM can be given a manifold structure which arises naturally from that on M. First note that TM is the union of sets of the form $\pi^{-1}U$ where U is the domain of a chart x on M. Now each $\mathbf{v} \in \pi^{-1}U$ can be written uniquely as $\mathbf{v} = \sum v^a (\frac{\partial}{\partial x^a})_m$ where $\pi(v) = m$ and so one has an injective map $\pi^{-1}U \to \mathbb{R}^{2n}$ given by $\mathbf{v} \to (x^1(m), ..., x^n(m), v^1, ..., v^n)$ whose range is the open subset $x(U) \times \mathbb{R}^n$ of $\mathbb{R}^{2n}$. It is thus a chart for TM and the set of all such charts is easily shown to be an atlas for TM which, since M is smooth, gives the tangent bundle a smooth manifold structure of dimension $2n$. The map π is then easily seen to be a *smooth* map between the manifolds TM and M. The consequent manifold topology on TM inherits many of the topological properties that may be possessed by M, for example, the properties of being Hausdorff, connected, second countable and paracompact (but not compactness).

4.7 Tensor Spaces and Tensor Bundles

With the notation of the previous section with M smooth and $\dim M = n$ let $m \in M$ and consider the *dual space* of the vector space T_mM, denoted by T_m^*M (see section 2.4). Let x be a chart whose domain contains m so that $(\frac{\partial}{\partial x^a})_m$ is a basis for T_mM and let $(dx^a)_m$ denote the corresponding dual basis in T_m^*M. Hence $(dx^a)_m((\frac{\partial}{\partial x^b}{}_m)) = \delta_b^a$ and each member $\mathbf{w}$ of T_m^*M may be written as $\mathbf{w} = \sum_{a=1}^{n} w_a (dx^a)_m$ where w_a are the *components* of $\mathbf{w}$ in the basis $(dx^a)_m$ of the chart x. Thus a new object (in fact an n-dimensional real vector space) T_m^*M is constructed at each $m \in M$ called the *cotangent space* to M at m. Its members are called *covectors* (sometimes *cotangent vectors*) or 1-*forms*. One can also construct the *cotangent bundle*

$$T^*M = \bigcup_{m \in M} T_m^*M$$

of M. Now if $\mathbf{w} \in T_m^*M$ and x and y are charts whose domains contain m then one may write

$$\mathbf{w} = \sum_{a=1}^{n} w_a (dx^a)_m = \sum_{a=1}^{n} w'_a (dy^a)_m \qquad \text{and} \qquad w_a = \mathbf{w}\left(\left(\frac{\partial}{\partial x^a}\right)_m\right)$$

It then follows from the work and notation of the previous section that

$$w'_a = \mathbf{w}\Big(\Big(\frac{\partial}{\partial y^a}\Big)_m\Big) = \sum_{b=1}^{n} A_a^b w_b = \sum_{b=1}^{n}\Big(\frac{\partial x^b}{\partial y^a}\Big)_m w_b, \tag{4.2}$$

$$w_a = \sum_{b=1}^{n}\Big(\frac{\partial y^b}{\partial x^a}\Big)_m w'_b \tag{4.3}$$

which gives the transformation law between the components of $\mathbf{w}$ in these two bases. It now follows by arguments essentially identical to those of the previous section that the cotangent bundle T^*M can be given the structure of a smooth $2n$-dimensional manifold. The projection map $T^*M \to M$ (also denoted by π unless some possible confusion neccessitates an alternative symbol) is defined by $\pi(\mathbf{w}) = m$ if $\mathbf{w} \in T_m^*M$ and it follows that $\pi^{-1}(m) = T_m^*M$ and that π is a smooth map between the manifolds T^*M and M.

It is pointed out that the members of T_mM and T_m^*M are sometimes called *contravariant vectors* and *covariant vectors*, respectively. It is also remarked that although the vector spaces T_mM and T_m^*M are isomorphic, there is no *natural* isomorphism between them (*i.e.* there is no *natural* way of pairing off tangent vectors and cotangent vectors) unless some other object defined on M intervenes to effect such an identification. This point will be discussed more fully later.

The idea involved in constructing the cotangent space can be generalised by considering the set of all *multilinear maps* (called *tensors of type* (r, s)) on the vector space $T_mM \oplus ... \oplus T_mM \oplus T_m^*M \oplus ... \oplus T_m^*M$ where there are s copies of T_mM and r copies of T_m^*M for non-negative integers r and s (see section 2.5). This set is a real vector space of dimension n^{r+s} called the *vector space of tensors of type* (r, s) *at* m and is denoted by $T_m{}_s^rM$. The union of all such vector spaces at all points of M (for fixed r and s) is the *bundle of tensors of type* (r, s) *on* M and is denoted by T_s^rM. The space of type $(0, 1)$ tensors at m is just T_m^*M whilst the space of type $(1, 0)$ tensors at m is the dual of T_m^*M and is thus naturally isomorphic to T_mM under an isomorphism which associates with $(\frac{\partial}{\partial x^a})_m$ in T_mM a member $\mathbf{e}_a$ in the dual of T_m^*M satisfying $\mathbf{e}_a((dx^b)_m) = (dx^b)_m((\frac{\partial}{\partial x^a})_m) = \delta_b^a$ (see section 2.4). It is usual to use the symbol $(\frac{\partial}{\partial x^a})_m$ also for $\mathbf{e}_a$ and so $(\frac{\partial}{\partial x^a})_m((dx^b)_m) = \delta_b^a$. The bundles of $(1, 0)$ and $(0, 1)$ tensors are then the tangent and cotangent bundles, respectively. If x is a chart whose domain contains m and if t is a type (r, s) tensor at m one has in the obvious basis (see section 2.5)

$$t = \sum_{a_1,\ldots,b_s=1}^{n} t^{a_1\ldots a_r}_{b_1\ldots b_s} (dx^{b_1})_m \otimes \ldots \otimes (dx^{b_s})_m \otimes (\frac{\partial}{\partial x^{a_1}})_m \otimes \ldots \otimes (\frac{\partial}{\partial x^{a_r}})_m$$

where $t^{a_1\ldots a_r}_{b_1\ldots b_s}$ are the *components* of t in the above basis (or in the chart x). Thus

$$t^{a_1\ldots a_r}_{b_1\ldots b_s} = t\Big(\Big(\frac{\partial}{\partial x^{b_1}}\Big)_m, \ldots, \Big(\frac{\partial}{\partial x^{b_s}}\Big)_m, (dx^{a_1})_m, \ldots, (dx^{a_r})_m\Big)$$

If y is another chart whose domain contains m and for which the components of t are $t'^{c_1\ldots c_r}_{d_1\ldots d_s}$ then one finds by similar arguments to those for tangent vectors and covectors

$$t'^{c_1\ldots c_r}_{d_1\ldots d_s} = \sum_{a_1,\ldots,b_s=1}^{n} \Big(\frac{\partial y^{c_1}}{\partial x^{a_1}}\Big)_m \cdots \Big(\frac{\partial y^{c_r}}{\partial x^{a_r}}\Big)_m \Big(\frac{\partial x^{b_1}}{\partial y^{d_1}}\Big)_m \cdots \Big(\frac{\partial x^{b_s}}{\partial y^{d_s}}\Big)_m t^{a_1\ldots a_r}_{b_1\ldots b_s} \tag{4.4}$$

When two tensors of the same type at m are added (or scalar multiplied) in their vector space structure one merely adds together (or scalar multiplies) their components in a given coordinate system. Again one concludes that $T^r_s M$ can be given the structure of a smooth manifold of dimension $n^{r+s}+n$ by copying the method above for TM and T^*M and the obvious projection map from this bundle to M is then smooth. Indices in the upper position are referred to as *contravariant indices* and those in the lower position as *covariant indices*. A tensor T is called *symmetric* (respectively, *skew-symmetric*) in two contravariant or two covariant indices a and b if in some (and, as is easily checked, any) coordinate system, $T^{\cdot a \cdot b \cdot}_{c \cdots d} = T^{\cdot b \cdot a \cdot}_{c \cdots d}$ (respectively, $T^{\cdot a \cdot b \cdot}_{c \cdots d} = -T^{\cdot b \cdot a \cdot}_{c \cdots d}$) and similarly for two covariant indices. A function $f : M \to \mathbb{R}$ can be regarded as a $(0, 0)$ tensor.

4.8 Vector and Tensor Fields

A vector field on a (smooth) manifold M of dimension n is essentially the attaching of a tangent vector to each point of M in a suitably differentiable way. More precisely, a *vector field* on M is a map $X : M \to TM$ such that $X(m) \in T_m M$ for each $m \in M$ (such a map is also called a *section* of TM). One usually asks, in addition, that X satisfies some differentiability requirement. Then X is called a *smooth vector field* if any of the following conditions (which are easily shown to be equivalent) are satisfied.

i) If x is any chart of M with domain U and if one writes $X(m) = \sum_{a=1}^{n} X^a(m)(\frac{\partial}{\partial x^a})_m$ at each $m \in U$, the n functions $X^a : U \to \mathbb{R}$ are smooth.
ii) If f is any smooth real valued function defined on some open subset V of M, the function $Xf : V \to \mathbb{R}$ given by $Xf(m) = X(m)(f)$ is smooth.
iii) The map $X : M \to TM$ is a smooth map when TM has the natural manifold structure described earlier (*i.e.* X is a *smooth section* of TM).

The definition here of a vector field X required it to be defined on the whole of M. This definition (and consequent smoothness criteria) are easily and obviously adapted to the case when X is defined on some open subset W of M. If X is defined on the whole of M it is called a *global vector field* on M. It is now clear that if x is a chart of M with domain U then the n vector fields $\frac{\partial}{\partial x^a}$ defined on U by $\frac{\partial}{\partial x^a}(m) = (\frac{\partial}{\partial x^a})_m$ are smooth vector fields on U. (If, more generally, M is a C^∞ manifold then a C^k $(k \geq 1)$ vector field may be defined by an obvious modification of the criteria above.)

If X and Y are global smooth vector fields on M and α, $\beta \in \mathbb{R}$ one may define a global smooth vector field $\alpha X + \beta Y$ by $(\alpha X + \beta Y)(m) = \alpha X(m) + \beta Y(m)$. Thus the set $F^1(M)$ of all global smooth vector fields on M can be given the structure of a real vector space (which, if M is Hausdorff, is not finite-dimensional). Here it is important to distinguish between vector space theoretic concepts applied to $F^1(M)$ and applied to $T_m M$ $(m \in M)$. In particular if X and Y are independent members of $F^1(M)$ then $X(m)$ and $Y(m)$ may not be independent members of $T_m M$. The manifold M is called *parallelisable* if there exist smooth vector fields $X^1, ..., X^n$ on M such that $X^1(m), ..., X^n(m)$ are independent members of the vector space $T_m M$ for each $m \in M$. Then the set $\{X^1, ..., X^n\}$ consists of n independent members of $F^1(M)$ called a *parallelisation* of M. It should also be noted that if $f : M \to \mathbb{R}$ is smooth and $X \in F^1(M)$ then the vector field fX defined by $fX(m) = f(m)X(m)$ is in $F^1(M)$.

One may define a *covector field* on M as a map $\omega : M \to T^*M$ such that $\omega(m) \in T_m^*M$ for each $m \in M$ (a *section* of T^*M). Then ω is called *smooth* if any of the following equivalent conditions are satisfied.

i) If x is any chart of M with domain U and if one writes $\omega(m) = \sum_{a=1}^{n} \omega_a(m)(dx^a)_m$ at each $m \in U$, the n functions $\omega_a : U \to \mathbb{R}$ are smooth.
ii) If X is any smooth vector field defined on some open subset V of M the map $\omega(X) : V \to \mathbb{R}$ given by $\omega(X)(m) = \omega(m)(X(m))$ is smooth.

iii) The map $\omega : M \to T^*M$ is smooth when T^*M has the natural manifold structure described earlier (*i.e.* ω is a *smooth section* of T^*M).

Again one would have obvious modifications for a covector field defined on some open subset W of M and ω is a *global covector field* if $W = M$. If x is a chart of M with domain U then the n covector fields dx^a defined on U by $dx^a(m) = (dx^a)_m$ are smooth on U. Degrees of differentiability other than C^∞ are also easily handled. Just as described earlier for vector fields, the set of all global smooth covector fields on M is a real vector space and if ω is such a field and $f : M \to \mathbb{R}$ is smooth then $f\omega$ is a smooth covector field on M in an obvious way. Regarding notation, what has here been called a covector field is sometimes referred to as a 1-*form field*.

The concept of a tangent vector at $m \in M$ can be motivated by the idea of a tangent vector to a path passing through m (as will be seen later). The concept of a covector at m can be motivated by the idea of the gradient of a function and leads to the following construction. Let f be a smooth function $M \to \mathbb{R}$. Then one can construct a smooth covector (1-form) field df defined by its action on any smooth vector field X defined on some open subset of M, $df(X) = X(f)$. Thus if x is a chart of M with domain U and by considering the smooth vector fields $\frac{\partial}{\partial x^a}$ on U one finds that, on U, $df = \sum_{a=1}^{n} (\frac{\partial f}{\partial x^a}) dx^a$. Also by choosing f to be, successively, the coordinate functions $x^1, ..., x^n$ one sees that if $m \in U$ then $dx^a(m)$ is exactly what was earlier called $(dx^a)_m$ and thus explains the latter notation. Again, the modifications when the original f was defined only on some open subset of M are clear. However it should be noted that whilst the smoothness of f guarantees the smoothness of df, the condition that f is C^k $(1 \le k < \infty)$ implies only that df is C^{k-1}.

A *tensor field* t of type (r, s) on M is a map $t : M \to T^r_s M$ such that $t(m) \in T_m{}^r_s$, $\forall m \in M$ (i.e. t is a section of $T^r_s M$). It is called *smooth* if any of the following equivalent conditions hold

i) In any chart x of M with domain U and if one writes

$$t(m) = \sum_{a_1,...,b_s} t^{a_1...a_r}_{b_1...b_s}(m)(\frac{\partial}{\partial x^{a_1}})_m \otimes ... \otimes (dx^{b_s})_m$$

at each $m \in U$ the n^{r+s} functions $t^{a_1...a_r}_{b_1...b_s} : U \to \mathbb{R}$ are smooth.

ii) If $X^1, ..., X^s$ are smooth vector fields and $\omega^1, ..., \omega^r$ are smooth covector fields each defined on the open subset V of M then the map $V \to \mathbb{R}$

given by $m \to t(m)(X^1(m), ..., X^s(m), \omega^1(m), ..., \omega^r(m))$ is smooth.

iii) The map $t : M \to T^r_s M$ is smooth when $T^r_s M$ has the manifold structure described earlier (*i.e.* t is a *smooth section* of $T^r_s M$).

Again one easily makes appropriate modifications when T is not defined on the whole of M or when the degree of differentiability is reduced. It should be noted that the above remarks about tensor fields in the case of type $(1,0)$ tensors reduce to those for vector fields above when the identification of a vector space and its double dual (section 2.4) is taken into account. If t and t' are smooth type (r,s) tensor fields on M and α, $\beta \in \mathbb{R}$ then the smooth type (r,s) tensor field $\alpha t + \beta t'$ is defined by $(\alpha t + \beta t')(m) = \alpha t(m) + \beta t'(m)$. If t is a type (r,s) and t' a type (p,q) smooth tensor field on M the *tensor product* of t and t' is the type $(r+p, s+q)$ smooth tensor field on M denoted by $t \otimes t'$ and defined in any chart U with coordinates x^a by the component relation

$$(t \otimes t')^{a_1 \dots a_{r+p}}_{b_1 \dots b_{s+q}} = t^{a_1 \dots a_r}_{b_1 \dots b_s} t'^{a_{r+1} \dots a_{r+p}}_{b_{s+1} \dots b_{s+q}}$$

where $t^{a_1 \dots a_r}_{b_1 \dots b_s}$ are the components of t in this chart and similarly for t'. If t is a smooth type (r,s) tensor field on M the *contraction* of t over the indices a_p and b_q (one contravariant and one covariant index) is the smooth type $(r-1, s-1)$ tensor field $\bar{t}$ with components

$$\bar{t}^{a_1 \dots a_{p-1} a_{p+1} \dots a_r}_{b_1 \dots b_{q-1} b_{q+1} \dots b_s} = \sum_{k=1}^{n} t^{a_1 \dots a_{p-1} k a_{p+1} \dots a_r}_{b_1 \dots b_{q-1} k b_{q+1} \dots b_s}$$

The *contraction* of a type (r,s) tensor field t (with $s \neq 0$) *with a vector field* k on M over the covariant index c of t is defined to be the contraction of the tensor $t \otimes k$ over k's contravariant index and the index c. Other types of tensor contractions can be similarly defined. The set of all smooth type (r,s) tensor fields on M, for fixed r and s, is a real vector space under the addition and multiplication defined at the beginning of this paragraph.

Given two smooth (respectively C^k, $1 \le k < \infty$) vector fields X and Y on M there is an important smooth (respectively C^{k-1}) vector field arising from X and Y on M called the *Lie bracket* (or just the *bracket*) of X and Y, denoted by $[X,Y]$ and defined by its action on a C^2 real valued function f (defined on some open subset of M) by

$$[X,Y]f = X(Y(f)) - Y(X(f)) \tag{4.5}$$

If X and Y have components X^a and Y^a in some chart x of M then the components of $[X, Y]$ in this chart are $\sum_{b=1}^{n}(Y^a{}_{,b}X^b - X^a{}_{,b}Y^b)$. This definition is easily adapted if X and Y are not global vector fields on M and then one easily checks that $[X, Y] = -[Y, X]$ and that for a vector field Z, $[X, [Y, Z]] + [Y, [Z, X]] + [Z, [X, Y]] = 0$. The latter is called the *Jacobi identity*.

It easily follows that the set of all smooth (but not the set of all C^k $(1 \leq k < \infty)$) global vector fields on M is a Lie algebra under the Lie bracket product. If X and Y are smooth vector fields on M and f and g smooth real valued functions on M then

$$[fX, gY] = fg[X, Y] + f(Xg)Y - g(Yf)X.$$

At this point the well known *Einstein summation convention* will be introduced whereby a twice repeated index is automatically summed over the (obvious) range, unless otherwise specified. There will be occasions, however, where it is convenient to retain the Σ symbol. As an example of this notation let T be a tensor at $m \in M$ of type (r, s) and $\{\overset{\alpha}{\mathbf{e}}\}$ a frame at m. Then the real numbers $T^{\alpha\cdots\beta}_{\gamma\cdots\delta} \equiv T^{a\cdots b}_{c\cdots d}\, \overset{\alpha}{e}{}^c \cdots \overset{\beta}{e}{}^d\, \underset{\gamma}{e}{}_a \cdots \underset{\delta}{e}{}_b$ are called the *frame components of* T (in the frame $\{\overset{\alpha}{\mathbf{e}}\}$).

4.9 Derived Maps and Pullbacks

Let M and M' be smooth manifolds and $\phi : M \to M'$ a smooth map. Let $m \in M$ and set $m' = \phi(m)$. If $\mathbf{v} \in T_m M$ there is a well defined member of $T_{m'}M'$ arising from $\mathbf{v}$ and ϕ, denoted by $\phi_{*m}\mathbf{v}$ and defined by

$$\phi_{*m}\mathbf{v}(f) = \mathbf{v}(f \circ \phi)$$

where f is any smooth function defined on some open neighbourhood V of m' (and then the domain of $f \circ \phi$ includes the neighbourhood $\phi^{-1}V$ of m). The map $\phi_{*m} : T_m M \to T_{m'}M'$ is easily checked to be a linear map between vector spaces. If x and y are charts whose domains contain m and m', respectively, the linear map ϕ_{*m} is completely determined by its action on the basis $(\frac{\partial}{\partial x^a})_m$ and is easily shown to be

$$\Big(\frac{\partial}{\partial x^a}\Big)_m \to \sum_{b=1}^{n}\Big(\frac{\partial(y^b \circ \phi)}{\partial x^a}\Big)_m \Big(\frac{\partial}{\partial y^b}\Big)_{m'} \qquad (n = \dim M')$$

where the matrix $(\frac{\partial(y^b \circ \phi)}{\partial x^a})_m$ is the *Jacobian* of the coordinate representative $y \circ \phi \circ x^{-1}$ of ϕ at m and its rank is called the *rank of ϕ at m*. This rank is clearly independent of the charts x and y. The map ϕ_{*m} is called the *derived linear function* of ϕ at m or the *differential* of ϕ at m. If M'' is another smooth manifold and $\phi' : M' \to M''$ is smooth then $(\phi' \circ \phi)_{*m} = \phi'_{*m'} \circ \phi_{*m}$.

With $\phi : M \to M'$ as above one also has a natural map $\phi_* : TM \to TM'$ defined by $\mathbf{v} \to \phi_* \mathbf{v} = \phi_{*m} \mathbf{v}$ where $\mathbf{v} \in T_m M$. This map is called the *differential* of ϕ and is a *smooth* map between the manifolds TM and TM'. With $\phi' : M' \to M''$ as above, it is clear that $(\phi' \circ \phi)_* = \phi'_* \circ \phi_*$.

With ϕ as above and $f : M' \to \mathbb{R}$ smooth define a smooth map $\phi^* f : M \to \mathbb{R}$ by $\phi^* f = f \circ \phi$, called the *pullback* of f under ϕ. Now let t be a type $(0, s)$ tensor at $m' = \phi(m)$. Then one can construct a type $(0, s)$ tensor at m, denoted by $\phi^*_m t$ and called the *pullback* of t under ϕ, by

$$\phi^*_m t(\mathbf{v}_1, ..., \mathbf{v}_s) = t(\phi_{*m}(\mathbf{v}_1), ..., \phi_{*m}(\mathbf{v}_s))$$

for $\mathbf{v}_1, ..., \mathbf{v}_s \in T_m M$. This construction enables one to give an equivalent definition of the map $\phi_{*m} : T_m M \to T_{m'} M'$ in that for $\mathbf{v} \in T_m M$ one defines $\phi_{*m} \mathbf{v}$ by its action on a member $\boldsymbol{\omega}$ of $T^*_{m'} M'$ by

$$\phi_{*m} \mathbf{v}(\boldsymbol{\omega}) = \mathbf{v}(\phi^*_m \boldsymbol{\omega})$$

This is easily checked to be equivalent to the original one and, further, allows one to *push forward* tensors of type $(r, 0)$. More precisely if t is a tensor of type $(r, 0)$ at $m \in M$ one defines a type $(r, 0)$ tensor at m' called the *pushforward* $\phi_{*m} t$ of t under ϕ by

$$\phi_{*m} t(\boldsymbol{\omega}_1, ..., \boldsymbol{\omega}_r) = t(\phi^*_m \boldsymbol{\omega}_1, ..., \phi^*_m \boldsymbol{\omega}_r)$$

for $\boldsymbol{\omega}_1, ..., \boldsymbol{\omega}_r \in T^*_{m'} M'$. However the following difference should be noted. If t is a *tensor field* of type $(0, s)$ on M' one can construct a *tensor field* of type $(0, s)$ on M denoted by $\phi^* t$, called the *pullback* of t under ϕ and defined in the above notation by

$$(\phi^* t)(m)(\mathbf{v}_1, ..., \mathbf{v}_s) = t(m')(\phi_{*m} \mathbf{v}_1, ..., \phi_{*m} \mathbf{v}_s).$$

The tensor field $\phi^* t$ is smooth if ϕ and t are. But there is no similar construction for type $(r, 0)$ tensor *fields* in the "direction" $M \to M'$. In particular although the differential ϕ_* of ϕ is a well defined map $TM \to TM'$ it may not map a vector field X on M to a vector field on M'. This is because one may have m_1, $m_2 \in M$ with $\phi(m_1) = \phi(m_2)$ but with $\phi_{*m_1} X(m_1) \neq \phi_{*m_2} X(m_2)$.

Suppose now that $\phi : M \to M'$ is a smooth *diffeomorphism*. Then the situation is much nicer. In fact if t is any smooth type (r, s) tensor field on M' then one has a smooth type (r, s) tensor field $\phi^* t$ on M defined by

$$\phi^* t(m)(\mathbf{v}_1, ..., \mathbf{v}_s, \boldsymbol{\omega}_1, ..., \boldsymbol{\omega}_r) = \\ t(m')(\phi_{m*}\mathbf{v}_1, ..., \phi_{m*}\mathbf{v}_s, {\phi^{-1}}^*_{m'}\boldsymbol{\omega}_1, ..., {\phi^{-1}}^*_{m'}\boldsymbol{\omega}_r)$$

for $\mathbf{v}_1, ..., \mathbf{v}_s \in T_m M$, $\boldsymbol{\omega}_1, .., \boldsymbol{\omega}_r \in T_m^* M$ and where $m' = \phi(m)$. Again $\phi^* t$ is called the *pullback* of t under ϕ. Since $\phi^{-1} : M' \to M$ is also a diffeomorphism one can move tensors in the opposite "direction" using ϕ^{-1}. Also, for $\phi : M \to M'$ a smooth diffeomorphism and X a smooth vector field on M one has a smooth vector field $\phi_* X$ on M' defined by $\phi_* X(m') = \phi_{*m} X(m)$ for $m \in M$ and $m' = \phi(m) \in M'$ [21]. In particular, if X and Y are smooth vector fields, then $[\phi_* X, \phi_* Y] = \phi_*[X, Y]$. Similar constructions can be done when the tensor fields and map ϕ are C^k $(1 \leq k \leq \infty)$ [21].

Returning to a smooth map $\phi : M \to M'$ let $m \in M$ and $m' = \phi(m)$. The behaviour of ϕ at points *close to* m is in a sense reflected by the action of ϕ_{*m} on $T_m M$. An important case where this is made precise is in the inverse function theorem for manifolds [19].

Theorem 4.5 The Inverse Function Theorem for Manifolds
Let M and M' be smooth manifolds and let $\phi : M \to M'$ be smooth. If $m \in M$ then ϕ_{*m} is a vector space isomorphism $T_m M \to T_{m'} M'$ if and only if there exists an open neighbourhood U of m in M such that the restriction $\phi_{|U}$ of ϕ to U is a diffeomorphism from the open subset U of M onto an open subset U' of M' containing m'.

The proof follows from the inverse function theorem given in section 4.2 for $\mathbb{R}^n$. If the conclusions of the theorem hold then M and M' have the same dimension.

4.10 Integral Curves of Vector Fields

Let M be a smooth n-dimensional manifold. A (smooth) *curve* or *path* in M is a smooth map $c : I \to M$ where I is an *open* interval of $\mathbb{R}$. Let t be the identity chart on $\mathbb{R}$, $m \in M$ and x a chart in M whose domain U contains m and intersects the range of c non- trivially. Then c, restricted to U, is often described by its *component functions* $c^a(t)$ where $c^a = x^a \circ c$ and t may be restricted to some open subset of I. Often, no distinction is drawn between the *image* of the curve c in M, $c(I)$, and the curve c itself (and a member of $c(I)$ is said to be *on* c). However, the map c contains more information, that is, its precise dependence on t, and to indicate this dependence t is referred to as the *parameter* of the curve. If t_0 is in the domain of the functions c^a the member $(\frac{\partial c^a}{\partial t})_{t_0}(\frac{\partial}{\partial x^a})_{c(t_0)}$ of $T_{c(t_0)}M$ with components $(\frac{\partial c^a}{\partial t})_{t_0}$ in the chart x is called the *tangent vector* (or said to be *tangent*) to the curve c at $c(t_0)$. To show this is a member of $T_{c(t_0)}M$ one must check that the change in its components satisfies (4.1) under a change of coordinates in M. This is easily done but will be achieved in a coordinate independent way below. Now suppose that J is another open interval of $\mathbb{R}$ and $f : J \to I$ is a bijective map such that f and f^{-1} are each smooth and where the derivative of f is nowhere zero. Define the curve c' in M by $c' = c \circ f$. Then c and c' have the same range in M but, in general, have different parameters and c' is called a *reparametrisation* of c. The tangent vectors to c and c' at the same point $m = c(t_0) = c'(t'_0)$ (where $f(t'_0) = t_0$) differ by a (non-zero) multiple $(\frac{\partial f}{\partial t'})_{t'_0}$ as is easily checked.

Let $\frac{\partial}{\partial t}$ be the vector field on $\mathbb{R}$ corresponding to the identity chart t. For the curve c in M define the map $\dot{c} = c_* \circ \frac{\partial}{\partial t}$, so that $\dot{c} : I \to TM$ is a *curve in* TM, that is, it associates with each $t_0 \in I$ a vector at $c(t_0)$ in M. This vector is precisely the *tangent vector* to the curve c at $c(t_0)$ since from section 4.9

$$\dot{c}(t_0) = \Big(c_* \circ \frac{\partial}{\partial t}\Big)(t_0) = c_*\Big(\frac{\partial}{\partial t}\Big)_{t_0} = \Big(\frac{d(x^a \circ c)}{dt}\Big)_{t_0}\Big(\frac{\partial}{\partial x^a}\Big)_{c(t_0)}$$

Now let X be a smooth vector field on M. A curve c in M is called an *integral curve* of X if its tangent vector at any point of the curve equals the value of X at that point, that is, if for each t_0 in the domain of c, $\dot{c}(t_0) = X(c(t_0))$ or, equivalently, if $\dot{c} = X \circ c$. If the domain of an integral curve c of X contains $0 \in \mathbb{R}$, c is called an *integral curve of x starting from* $m = c(0)$. Finding integral curves involves solving first order differential equations. In fact, within a chart domain U, if $X = X^a\frac{\partial}{\partial x^a}$ for smooth functions $X^a : U \to \mathbb{R}$, a curve c is an integral curve of X if

$$\frac{d}{dt}[c^a(t)] = \frac{d}{dt}[x^a(c(t))] = X^a(x^1(c(t)), ..., x^n(c(t))) \qquad (4.6)$$

an expression which is often loosely written as $\frac{dx^a}{dt} = X^a$. The arbitrary constant that occurs in solving first order equations such as (4.6) can, in general, be loosely identified with the choice of starting point along a particular integral curve of X. In fact, if $c : I \to M$ is an integral curve of X

where $0 \in I$ (and so c starts from $c(0) = m$) and if $t_0 \in I$ let $f : \mathbb{R} \to \mathbb{R}$ denote the *translation* $t \to t + t_0$. Then it can be checked that $c' = c \circ f$ is also an integral curve of X with domain $f^{-1}(I)$ starting from $c'(0) = c(t_0)$.

An important result regarding the existence of integral curves of a given smooth vector field on M can now be stated [19].

Theorem 4.6 *Let X be a smooth vector field on M. Then given any $m_0 \in M$ there exists an open neighbourhood U of m_0 and an open interval I of $\mathbb{R}$ containing 0 such that there is an integral curve of X with domain I starting from any point $m \in U$. Any other integral curve of X starting from m coincides with this curve on some neighbourhood of* 0.

Let X be a smooth vector field on M and suppose c_1 and c_2 are integral curves of X with domains I_1 and I_2 each of which contains 0 and which start from the same point of M. Then c_1 and c_2 need not coincide on $I_1 \cap I_2$ (an example being easily constructed - see, *e.g.* [19]). However, c_1 and c_2 will coincide on $I_1 \cap I_2$ if M is *Hausdorff* [19]. For such manifolds the union of the domains of all integral curves of X which start from a given point $m \in M$ is an open interval of $\mathbb{R}$ containing 0 on which is defined an integral curve of X called the *maximal integral curve of X starting from m*. Again, with M Hausdorff, if the domain of the maximal integral curve of a smooth vector field X through any point $m \in M$ is always $\mathbb{R}$ then X is called *complete*.

There is an instructive geometrical interpretation of the derived linear function which may now be briefly commented on. Let M and M' be smooth manifolds, $\phi : M \to M'$ a smooth map and $c : I \to M$ a smooth curve in M. Then $\phi \circ c : I \to M'$ is a smooth curve in M'. Further one may compute $(\dot{\phi \circ c})$ to find

$$(\dot{\phi \circ c})(t_0) = \phi_{*c(t_0)}\left(c_*\left(\frac{\partial}{\partial t}\right)_{t_0}\right)$$

thus showing that if $\mathbf{v} \in T_m M$ is tangent to the curve c at $m \in M$ then $\phi_{*m}\mathbf{v}$ is tangent to the curve $\phi \circ c$ at $\phi(m) \in M'$.

4.11 Submanifolds

In this section the important idea of a submanifold of a manifold is discussed. Let M be a smooth manifold of dimension n and M' a subset of M. The requirement that M' be a (smooth) submanifold of M, roughly speaking, consists of two points: (i) that M' be a (smooth) manifold in its own right and (ii) that it be contained in M in a *smooth way*. One further definition is required to achieve this. Let $\bar{M}$ and M be smooth manifolds of dimensions $\bar{n}$ and n and let $\phi : \bar{M} \to M$ be smooth. The map ϕ is called an *immersion* if for each $\bar{m} \in \bar{M}$ the rank of ϕ at $\bar{m}$ equals $\bar{n}$, that is, the dimension of $\bar{M}$ (and so necessarily $\phi_{*\bar{m}}$ is injective for each $\bar{m} \in \bar{M}$ and $\dim \bar{M} \leq \dim M$). Now, returning to the original manifold M and subset M' of M, suppose M' can be given a manifold structure in such a way that the natural inclusion map $i : M' \to M$ is an immersion. Then M' is called a *submanifold* of M (and necessarily $\dim M' \leq \dim M$).

There are a number of points regarding submanifolds which should be discussed. It will be recalled that a subset of a group (or a vector or topological space) can be a subgroup (or a vector or topological subspace) in at most one way since its structure is imposed upon it by the containing group (or vector or topological space). However, the definition of the submanifold M' above contained no such imposition on it from M other than the requirement that i be an immersion. One consequence of this weakening of the hold of the mother set over its subset is that often a subset M' of M can be given the structure of a submanifold of M in more than one way. Furthermore, different submanifold structures on the subset M' are not necessarily of the same dimension. Thus one should be careful about presupposing that a particular submanifold structure for a subset M' of M is in any sense special. Next, if the subset M' has a submanifold structure then it has a natural topology arising from that structure as described earlier. But M' also has a natural topology arising from the fact that it is a subset of M and thus inherits subspace topology from the natural topology on M which arises from the latter's manifold structure. *These two topologies are not necessarily equal.* Of course, the map $i : M' \to M$, being smooth, is continuous (when M' and M have their respective manifold topologies) and it easily follows that any subset U of M' which is open in subspace topology on M' is open in the manifold topology on M' but the converse may fail. It may at this point seem prudent to strengthen the definition of submanifold in order to force the equality of these topologies. However, submanifolds, as defined above, occur naturally in many areas of manifold theory and so the original definition will be retained (*cf.* the earlier remarks concerning the Hausdorff property for manifolds). A submanifold M' of M for which the above two topologies coincide is sufficiently special to be distinguished by a name and is called (following [19]) a *regular submanifold of M*. However it should be noted that this nomenclature is by no means universal. In fact, the terms *immersed* and *embedded* submanifolds are sometimes used, respectively, for what are here called submanifolds and regular submanifolds. A submanifold of M which is a closed subset of M is called a *closed submanifold* of M. Topological statements about submanifolds will always refer to their manifold topologies (and may differ from topological statements about them as *subspaces* of their containing manifold).

Some examples and properties of submanifolds can now be given for which [19] is a general reference.

i) Let U be an open subset of M and $i : U \to M$ the inclusion map. If x is a chart of M with domain $V \subseteq U$ then $x \circ i$ is a chart for U. These charts are easily shown to lead to a natural smooth manifold structure for U making it a submanifold of M of the same dimension as M and then U is called an *open submanifold* of M. An open submanifold of M is necessarily regular and any submanifold of M with dimension equal to that of M is necessarily an open submanifold of M. As an example, use of the determinant function shows that $GL(n, \mathbb{R})$ *is an open submanifold of* $M_n\mathbb{R}$.
ii) The set $M_n\mathbb{R}$ and its subsets of symmetric and skew-symmetric members were each given manifold structures in section 4.3. With these structures the latter two subsets are regular submanifolds of the manifold $M_n\mathbb{R}$.
iii) If V is an n-dimensional real vector space and W an m-dimensional subspace of V then W and V acquire natural smooth manifold structures of dimension n and m, respectively, (example (i) of section 4.3) and then W is an m-dimensional regular closed submanifold of V.
iv) If $m \in M$ and T_mM is given its natural manifold structure as a real vector space then it becomes an n-dimensional closed regular submanifold of TM.
v) Let M and $\bar{M}$ be smooth manifolds of dimensions n and $\bar{n}$, respectively, $(n > \bar{n})$ and let $f : M \to \bar{M}$ be smooth. Let $\bar{m} \in \bar{M}$ and suppose f has rank $\bar{n}$ at each point of the subset $f^{-1}\{\bar{m}\} \equiv \{m \in M : f(m) = \bar{m}\} \equiv M'$ of M. Then M' can be given the structure of a closed regular submanifold of M of dimension $n - \bar{n}$. A special case of this occurs when $\bar{M} = \mathbb{R}$ (and $\bar{n} = 1$) and then M' is an $(n-1)$-dimensional *hypersurface* of M. In this case an obvious choice of f reveals that $S^n \equiv \{\mathbf{x} \in \mathbb{R}^{n+1} : \sum_{i=1}^{n+1} x_i^2 = 1\}$ is an n-dimensional closed regular submanifold of $\mathbb{R}^{n+1}$.
vi) If M_1, M_2, M'_1, M'_2 are smooth manifolds with M'_1 a submanifold of M_1 and M'_2 a submanifold of M_2 then the product manifold $M'_1 \times M'_2$

is a submanifold of $M_1 \times M_2$. The result is still true if submanifold is replaced throughout by regular submanifold.

vii) Suppose M, M_1 and M_2 are smooth manifolds with $M_2 \subseteq M_1 \subseteq M$. If M_2 is a submanifold of M_1 and M_1 is a submanifold of M then M_2 is a submanifold of M. The same is true if submanifold is everywhere replaced by regular submanifold.

viii) If a subset M' of a smooth manifold M admits the structure of a regular submanifold of M of dimension n' then M' admits no other submanifold structure of dimension n' and no other regular submanifold structure of any dimension.

ix) If M' is a regular submanifold of M and M'' is a submanifold of M satisfying $M'' \subseteq M'$ then M'' is a submanifold of M'.

Suppose M' is an n'-dimensional submanifold of M and let $i : M' \to M$ be the inclusion map. If $m \in M'$ the map $i_{*m} : T_m M' \to T_m M$ is an injective map whose range is an n'-dimensional subspace of $T_m M$ called the *subspace of* $T_m M$ *tangent to* M' and its members are said to be *tangent to* M' at m. If X is a smooth vector field on M it is said to be *tangent to* M' if for each $m \in M'$, $X(m)$ is tangent to M'. This is the case if and only if there exists a smooth vector field $\tilde{X}$ on M' such that at each $m \in M'$, $i_{*m}\tilde{X}(m) = X(m)$. In this sense vector fields on M which are tangent to M' induce a corresponding vector field on M'. Further, if c' is an integral curve of $\tilde{X}$ then $c = i \circ c'$ is an integral curve of X on an appropriate domain.

Because of the fact that the natural topology of a submanifold need not be its topology as a subspace, some care is needed in handling submanifolds. The following theorem and remarks show the importance of regular submanifolds (see, *e.g.* [19]).

Theorem 4.7 *Let M_1 and M_2 be smooth manifolds and let $f: M_1 \to M_2$ be smooth. Suppose M'_1 and M'_2 are submanifolds of M_1 and M_2 respectively. Then*

i) the restriction of f to M'_1 is a smooth map $M'_1 \to M_2$,

ii) if $f(M_1) \subseteq M'_2$ and M'_2 is a regular submanifold of M_2 then f is a smooth map $M_1 \to M'_2$,

iii) if $f(M'_1) \subseteq M'_2$ and if the induced map $f' : M'_1 \to M'_2$ is smooth then if $\mathbf{v} \in TM_1$ is tangent to M'_1, $f_*\mathbf{v}$ is tangent to M'_2.

In fact (i) is trivial since if $i : M'_1 \to M_1$ is the inclusion map the restriction of f to M'_1 is just the map $f \circ i$. Part (ii) may fail if the word regular is dropped. (But if the word regular is dropped and if the map $f : M_1 \to M'_2$ is continuous then it is necessarily smooth as a well known result on immersions shows [19].) Part (iii) may fail if f' is not smooth (but f' is necessarily smooth if M_2' is a regular submanifold of M_2).

The following example exhibits some of the unexpected behaviour of non-regular submanifolds alluded to in the above results. Let $M = \mathbb{R}^3$ with its usual smooth manifold structure and let S be the subset $z = 0$ of M. The set S can be given the following two distinct structures as a submanifold of M. Firstly S may be given a natural structure as a smooth 2-dimensional manifold M' diffeomorphic to $\mathbb{R}^2$. In the standard global charts the inclusion map $i : M' \to M$ is $(x, y) \to (x, y, 0)$. Secondly, for each $y \in \mathbb{R}$ define the map $\psi_y : (x, y, 0) \to x$. Each ψ_y is a chart for S and these charts give rise to a 1-dimensional manifold structure for S (*cf.* example (vii), section 4.3). The resulting manifold is denoted by M'' and the inclusion map $M'' \to M$ in the domain of the chart ψ_y is $x \to (x, y, 0)$. It is then easy to check that M' and M'' are each submanifolds of M, the former being regular (and the latter, by consideration of its chart domains, clearly not). Now the inclusion map $i : M' \to M$ given above is a smooth map and its range is equal to the submanifold M''. But i is not a smooth (or even continuous) map $M' \to M''$ (*cf.* part (ii) of theorem 4.7). Also this map $i : M' \to M''$ can be regarded as the restriction to M' of the (smooth) identity map $M \to M$. Now let $m \in M'$ and let $\mathbf{v} \in T_m M$ be the tangent vector with components $(1, 1, 0)$ in the standard chart on M. Thus $\mathbf{v}$ is tangent to M' at m but $i_{*m}\mathbf{v}$ is not tangent to M'' at m. (Since $i : M \to M$ is the identity map, i_{*m} is the identity map $T_m M \to T_m M$ for each $m \in M$ and, in the above coordinates, members of $T_m M$ tangent to M'' have the form $(a, 0, 0)$ $(a \in \mathbb{R})$ (*cf.* part (iii) of theorem 4.7).

The above example shows, in particular, that a subset of a manifold can sometimes be given different structures as a submanifold of M. However, it follows from example (viii) of this section that if one has a topological subspace M' of M and one wishes to find a submanifold structure for M' whose corresponding topology coincides with the subspace topology then there is at most one such structure.

Some topological properties of submanifolds are contained in the following theorem.

Theorem 4.8 *Let M be a smooth manifold and M' a submanifold of M.*

i) If M' is compact its underlying set is a compact subset of M (and M' is necessarily regular if M is Hausdorff).

ii) If M' is connected its underlying subset is a connected subset of M.

iii) If M is Hausdorff then M' is Hausdorff.

iv) If M is second countable and M' is either a connected or a regular submanifold then M' is second countable.

v) Any component of M is a connected open submanifold of M.

vi) If M' is regular and $\dim M' < \dim M = n$ (or, equivalently, M' is regular and is not an open submanifold of M) then M' is not a dense subset of M and $M \setminus M'$ is a dense subset of M.

Proof. The proofs of the first part of (i) and of (ii), (iii), (v) and the second option in (iv) are clear. The proof of the second part of (i) uses the straightforward result that a continuous bijective map from a compact topological space to a Hausdorff topological space is a homeomorphism (section 3.8). The proof of the first option in (iv) and the first part of (vi) can be found in [19]. For the last part of (vi) let U be an open subset of M such that $\emptyset \neq U \subseteq M'$. Then U is an open submanifold of M and $\dim U = n$. Also, since $U \subseteq M'$ and M' is regular, U is a submanifold of M' and the contradiction $\dim M' = n$ follows (see result (ix) of this section)□

The concept of a submanifold enables one to ask the question whether any n-dimensional manifold M is essentially just a regular submanifold of $\mathbb{R}^N$ (with its usual manifold structure) for some N. Clearly the answer is no without (at least) the clauses that M be second countable and Hausdorff as theorem 4.8 shows. Some years ago Whitney [26] showed that *if M is second countable and Hausdorff then M is diffeomorphic to a regular submanifold of $\mathbb{R}^{2n}$.*

If M' is a submanifold of M the fact that the inclusion map $i : M' \to M$ is an immersion leads to the following result which links the charts in M' and M [19].

Theorem 4.9 *Let M' be an n'-dimensional submanifold of the n-dimensional manifold M with inclusion $i : M' \to M$ and let $m \in M'$. Then there exist charts x' in M' and x in M whose domains U' and U each contain m such that the corresponding coordinate representative $x \circ i \circ x'^{-1}$ of i is just the map $(z_1, ..., z_{n'}) \to (z_1, ..., z_{n'}, 0, ..., 0)$ from some open subset of $\mathbb{R}^{n'}$ into $\mathbb{R}^n$. In addition if M' is regular, one can choose U and U' such that $U' = U \cap M'$.*

The proof of the first part is a standard property of immersions. For the final sentence one can use the continuity of i to see that $i^{-1}U$ is open and hence that one can restrict U' so that $U' \subseteq U$. Then since M' is regular there is an open subset W in M such that $U' = M' \cap W$. Thus by restricting U to $U \cap W$ one has $M' \cap (U \cap W) = U' \cap U = U'$ to complete the result.

There is another type of submanifold which is intermediate between a general submanifold and a regular one and which will be of use later. It can be motivated by pointing out that the potential failure of theorem 4.7(ii) when M_2' is not regular is an irritating feature of such submanifolds. Let M' be a submanifold of a smooth manifold M. Then M' is called a *leaf* (of M) [27] if it is connected and if whenever A is a locally connected topological space and $f : A \to M$ a continuous map whose range lies inside M' then the associated map $A \to M'$ is continuous. The conclusion of theorem 4.7(ii) now holds if "regular submanifold" is replaced by "leaf" in that if M_1 is a smooth manifold and $f : M_1 \to M$ is a smooth map whose range lies in the leaf M' of M then the associated map $M_1 \to M'$ is continuous (theorem 4.3(i)) and hence smooth (see the remark following theorem 4.7). Just as for regular submanifolds, if a subset of M admits the structure of a leaf of M, then this structure is unique. It is remarked that a (connected) submanifold need not be a leaf and a leaf need not be a regular submanifold. In fact the so called "irrational wrap" on the torus $S^1 \times S^1$ (see, *e.g.* [22]) is a leaf [27] but not a regular submanifold of it whereas the (connected) "figure eight" submanifold of $\mathbb{R}^2$ (see, *e.g.* [19]) is easily seen not to be a leaf.

If M_1 and M_2 are smooth manifolds of dimension m and n, respectively, let M be the $(m+n)$-dimensional smooth manifold given by $M = M_1 \times M_2$ and p_1 and p_2 the usual (smooth) projections. Then, for each $a \in M_1$, $b \in M_2$ the subsets $b\tilde{M}_1 = \{(m_1, b) : m_1 \in M_1\}$ and $a\tilde{M}_2 = \{(a, m_2) : m_2 \in M_2\}$ can be given the structure of regular submanifolds of $M_1 \times M_2$ with each $b\tilde{M}_1$ (respectively, $a\tilde{M}_2$) diffeomorphic to M_1 (respectively, M_2). The map $f : T(M_1 \times M_2) \to TM_1 \times TM_2$ given by $f = (p_{1*}, p_{2*})$ can be shown to be a diffeomorphism [19] and then if X is a smooth vector field and O_1 the zero vector field on M_1 (and similarly for Y and O_2 on M_2), $f^{-1} \circ (X, O_2)$ and $f^{-1} \circ (O_1, Y)$ are, in an obvious notation, smooth vector fields on $M_1 \times M_2$ tangent, respectively, to each $b\tilde{M}_1$ and to each $a\tilde{M}_2$.

This section is concluded by remarking that, with the above notation, tensors of type $(0, r)$ on M may be *pulled back* to M' using the map i^*.

4.12 Quotient Manifolds

Let M and M' be manifolds and $f : M \to M'$ be a smooth map. Then f is called a *submersion* if at each $m \in M$, *rank* f = *dim* M' (and necessarily $\dim M' \leq \dim M$ and f_{*m} is surjective for each $m \in M$). The projection maps $p_i : \mathbb{R}^n \to \mathbb{R}$ are simple examples of submersions when $\mathbb{R}^n$ and $\mathbb{R}$ have their standard manifold structures.

Now let $\sim$ be an equivalence relation on M and let $M/\sim$ denote the associated quotient set. If $M/\sim$ can be given a manifold structure such that the natural map $\mu : M \to M/\sim$, which sends $m \in M$ to the equivalence class in $M/\sim$ containing m, is a *submersion* then $M/\sim$ is called a *quotient manifold of* M. For example, the equivalence relation $\sim$ on $\mathbb{R}^n$ given by $(x^1, ..., x^n) \sim (x'^1, ..., x'^n)$ if $x^i = x'^i$ for a particular i, $(1 \leq i \leq n)$ gives rise to the quotient set $M/\sim = \mathbb{R}$. When $\mathbb{R}$ has its standard manifold structure the associated natural map $M \to M/\sim$ is just p_i, which is then a submersion, and so $M/\sim$ is a quotient manifold of $\mathbb{R}^n$.

Although conceptually more difficult than a submanifold structure, a quotient manifold structure has some simpler mathematical properties. For example [19], the topology arising on a quotient manifold $M/\sim$ from its manifold structure is the same as the quotient topology on $M/\sim$ arising from M. Also, if a quotient set $M/\sim$ admits the structure of a quotient manifold then it does so in only one way. However, a quotient manifold $M/\sim$ of a Hausdorff manifold M may not be Hausdorff (*cf.* a remark in section 4.5).

4.13 Distributions

A vector field on a manifold M selects a member of T_mM at each $m \in M$. Sometimes there naturally arises on M not a well defined vector field but rather a well defined *direction field*, that is, a well defined 1-dimensional subspace of T_mM for each $m \in M$. More generally, one may have a well defined r-dimensional subspace of T_mM at each $m \in M$ picked out by some construction on M. This idea leads to the concept of a *distribution* on M (sometimes referred to as a *distribution in the sense of Frobenius* to distinguish it from other types of distributions). A distribution is essentially the selection of an r-dimensional subspace of T_mM for each $m \in M$ (and with r independent of $m \in M$) in a *smooth* manner. To be more precise, an *r-dimensional distribution on* M is a map D which associates with each

$m \in M$ an r-dimensional subspace of T_mM $(0 < r \leq n)$ such that for each $m \in M$ there is an open neighbourhood U of m and r smooth vector fields $X_1, ..., X_r$ defined on U such that $X_1(p), ..., X_r(p)$ span $D(p)$ for each $p \in U$.

It is clear in this definition how the smoothness of D is provided by the *fitting* of the vector fields $X_1, ..., X_r$ to D.

Because of the constancy of r and the smoothness of $X_1, ..., X_r$ one might ask if there exist, locally, r-dimensional submanifolds of M in which these vector fields lie (*i.e.* to which they are tangent). The answer in general is no and it is, therefore, of importance to have a theorem saying precisely when such submanifolds exist in terms of some property of the distribution. Let D be an r-dimensional distribution on M. Then a submanifold M' of M is an *integral manifold of D* if the natural inclusion map $i : M' \to M$ satisfies the condition that the range of the map $i_{m*} : T_mM' \to T_mM$ is exactly $D(m)$ for each $m \in M'$, that is, $D(m)$ is the subspace of T_mM tangent to M'. Thus if M' exists it is necessarily r-dimensional. A distribution D is then called *integrable* if every $m \in M$ is contained in an integral manifold of D. Also, a smooth vector field X on M (or some open subset of M) is said to *belong* to a distribution D if $X(m) \in D(m)$ for each m in the domain of X. It follows that if X belongs to D and if M' is an integral manifold of D then there exists a smooth vector field X' on M' (or some open subset of M') such that at each $m \in M'$ $i_{*m}X'(m) = X(m)$ (*i.e.* X is tangent to M'). The well known theorem due to Frobenius (see, *e.g.* [19]) which decides when a distribution D is integrable can now be stated and is done so in terms of a condition on vector fields like $X_1, ..., X_r$ in the first paragraph of this section which belong to D and span it at each point of the appropriate domain.

Theorem 4.10 (Frobenius)

Let M be a smooth manifold and D an r-dimensional distribution on M. Then D is integrable if and only if $[X, Y]$ belongs to D whenever X and Y belong to D, where X and Y are smooth vector fields on some open subset of M. (This latter condition on D is expressed by saying D is involutive)

An alternative, useful version of this theorem arises in the following situation. Let $X_1, \ldots, X_r$ be r smooth vector fields defined on M with the property that at any point $m \in M$, $X_1(m), \ldots, X_r(m)$ are *independent* members of T_mM (so necessarily $r \leq dim\ M$). This set of vector fields leads to an r-dimensional distribution D on M (called the distribution *spanned* by (or *arising from*) $X_1, \ldots, X_r$) which associates with $m \in M$ the subspace of T_mM spanned by $X_1(m),\ldots,X_r(m)$ (and then $X_1,\ldots,X_r$ belong to D). One asks if these vector fields have the *surface forming condition* namely that each $m \in M$ lies in a (necessarily r-dimensional) submanifold M' of M such that the subspace of T_mM tangent to M' is precisely the span of $\{X_1(m),\ldots,X_r(m)\}$, that is, of $D(m)$ (*i.e.* D is integrable).

Theorem 4.11 *Let M be a smooth manifold of dimension n and let $X_1,\ldots,X_r$ be smooth global vector fields on M ($r \leq n$) such that $X_1(m),\ldots,X_r(m)$ are independent members of T_mM for each $m \in M$. Then the distribution D spanned by these vector fields is integrable if and only if there exists smooth functions a^k_{ij} on M such that these vector fields satisfy*

$$[X_i, X_j] = \sum_{k=1}^{r} a^k_{ij} X_k \qquad (1 \leq i,j \leq r) \tag{4.7}$$

Proof. Using (4.7), the properties of the Lie bracket and theorem 4.10 the theorem is proved provided that whenever X is a smooth vector field on some open subset W of M belonging to D, so that $X = \sum\limits_{i=1}^{r} \alpha_i X_i$, for functions $\alpha_i : W \to \mathbb{R}$ then these functions are smooth. To establish this let x be a chart of M whose domain $U \subseteq W$ contains $m \in M$. Then, in components, one has $X^a = \sum_{i=1}^r \alpha_i X^a_i$ and the matrix X^a_i has rank r at each point of U. After (possibly) renumbering the coordinate functions one may assume that the $r \times r$ matrix X^a_i $(1 \leq i, a \leq r)$ is non-singular at m and hence on some open coordinate domain $V \subseteq U$ with coordinates x^a (see section 3.11). Thus in V

$$\begin{pmatrix} X^1 \\ \vdots \\ X^r \end{pmatrix} = \begin{pmatrix} X^1_1 \ldots X^1_r \\ \vdots \\ X^r_1 \ldots X^r_r \end{pmatrix} \begin{pmatrix} \alpha_1 \\ \vdots \\ \alpha_r \end{pmatrix}$$

Cramer's rule then gives the unique solution for each of the α_i and which are thus smooth since X and $X_1,\ldots,X_r$ are. This completes the proof. □

Some other useful results can be collected together in the next theorem (see, *e.g.* [19]).

Theorem 4.12 *Let M be a smooth manifold of dimension n and let $m \in M$.*

i) *If X is a smooth vector field on M and $X(m) \neq 0$ there exists a chart x whose domain U contains m and such that, in U, X has components $X^1 = 1, X^2 = \ldots = X^n = 0$ (i.e. $X = \frac{\partial}{\partial x^1}$ on U).*

ii) *If $X_1, \ldots, X_r$ are smooth vector fields on M such that $X_1(m), \ldots, X_r(m)$ are independent members of T_mM for each $m \in M$ and also $[X_i, X_j] = 0$ $(1 \leq i, j \leq r)$ then there exists a chart x with domain U containing m in which $X_1 = \frac{\partial}{\partial x^1}, \ldots, X_r = \frac{\partial}{\partial x^r}$.*

iii) *Every 1-dimensional distribution on M is integrable.*

If D is an integrable distribution on M there exists through each $m \in M$ a *unique maximal connected integral manifold* of D. This is a connected integral manifold of D through m and is maximal in that it contains all other connected integral manifolds of D through m. It is also remarked here that a 1-dimensional distribution on M is sometimes called a *line element field* or a *field of directions* on M. There are obvious geometrical comparisons to be made between integral manifolds of an (integrable) distribution and integral curves of a vector field. In fact if X is a nowhere zero vector field on a manifold M and if $c : I \to M$ is an integral curve of X then $c(I)$ can be given the structure of a (1-dimensional) integral manifold of the 1-dimensional distribution on M spanned by X [19].

The concept of a distribution D defined above required the constancy of *dim* $D(m)$ over M. In practice, one often encounters a structure on M similar to a distribution but with this constancy of dimension dropped. To avoid possible confusion such a structure will be here called a generalised distribution and will be described less generally (but in a manner sufficient for the present purposes) in terms of a certain family of smooth vector fields on M. Let S be a *real vector space* of *global smooth* vector fields on a smooth *connected paracompact* manifold M under the usual addition and scalar multiplication of vector fields on M. For each $m \in M$ let S_m be the subspace of T_mM consisting of the values of all members of S at m, that is, $S_m = \{X(m) : X \in S\}$. The map $m \to S_m$ is called the *generalised distribution* on M arising from S. The dimension of S_m may vary with m (and then S is sometimes called *singular*). If $\dim S_m$ is constant over m (the *non-singular* case), the map $m \to S_m$ is a *distribution* on M. Let V_p denote the subset of M at each point m of which, $\dim S_m = p$ $(0 \leq p \leq n)$, that is, $V_p = \{m \in M : \dim S_m = p\}$. Then V_n is an open subset of M (possibly empty) and one has a *disjoint* decomposition of M in the form

$$M = V_n \cup \operatorname{int} V_{n-1} \cup \ldots \cup \operatorname{int} V_0 \cup V \tag{4.8}$$

where the disjointness of the decomposition defines the (necessarily closed) subset V. Now if U is a non-empty open subset of M contained in V it follows by disjointness that $U \cap V_n = \emptyset$. If for some $m \in U$ $dim\ S_m = n-1$ then (section 3.11) there exists an open neighbourhood U' of m such that $m \in U' \subseteq U$ and $U' \subseteq V_{n-1}$. But this gives the contradiction (by disjointness) that $m \in \operatorname{int} V_{n-1}$. Proceeding with this argument (*i.e.* using the result, from section 3.11, that $\bigcup_{k=p}^{n} V_k$ is open for each p, $1 \leq p \leq n$) gives the contradiction that no point of U is in $V_n, V_{n-1}, \ldots,$ or V_0 and hence U is empty. It follows that V is closed and has empty interior and is hence *nowhere dense* in M (section 3.3). Thus $V_n \cup \operatorname{int} V_{n-1} \cup \ldots \cup \operatorname{int} V_0$ is an *open dense* subset of M.

With the notation of the previous paragraph one can still ask about the existence of integral and maximal integral manifolds for S where now a submanifold M' of M is called an *integral manifold* of S if for each $m' \in M'$ the subspace of $T_{m'}M$ tangent to M' is the subspace $S_{m'}$ (so that the members of S are *tangent to* M') and a *maximal integral manifold* of S if it is a connected integral manifold of S which is not properly contained in any other connected integral manifold of S. After another definition a powerful result based on the work of Hermann, Stefan and Sussman [27]-[30] which generalises the theorem of Frobenius can be stated. The vector space S will be called *locally finitely generated* if for each $m \in M$ there exists an open neighbourhood U of m and a finite subset $X_1, ..., X_r \in S$ such that for any $X \in S$, $X = \sum_{i=1}^{r} \alpha_i X_i$ *on* U for smooth functions $\alpha_i : U \to \mathbb{R}$. Now suppose S is a *Lie algebra* with respect to its assumed vector space structure and the Lie bracket operation (in fact, a Lie subalgebra of the Lie algebra of *all* smooth vector fields on M under the bracket operation).

Theorem 4.13 *Let S be a real Lie algebra of global smooth vector fields on a smooth connected paracompact manifold M. Then if S is locally finitely generated (including the special case when S is finite-dimensional) there exists a unique maximal integral manifold of S through each $m \in M$ for which* $\dim S_m \geq 1$ *and this integral manifold is a leaf of M.*

The maximal integral manifolds guaranteed by this theorem need not be regular submanifolds of M but since they are leaves, the conclusion of theorem 4.7(ii) when M_2' is such a submanifold is guaranteed. It is also remarked here that in the case of a (Fröbenius) distribution the conclusion of theorem 4.7(ii) for M_2' a maximal integral manifold has been known for some time since, because M is paracompact and connected, it is second countable (theorem 4.4) and since M_2' is connected it is also second countable (theorem 4.8(iv)). The result now follows from a result in [19].

4.14 Curves and Coverings

In chapter 3 the idea of a *curve* in a topological space as a certain *continuous* map was discussed whereas in this chapter the curves of importance have been smooth curves in manifolds. Similarly, the concepts of the fundamental group and of coverings were discussed topologically, that is, using continuous maps, and one may similarly enquire about their smooth counterparts when applied to manifolds. This section provides a brief discussion of some of these points.

Let M be a smooth Hausdorff manifold and let $p, q \in M$. From chapter 3 a curve c in M from p to q is a continuous map $c : [0,1] \to M$ with $c(0) = p$, $c(1) = q$. *A piecewise C^k curve* ($k \geq 1$) in M from p to q is a map c' defined on the closed interval $[a, b]$ such that $c'(a) = p$, $c'(b) = q$ and such that the interval $[a, b]$ can be divided into finitely many closed subintervals $[a, s_1], [s_1, s_2], \ldots, [s_{m-1}, b]$ with $a < s_1 < \ldots < s_{m-1} < b$ on each of which c' agrees with a C^k map into M from an open interval of $\mathbb{R}$ containing that subinterval [19]. It is now straightforward to show that for $p, q \in M$, the relation $p \sim q \Leftrightarrow$ there exists a piecewise C^k curve from p to q, is an equivalence relation on M each equivalence class of which is open and closed in M (and hence equal to M if M is connected, equivalently, path connected from theorem 4.3(iv) [19]. Hence if one can find a (continuous) curve between two points of M one can find a piecewise C^k curve between them (and clearly conversely from the material in section 3.10). Hence the definition of path connectedness (or local path connectedness) could be stated in terms of (continuous) curves or piecewise C^k curves ($k \geq 1$). In fact, if one can find a curve from p to q in M one can, by smoothing procedures, find a smooth curve from p to q and so the definition of path connectedness could be phrased in terms of smooth curves.

Since a manifold is a topological space one can discuss its fundamental group in terms of (continuous) curves. It can be shown that the results obtained are the same as if one had used piecewise smooth curves or even smooth curves together with (in the obvious sense) smooth homotopies. In fact the only new results required are, firstly, that if c is a closed curve at p then there is a *smooth* closed curve c' at p which is *continuously* homotopic (*i.e.* in the sense of section 3.10) to c and, secondly, if c_1 and c_2 are two closed *smooth* curves at p which are *continuously* homotopic to each other then they are *smoothly* homotopic to each other.

Now let M, $\tilde{M}$ be smooth connected Hausdorff manifolds. Regarding them firstly as topological spaces it makes sense to ask if $\tilde{M}$ is a covering *space* of M, that is, if a *covering* map $p : \tilde{M} \to M$ exists as described in section 3.10. Now define a *smooth covering* (as opposed to a *covering* which is a continuous map between topological spaces) $p' : \tilde{M} \to M$ as a *smooth* surjective map such that for each $m \in M$ there exists a connected open subset U containing m such that each component of $p'^{-1}U$ is diffeomorphic to U under (the appropriate restriction of) p' and when these open subsets have their usual open submanifold structure. It follows that the dimensions of M and $\tilde{M}$ and the rank of p' at any $m \in \tilde{M}$ are equal. Next let $\tilde{M}$ be a *covering* of M with covering map $p : \tilde{M} \to M$ (so that only the manifold topologies of M and $\tilde{M}$ are relevant). Then it does not necessarily follow that p is a smooth covering since p may not actually be smooth (just consider one of those sets upon which different manifold structures can be put but which have identical manifold topologies (see *e.g.* [19]) and consider the identity map on the set). However one does have the result that, for topological spaces M and $\tilde{M}$ and a covering $p : \tilde{M} \to M$ and given a manifold structure for M whose manifold topology equals the original topology on M, there exists a unique manifold structure for $\tilde{M}$ whose manifold topology equals the original topology on $\tilde{M}$ and for which p is a smooth covering [18]. The manifold $\tilde{M}$ is called a *covering manifold of* M. If $\tilde{M}$ is simply connected it is called the *universal covering manifold* of M. Every smooth connected Hausdorff manifold has a unique universal covering manifold in the sense that if $\tilde{M}_1$ and $\tilde{M}_2$ are universal covering manifolds of M with smooth covers $p_1 : \tilde{M}_1 \to M$ and $p_2 : \tilde{M}_2 \to M$ there exists a smooth diffeomorphism $h : \tilde{M}_1 \to \tilde{M}_2$ such that $p_2 \circ h = p_1$. To see this note that section 3.10 guarantees a homeomorphism $h' : \tilde{M}_1 \to \tilde{M}_2$ such that, as continuous maps, $p_2 \circ h' = p_1$. Now p_1 and p_2 are immersions and a standard result [19] now shows that h and h^{-1} are smooth and the result follows. Further, if one defines an equivalence relation $\sim$ on $\tilde{M}$ by $m_1 \sim m_2 \Leftrightarrow p(m_1) = p(m_2)$ then any smooth connected Hausdorff manifold is diffeomorphic to a quotient manifold of a simply connected manifold (*i.e.* to $\tilde{M}/\sim$). The proof is similar to that for topological spaces in 3.10. One has a natural bijection $f : M \to \tilde{M}/\sim$ from which a natural manifold structure for $\tilde{M}/\sim$ results making f a diffeomorphism. The natural map $i : \tilde{M} \to \tilde{M}/\sim$ then satisfies $i = f \circ p$ and is smooth with rank equal to the rank of p which equals $\dim \tilde{M}$ at each point of $\tilde{M}$. Thus i is a submersion (section 4.12) and the result follows.

4.15 Metrics on Manifolds

From the point of view of this book one of the most important constructions on a manifold is that of a metric. Let M be a smooth manifold and let $m \in M$. A *metric at* m is an inner product on $T_m M$, that is, a symmetric non-degenerate bilinear form on $T_m M$. In other words, a metric at p is a symmetric type $(0,2)$ tensor at p whose components constitute an $n \times n$ non-singular symmetric matrix at p. A *smooth metric on* M is a global smooth tensor field of type $(0,2)$ on M which is a metric at m when evaluated at any $m \in M$. If in the above definitions one replaces inner product by positive definite inner product the resulting metric on M (or metric at $m \in M$) is called *positive definite*. If, however, one replaces inner product by inner product of Lorentz signature one arrives at a (smooth) *Lorentz metric on* M (or at $m \in M$). In fact the *signature* of a metric on a *connected* manifold M is the signature of any of the inner products representing it (it being straightforward (see section 4.16 after (4.39)) then to prove that this signature is the same at each point of M). All concepts of orthogonality for a metric are as defined for inner products in section 2.5.

Let g_m be a metric at $m \in M$ and let x be a chart whose domain contains m. Then at m one has (recalling the Einstein summation convention)

$$g_m = g_{ab}(dx^a)_m \otimes (dx^b)_m \quad \left[\text{i.e. } g_{ab} = g_m\left(\left(\frac{\partial}{\partial x^a}\right)_m, \left(\frac{\partial}{\partial x^b}\right)_m\right)\right] \tag{4.9}$$

where the components g_{ab} $(= g_{ba})$ of g_m in the chart x form a symmetric matrix with $\det(g_{ab}) \neq 0$. Under a change of coordinates the general law (4.4) shows that, at m, $g'_{ab} = A_{ac} g_{cd} A^{\mathrm{T}}_{db}$ where $A_{ab} = \left(\frac{\partial x^b}{\partial y^a}\right)_m$. It now follows from section 2.5 (and, in particular, from Sylvester's law) that one may change coordinates about m so that, *at* m, the matrix g'_{ab} assumes the Sylvester form (2.1) appropriate to its signature. Let g^{ab} be the matrix inverse to g_{ab} and define the $(2,0)$ tensor ${g_m}^{-1} = g^{ab}(\frac{\partial}{\partial x^a})_m \otimes (\frac{\partial}{\partial x^b})_m$ at m (this definition being easily shown to depend only on g_m and not on the chart x in which the components were computed). A metric g_m at $m \in M$ distinguishes a vector space isomorphism from $T_m M$ to its dual space $T^*_m M$ where, hitherto, no such natural isomorphism existed. This isomorphism, denoted by f_{g_m} (to highlight its dependence on g_m) is defined by $\mathbf{v} \rightarrow f_{g_m}\mathbf{v} \in T^*_m M$ where $f_{g_m}\mathbf{v}(\mathbf{u}) = g_m(\mathbf{v}, \mathbf{u})$ $(\mathbf{u}, \mathbf{v} \in T_m M)$. If in some chart x

whose domain contains m, $\mathbf{v} = v^a(\frac{\partial}{\partial x^a})_m$, then $f_{g_m}\mathbf{v} = (g_{ab}v^b)(dx^a)_m$. One traditionally defines $v_a \equiv g_{ab}v^b$ so that $f_{g_m}\mathbf{v} = v_a(dx^a)_m$ (and provided, of course, that no ambiguity is involved *i.e.* that the metric g_m is *understood*). Thus one says that the metric g_m *lowers indices* at $m \in M$. The process of *raising indices* can be accomplished by using the map $f_{g_m^{-1}} : T_m^*M \to T_mM$ defined for $\boldsymbol{\omega} \in T_m^*M$ by $f_{g_m^{-1}}\boldsymbol{\omega}(\mathbf{t}) = g_m^{-1}(\boldsymbol{\omega}, \mathbf{t})$ ($\mathbf{t} \in T_m^*M$) and in the chart x the effect, in components, is $t_a \to t^a \equiv g^{ab}t_b$. The definition of g_m^{-1} shows that $g_{ac}g^{cb} = \delta_a^b$ and it is then easily checked that raising and lowering indices are *inverse* operations. If M admits a smooth metric g then the above shows that any smooth vector field on some open subset U of M has a naturally *associated covector field* on U associated with it (and vice versa). If X has components X^a in some chart of M contained in U then the associated covector field has components $X_a = g_{ab}X^b$ in U.

One may extend index raising and lowering to general tensors. Omitting the details (which are similar to those described in the previous paragraph—see *e.g.* [24]) one can lower (or raise), using a given metric g_m, a particular index of a type (r, s) tensor T at m turning it into a type $(r-1, s+1)$ (or a type $(r+1, s-1)$ tensor at m. The same symbol T would still be used for the resulting tensor. Thus, for example, one would have in components $T_{abcd} \to T^a{}_{bcd} \equiv g^{ae}T_{ebcd}$, $T^{abcd} \to T_{ab}{}^{cd} \equiv g_{ae}g_{bf}T^{efcd}$. Although the order of the *up* (contravariant) indices and that of the *down* (covariant) indices is important there is no such significance between the ordering of *up* relative to *down* indices. In practice one does allow significance to the relative position of an up and a down index but only to indicate relative position when the up index is lowered or the down index raised.

One can always construct a metric of any signature (r, s) at a point $m \in M$ subject to $r + s = \dim M$ (and, indeed, one may construct a smooth metric of this signature on the domain U of any chart x of M by choosing an appropriate non-singular matrix g_{ab} and defining the metric to be $g_{ab}dx^a \otimes dx^b$ on U). However, one cannot necessarily define a global smooth metric on a given manifold M. The following theorems summarise the situation in the positive definite and Lorentz cases.

Theorem 4.14 *Let M be a smooth connected Hausdorff manifold. Then the following conditions are equivalent.*

i) M admits a global smooth positive definite metric.
ii) M is paracompact.
iii) M is second countable.
iv) M admits a partition of unity.

v) The manifold topology of M is metrisable.

Proof. A brief sketch of how the proof proceeds is easily given. One notes first that the equivalence of (ii), (iii) and (iv) is just the statement of theorem 4.4. Next, the existence of a partition of unity on M and the existence of local smooth positive definite metrics in the chart domains of M together allow a *smooth globalising* procedure to be carried out which results in a smooth global positive definite metric on M and so (i) follows from either of the equivalent conditions (ii), (iii) or (iv) [19]. If, however, (i) is assumed to hold then the metric g which arises can be used to construct the *magnitude* (or *size*) $g(\mathbf{v},\mathbf{v})^{\frac{1}{2}}$ of any tangent vector $\mathbf{v}$ at any point of m. Now since M is connected there is a piecewise smooth curve $c : [a,b] \to M$ joining any two points p, $q \in M$. Thus $[a,b]$ can be divided up according to $a = t_0 < t_1 < \ldots < t_m = b$ and so that c agrees with a smooth curve c_a on $[t_{a-1}, t_a]$ $(a = 1, \ldots, m)$. The length of c is defined to be $\sum_{a=1}^{m} \int_{t_{a-1}}^{t_a} g(v(t), v(t))^{\frac{1}{2}} dt$ where $v(t)$ is the tangent vector to the relevant curve c_a. Such lengths are non-negative real numbers and allow a map $d : M \times M \to \mathbb{R}$ to be defined where for p, $q \in M$, $d(p,q)$ is the infimum of the lengths of all piecewise smooth curves from p to q. *The map d turns out to be a metric for M* in the sense of section 3.2 and the consequent metric topology on M coincides with the manifold topology for M. Thus (v) follows. Finally if (v) holds then (ii) (and hence (iii) and (iv)) follow from Stone's theorem (see section 3.8) which says that any metric space is paracompact. □

Theorem 4.15 *Let M be a smooth connected paracompact n-dimensional manifold ($n \geq 2$). Then M admits a (smooth) Lorentz metric if and only if it admits a 1-dimensional distribution.*

Proof. To prove this one notes that the conditions of the theorem together with the previous theorem ensure the existence of a global smooth positive definite metric γ on M. If M also admits a 1-dimensional distribution then in some open coordinate neighbourhood U of any $m \in M$ there exists a smooth vector field t which spans the distribution. Clearly one may assume t is everywhere of unit size with respect to γ (*i.e.* $\gamma(t,t) = 1$ everywhere on U) since γ is smooth and so t is determined up to a sign at each point of U. It follows that the smooth tensor $t \otimes t$ of type $(2,0)$ is uniquely determined on U. Then by using γ to lower indices one obtains a smooth type $(0,2)$ tensor on U with components $t_a t_b$ ($t_a = \gamma_{ab} t^b$). By covering M with such neighbourhoods and their corresponding type $(0,2)$

tensors the latter clearly give rise to a global type $(0,2)$ tensor h on M (the ambiguity in the sign of the vector spanning the distribution at any point being cancelled in the definition of h). One then constructs the global $(0,2)$ smooth tensor $g \equiv \gamma - 2h$ on M which is, in fact, a smooth Lorentz metric on M. To see this let $\mathbf{t}, \mathbf{v}_1, ..., \mathbf{v}_{n-1}$ be an orthonormal basis (with respect to γ) at $m \in M$ with t spanning the distribution at m. In this basis g has components $g_{ab} = \text{diag}(-1, 1, \ldots, 1)$ and so is in the Sylvester form for a Lorentz metric at m.

Conversely let γ be a global smooth positive definite metric on M (which exists from the paracompactness of M by the last theorem) and let g be a Lorentz metric for M with signature $(-1, 1, \ldots, 1)$. Let x be a chart whose domain contains m and using components in this chart at m consider the eigenvector-eigenvalue problem $g_{ab}k^b = \lambda\gamma_{ab}k^b$ for $\mathbf{k} \in \mathbb{R}^n$, $\lambda \in \mathbb{C}$. By adjusting the chart so that $\gamma_{ab} = \delta_{ab}$ at m and using the symmetry of g one sees (section 2.6) that all such eigenvalues are real and that g_{ab} is diagonalisable over $\mathbb{R}$. Also, $\det g_{ab} < 0$ and so at least one such eigenvalue, say λ_0, must be negative. Denote by u^a a corresponding eigenvector so that at m, $g_{ab}u^b = \lambda_0\gamma_{ab}u^b$. A contraction with u then shows that $g_{ab}u^a u^b < 0$. If there exists another negative eigenvalue $\lambda_1 \neq \lambda_0$ and corresponding eigenvector v then again one finds $g_{ab}v^a v^b < 0$ and, since $\lambda_0 \neq \lambda_1$, one easily finds $g_{ab}u^a v^b = 0$. This latter equation and the previous two inequalities are easily checked to be inconsistent with the Lorentz signature of g and so there is a unique negative eigenvalue $\lambda_0(m)$ at m. The diagonalisability of g_{ab} then shows that the $\lambda_0(m)$-eigenspace is a 1-dimensional subspace of $T_m M$. But $\lambda_0(m)$ is a simple root of the polynomial $\det(g_{ab} - \lambda\gamma_{ab}) = 0$ at each $m \in M$ and hence depends smoothly on the smooth coefficients of the polynomial (see, *e.g.* [31]). Hence the numbers $\lambda_0(m)$ determine a global smooth function $M \to \mathbb{R}$ (also denoted by λ_0). Now consider the global smooth tensor field A of type $(1,1)$ defined in the domain of any chart by the components $A^a{}_b = g^{ac}(g_{cb} - \lambda_0\gamma_{cb})$. At each $m \in M$ the matrix $A^a{}_b$ has rank $n-1$ and the $(n-1)$-dimensional subspaces of $T_m M$ given at each $m \in M$ in components by $\{A^a{}_b v^b : v \in T_m M\}$ give rise to a smooth $(n-1)$-dimensional distribution on M [31]. The orthogonal complement with respect to g of this distribution is then itself a (1-dimensional) smooth distribution [31] and agrees with the λ_0-eigenspace at each $m \in M$. Thus M admits a 1-dimensional distribution. □

It may be wondered whether the paracompactness clause in the proceeding theorem is really necessary. It turns out that in the case of most importance for this book, *i.e.* $\dim M = 4$, it is in the sense that if a smooth connected Hausdorff 4-dimensional manifold admits a smooth Lorentz metric it is necessarily paracompact [32].

Now suppose that M is a smooth manifold admitting a metric g and let M' be a submanifold of M with inclusion $i : M' \to M$. Then the pullback i^*g is a smooth symmetric tensor of type $(0, 2)$ on M' but it is not necessarily a metric for M' since it may fail to be non-degenerate at some (or all) points of M'. It is, however, a standard result that if g is positive definite then i^*g *is* a positive definite metric for M' called the *metric induced on* M' *by* g [20]. For Lorentz metrics one has the following theorem.

Theorem 4.16 *Let* M *be a smooth manifold with Lorentz metric* g *of signature* $(-1, 1, ..., 1)$ *and let* M' *be a submanifold of* M *with inclusion* i. *If* $\dim M' = \dim M$ *then* M' *is an open submanifold of* M *and* i^*g *is a Lorentz metric for* M' *(the restriction of* g *to* M'*). Otherwise*

i) if for each $m \in M'$ *and* $\mathbf{v} \in T_mM$, $\mathbf{v} \neq 0$ *with* $\mathbf{v}$ *tangent to* M', $g_m(\mathbf{v}, \mathbf{v}) > 0$ *then* i^*g *is a positive definite metric for* M',
ii) if for each $m \in M'$ *there exists* $\mathbf{v} \in T_mM$ *with* $\mathbf{v}$ *tangent to* M' *and* $g_m(\mathbf{v}, \mathbf{v}) < 0$ *then* i^*g *is a Lorentz metric for* M'.

The proof of (i) is clear since if $\mathbf{u} \in T_mM'$, $\mathbf{u} \neq 0$, then $(i^*g)_m(\mathbf{u}, \mathbf{u}) > 0$. For (ii) the proof will be clear after the algebraic discussion of Lorentz metrics in the next chapter. Clearly, for a general submanifold M' of M, the nature (degenerate, positive-definite or Lorentz) of i^*g will vary over M'.

A smooth manifold M together with a metric g on M is often written as the pair (M, g) or simply as M if g is clear. If g is positive definite, (M, g) or M is sometimes called *Riemannian*. However this notation is not universal, some authors preferring the latter term for any manifold with a metric (of arbitrary signature). Hence it will not be used in this text. If g has Lorentz signature, (M, g) or M is sometimes called *Lorentz* or *Lorentzian*. If $M = \mathbb{R}^n$ and g is a smooth metric on M which in the usual global chart has components g_{ab}, where g_{ab} equals the appropriate Sylvester matrix for g, at each $m \in M$ then (M, g) is called *pseudo-Euclidean* (and *Euclidean* if g is positive definite). If each point of M has a coordinate neighbourhood in which the metric takes constant values, as in the last sentence, it is called *locally pseudo-Euclidean* or *locally Euclidean*. If (M, g) and (M', g') are smooth manifolds with smooth metrics g and g' and if $f : M \to M'$

is a diffeomorphism such that $f^*g' = g$ then f is called an *isometry* and (M, g) and (M', g') are then *isometric*. Again if (M, g) and (M', g') are manifolds with metrics let $\tilde{M} = M \times M'$ be the smooth product manifold and $i : \tilde{M} \to M$, $j : \tilde{M} \to M'$ the natural projections. There is a natural smooth metric $g \otimes g'$ on $\tilde{M}$ defined by $g \otimes g' \equiv i^*g + j^*g'$ and called the *product* of g and g' (see *e.g.* [24]) and one refers to the *metric product* of M and M'.

A notational remark can now be made regarding metrics on manifolds. An old classical notation often employed in relativity theory is that which describes a metric g in the domain U of a chart x by writing $ds^2 = g_{ab}dx^a dx^b$. Although formally similar to (4.9) it had quite a different meaning, classically, being thought of as the *square of the distance* between two points whose coordinate separation was dx^a (at least in the positive definite case)! The modern approach to metrics given here describes them as giving *size* to tangent vectors. The above classical notation is still usefully employed but only in the sense of being an alternative (possibly more picturesque) way of actually defining g in that chart.

4.16 Linear Connections and Curvature

One of the most important concepts in differential geometry and theoretical physics is that of a connection. More details will be provided in a later chapter when holonomy theory is considered. In this section the basic idea of a *linear connection* (sometimes shortened to *connection*) is considered.

A *(linear) connection* ∇ on M is a map which associates with two smooth vector fields X and Y defined on open subsets U and V of M, respectively, a third smooth vector field denoted by $\nabla_X Y$ defined on $U \cap V$ such that for smooth real valued functions f, g and a smooth vector field Z defined on appropriate open subsets of M and a, $b \in \mathbb{R}$

$$\nabla_Z(aX + bY) = a\nabla_Z X + b\nabla_Z Y \tag{4.10}$$

$$\nabla_{fX+gY} Z = f\nabla_X Z + g\nabla_Y Z \tag{4.11}$$

$$\nabla_X(fY) = f\nabla_X Y + (X(f))Y \tag{4.12}$$

where all appropriate domains are assumed to be non-empty open subsets of M. A connection as defined above will be called *smooth*. Rather than repeat statements about domains of definition of various constructions it will be assumed, unless stated otherwise, that they are open subsets of M or the obvious intersections of open subsets of M. It is convenient, for f

and X as above, to define $\nabla_X f \equiv X(f)$ so that (4.12) assumes a Leibniz form.

The quantity $\nabla_X Y$ is referred to as the *covariant derivative of the vector field* Y *along the vector field* X (and, similarly, $\nabla_X f$ is the covariant derivative of f along X). As the form of the above axioms suggest, they are an attempt to place a structure on M which allows a sensible derivative of a vector field Y along the integral curves of another vector field X to be defined. By implication one thus has the concept of a *constant* vector field along the integral curves of a vector field Y and, it turns out, along any smooth curve in M. Thus if M is connected (and hence path connected) one has a notion of *parallel displacement* or *transport* (movement "without change") of tangent vectors between any two points of M but along a particular curve connecting these points (and which may depend on the curve chosen).

A manifold M may not admit a connection. However, any manifold admits one *locally* by virtue of it being *locally like* $\mathbb{R}^n$. To see this let x be a chart of M with domain U and let X and Y be any smooth vector fields on U with $X = X^a \frac{\partial}{\partial x^a}$ and $Y = Y^a \frac{\partial}{\partial x^a}$ for smooth functions X^a and Y^a. Then define ∇ on the open submanifold U of M by

$$\nabla_X Y = X(Y^a)\frac{\partial}{\partial x^a} = (Y^a_{,b}X^b)\frac{\partial}{\partial x^a} \tag{4.13}$$

It is easily checked that ∇ is a connection on U (which is dependent on the chart x). The same construction reveals a standard global linear connection on $\mathbb{R}^n$ from its usual global chart.

Turning to the general existence of connections on a manifold M the following theorem (see *e.g.* [19]) is fundamental.

Theorem 4.17 *Any paracompact smooth manifold admits a smooth (linear) connection.*

The proof is similar in essence to that of theorem 4.14 for positive definite metrics. One uses a partition of unity on M (which exists from theorem 4.4) to spread out the above *local* connections smoothly on M. Henceforth M will be assumed to be *paracompact*.

Let M be a (smooth paracompact) manifold which admits a connection ∇. There are associated with ∇ two constructions on M of particular importance. Let X and Y be smooth vector fields and consider the smooth vector field $\tilde{T}(X,Y)$ defined by

$$\tilde{T}(X,Y) = \nabla_X Y - \nabla_Y X - [X,Y] \qquad (= -\tilde{T}(Y,X)) \tag{4.14}$$

$\tilde{T}$ is called the *torsion* of ∇. *All the connections considered in this text will have zero torsion in the sense that $\tilde{T}(X,Y)$ will always be the zero vector field* and then the connection ∇ is called *symmetric*. This will be discussed further later. Of more importance here is the *curvature structure* $\tilde{R}$. Let X, Y, Z be smooth vector fields and define a smooth vector field by

$$\tilde{R}(X,Y)Z = \nabla_X(\nabla_Y Z) - \nabla_Y(\nabla_X Z) - \nabla_{[X,Y]}Z \quad (= -\tilde{R}(Y,X)Z) \tag{4.15}$$

it is easily seen that $\tilde{R}$ is linear in each of its arguments in the sense that, for example, for vector fields X_1 and X_2 and $a, b \in \mathbb{R}$

$$\tilde{R}(aX_1 + bX_2, Y)Z = a\tilde{R}(X_1,Y)Z + b\tilde{R}(X_2,Y)Z \tag{4.16}$$

and from (4.15) one can also show [33] that if f, g and h are smooth functions

$$\tilde{R}(fX, gY)hZ = fgh\tilde{R}(X,Y)Z \tag{4.17}$$

The curvature structure $\tilde{R}$ leads to a global type $(1,3)$ smooth tensor field also denoted $\tilde{R}$ on M called the *curvature* (or *Riemann*) *tensor* (associated with the connection ∇). To see this let $m \in M$ and let x be a chart whose domain U contains m. Let $\mathbf{X'}_1$, $\mathbf{X'}_2$, $\mathbf{X'}_3 \in T_mM$ and $\boldsymbol{\omega}' \in T_m^*M$ be given by $\mathbf{X'}_k = X_k^a(\frac{\partial}{\partial x^a})_m$ $(k = 1,2,3)$, $\boldsymbol{\omega}' = \omega_a(dx^a)_m$ $(X_k^a, \omega_a \in \mathbb{R})$ and extend them smoothly to U by $X_k = X_k^a \frac{\partial}{\partial x^a}$ $(k = 1,2,3)$, $\omega = \omega_a dx^a$ for smooth functions X_k^a and ω_a on U. Then define

$$\tilde{R}(m)(\boldsymbol{\omega}', \mathbf{X'}_1, \mathbf{X'}_2, \mathbf{X'}_3) \equiv [(\tilde{R}(\mathbf{X}_2, \mathbf{X}_3)\mathbf{X}_1)(\boldsymbol{\omega})]_m \tag{4.18}$$

It is straightforward to check that this definition is independent of the choice of X_k and ω provided they yield X'_k and ω' on evaluation at m. It also follows from the definition that $\tilde{R}$ is smooth. In components in U one has

$$\tilde{R} = R^a{}_{bcd}\frac{\partial}{\partial x^a} \otimes dx^b \otimes dx^c \otimes dx^d \tag{4.19}$$

where the smooth functions $R^a{}_{bcd}$ are the *curvature tensor components*. By virtue of (4.15) they satisfy

$$R^a{}_{bcd} = -R^a{}_{bdc} \tag{4.20}$$

Since ∇ is symmetric, $T \equiv 0$ and (4.14) and (4.15) give

$$\tilde{R}(X,Y)Z + \tilde{R}(Y,Z)X + \tilde{R}(Z,X)Y = 0 \tag{4.21}$$

which in terms of components is the (algebraic) Bianchi identity

$$R^a{}_{bcd} + R^a{}_{cdb} + R^a{}_{dbc} = 0 \tag{4.22}$$

Again in the chart x with domain U define smooth functions Γ^a_{bc} on U by the relations

$$\nabla_{\frac{\partial}{\partial x^b}} \frac{\partial}{\partial x^c} = \Gamma^a_{bc} \frac{\partial}{\partial x^a} \tag{4.23}$$

The functions Γ^a_{bc} are called the *coefficients* of the connection ∇ in the chart x and (4.14) shows that ∇ is symmetric if and only if $\Gamma^a_{bc} = \Gamma^a_{cb}$ in each coordinate domain. They are *not* the components of any tensor at any point of U since if y is another chart of M whose domain intersects U then by writing out (4.23) in the y coordinates, using the symbol Γ'^a_{bc} for the connection coefficients in these coordinates, one easily finds on the intersection of these coordinate domains that

$$\Gamma'^a_{bc} = \frac{\partial x^e}{\partial x'^b} \frac{\partial x^f}{\partial x'^c} \frac{\partial x'^a}{\partial x^d} \Gamma^d_{ef} + \frac{\partial^2 x^d}{\partial x'^b \partial x'^c} \frac{\partial x'^a}{\partial x^d} \tag{4.24}$$

It should be remarked at this point that if ∇ and $\bar{\nabla}$ are two connections on M whose coefficients are Γ^a_{bc} and $\bar{\Gamma}^a_{bc}$ in some coordinate domain U of M then the functions $\Gamma^a_{bc} - \bar{\Gamma}^a_{bc}$ do define a smooth tensor field on U of type $(1,2)$. Also, if one specifies smooth functions Γ^a_{bc} in each domain of an atlas of M which satisfy (4.24) on the intersection of any two such domains then one has, in fact, defined a connection on M through (4.23) whose coefficients are the functions Γ^a_{bc}. It is symmetric if and only if $\Gamma^a_{bc} = \Gamma^a_{cb}$ in each coordinate domain (and it is noted how this property is preserved by (4.24)). A convenient expression for the curvature tensor components in any chart domain can be obtained from (4.18), (4.19) and (4.23)

$$R^a{}_{bcd} = \Gamma^a_{db,c} - \Gamma^a_{cb,d} + \Gamma^e_{db}\Gamma^a_{ce} - \Gamma^e_{cb}\Gamma^a_{de} \tag{4.25}$$

Now let X be a smooth vector field. From the connection ∇ and X one can define a type $(1,1)$ tensor field ∇X on M as that unique $(1,1)$ tensor field such that the (only possible) contraction of ∇X with any smooth vector field Y is the vector field $\nabla_Y X$. Now by using (4.11), (4.12) and (4.23) one easily finds that in any coordinate domain U of M and for vector fields X and Y defined on U

$$\nabla_Y X = (X^a{}_{;b} Y^b) \frac{\partial}{\partial x^a} \tag{4.26}$$

where

$$X^a{}_{;b} = X^a{}_{,b} + \Gamma^a_{bc} X^c \tag{4.27}$$

Hence (since, from (4.24), $X^a{}_{;}b$ are the components of a type $(1,1)$ tensor on U) one has on U

$$\nabla X = (X^a{}_{;b}) \frac{\partial}{\partial x^a} \otimes dx^b \tag{4.28}$$

The tensor ∇X (with components $X^a_{;b}$) is called the *covariant derivative of* X (with respect to the connection ∇). Extending this idea to a function f on (some open subset of) M, and recalling that $\nabla_X f = X(f)$, the above concept yields the definition $\nabla f = df$ for the covariant derivative of f and which is consistent with the Leibniz rule for $\nabla(fX)$.

The notion of taking the covariant derivative along a vector field can be extended from vector fields to arbitrary tensor fields by noting that there is for any given smooth vector field X with domain U a unique operator ∇_X which maps a smooth tensor field of type (r, s) on U to a tensor field of the same type on U, which coincides with $\nabla_X Y$ defined above when it operates on a smooth vector field Y on U and which satisfies the following conditions for smooth tensors S and T on U of the same type, for a smooth function f on U and for a, $b \in \mathbb{R}$ (see *e.g.* [20],[34])

(1) $\nabla_X f = X(f)$
(2) $\nabla_X(aS + bT) = a\nabla_X S + b\nabla_X T$
(3) $\nabla_X(S \otimes T) = \nabla_X S \otimes T + S \otimes \nabla_X T$
(4) ∇_X commutes with the contraction operator
(5) $\nabla_{fX} S = f\nabla_X S$
(6) $\nabla_{X+Y} Z = \nabla_X Z + \nabla_Y Z$.

This extension of ∇_X to arbitrary tensor fields leads to a similar extension of the covariant derivative operator ∇ to arbitrary tensor fields. In fact if T is a smooth tensor field of type (r, s) defined on some open subset U of M then the smooth tensor field ∇T of type $(r, s+1)$ is defined as that tensor which when contracted over the extra covariant index with any such X gives $\nabla_X T$. It can then be shown that, in any coordinate domain where T has components $T^{a_1,\ldots,a_r}_{b_1,\ldots,b_s}$

$$\nabla T \equiv T^{a_1\ldots a_r}_{b_1\ldots b_s;b} \frac{\partial}{\partial x^{a_1}} \otimes \ldots \otimes \frac{\partial}{\partial x^{a_r}} \otimes dx^{b_1} \otimes \ldots \otimes dx^{b_s} \otimes dx^b \tag{4.29}$$

where the components of ∇T are given by

$$\begin{aligned} T^{a_1\dots a_r}_{b_1\dots b_s;b} = \frac{\partial}{\partial x^b} T^{a_1\dots a_r}_{b_1\dots b_s} + \Gamma^{a_1}_{bc} T^{ca_2\dots a_r}_{b_1\dots b_s} + \dots + \Gamma^{a_r}_{bc} T^{a_1\dots a_{r-1}c}_{b_1\dots b_s} \\ - \Gamma^{c}_{bb_1} T^{a_1\dots a_r}_{cb_2\dots b_s} - \dots - \Gamma^{c}_{bb_s} T^{a_1\dots a_r}_{b_1\dots b_{s-1}c} \end{aligned} \tag{4.30}$$

and that, if $X = X^a \frac{\partial}{\partial x^a}$, the components of $\nabla_X T$ are $T^{a_1,\dots,a_r}_{b_1,\dots,b_s;b} X^b$.

Now let x be a chart of M with domain U and let T be a smooth type (r,s) tensor field and X a smooth vector field defined on U. For $m \in U$ let c be an integral curve of X passing through m and with t the parameter of c and let $c^a = x^a \circ c$ so that $X^a = \frac{dc^a}{dt}$ on the range of c in U. The condition $\nabla_X T = 0$ is then

$$T^{a_1\dots a_r}_{b_1\dots b_s;b} = \frac{d}{dt} T^{a_1\dots a_r}_{b_1\dots b_s} + \Gamma^{a_1}_{bc} T^{ca_2\dots a_r}_{b_1\dots b_s} \frac{dc^b}{dt} + \dots - \Gamma^{c}_{bb_s} T^{a_1\dots a_r}_{b_1\dots b_{s-1}c} \frac{dc^b}{dt} = 0 \tag{4.31}$$

and if this is the case one says that T is *covariantly constant along* c or that T is *parallely transported along* c. Thus the connection ∇ allows a concept of "moving" tensors along curves "without change" because if one specifies the value of T at m the first order differential equation (4.31) will yield a unique solution for T at least at points on the image of c in U in some neighbourhood of m. Again one refers to this solution as the *parallel transport* (or *displacement*) or *transfer* of $T(m)$ along c. It can be shown [20] that if $p, q \in M$ lie on a smooth curve c then given any tensor at p it may be parallely transported uniquely along c to q. Further, parallel transport gives rise in this way to a (curve dependent) isomorphism of the vector spaces $T_p{}^s_r \to T_q{}^s_r$.

If a smooth tensor T satisfies $\nabla T = 0$ on M (or on some open subset U of M) T is called *covariantly constant* on M (or U). If T is nowhere zero and satisfies $\nabla T = T \otimes w$ on M (or U) it is called *recurrent* on M (or U) and the smooth 1-form field w is called the *recurrence 1-form* of T.

Now let c be a smooth curve in M and let $m \in M$ be a point on c. Let $\mathbf{X} \in T_m M$, $\mathbf{X} \neq 0$, agree with the tangent vector to c at m. If the parallel transport of $\mathbf{X}$ along c is a multiple of the tangent vector to c at each point of c then c is called a *geodesic* (of the connection ∇ through m with initial value $\mathbf{X}$). A standard argument then shows that for some *reparametrisation* c' of c the parallel transport of $\mathbf{X}$ along c' *agrees* with the tangent vector to c' at each point of c'. In this case the parameter of c' is called an *affine parameter* and c' an *affinely parametrised geodesic* (of ∇). If c_1 and c_2 are reparametrisations of c with parameters t_1 and t_2, respectively, and if each is an affinely parametrised geodesic, the affine parameters t_1 and t_2 are related by $t_2 = at_1 + b$ ($a, b \in \mathbb{R}$, $a \neq 0$). Suppose now that c is an affinely parametrised geodesic with affine parameter t. Then since ∇ is smooth it follows that, provided c is C^1, it is necessarily smooth [20] and the components $c^a = x^a \circ c$ in a chart x satisfy the *geodesic equation*

$$\frac{d^2c^a}{dt^2} + \Gamma^a_{bd}\frac{dc^b}{dt}\frac{dc^d}{dt} = 0 \tag{4.32}$$

Conversely if c satisfies (4.32) it is an affinely parametrised geodesic.

An *affinely parametrised* geodesic $c : (a,b) \to M$ ($a, b \in \mathbb{R}$) is called *maximal* if c cannot be extended as an affinely parametrised geodesic to some open interval of $\mathbb{R}$ properly containing (a,b). An affinely parametrised geodesic is called *complete* if its domain is the whole of $\mathbb{R}$. If every affinely parametrised geodesic in M is complete, ∇ (or M if ∇ is understood) is called *geodesically complete* (or just *complete* if the context is clear).

Theorem 4.18 *Let M be a smooth paracompact manifold with smooth symmetric connection ∇ and let $m \in M$ and $\mathbf{v} \in T_mM$.*

i) There is a unique affinely parametrised maximal geodesic c such that $c(0) = m$, $\dot{c}(0) = \mathbf{v}$

ii) If c' is an affinely parametrised geodesic satisfying $c'(0) = m$, $\dot{c}'(0) = \mathbf{v}$ then c' is defined on some open subinterval of the domain of c and agrees with it there.

iii) If c' is an affinely parametrised geodesic satisfying $c'(0) = m$, $\dot{c}'(0) = \lambda\mathbf{v}$, $(0 \neq \lambda \in \mathbb{R})$ then c' is a reparametrisation of c on the intersection of their domains.

Part (i) of this theorem says that given $m \in M$ and $\mathbf{v} \in T_mM$ there is always a geodesic starting from m with initial tangent vector $\mathbf{v}$ and part (ii) says that, roughly speaking, it is *as unique as it can be* [18]. Part (iii), again loosely speaking, says that such a geodesic is determined by its initial starting point and *direction*.

There is an important map arising on such a manifold M with a connection. First note that if an affinely parametrised geodesic $c : (-a, a) \to M$ satisfies $c(0) = m$, $\dot{c}(0) = \mathbf{v}$ with $m \in M$, $\mathbf{v} \in T_mM$ and $a \in \mathbb{R}$ then the affinely parametrised geodesic $c' = c \circ f$, where $f : (-a\lambda^{-1}, a\lambda^{-1}) \to (-a, a)$ is defined by $f(x) = \lambda x$ $(0 \neq \lambda \in \mathbb{R})$, satisfies $c'(0) = m$, $\dot{c}' = \lambda\mathbf{v}$. Then let $W \subseteq T_mM$ be that subset of T_mM such that if $\mathbf{v} \in W$ and c is an affinely parametrised geodesic with $c(0) = m$ and $\dot{c}(0) = \mathbf{v}$ then $c(1)$ is defined. Thus one has a map $W \to M$ called the *exponential map* at m (arising from the connection) and denoted by $\exp_m$ and defined by $\exp_m \mathbf{v} = c(1)$. It follows that, if $t\mathbf{v} \in W$, $\exp_m(t\mathbf{v}) = c(t)$ and that the maximal geodesic starting from m with initial tangent $\mathbf{v}$ is the map $t \to \exp_m(t\mathbf{v})$ for appropriate t. The following theorem now holds [18],[19],[20].

Theorem 4.19 *Let M be a smooth paracompact manifold admitting a smooth symmetric connection and let $m \in M$. Then there exists an open subset V of T_mM containing 0 and an open subset U of M containing m such that $\exp_m : V \to U$ is a (smooth) diffeomorphism between the obvious open submanifold structures on V and U.*

Henceforth, the exponential map will always be understood to have an *open* domain and range as described in theorem 4.19.

Now let $\mathbf{X}_1, ..., \mathbf{X}_n$ be a basis for T_mM and, with the notation of the previous theorem, define a smooth map $U \to \mathbb{R}^n$ with components x^a by $x^a(\exp_m(\mathbf{v})) = v^a$ given that $\mathbf{v} = v^a\mathbf{X}_a$. This defines a chart x for M with domain U as is easily checked. The corresponding coordinate system is called a *normal coordinate system for M at m*. The coordinate expression for an affinely parametrised geodesic c in U satisfying $c(0) = m$ and $\dot{c}(0) = \mathbf{v}$ is then $c^a(t) = tv^a$ where $c^a = x^a \circ c$ and any curve of this form is an affinely parametrised geodesic (lying in U). The point m is the *origin* of the above normal coordinate system and its coordinates are $(0, ..., 0)$. It follows that for these normal coordinates $\left(\frac{\partial}{\partial x^a}\right)_m = \mathbf{X}_a$. It also follows from (4.32) and the above discussion that, in these coordinates, the coefficients Γ^a_{bc} vanish at m. Hence any $m \in M$ admits a coordinate domain in which $\Gamma^a_{bc}(m) = 0$.

There are a number of differential identities satisfied by the curvature tensor on M arising from a (symmetric) connection ∇. Let x be a chart of M with domain U. Covariant differentiation of tensors on U is not necessarily commutative and a series of identities, the *Ricci identities*, point this out. They are given here in component form for a vector field X, a covector field ω and a type $(0,2)$ tensor field T on U. They can be derived, somewhat tediously but easily, from (4.25) and (4.30). A double covariant derivative will be denoted by two indices after the semi-colon rather than two semi-colons. Thus $X^a{}_{;b;c}$ will be written $X^a{}_{;bc}$.

$$X^a{}_{;bc} - X^a{}_{;cb} = X^d R^a{}_{dcb} \tag{4.33}$$

$$\omega_{a;bc} - \omega_{a;cb} = \omega_d R^d{}_{abc} \tag{4.34}$$

$$T_{ab;cd} - T_{ab;dc} = T_{eb} R^e{}_{acd} + T_{ae} R^e{}_{bcd} \tag{4.35}$$

The next identity concerns only the curvature and connection and is referred to as the (differential) *Bianchi identity*. It is

$$R^a{}_{bcd;e} + R^a{}_{bde;c} + R^a{}_{bec;d} = 0 \tag{4.36}$$

There is a very important tensor which derives immediately from the curvature tensor by contraction. It is called the *Ricci tensor* (sometimes denoted by *Ricc*) and is of type $(0,2)$. In components it is defined by

$$R_{ab} \equiv R^c{}_{acb} \tag{4.37}$$

If the connection ∇ on M is such that the curvature tensor is identically zero on M then ∇ is called a *flat connection* (or sometimes, if ∇ is understood, M is called *flat*). Another term, which is non-standard but will be useful later on is *non-flat* (not to be confused with *not flat* which is the opposite of *flat*). A manifold M with connection ∇ will be called *non-flat* if the curvature tensor does not vanish over any non-empty open subset of M.

Now let M be a smooth paracompact manifold admitting a smooth connection ∇ and a smooth metric g. The connection and metric are, in general, independent of each other but it is often convenient to impose some compatibility requirement on them jointly. Perhaps the most important one and certainly the most important for this text is the requirement that when a tangent vector X undergoes parallel transport with respect to ∇ along a smooth curve in M its inner product with itself with respect to g is constant along that curve. Thus if x is a chart on M with domain U and if c is a smooth curve in U then $g(X,X)$ is constant along c. Then using the condition for the parallel transport of X along c and the requirement that the condition applies to all vectors along all curves one easily sees that $\nabla g = 0$ or, in components, $g_{ab;c} = 0$. Thus

$$\frac{\partial g_{ab}}{\partial x^c} - \Gamma^d_{ca} g_{db} - \Gamma^d_{cb} g_{ad} = 0 \tag{4.38}$$

A simple permuting of the indices a, b, c then leads to a convenient expression for the coefficients of the connection or, as they are sometimes called in this case, the *Christoffel Symbols* Γ^a_{bc}, namely

$$\Gamma^a_{bc} = \frac{1}{2} g^{ad} \left(\frac{\partial g_{db}}{\partial x^c} + \frac{\partial g_{dc}}{\partial x^b} - \frac{\partial g_{bc}}{\partial x^d} \right) \tag{4.39}$$

For a given metric g on M the uniquely determined symmetric connection whose coefficients are given by (4.39) is called the *Levi-Civita connection* associated with g. A connection with this property is called a (symmetric)

metric connection and, in particular, a *metric connection compatible with the metric g* (and if ∇ is flat, a metric compatible with ∇ is called a *flat metric*). The compatibility condition $\nabla g = 0$ ensures that the inner product with respect to g of any two vectors parallely transported along any curve c is constant along c (and hence the constancy of the signature of a metric on a connected manifold stated in section 4.15). Also one can easily show that the object with components δ^a_b ($= 1$ if $a = b$ and zero otherwise) at $m \in M$ and in every coordinate system whose domain contains m is a $(1,1)$ tensor at m. This then gives rise to an obvious tensor field δ on M satisfying $\nabla\delta = 0$, *i.e.* $\delta^a_{b;c} = 0$. It then follows from the relation $g_{ab}g^{bc} = \delta^c_a$ that ${g^{ab}}_{;c} = 0$. Hence the rules of raising and lowering indices *commute* with covariant differentiation in the sense that, for example,

$$T_{ab} \equiv {T_a}^c g_{cb} \Rightarrow T_{ab;d} = {T_a}^c{}_{;d} g_{cb} \tag{4.40}$$

The curvature tensor of a connection naturally arose as a $(1,3)$ tensor with components ${R^a}_{bcd}$. One can now define a global smooth type $(0,4)$ tensor on M also denoted by $\tilde{R}$ and also called the *curvature tensor* (or the *Riemann tensor*) and defined in components by

$$R_{abcd} = g_{ae}{R^e}_{bcd} \tag{4.41}$$

Then one has the extra algebraic symmetry relations

$$R_{abcd} = -R_{abdc} = -R_{bacd}, \qquad R_{abcd} = R_{cdab} \tag{4.42}$$

These algebraic relations between the curvature components can then easily be used to show that the Ricci tensor defined in (4.37) is a *symmetric* type $(0,2)$ tensor. Further, one can now define an important smooth real valued function on M denoted by R and called the *Ricci scalar*. In components it is given by

$$R = R_{ab}g^{ab} \tag{4.43}$$

If $n = \dim M \geq 3$ and if in every coordinate system on M the curvature and metric tensor components satisfy

$$R_{abcd} = \alpha(g_{ac}g_{bd} - g_{ad}g_{bc}) \tag{4.44}$$

then the function α is given by $\alpha = \frac{R}{n(n-1)}$, $R_{ab} = \frac{R}{n}g_{ab}$ and the Bianchi identity (4.36) forces α to be constant (Schur's theorem - see, *e.g.* [35]). Such a manifold is said to have *constant curvature* α. If (4.44) holds at some $m \in M$ then M is said to be of *constant curvature at m* (a poor notation it has to be admitted!). If $n = 2$, (4.44) necessarily holds but α need not be constant.

With the notation of the previous paragraph if the Ricci and metric tensors are proportional so that, in coordinates, $R_{ab} = \beta g_{ab}$ for some function β, then $\beta = \frac{R}{n}$ and M is called an *Einstein space* (and so if M has constant curvature it is an Einstein space). M is a *proper* Einstein space if $\beta \neq 0$. If $n \geq 3$, suitable contractions of (4.36) show that R and hence β are constant. If M is flat it can be shown that each $m \in M$ admits a coordinate neighbourhood U in which the metric components g_{ab} are constant (and then a further coordinate transformation will ensure that the components g_{ab} take the appropriate Sylvester form everywhere in U).

Let g and g' be global smooth metrics on M. Suppose there exists a smooth function $\phi : M \to \mathbb{R}$ such that ϕ is nowhere zero on M and such that $g' = \phi g$. Then g and g' are said to be *conformally related* (or just *conformal*). If $\phi(m) > 0$ at each $m \in M$ then g and g' have the same signature at each $m \in M$ whereas if $\phi(m) < 0$ at each $m \in M$ the signature of g' is, in an obvious sense, the reverse of that of g at each $m \in M$. There is an important (smooth) tensor that derives from the metric tensor on M but which is unchanged if it is recalculated using a conformally related metric. This is the *Weyl tensor* (or the *conformal curvature tensor*) which is denoted by C and defined in components (and only in the case that $n = dim\ M \geq 3$) by

$$C^a{}_{bcd} = R^a{}_{bcd} - \frac{1}{n-2}(R^a{}_c g_{bd} + R_{bd}\delta^a_c - R_{bc}\delta^a_d - R^a{}_d g_{bc}) - \frac{R}{(n-1)(n-2)}(g_{bc}\delta^a_d - g_{bd}\delta^a_c) \tag{4.45}$$

The Weyl tensor (which is identically zero when $n = 3$) has some pleasant algebraic properties which can be calculated together here in terms of the components of the related type $(0,4)$ (Weyl) tensor $C_{abcd} = g_{ae}C^e{}_{bcd}$.

$$C_{abcd} = -C_{bacd} = -C_{abdc} = C_{cdab} \tag{4.46}$$

$$C_{abcd} + C_{acdb} + C_{adbc} = 0 \tag{4.47}$$

$$C^c{}_{acb} \equiv g^{cd}C_{cadb} = 0 \tag{4.48}$$

However it is the Weyl tensor of type $(1,3)$ given in (4.45) and not the type $(0,4)$ version which has the *conformally invariant* property of being the same for conformally related metrics. The converse is false in the sense that two metrics giving rise to the same Weyl tensor as in (4.45) are not necessarily conformally related. A counter-example will be given in chapter 9.

A smooth manifold M of dimension $n \geq 4$ and admitting a smooth metric g is called *conformally flat* if its associated Weyl tensor vanishes identically on M. If this is the case then for each $m \in M$ there is an open neighbourhood U of M, a flat metric h on U and a smooth function $\sigma : U \to \mathbb{R}$ such that, on U, $g = \sigma h$. Conversely if such neighbourhoods exist about each $m \in M$ the conformally invariant property of the Weyl tensor, together with the easily established fact that the Weyl tensor on a flat manifold vanishes identically, show that M is conformally flat. The manifold M is called *non-conformally flat* if the Weyl tensor does not vanish over some non-empty open subset of M (and the term *not conformally flat* means the opposite of conformally flat). If $dim\ M = 3$ the Weyl tensor (4.45) vanishes identically on M whilst if $dim\ M = 2$ neighbourhoods like U above exist about each $m \in M$. For such manifolds of dimension 3 there is another tensor which plays the role of the Weyl tensor in determining the existence of neighbourhoods such as U above. These results are discussed in [35].

Some other topics can be briefly dealt with here. Let X be a smooth covector field on the smooth Hausdorff manifold M. Then X is called *exact* (or a *global gradient* or *normal*) if there is a smooth function $f : M \to \mathbb{R}$ such that $X = df$ and X is called *closed* if, in any cordinate system on M, $X_{a,b} = X_{b,a}$. Clearly if X is exact it is closed and, *if M is simply connected, X is closed if and only if it is exact* [25]. For any manifold M, if X is closed then each $m \in M$ admits a connected coordinate neighbourhood U and a smooth function $\phi : U \to \mathbb{R}$ such that $X = d\phi$ (*i.e.* $X_a = \phi_{,a}$) on U so that X is locally a gradient. The above terms are sometimes applied to a vector field X on a manifold M admitting a metric g if they apply to its associated covector field (section 4.15). There is a similar but more general condition on the covector field X (or its associated vector field where appropriate). Suppose each $m \in M$ admits an open connected neighbourhood U on which $X = \psi d\phi$ where $\psi, \phi : U \to \mathbb{R}$ are smooth. Then X is called *hypersurface orthogonal*. The name is unfortunate since no metric need be involved. It arises from the fact that if M admits a metric then on any open neighbourhood U on which $d\phi$ never vanishes the vector field associated with X is everywhere *orthogonal to the submanifolds* (hypersurfaces) of constant ϕ in U (which exist from section 4.11, example (v)), that is, everywhere orthogonal to the vectors tangent to these submanifolds. Thus this definition for a covector field X is a statement about X but for a vector field (when M has a metric g) it is a statement about X and g. [For example, let X be a covector field on M such that X is nowhere zero on some coordinate neighbourhood U of M and is *not* hypersurface orthogonal on U. Let g be a metric on U and consider the vector field on U with components $X^a \equiv g^{ab}X_b$. Suppose U is chosen (as it can be) such that $X^a = \delta_1^a$ on U. Then for this *vector field* on U the associated covector field $X_a = g_{ab}X^b$ is *not* hypersurface orthogonal (by definition) but, if h is the Euclidean metric on U, the covector field $h_{ab}X^b = \delta_a^1$ *is* hypersurface orthogonal.] If a covector field X is hypersurface orthogonal and $X(m) \neq 0$ then, whilst not necessarily a gradient on some open neighbourhood of m, it may be scaled so that it is. Of course a closed covector field is hypersurface orthogonal.

It is convenient here to mention a point of notation which involves the use of round and square brackets in coordinate expressions to represent complete symmetrisation and skew-symmetrisation, respectively. Thus, for example, if T_{ab} and T_{abc} are tensor components at some point of a manifold then

$$T_{[ab]} \equiv \frac{1}{2}(T_{ab} - T_{ba}) \qquad T_{(ab)} \equiv \frac{1}{2}(T_{ab} + T_{ba})$$
$$T_{[abc]} \equiv \frac{1}{6}(T_{abc} + T_{bca} + T_{cab} - T_{bac} - T_{cba} - T_{acb}).$$

As an example, if X is a *nowhere zero* covector field on M then the condition for it to be hypersurface orthogonal is that in any coordinate system, $X_{[a,b}X_{c]} = 0$. This follows from the "form" version of the Fröbenius (distribution) theorem (see, *e.g.* [22]).

4.17 Grassmann and Stiefel Manifolds

Consider the vector space $\mathbb{R}^n$ and let $V \subseteq \mathbb{R}^n$ be an m-dimensional subspace of $\mathbb{R}^n$. Suppose V is such that it may be spanned by m vectors in $\mathbb{R}^n$ which, in the standard basis for $\mathbb{R}^n$, have components

$$(1, 0, \ldots, 0, x_{m+1}, \ldots, x_n), \ldots, (0, \ldots, 0, 1, y_{m+1}, \ldots, y_n).$$

Then these vectors are *uniquely* determined by V and such m-dimensional subspaces may then be identified with the member $(x_{m+1}, \ldots, x_n, \ldots, y_{m+1}, \ldots, y_n)$ of $\mathbb{R}^{m(n-m)}$. This gives a one-to-one map from a *subset* of the set $G(m, \mathbb{R}^n)$ of all m-dimensional subspaces of $\mathbb{R}^n$

onto $\mathbb{R}^{m(n-m)}$. By extending this idea but choosing the $0's$ and $1's$ in slots other than the first m slots one obtains a set of $m(n-m)$-dimensional charts for $G(m, \mathbb{R}^n)$ which can be shown [19] to yield a smooth atlas for $G(m, \mathbb{R}^n)$. The resulting manifold $G(m, \mathbb{R}^n)$ is then called a *Grassmann manifold* (in this case of all m-dimensional subspaces of $\mathbb{R}^n$) and has dimension $m(n-m)$.

There is an alternative approach to the Grassmann manifolds which involves the introduction of a Stiefel manifold. An *m-frame* in $\mathbb{R}^n$ is an ordered set $v = (\mathbf{v}_1, \ldots, \mathbf{v}_m)$ of m linearly independent vectors in $\mathbb{R}^n$. Again with components in the usual basis for $\mathbb{R}^n$ one can arrange the members of the m-frame v to get a bijective map from the set $V(m, \mathbb{R}^n)$ of m-frames in $\mathbb{R}^n$ *onto* the set of $n \times m$ real matrices of rank m. An application of the ideas in section 3.11 shows that the latter set is an open submanifold of the mn-dimensional smooth manifold $M_{n\times m}\mathbb{R}$. Hence the set $V(m, \mathbb{R}^n)$ becomes an mn-dimensional smooth manifold (with a global chart) called the *Stiefel manifold of m-frames in* $\mathbb{R}^n$. If one then maps the frame v into the m-dimensional subspace of $\mathbb{R}^n$ that its members span one obtains a map f from $V(m, \mathbb{R}^n)$ *onto* $G(m, \mathbb{R}^n)$ which can be shown to be a submersion [19]. Defining the equivalence relation $\sim$ on $V(m, \mathbb{R}^n)$ by $x \sim y \Leftrightarrow f(x) = f(y)$ one sees that $G(m, \mathbb{R}^n)$ is (diffeomorphic to) a quotient manifold of $V(m, \mathbb{R}^n)$.

Each Grassmann manifold is Hausdorff, second countable and compact [19]. Also $G(m, \mathbb{R}^n)$ is connected as will be shown in section 5.7. The Grassmann manifold $G(m, \mathbb{R}^n)$ *is diffeomorphic to* $G(n-m, \mathbb{R}^n)$ as can be shown by employing the standard metric on $\mathbb{R}^n$ (defined by $(\mathbf{x}, \mathbf{y}) \to x^i y^i$ with $\mathbf{x} = (x^1, \ldots, x^n)$ and $\mathbf{y} = (y^1, \ldots, y^n) \in \mathbb{R}^n$) and mapping a member V of $G(m, \mathbb{R}^n)$ into its orthogonal complement W with respect to this metric. The Grassmann manifold $G(1, \mathbb{R}^{n+1})$ is *real projective space*, usually denoted by $P^n\mathbb{R}$, and has dimension n.

Chapter 5

Lie Groups

5.1 Topological Groups

Let G be a set. It has been seen in earlier chapters how G may be given the structure of a group with multiplication represented by $\cdot$ and how G may be given a topological structure $\mathcal{T}$. These structures need have no relationship to each other. But given that G has these structures one can arrange a compatibility requirement between them by insisting that for $a, b \in G$, the *group function* $\phi_1 : G \times G \to G$ defined by $\phi_1(a, b) = a \cdot b$ (or more simply ab) and the *inverse operator* $\phi_2 : G \to G$ defined by $\phi_2(a) = a^{-1}$ are *continuous* maps with respect to $\mathcal{T}$ and the corresponding product topology on $G \times G$. If such is the case the set G together with its group and topological structure is called a *topological group* and will simply be labelled G given that $\cdot$ and $\mathcal{T}$ are understood. It is not difficult to see that for a given group structure on G there is always a topology on G (for example, the discrete topology) which makes it a topological group, but it is not necessarily unique. However, given a topology on a set G, the next paragraph shows that there may not exist a topological group structure for G (and if there does, it is not necessarily unique as the situation when a set is given two distinct group structures each with discrete topology shows). If G_1 and G_2 are topological groups, a map $f : G_1 \to G_2$ which is continuous *and* a group homomorphism is called a *topological group homomorphism*. If f is, in addition, a homeomorphism then it is called a *topological group isomorphism* and G_1 and G_2 are *isomorphic topological groups*.

A subgroup H of a topological group G together with its subspace topology is itself a topological group and is called a *topological subgroup* of G. If H is an open (respectively closed) subset of G it is called an *open* (respectively *closed*) *subgroup* of G. If $a \in G$ the maps L_a and $R_a : G \to G$

given for $g \in G$ by $L_a(g) = ag$ and $R_a(g) = ga$ are called *left* and *right translations*. They are bijective and their inverses are also left and right translations respectively, *i.e.* $L_a^{-1} = L_{a^{-1}}$, $R_a^{-1} = R_{a^{-1}}$. Although they are not necessarily (group) isomorphisms they are (topological) homeomorphisms and, as a consequence, if H is an open subgroup of G, any right or left coset of H in G is an open subset of G. Hence if H is an open subgroup of G it is also a closed subgroup of G (since its complement is a union of such open cosets). The facts that L_a and R_a are homeomorphisms show that not any topological space can be given a group structure to make it a topological group since a high degree of "topological homogeneity" is required.

The topological group G may be connected. If it is not, the component of G containing the identity e, G_e, and called the *identity component*, is particularly important. In fact G_e is a *normal closed subgroup* of G (and the other components of G are the cosets of G_e in G, there being no distinction between left and right cosets here since G_e is normal - see section 2.2) [19].

The following result will be important later [19]. If G is a topological group then G_e (which inherits the structure of a connected topological group) is such that *it is generated by any open subset U in G_e which contains e.*

Any real vector space is a group and also a topological space (inherited through its manifold structure - see chapter 4). With these structures it is easily seen to be a topological group. Similarly, $GL(n, \mathbb{R})$ is a topological group. More examples of topological groups can be found in the next section and in the standard texts (see, *e.g.* [36]). In particular, if $G_1, \ldots, G_n$ are topological groups then so is $G_1 \times \cdots \times G_n$ with its product group and topology structures.

It is remarked that for a set G with both a group and a topology structure, the continuity of ϕ_1 does not imply the continuity of ϕ_2, and vice versa (but the continuity of *both* is insisted upon in the definition of a topological group). In fact the family of subsets $\{[a, b) : a, b \in \mathbb{R}\}$ is a basis for a topology $\mathcal{T}$ on $\mathbb{R}$ such that, with respect to the usual additive group structure on $\mathbb{R}$, ϕ_1 is continuous but ϕ_2 not. On the other hand the group G with three members $\{e, x, y\}$ and identity e ($\Rightarrow xx = y,\ yy = x,\ xy = yx = e$) and when given the topology $\mathcal{T} = \{\phi, \{e\}, G\}$ is such that ϕ_2 is continuous but ϕ_1 not. However, the continuity of the *single* map $G \times G \to G$ given by $(a, b) \to ab^{-1}$ *is* equivalent to the topological group condition (just combine this latter map with the (continuous) map $a \to (e, a)$ to get the continuity of ϕ_2 and then with the (now continuous) map $(a, b) \to (a, b^{-1})$ to see that

ϕ_1 is continuous, the converse being clear).

5.2 Lie Groups

Let G be a group. Suppose that G can also be given the structure of a smooth manifold of dimension n. Again, these structures need not be related. Suppose, however, that one insists on the compatibility requirement that the group function $\phi_1 : G \times G \to G$ is smooth when $G \times G$ is given the manifold product structure from G. Then G, together with its group and manifold structure, is called a *Lie Group* (of dimension n). It should be noted that it is not necessary to insist also that the inverse operator $\phi_2 : G \to G$ is smooth. In fact it follows automatically for a Lie group that ϕ_2 is a (smooth) diffeomorphism on G [19]. Hence a Lie group with its given group structure and manifold topology is a topological group. The extra power in the manifold structure of a Lie group forces the smoothness of ϕ_2 from that of ϕ_1 whereas the continuity of both ϕ_1 and ϕ_2 must be assumed for a topological group.

Some elementary properties of Lie groups can now be briefly reviewed, further details being available elsewhere [19], [20], [22] (and, in fact, [19] will be used as a general reference for much of this chapter). All topological references will always be to the underlying manifold topology. If G is a Lie group the left and right translations L_a and R_a ($a \in G$) and the inverse operator ϕ_2 are diffeomorphisms $G \to G$. The identity component G_e of G, where e is the identity of G, is an *open* (and closed) subset of G and hence an open submanifold of G. From the previous section G_e is also a normal subgroup of G and its cosets in G are the (open and closed) components of G. These cosets are (smoothly) diffeomorphic to G_e. Then G_e inherits the structure of a connected Lie group which is generated by any open subset of G_e containing e. *Hence there are no open subgroups of G properly contained in G_e*. If $G_1, \ldots, G_n$ are Lie groups then $G_1 \times \cdots \times G_n$ is a Lie group with the product group and manifold structures. It can be shown that if a *topological group* satisfies the T_1 property (section 3.8) then it is necessarily Hausdorff. However, the topology of any manifold is T_1 (theorem 4.3(*ii*)) and so *every Lie group is Hausdorff*. Also it can be shown that every *connected* Lie group is second countable (and, in particular, G_e is second countable for any Lie group G). It follows that any *connected* Lie group is paracompact (theorem 4.4). With the obvious natural structures $\mathbb{R}^n$, $GL(n, \mathbb{C})$ and $GL(n, \mathbb{R})$ are Lie groups. If G_1 and G_2 are Lie groups and $f : G_1 \to G_2$ is smooth and a group homomorphism then f is called a *Lie group homomorphism*. If, in addition, f is bijective and f^{-1} is smooth it is called a *Lie group isomorphism* and G_1 and G_2 are *isomorphic as Lie groups* (*Lie isomorphic*). It can be shown [22] that if $f : G_1 \to G_2$ is a topological group homomorphism then it is a Lie group homomorphism and hence that *if G_1 and G_2 are isomorphic as topological groups they are isomorphic as Lie groups.*

5.3 Lie Subgroups

A subgroup of a topological group G has a natural subspace topology and hence a natural structure as a topological subgroup of G. For a Lie group the situation is a little more complicated. Let G be a Lie group and H a subset of G. Suppose that H is both a subgroup of G and a submanifold of G and, that with these two structures, is itself a Lie group. Then H is called a *Lie subgroup* of G of dimension equal to that of H as a manifold. The complication is that the topology on H may not be subspace topology and in this sense H may not be a *topological* subgroup of G! It is, in fact, true that if G is a Lie group and *H is a subset of G which is simultaneously a subgroup and a submanifold of G then H is a Lie subgroup of G.* In other words the compatibility of these structures needed to give a Lie group structure on H follows automatically. This result is not obvious (see *e.g.* [22]) because the restriction of the group function $\phi_1 : H \times H \to H$ is not obviously a smooth map between the manifolds $H \times H$ and H (see *e.g.* theorem 4.7). It is obvious, of course, if H is a *regular* submanifold of G but this need not be the case. In what is to follow, and given that G is a Lie group and $H \subseteq G$, the statement that H is a subgroup of G means exactly what it says in the group theoretic sense whereas the statement H is a topological subgroup of G again means exactly what it says but with G regarded as a topological group. The statement that H is a Lie subgroup of G is as defined above.

A group usually becomes more interesting when it admits a manifold structure which then makes it a Lie group. Similar remarks apply when a subgroup of a Lie group is found to admit the structure of a Lie subgroup. In this respect, the following so called 'closed subgroup theorem' is important. A detailed discussion and proof can be found in [19].

Theorem 5.1 *Let G be a Lie group and H a subgroup of G. If H is a closed subset of G which is not discrete then H admits a unique structure as* *a regular submanifold of G and is then a Lie subgroup of G. If, however, H is closed but not open then the set $L(G,H)$ of left cosets of H in G admits the structure of a Hausdorff quotient manifold of G.*

Another useful result in this direction is

Theorem 5.2 *Let G be a Lie group of dimension n and M a manifold of dimension $m < n$. Let $f : G \to M$ be a smooth map which has rank m at the identity e of G and let $H = f^{-1}\{f(e)\}$. Suppose also that $f(gh) = f(g)$ for each $h \in H$, $g \in G$. Then H can be given the structure of a (closed) regular submanifold of G of dimension $n - m$ and then the structure of a (closed) Lie subgroup of G of dimension $n - m$.*

It is remarked that the "compatibility" of the structures on H between these theorems follows from section 4.11, example (viii). It is also remarked that it may be possible to give H in theorem 5.1 a non-regular manifold structure so that it is then a Lie subgroup of G. As an application of these results consider the n^2-dimensional Lie group $GL(n,\mathbb{R})$. This group has two components: the identity component and its coset and which contain those non-singular matrices with positive determinant and negative determinant, respectively. Important subgroups of it are $O(n)$ and $SL(n,\mathbb{R})$. In fact the continuous maps $f : GL(n,\mathbb{R}) \to S(n,\mathbb{R})$ and $g : GL(n,\mathbb{R}) \to \mathbb{R}$ given by $f(A) = AA^{\mathrm{T}}$ and $g(A) = \det A$ show that they are the subsets $f^{-1}(I_n)$ and $g^{-1}(1)$, respectively, and are hence *closed* subgroups of $GL(n,\mathbb{R})$. Since neither is discrete it follows from theorem 5.1 that each can be given a regular submanifold structure which then makes it a Lie subgroup of $GL(n,\mathbb{R})$. The dimension of $SL(n,\mathbb{R})$ is $n^2 - 1$ and of $O(n)$ is $\frac{1}{2}n(n-1)$ and this follows from theorem 5.2 by again employing the maps f and g. An application of these results to the Lorentz group will be given in section 6.3.

A Lie subgroup H of a Lie group G is called a *connected Lie subgroup* if H is connected in its manifold topology. It is then necessarily a connected subspace of G (but the obvious converse may fail and a counter-example can be constructed from the example following the proof of theorem 4.7). It can be shown that if a subset H of a Lie group G can be given two structures as a *connected* Lie subgroup of G then these structures are Lie isomorphic. Hence if, in addition, H is a *closed* subset of G this unique Lie subgroup structure on H makes H a *regular* submanifold of G from theorem 5.1. The Lie subgroup $SL(n,\mathbb{R})$ is connected but $O(n)$ is not. The latter has two components: the identity component denoted by $SO(n)$ and its coset, consisting of those matrices with determinant $+1$ and -1, respectively. It is noted here that, from the above and section 4.11, example (vii), $O(n)$ and $SO(n)$ are regular submanifolds of $GL(n, \mathbb{R})$ and, clearly, the latter is an open submanifold of $\mathbb{R}^{n^2}$. Thus, since $O(n)$ and $SO(n)$ are bounded (and closed) *subspaces* of $\mathbb{R}^{n^2}$ they are compact (section 3.8).

5.4 Lie Algebras

There is an important vector space associated with a Lie group and which is called its Lie algebra. Before introducing this a few remarks on vector fields are required.

Let M be a smooth manifold and X a global smooth vector field on M and let $f : M \to M$ be a smooth diffeomorphism on M. The vector field X is called f-invariant if $f_* \circ X = X \circ f$ (*i.e.* X is preserved when "pushed forward" by f_* - see section 4.9). It is not hard to check that if X and Y are smooth vector fields on M which are f-invariant then so also is $[X, Y]$.

Now let G be a Lie group of dimension n with identity e and consider the left and right translations L_a and R_a for each $a \in G$, all of which are smooth diffeomorphisms of G. A global smooth vector field X on G is called *left* or *right invariant* if it is, respectively, L_a-invariant or R_a-invariant for each $a \in G$. Now let $\mathbf{v}$ be a vector in the tangent space T_eG to G at e and define a global vector field X on G by $X(a) = L_{a*}(\mathbf{v})$ for each $a \in G$. Then it can be shown that X is a left-invariant vector field on G and that any left-invariant vector field X on G can be obtained in this manner from the vector $X(e) \in T_eG$. It is then clear that the set of left invariant vector fields on G is a real vector space isomorphic to T_eG under the isomorphism $X(e) \to X$ and is hence of dimension n. (Thus any Lie group is parallelisable - see section 4.8.)

This vector space of left-invariant vector fields can be given the structure of a *Lie algebra* (see section 2.7) under the (Lie) bracket operation which associates with two such vector fields X and Y the global smooth vector field $[X, Y]$ (since then $[X, Y]$ is left-invariant). This Lie algebra structure is then transferred to T_eG by the above isomorphism so that if $\mathbf{u}, \mathbf{v} \in T_eG$ arise from left-invariant vector fields X and Y, respectively, on G (so that $X(e) = \mathbf{u}$, $Y(e) = \mathbf{v}$) then one defines $[\mathbf{u}, \mathbf{v}] \equiv [X, Y](e)$. If $\{\mathbf{v}_i\}$ is a basis for T_eG which arises, respectively, from left-invariant vector fields $X_1, \ldots, X_n$ on G it easily follows that $[X_a, X_b] = \sum_{c=1}^{n} c^c_{ab} X_c$ where the $c^c_{ab} \in \mathbb{R}$ are the *structure constants* in this basis. It follows that $[\mathbf{v}_a, \mathbf{v}_b] = \sum_{c=1}^{n} c^c_{ab} \mathbf{v}_c$. The vector space T_eG with this Lie algebra structure is referred to as the *Lie algebra of G*. It will be generically labelled LG although more specific symbols are available for the more commonly used Lie groups. It is remarked that analogous results hold for the right-invariant vector fields on G and the vector space T_eG with its resulting Lie algebra structure is denoted by RG. The bracket operations in LG and RG are not the same but differ only in sign.

In order to see the role played by the Lie algebra of G consider the set of d-dimensional distributions on G $(1 \leqslant d \leqslant n)$ which are left-invariant in the sense that they are preserved by all the left translations. More precisely a distribution D on G is *left-invariant* if for each $a, g \in G$, $D(ag) = L_{a*}D(g)$. Clearly any such left-invariant distribution D uniquely determines a subspace $D(e)$ of T_eG and conversely every subspace U of T_eG uniquely determines a left invariant distribution D on G according to $D(g) = L_{g*}(U)$ $\forall g \in G$ (and so $D(e) = U$). The question now arises as to the integrability of such left-invariant distributions and to the relations between left-invariant distributions in G, subalgebras of LG and Lie subgroups of G. To this and other ends, the following theorem is important.

Theorem 5.3 *Let G be a Lie group with Lie algebra LG.*

i) *Let U be a subspace of T_eG and D the associated left-invariant distribution on G (so that $D(e) = U$). Then D is integrable if and only if U is a subalgebra of LG.*

ii) *Let H be a Lie subgroup of G with Lie algebra LH and let $i : H \to G$ be the (smooth) inclusion map. Then $i_{*e} : LH \to LG$ is a Lie algebra isomorphism between LH and a subalgebra of LG. Also H is an integral manifold of the left-invariant distribution on G determined by the subspace $i_{*e}(T_eH)$.*

iii) *Let U be a subalgebra of LG. Then there exists a unique connected Lie subgroup H of G such that, if i is the inclusion map $H \to G$, i_{*e} is a Lie algebra isomorphism between LH and U.*

Roughly speaking, theorem 5.3 says that each Lie subgroup of G picks out a unique Lie subalgebra of LG whereas each Lie subalgebra of LG is the Lie algebra of some (not necessarily unique) Lie subgroup of G and is the Lie algebra of a *unique connected* Lie subgroup of G. It should also be remarked that the Lie algebra U in part (iii) which was ultimately identified with the Lie algebra of a Lie group (*i.e.* the Lie algebra LH of H) was assumed itself to arise as a subalgebra of the Lie algebra of a Lie group (*i.e.* G). More generally one can show that *any* finite dimensional (real) Lie algebra is the

Lie algebra of a connected Lie subgroup of $GL(n,\mathbb{R})$ for some n. This is a consequence of Ado's theorem (see *e.g.* [19]). It also follows that the Lie algebra of G is identical to that of G_e.

As an example consider the Lie group $GL(n,\mathbb{R})$. In section 4.11 it was shown that $GL(n,\mathbb{R})$ is an open submanifold of $M_n\mathbb{R}$ and since $M_n\mathbb{R}$ admits a global chart so does $GL(n,\mathbb{R})$. Let the n^2 coordinate functions for this chart on $GL(n,\mathbb{R})$, in an obvious notation, be x^{ab} $(1 \leqslant a,b, \leqslant n)$. The left-invariant vector fields on G can be obtained from the basis members $\left(\frac{\partial}{\partial x^{ab}}\right)_e$ of $T_eGL(n,\mathbb{R})$ by employing the left translations in G. Thus one obtains left-invariant vector fields X_{ab} defined at g (where g is the matrix g_{ab}) by $X_{ab}(g) = L_{g*}\left(\frac{\partial}{\partial x^{ab}}\right)_e$ and which take the general form $X_{ab} = P_{ab}{}^{cd}\left(\frac{\partial}{\partial x^{cd}}\right)$ on G (recalling the summation convention). The functions P can be calculated for each choice of a and b according to

$$\begin{aligned} P_{ab}{}^{cd}(g) &= X_{ab}(g)(x^{cd}) = L_{g*}\left(\frac{\partial}{\partial x^{ab}}\right)_e (x^{cd}) \\ &= \left(\frac{\partial}{\partial x^{ab}}\right)_e (x^{cd}\circ L_g) = \left(\frac{\partial}{\partial x^{ab}}\right)_e (g_{ce}x^{ed}) = g_{ca}\delta_{bd} \end{aligned}$$

Hence $X_{ab}(g) = g_{ca}\left(\frac{\partial}{\partial x^{cb}}\right)_g$ and in the above coordinates, $X_{ab} = x^{ca}\frac{\partial}{\partial x^{cb}}$. If one writes $[X_{ab}, X_{cd}] = \alpha^{pq}\frac{\partial}{\partial x^{pq}}$ then

$$\begin{aligned} \alpha^{pq} &= X_{ab}\left(X_{cd}(x^{pq})\right) - X_{cd}\left(X_{ab}(x^{pq})\right) \\ &= X_{ab}\left(x^{rc}\delta_{pr}\delta_{qd}\right) - X_{cd}\left(x^{ra}\delta_{pr}\delta_{qb}\right) \\ &= \left(x^{pa}\delta_{qd}\delta_{cb} - x^{pc}\delta_{qb}\delta_{ad}\right) \end{aligned}$$

and so

$$[X_{ab}, X_{cd}] = \delta_{cb}X_{ad} - \delta_{ad}X_{cb} \tag{5.1}$$

Now the Lie algebra of $GL(n,\mathbb{R})$ must on grounds of dimension be isomorphic (as a vector space) to $M_n\mathbb{R}$ and so the product (5.1) may be transferred to $M_n\mathbb{R}$ by means of the basis E_{ab} for $M_n\mathbb{R}$, where E_{ab} is the matrix with 1 in the ab position and zeros elsewhere, according to

$$[E_{ab}, E_{cd}] = \delta_{cb}E_{ad} - \delta_{ad}E_{cb} \tag{5.2}$$

Thus if $A = \sum A_{ij}E_{ij}$ and $B = \sum B_{ij}E_{ij}$ are in $M_n\mathbb{R}$ the Lie algebra binary operation can be found from 5.2 to be $[A,B] = \sum A_{ij}B_{mn}[E_{ij}, E_{mn}] = AB - BA$ (the *commutator* of A and B).

Returning to theorem 5.2 it can be shown, in the notation of that theorem, that the subalgebra of LG corresponding to the Lie subgroup H

(identified with M) is the subspace $(f_*)^{-1}(\mathrm{O})$ where O is the zero vector of $T_{f(e)}M$. Since the appropriate function f and manifold M are known (and were given earlier) for the Lie subgroups $O(n)$ and $SL(n,\mathbb{R})$ of $GL(n,\mathbb{R})$ one can calculate their Lie algebras as subalgebras of the Lie algebra of $GL(n,\mathbb{R})$ given above. For $O(n)$ the map considered was $f : GL(n,\mathbb{R}) \to S(n,\mathbb{R})$ given by $f(A) = AA^{\mathrm{T}}$ which in the usual global charts x^{ab} for $GL(n,\mathbb{R})$ and y^{ab} for $S(n,\mathbb{R})$ (section 4.3, example (iv)) is $y^{ab} \circ f = x^{ac}x^{bc}$. Now $f_* \left(\frac{\partial}{\partial x^{ab}}\right)_e = \sum_{c \leq d}^{n} A_{ab}^{cd} \left(\frac{\partial}{\partial y^{cd}}\right)_{f(e)}$ where

$$\begin{aligned}
A_{ab}^{cd} &= f_* \left(\frac{\partial}{\partial x^{ab}}\right)_e (y^{cd}) = \left(\frac{\partial}{\partial x^{ab}}\right)_e (y^{cd} \circ f) \\
&= \left(\frac{\partial}{\partial x^{ab}}\right)_e (x^{ce}x^{de}) = \delta_a^c \delta_b^d + \delta_a^d \delta_b^c
\end{aligned}$$

So for a general member $\mathbf{v} = B^{ab} \left(\frac{\partial}{\partial x^{ab}}\right)_e$ of $T_e GL(n,\mathbb{R})$ one has

$$f_* \mathbf{v} = \sum_{a \leqslant b}^{n} (B^{ab} + B^{ba}) \left(\frac{\partial}{\partial y^{ab}}\right)_{f(e)} \tag{5.3}$$

Then the subalgebra of $GL(n,\mathbb{R})$ associated with the Lie subgroup $O(n)$ is $(f_*)^{-1}(\mathrm{O})$, that is, *the Lie algebra of $O(n)$ is the vector space of skew-symmetric members of $M_n\mathbb{R}$* (obtained by setting the expression in (5.3) equal to zero) under the binary operation induced by (5.2). A similar calculation for the Lie subgroup $SL(n,\mathbb{R})$ using the determinant function shows that *the Lie algebra of $SL(n,\mathbb{R})$ is the vector space of tracefree matrices* under the binary operation induced by (5.2).

5.5 One Parameter Subgroups and the Exponential Map

Let G be a Lie group with identity e and let X be a left-invariant vector field on G. Suppose c is an integral curve of X with domain some open interval I of $\mathbb{R}$ containing 0 and which starts at e (so that $c(0) = e$ and $\dot{c} = X \circ c$). For $a \in G$ consider the curve $c_a = L_a \circ c$ with domain I starting from $a \in G$. Then with the usual coordinate t on the curve c

$$\begin{aligned}
\dot{c}_a = c_{a*} \circ \frac{\partial}{\partial t} &= L_{a*} \circ c_* \circ \frac{\partial}{\partial t} = L_{a*} \circ \dot{c} \\
&= L_{a*} X \circ c = X \circ L_a \circ c = X \circ c_a
\end{aligned}$$

and so c_a is also an integral curve of X with the same domain I as c. Because there exists an integral curve of X with the same domain I through any point of G and since G is Hausdorff it can be shown that X is a *complete vector field* [19]. Hence *each left-invariant vector field on G is complete.*

This completeness feature turns out to be rather important for several reasons, the first of which can be described now. Let $\mathbf{v} \in T_eG$, let X be the left-invariant vector field determined by $\mathbf{v}$, so that $X(e) = \mathbf{v}$, and let c be the maximal integral curve of X starting at e. One can then define a map called the *exponential mapping for G*, denoted by exp and defined as a map $\exp : T_eG \to G$ by $\exp \mathbf{v} = c(1)$. Sometimes one writes $e^{\mathbf{v}}$ for $\exp \mathbf{v}$. This map is defined on the whole of T_eG (by the completeness described above) and is smooth when T_eG has its standard manifold structure. Care must be taken not to confuse this exponential map with the exponential map arising from a connection (section 4.16). In fact, the exponential map described here can also be thought of as arising from a connection on G and its smoothness can be deduced from this fact. The name "exponential" is suggested by the fact that if $\alpha, \beta \in \mathbb{R}$ and $\mathbf{v} \in T_eG$ then

$$\exp(\alpha + \beta)\mathbf{v} = \exp \alpha\mathbf{v} \exp \beta\mathbf{v} \tag{5.4}$$

The real line $\mathbb{R}$ with its usual (identity global) chart and binary operation of addition is a 1-dimensional Lie group. Let G be any Lie group and let $f : \mathbb{R} \to G$ be a group homomorphism which is also a smooth map. Then f is called a *1-parameter subgroup of G*. It follows from (5.4) that for $\mathbf{v} \in T_eG$ the map $\phi_{\mathbf{v}} : t \to \exp t\mathbf{v}$ is a 1-parameter subgroup of G. But one can say more. In fact, *every* 1-parameter subgroup of G is of the form $\phi_{\mathbf{v}}$ for some (unique) $\mathbf{v} \in T_eG$. Now the range of ϕ_0 is $\{e\}$ and, if $\mathbf{v} \neq 0$, the range of $\phi_{\mathbf{v}}$ can be given the structure of a 1-dimensional submanifold of G (because $\phi_{\mathbf{v}}$ is an integral curve of the nowhere zero vector field $g \to L_{g*}\mathbf{v}$ on G - see section 4.13) and is also a subgroup. It follows from section 5.3 of this chapter that the *range of $\phi_{\mathbf{v}}$ is a 1-dimensional Lie subgroup of G* (and is often identified with the 1-parameter subgroup above).

One of the important aspects of the exponential map is the way in which it generates members of G from the Lie algebra LG of G. Since $GL(n, \mathbb{R})$ and certain of its subgroups play an important role in this book, this Lie group will be used now as an example of this feature. In the notation employed in the previous section and regarding $GL(n, \mathbb{R})$, and now denoting its Lie algebra by $g\ell(n, \mathbb{R})$, let $\mathbf{v} = V^{ab}\left(\frac{\partial}{\partial x^{ab}}\right)_e \in g\ell(n, \mathbb{R})$ ($V \in M_n\mathbb{R}$). The corresponding left-invariant vector field X on G is then

$X : g \to L_{g*}\mathbf{v}$ where

$$X(g) = L_{g*}\left[V^{ab}\left(\frac{\partial}{\partial x^{ab}}\right)_e\right] = V^{ab}g_{ca}\left(\frac{\partial}{\partial x^{cb}}\right)_g \tag{5.5}$$

and so (replacing g_{ab} by x^{ab})

$$X = V^{ab}x^{ca}\frac{\partial}{\partial x^{cb}} \tag{5.6}$$

Now an integral curve of this vector field is $\phi_{\mathbf{v}}$ and so $\dot{\phi}_{\mathbf{v}} = X \circ \phi_{\mathbf{v}}$. Now if $c^{ab} = x^{ab} \circ \phi_{\mathbf{v}}(t) = x^{ab} \circ \exp t\mathbf{v}$ then from (5.6) one has

$$\frac{dc^{ab}}{dt} = V^{cb}x^{ac}(\phi_{\mathbf{v}}(t)) = V^{cb}c^{ac} \tag{5.7}$$

with initial condition $c^{ab}(0) = x^{ab}(e) = \delta_{ab}$. The unique solution is

$$c^{ab} \equiv x^{ab} \circ \exp tV = \delta_{ab} + \sum_{s=1}^{\infty}\frac{t^s}{s!}(V^s)^{ab} \tag{5.8}$$

where

$$\exp tV = I_n + \sum_{s=1}^{\infty}\frac{t^s}{s!}V^s \tag{5.9}$$

This notation is standard and no confusion should arise between it and the $\exp_m$ notation of section 4.16.

A few remarks may be made here regarding the exponential map. First suppose H is a Lie subgroup of a Lie group G. Then one must distinguish between the exponential map associated with H (and denoted by $\exp_H$) so that $\exp_H : T_eH \to H$ and the exponential map exp of G when restricted to T_eH. Fortunately the relation between them is pleasant in the sense that if $i : H \to G$ is the natural inclusion map then $\exp \circ\, i_{*e} = i \circ \exp_H$. Thus the range of the curve $\phi_{\mathbf{v}} : t \to \exp t\mathbf{v}$ for $\mathbf{v} \in i_* T_eH$ lies in H. There is a converse result if H has a second countable topology (and note that from section (5.2) every connected Lie group is second countable) in that any smooth curve c in G whose range lies in H is such that every tangent vector $\dot{c}(t)$ is tangent to H. Second, one might ask whether the exponential map $\exp : T_eG \to G$ for a Lie group G is *onto* or not. If it is onto (that is every $g \in G$ equals $\exp(\mathbf{v})$ for some $\mathbf{v} \in T_eG$) the Lie group is called *exponential*. Clearly, for a Lie group to be exponential it must be connected but this is not sufficient. Less obvious is the result that a connected Lie subgroup of an exponential Lie group need not be exponential. This feature will be seen later when the Lorentz group is introduced. Clearly (the range of) any 1-parameter subgroup of a Lie group G is an exponential Lie subgroup of G. It can be shown that any compact connected Lie group is exponential [37] and that $SL(2,\mathbb{R})$ is not. However every connected Lie group is "almost" exponential as the following theorem shows.

Theorem 5.4 *Let G be a connected Lie group. Then every $g \in G$ is the product of finitely many members of G each of which is the exponential of some member of the Lie algebra of G.*

Proof. This follows from the fact mentioned in section (5.2) that G is generated by *any* open subset of G containing the identity e of G. But the range of exp is such an open subset and this completes the proof. □

It is convenient at this point to mention briefly some properties of the exponential map. First consider the metric space $\mathbb{R}^{n^2}$ for some positive integer n with the usual Euclidean metric. The set $M_n\mathbb{R}$ may be identified with this metric space and then one can talk sensibly about convergence of sequences of such matrices. If $P \in M_n\mathbb{R}$ then the series (5.9) for $\exp P$ can be shown to be convergent to a member of $GL(n,\mathbb{R})$ and one has the following properties for $B, P, Q \in M_n\mathbb{R}$ and B non-singular.

i) If $Q = BPB^{-1}$ then $\exp Q = B(\exp P)B^{-1}$.
ii) If $PQ = QP$ then $\exp(P+Q) = \exp P \exp Q = \exp Q \exp P$.
iii) $\exp(-P) = (\exp P)^{-1}$.
iv) If $\mathbf{x}$ is a (real or complex) eigenvector of P with (real or complex) eigenvalue λ then $\mathbf{x}$ is an eigenvector of $\exp P$ with eigenvalue e^{λ}.
v) If $P \in M_n\mathbb{R}$ the map $\mathbb{R} \to M_n\mathbb{R}$ given by $t \to e^{tP}$ is differentiable and

$$\frac{d}{dt}(\exp tP) = P \exp tP = (\exp tP)\, P.$$

vi) If $O \in M_n\mathbb{R}$ is the zero matrix, $\exp O = I_n$.
vii) If $P \in M_n\mathbb{R}$ and if $\mathbf{v}$ $(0 \neq \mathbf{v} \in \mathbb{C}^n)$ is an eigenvector of $\exp tP$ with eigenvalue $\lambda(t) \in \mathbb{C}$ for each $t \in \mathbb{R}$ then $\lambda(t) = e^{\lambda t}$ $(\lambda \in \mathbb{C})$ and $\mathbf{v}$ is an eigenvector of P with eigenvalue λ.

The proofs of (i)–(vi) can be found in [38],[22]. For (vii) one has $\mathbf{v} \exp tP = \lambda(t)\mathbf{v}$ and so $\lambda(t)$ is a smooth function on $\mathbb{R}$. Also setting $t = 0$ gives $\lambda(0) = 1$ and, since $\exp tP$ is non-singular, $\lambda(t) \neq 0$, $\forall t \in \mathbb{R}$. Now differentiate and use (v) above to get $(\mathbf{v} \exp tP)\, P = \dot{\lambda}\mathbf{v}$ or $\mathbf{v}P = \dot{\lambda}\lambda^{-1}\mathbf{v}$ where $\dot{\lambda} \equiv \frac{d\lambda}{dt}$. So $\dot{\lambda}\lambda^{-1}$ is constant and it follows that $\lambda(t) = e^{\lambda t}$ $(\lambda \in \mathbb{C})$ and hence that $\mathbf{v}P = \lambda\mathbf{v}$.

5.6 Transformation Groups

Let G be a group and M a smooth manifold. Then G is said to act on M as a *transformation group* if there is a map $\phi : G \times M \to M$ with the properties that

i) given $g \in G$ the function $\phi_g : M \to M$ given by $\phi_g(m) = \phi(g, m)$ is a smooth diffeomorphism of M,
ii) if $g, h \in G$, $\phi_g \circ \phi_h = \phi_{gh}$.

If e is the identity of G then the map ϕ_e is necessarily the identity map on M. This follows because if $m \in M$ and $m' = \phi_e^{-1}(m)$ then $\phi_e(m) = \phi_e(\phi_e(m')) = \phi_e(m') = m$. However e may not be the only member of G which gives rise in this way to the identity map on M. If e is the only such member of G, G is said to act *effectively* (or to be an *effective action*) on M. If ϕ_e is the only such map to fix *any* point of M then G is said to act freely (or to be a *free action*) on M. Clearly a free action is effective. Now let $g \in G$. Then $gg^{-1} = e$ and so $\phi_g \circ \phi_{g^{-1}} = \phi_e$. It follows that $\phi_g^{-1} = \phi_{g^{-1}}$. Since each ϕ_g is a diffeomorphism it follows that the map $\phi : G \times M \to M$ is *surjective*. The action of G on M is said to be *transitive* if given $m_1, m_2 \in M$ $\exists\, g \in G$ such that $\phi_g(m_1) = m_2$.

The above action $\phi : G \times M \to M$ is, for obvious reasons, sometimes referred to as a *left action* and G is said to act *(on M) on the left*. A *right action* of G on M can be defined as a map $\psi : M \times G \to M$ such that (i)$'$ for $g \in G$, the map $\psi_g : m \to \psi(m, g)$ is a diffeomorphism of M and (ii)$'$ for $g, h \in G$, $\psi_g \circ \psi_h = \psi_{hg}$. It should be pointed out that there is nothing special about G and M that produces a left action rather than a right action, or vice versa, because if ϕ is a left action of G on M then the map $\psi : M \times G \to M$ given by $\psi(m, g) = \phi(g^{-1}, m)$ is a right action, and vice versa. However, unless otherwise specified, a transformation group action will always refer to a left action as defined above

5.7 Lie Transformation Groups

Let G be a Lie group and M a smooth manifold. Then G is said to act on M as a *Lie transformation group* if there is a *smooth* map $\phi : G \times M \to M$ satisfying the conditions (i) and (ii) given in the previous section. [Note that (ii) may be rewritten as $\phi(g, \phi(h, m)) = \phi(gh, m)$ for each $m \in M$ and $g, h \in G$.] Thus G acts on M as a *transformation group* and the notation of the previous section will be employed here. However, for a *Lie* transformation group the conditions (i) and (ii) can be replaced by an equivalent set (i)″ and (ii) where (i)″ is the condition that the map ϕ is *surjective* (see, *e.g.* [19]). Clearly (i) and (ii) imply (i)″ and (ii). For the converse one notes that if $m \in M$ then, from (i)″, there exists $m' \in M$ and $h \in G$ such that $\phi_h(m') = m$ and so $\phi_e(m) = \phi_e(\phi_h(m')) = \phi_h(m') = m$ and so ϕ_e is still the identity map on M. It is then immediate that for $g \in G$, ϕ_g has an inverse function $\phi_{g^{-1}}$ and that the smoothness of ϕ_g follows from that of ϕ. Hence each ϕ_g is a diffeomorphism of M. The terms *effectively* and *freely* will be used with the same meaning as in the previous section. Also the remarks about left and right actions in the previous section apply also to Lie transformation groups and although the results to follow are stated for left actions they apply equally well for right actions when appropriately rewritten.

In the following examples of Lie transformation groups (i), (iv) and (v) are right actions, (ii) and (iii) are left actions and (vi) is either.

i) $\phi : \mathbb{R}^n \times GL(n,\mathbb{R}) \to \mathbb{R}^n$ where $\phi(\mathbf{v}, A) = \mathbf{v}A$.
ii) $\phi : GL(n,\mathbb{R}) \times S(n,\mathbb{R}) \to S(n,\mathbb{R})$ where $\phi(A, S) = ASA^T$.
iii) $\phi : GL(n,\mathbb{R}) \times M_n\mathbb{R} \to M_n\mathbb{R}$ where $\phi(A, B) = ABA^{-1}$.
iv) $\phi : \mathbb{R}^n \times O(n) \to \mathbb{R}^n$ where $\phi(\mathbf{v}, A) = \mathbf{v}A$.
v) $\phi : S^{n-1} \times O(n) \to S^{n-1}$ where $\phi(\mathbf{v}, A) = \mathbf{v}A$.
vi) Let X be a *complete* vector field on a smooth *Hausdorff* manifold M (section 4.10). Then the map $\phi : \mathbb{R} \times M \to M$ given by $\phi(r, m) = c_m(r) \equiv \phi_r(m)$ where c_m is the maximal integral curve of X starting at m, is smooth and satisfies the condition that $\phi_r \circ \phi_s = \phi_{r+s}$ $(r, s \in \mathbb{R})$. This shows that the Lie group $\mathbb{R}$ acts on M as a Lie transformation group.

It is remarked that modifications of examples (ii) and (iii) to $(S, A) \to A^T SA$ and $(B, A) \to A^{-1}BA$ produces right rather than left actions. The action represented by example (iii) is not effective whereas (i) and (iv) are effective but not free.

Let G act on M as a Lie transformation group as described above. If H is a Lie subgroup of G then, by restricting the map ϕ above to a map $H \times M \to M$, one sees that H *also acts on* M *as a Lie transformation group* (see examples (i) and (iv) above). Also, a subset A of M is *invariant* under G if $\phi(G \times A) \subseteq A$. An equivalence relation $\sim$ on M is said to be *preserved* by G if for $m_1, m_2 \in M$ and $g \in G$, $m_1 \sim m_2 \Rightarrow \phi_g(m_1) \sim \phi_g(m_2)$. One can now show that if N is a *regular* submanifold of M which is invariant under G then G acts (in an obvious way) as a Lie transformation group on N. (The problem here for a general submanifold N of M is the smoothness of the resulting map $G \times N \to N$ (see section 4.11). This problem is avoided if N is regular.)

Theorem 5.5 *Let G be a Lie group acting as a Lie transformation group on the smooth manifold M and let $\sim$ be an equivalence relation on M which leads to a quotient manifold $M/\sim$ of M and which is preserved by G. Then G acts naturally on $M/\sim$ as a Lie transformation group.*

This theorem has the following consequence. Suppose that H is a subgroup of G and $L(G, H)$ the set of left cosets of H in G. Suppose also that $L(G, H)$ can be given the structure of a quotient manifold of G. Then the group function $\phi_1 : G \times G \to G$ shows that G acts on itself as a Lie transformation group and clearly leads to a well defined map $G \times L(G, H) \to L(G, H)$. So by the previous theorem G acts on $L(G, H)$ on the left as a Lie transformation group with action $(g, aH) \to (ga)H$. The same results hold with *left* everywhere replaced by *right*. Another consequence is that the action of $GL(n, \mathbb{R})$ in example (i) above leads to an obvious action on the Stiefel manifold $V(m, \mathbb{R}^n)$ and hence on the Grassmann manifold $G(m, \mathbb{R}^n)$ which is a quotient manifold of $V(m, \mathbb{R}^n)$ (see section 4.17). Restricting this action to the identity component of $GL(n, \mathbb{R})$ and to a single member of $G(m, \mathbb{R}^n)$ and noting that the range of this restricted action is connected and equal to $G(m, \mathbb{R}^n)$ shows that $G(m, \mathbb{R}^n)$ is connected.

5.8 Orbits and Isotropy Groups

Let a Lie group G act on a smooth *Hausdorff* manifold M as a Lie transformation group. The Hausdorff assumption will be needed later in this section and is imposed now as a blanket statement for the whole section. Define an equivalence relation $\sim$ on M by $m_1 \sim m_2$ if there exists $g \in G$ such that $\phi_g(m_1) = m_2$ $(m_1, m_2 \in M)$. The associated equivalence class containing $m \in M$ is $\{\phi_g(m) : g \in G\}$ and is denoted by O_m and called the *orbit* of m (under G). If for $m \in M$ one defines the map $\phi_m : G \to M$ by $\phi_m(g) = \phi(g, m)$ then ϕ_m is smooth and its range is the orbit O_m of m. If the action of G on M is transitive there is only one orbit, which is M. If G is connected it follows that each orbit is a connected subset of M.

Again for $m \in M$ consider the subset of G whose members g "fix" m. This set is $\{g \in G : \phi_g(m) = m\}$ and is easily seen to be a *subgroup* of

G. It is also equal to the set $\phi_m^{-1}\{m\}$ and is denoted by I_m and called the *isotropy group* at m (under G). If m_1 and m_2 are members of the same orbit (so that there exists $g \in G$ such that $\phi_g(m_1) = m_2$) then if $g' \in I_{m_2}$ it follows that $\phi_g^{-1} \circ \phi_{g'} \circ \phi_g(m_1) = m_1$ and so $g^{-1}g'g \in I_{m_1}$. A similar argument in reverse then shows that I_{m_1} and I_{m_2} are *conjugate* (and hence isomorphic) subgroups of G.

The isotropy group I_m at m is a subgroup of G and leads to the set $L(G, I_m)$ of left cosets of I_m in G. If g_1, g_2 are in the same left coset then it follows that $\phi_m(g_1) = \phi_m(g_2)$ and so ϕ_m gives rise to a map $\psi_m : L(G, I_m) \to M$ given by $\psi_m(gI_m) = \phi_g(m)$ and whose range is the orbit O_m of m. Also if g_1I_m and g_2I_m are members of $L(G, I_m)$ then the condition $\psi_m(g_1I_m) = \psi_m(g_2I_m)$ implies that $\phi_{g_1}(m) = \phi_{g_2}(m)$. Thus $\phi_{g_1^{-1}g_2}(m) = m$ and so $g_1^{-1}g_2 \in I_m$. Hence $g_1I_m = g_2I_m$ and so ψ_m is one-to-one. Now $I_m = \phi_m^{-1}\{m\}$ and so since M is T_1 it follows that I_m *is a closed subgroup of* G. Thus either I_m is an open subset of G or $L(G, I_m)$ admits the structure of a quotient manifold of G *and* either I_m is discrete or it admits a unique structure as a regular submanifold of G and is then a Lie subgroup of G (theorem 5.1). If m_1 and m_2 lie on the same orbit, then I_{m_1} and I_{m_2} are either both discrete or diffeomorphic. One now has the following theorem.

Theorem 5.6 *Let G be a Lie group which acts as a Lie transformation group on the smooth Hausdorff manifold M. Then for $m \in M$ one has the following results.*

i) If I_m is not open in G the map $\psi_m : L(G, I_m) \to M$ is a smooth one-to-one immersion whose range is O_m. Hence the orbit O_m of m can be given the structure of a Hausdorff submanifold of M diffeomorphic to $L(G, I_m)$ under the map ψ_m. If I_m is discrete, $\dim O_m = \dim G$ and if not, $\dim O_m = \dim G - \dim I_m$.

ii) If G is second countable and acts transitively on M then $L(G, I_m)$ (which cannot now be discrete) is diffeomorphic to M. If also the action is free then G and M are diffeomorphic.

It should be noted in conjunction with part (ii) that any *connected* Lie group (*e.g.* the identity component of any Lie group) is second countable.

Under the conditions of part (i) of the above theorem the orbit O_m has a submanifold structure and the original action $\phi : G \times M \to M$ induces an obvious action $\phi' : G \times O_m \to O_m$ on O_m. This would immediately be smooth if O_m were a regular submanifold of M. Although O_m need not be a regular submanifold of M the smoothness of ϕ' can still be established as the next theorem shows.

Theorem 5.7 *Let G be a Lie group which acts as a Lie transformation group on the smooth Hausdorff manifold M. Let $m \in M$ and suppose I_m is not open in G. Then G acts on O_m as a Lie transformation group.*

Proof. Let ϕ' be the restriction $G \times O_m \to O_m$ of the original action of G on M. Clearly ϕ' is surjective. The previous theorem guarantees that the map ψ_m regarded as a map $\psi_m : L(G, I_m) \to O_m$ is a diffeomorphism. One can also define a diffeomorphism $\mu : G \times L(G, I_m) \to G \times O_m$ by the map $(g, q) \to (g, \psi_m(q))$ and a smooth map $\nu : G \times L(G, I_m) \to L(G, I_m)$ by the map $(g, g_1 I_m) \to g g_1 I_m$ (see end of last section). The proof is completed by noting that $\phi' = \psi_m \circ \nu \circ \mu^{-1}$. □

It is remarked that for $m \in M$, if I_m is not an open subgroup of G (so that $L(G, I_m)$ is a quotient manifold of G), then if $\lambda : G \to L(G, I_m)$ is the natural submersion one has $\phi_m = \psi_m \circ \lambda$ with ψ_m now regarded as a smooth map $L(G, I_m) \to O_m$ and so ϕ_m can be regarded as a smooth map from G onto O_m. *Hence if G is connected, O_m is a connected submanifold of M.* It is also pointed out that if M is paracompact (as will usually be the case in this book) and G connected then O_m is connected, paracompact and second countable. This follows by noting that the component M_0 of M containing O_m is closed and hence paracompact (section 3.8) and so admits a global smooth positive definite metric. This metric induces a similar one on O_m and so O_m is paracompact and second countable (theorem 4.14).

As an application of theorem 5.6 consider the left action (example (ii) section 5.7) of $GL(4, \mathbb{R})$ on $S(4, \mathbb{R})$ given by $\phi(A, S) = ASA^{\mathrm{T}}$ and let $S_0 = \operatorname{diag}(-1, 111) \in S(4, \mathbb{R})$. Then I_{S_0} is the *Lorentz group* and, since not discrete, it is a Lie subgroup of $GL(4, \mathbb{R})$ of dimension six (see chapter 6) and so is not an open submanifold of $GL(4, \mathbb{R})$. The corresponding orbit O_{S_0} is then, by theorem 5.6, a submanifold of $S(4, \mathbb{R})$ diffeomorphic to $L(GL(4, \mathbb{R}), I_{S_0})$ and $\dim O_{S_0} = \dim GL(4, \mathbb{R}) - \dim I_{S_0} = 16 - 6 = 10$. Since $\dim S(4, \mathbb{R}) = 10$, O_{S_0} is an open submanifold of $S(4, \mathbb{R})$. Similar results apply if S_0 is changed to $\operatorname{diag}(1111)$ (and then I_{S_0} is the orthogonal group $O(4)$).

5.9 Complete Vector Fields

Return now to the Lie group G with identity e and let X be a left-invariant vector field on G with $X(e) = \mathbf{v} \in T_e G$. It was mentioned earlier that the integral curve of X starting at e is the curve $\phi_{\mathbf{v}} : t \to \exp t\mathbf{v}$. For any other $a \in G$ it was shown at the beginning of section 5.5 that the integral curve of X starting at a is the curve $t \to L_a(\exp t\mathbf{v})$. So the integral curves of X are obtained by left translations from the curve $\phi_{\mathbf{v}}$. If Y is a right-invariant vector field on G satisfying $Y(e) = \mathbf{u} \in T_eG$ the integral curve of Y starting at e can also be shown to be the curve $\phi_{\mathbf{u}} : t \to \exp t\mathbf{u}$ and an argument similar to that for left-invariant vector fields shows that for $a \in G$ the integral curve of Y starting at a is the curve $t \to R_a(\exp t\mathbf{u})$. Thus the integral curves of Y are obtained by right translating the curve $\phi_{\mathbf{u}}$. Let $H_{\mathbf{v}}$ be the subgroup of G defined by the range of $\phi_{\mathbf{v}}$. Then if X and Y are left- and right-invariant vector fields, respectively, on G satisfying $X(e) = Y(e) = \mathbf{v} \in T_eG$ the ranges of their integral curves in G are, respectively, the left and right cosets of $H_{\mathbf{v}}$ in G.

The *right-invariant* vector fields on G can be mirrored in M by an important family of vector fields on M. Let $\mathbf{v} \in T_eG$ and $m \in M$. One can define a vector in T_mM using $\mathbf{v}$ and the map $\phi_m : G \to M$, $\phi_m(g) = \phi(g, m)$, namely the vector $\phi_{*m}\mathbf{v}$. The collection of all such vectors, one at each $m \in M$, determined by $\mathbf{v}$ can be shown to give rise to a smooth vector field $\tilde{X}$ on M. The set of all such vector fields obtained in this way from each $\mathbf{v} \in T_eG$ is then a real vector space of smooth vector fields on M, denoted by $R(G, M)$ (for reasons that will become clear later) and the map $f : \mathbf{v} \to \tilde{X}$ is a linear map from T_eG *onto* $R(G, M)$. Hence $R(G, M)$ is finite-dimensional with dimension $\leq \dim G$. Further, if I_m is not open in G and since ϕ_m can be regarded as a smooth map $G \to O_m$, it follows from the definition of $\tilde{X}$ and theorem 7.7(iii) that *each member of $R(G, M)$ is tangent to O_m*. It can also be shown [19] that the rank of ϕ_{*m} equals the dimension of O_m and hence that *each orbit is an integral manifold of the distribution on M arising from $R(G, M)$*. Now let X be the *right-invariant* vector field on G determined by $\mathbf{v} \in T_eG$, so that $X(e) = \mathbf{v}$. Then for $a \in G$ and $m \in M$

$$\phi_{*m}X(a) = \phi_{*m}(R_{a*}\mathbf{v}) = (\phi_m \circ R_a)_*\mathbf{v} \tag{5.10}$$

Now let $g \in G$ and put $\phi_m(a) = m'$. Then

$$\phi_m \circ R_a(g) = \phi_m(ga) = \phi(ga, m) = \phi(g, \phi(a, m)) = \phi_{m'}(g) \tag{5.11}$$

It follows that $\phi_{*m}X(a) = \phi_{*m'}\mathbf{v} = \tilde{X}(m')$ and so the right-invariant vector field X on G arising from $\mathbf{v} \in T_eG$ is related to the vector field $\tilde{X}$ on M arising from $\mathbf{v}$ in a natural way with respect to the map ϕ_m for any $m \in M$.

It should be noted here how the original *left* action of G on M represented by the map ϕ here leads naturally to consideration of *right-invariant* vector fields on G. Continuing, let X_1 and X_2 be right-invariant vector fields on G and $\tilde{X}_1$ and $\tilde{X}_2$ the corresponding vector fields on M related to X_1 and X_2 under the map ϕ_{*m} for each $m \in M$. It can then be shown that $[\tilde{X}_1, \tilde{X}_2]$ is similarly related to $[X_1, X_2]$. Evaluating this relation at e then shows that $R(G, M)$ is a (finite dimensional) Lie algebra of smooth vector fields on M under the bracket operation and that the map f above is a linear map between the Lie algebras RG and $R(G, M)$ which preserves the algebra structures, *i.e.* $[f(\mathbf{v}_1), f(\mathbf{v}_2)] = f[\mathbf{v}_1, \mathbf{v}_2]_R$ for $\mathbf{v}_1, \mathbf{v}_2 \in RG$ and where $[\ \ ]_R$ denotes the algebra operation in RG. The notation $R(G, M)$ is now clear.

Let $\tilde{X} = f(\mathbf{v})$ be the vector field on M arising from $\mathbf{v} \in RG$ as described above. The relation between the right-invariant vector field X on G arising from $\mathbf{v}$, and $\tilde{X}$, under the map ϕ_m for $m \in M$ shows that the map $t \to \phi_m(\exp t\mathbf{v}) = \phi_{\exp t\mathbf{v}}(m)$ is an integral curve of $\tilde{X}$ starting from m and defined on $\mathbb{R}$. This together with the arbitrariness of m shows that $\tilde{X}$ is a *complete* vector field on M.

Theorem 5.8 *If the Lie group G acts on a smooth Hausdorff manifold M as a Lie transformation group then $R(G, M)$ is a finite-dimensional (in fact of dimension $\leqslant \dim G$) Lie algebra of smooth complete vector fields on M. If the action of G on M is effective $R(G, M)$ is Lie isomorphic to RG under f.*

Proof. Only the last sentence has not been established. So suppose $\mathbf{v} \in RG$ and $f(\mathbf{v})$ is the zero vector field on M. Consideration of the integral curves discussed above now shows that $\phi_{\exp t\mathbf{v}}(m) = m$, $\forall m \in M$ and $\forall t \in \mathbb{R}$. The effective assumption then shows that $\exp t\mathbf{v} = e$, $\forall t \in \mathbb{R}$ and so $\mathbf{v} = 0$. Thus f is one-to-one and hence bijective. □

The above results help to give an elegant geometrical interpretation of the action of G on M. For example the action of the members of the subgroup $H_{\mathbf{v}} = \{\exp t\mathbf{v} : t \in \mathbb{R}\}$ of G on $m \in M$ simply traces out the integral curve starting at m of the vector field $\tilde{X}$ on M associated with $\mathbf{v} \in T_eG$. Also for $g \in G$, $m \in M$ let $K = \phi_m(H_{\mathbf{v}}g)$ be the image in M under ϕ_m of the right coset $H_{\mathbf{v}}g$ of $H_{\mathbf{v}}$ in G. Then

$$K = \{\phi_{hg}(m) : h \in H_{\mathbf{v}}\} = \{\phi_h(\phi_g(m)) : h \in H_{\mathbf{v}}\}$$

and so K (which is the range of the integral curve of $\tilde{X}$ starting from $\phi_m(g)$) can be thought of either as the set of all images of m under the action of all members of $H_{\mathbf{v}}g$ or as the set of all images of $\phi_g(m)$ under the action of all members of $H_{\mathbf{v}}$. Theorem (5.4) has the following interpretation.

Theorem 5.9 *Let G be a connected Lie group which acts on a smooth Hausdorff manifold M as a Lie transformation group. Then for any $g \in G$ there exists a positive integer k such that the associated map $\phi_g : M \to M$ can be described by*

$$\phi_g(m) = \phi_{\exp \mathbf{v}_1}(\phi_{\exp \mathbf{v}_2}(\cdots \phi_{\exp \mathbf{v}_k}(m) \cdots))$$

for members $\mathbf{v}_1, \mathbf{v}_2, \ldots, \mathbf{v}_k$ of T_eG.

Proof. The proof follows from theorem 5.4 by writing $g = g_1 g_2 \cdots g_k$ where $g_1 = \exp \mathbf{v}_1, g_2 = \exp \mathbf{v}_2, \ldots, g_k = \exp \mathbf{v}_k$. □

The interpretation is that the map ϕ_g can be regarded in k stages each of which has the form $\phi_{\exp \mathbf{v}}$ for some $\mathbf{v} \in T_eG$ and which moves the appropriate point a parameter distance unity along the integral curve starting at that point of the member of $R(G, M)$ associated with $\mathbf{v}$ under f.

Thus the action of G as a Lie transformation group on a smooth Hausdorff manifold M leads to a finite-dimensional Lie algebra of smooth, complete vector fields $R(G, M)$ on M. These vector fields facilitate a geometrical interpretation of the action of G on M. One naturally asks the question whether such a Lie algebra of vector fields on M necessarily gives rise in this way to a Lie group acting as a Lie transformation group on M. The answer is provided by the beautiful theorem of Palais [39].

Theorem 5.10 *Let S be a non-trivial finite-dimensional Lie algebra of smooth complete vector fields on a Hausdorff manifold M. Then there is a connected Lie group G which acts effectively on M as a Lie transformation group and is such that the Lie algebra $R(G, M)$ associated with this action equals S.*

5.10 Groups of Transformations

So far the discussion has concerned an abstract action of a Lie group G on a manifold M described by a function $\phi : G \times M \to M$. This action then gives rise to a diffeomorphism $\phi_g : M \to M$ for each $g \in G$. Suppose, however, one is not given this action of a Lie group G on M but rather a *set* G of diffeomorphisms of M which form a group under the usual rules of composition, identity and inversion of such maps. Can one find a Lie group structure for G such that the map $G \times M \to M$ given by $(g, m) \to g(m)$ is smooth? This, from section (5.7), would mean that G now acts on M as a Lie transformation group. However there may be more than one such action and an extra condition is needed to guarantee uniqueness (see, *e.g.* [19]).

Suppose then that G is a group of diffeomorphisms of a Hausdorff manifold M under composition, identity and inversion, as described above. If X is a smooth complete vector field on M such that each of its associated diffeomorphisms $\phi_r : M \to M$ (section (5.7) example (vi)) belongs to G then X is said to be *tangent to G*. Such a group G is called a *Lie group of transformations of M* if it admits a Lie group structure such that

i) the map $\phi : G \times M \to M$ given by $\phi(g, m) = g(m)$ is smooth,
ii) for any smooth complete vector field X on M which is tangent to G the map $\mathbb{R} \to G$ given by $r \to \phi_r$ is a smooth group homomorphism (*i.e.* a one-parameter subgroup of G).

The following theorem now summarises the situation [39] where the distinction between a *Lie group of transformations* and a *Lie transformation group* should be noted.

Theorem 5.11 *A group G of diffeomorphisms of a Hausdorff manifold M admits at most one structure as a Lie group of transformations of M. The map ϕ then shows that, given such a structure, G acts on M as a Lie transformation group and the action is effective. G admits the structure of a Lie group of transformations of M if and only if the set S of smooth complete vector fields on M which are tangent to G is a finite-dimensional Lie algebra under the bracket operation. If such is the case then G acts effectively on M as a Lie transformation group and $S = R(G, M)$ (in the notation of section (5.9)).*

A curious fact emerges in the proof of this result [39]. If X and Y are smooth *complete* vector fields on a Hausdorff manifold M then neither $X + Y$ nor $[X, Y]$ need be complete. Hence if S is the Lie algebra of all smooth vector fields on M and S' is a subset of *complete* members of S, the smallest subalgebra S'' of S which contains S' (the subalgebra of S *generated* by S') need not consist only of complete vector fields. However, *if S'' is finite-dimensional, every member of S'' is complete.*

5.11 Local Group Actions

In the last five sections the transformations on M which have been discussed have been diffeomorphisms of M. Often one is led, directly or indirectly, to transformations which are only defined on part of M and to this situation attention can now be turned.

Consider a smooth vector field X on a smooth Hausdorff manifold M. Then given $m \in M$ there exists $\varepsilon > 0$ and an open subset U of M containing m such that for any $p \in U$ an integral curve c_p of X exists which is defined on $(-\varepsilon, \varepsilon)$ and starts at p (theorem 4.6). This gives rise, for each $t \in (-\varepsilon, \varepsilon)$, to a map $\phi_t : U \to \phi_t(U)$ defined by $p \to c_p(t)$ for each $p \in U$. So each point of U is moved a parameter distance t along the integral curve of X through that point. Each map ϕ_t is smooth and is, in fact, a diffeomorphism between the open submanifolds U and $\phi_t(U)$ of M. However, for $t \neq 0$, ϕ_t may not be defined on the whole of M and is so defined if and only if there exists an integral curve of X defined on the *same* open interval of $\mathbb{R}$ starting at any point of M. This is equivalent to X being *complete* [19] and then *every* such map ϕ_t constructed above is a *diffeomorphism* of M. Thus whereas smooth *complete* vector fields on M lead to an action of $\mathbb{R}$ on M as a Lie transformation group (section (5.7) example (vi)), incomplete smooth vector fields on M lead to "local diffeomorphisms" of M. This leads to the idea of a *flow box* associated with X [21],[22].

Theorem 5.12 *Let X be a smooth vector field on a Hausdorff manifold M. For each $m \in M$ there exists a triple (U, ε, ϕ), called a flow box of X at m, where*

i) U is an open submanifold of M containing m, $\varepsilon \in \mathbb{R}$ and $\varepsilon > 0$ or $\varepsilon = \infty$,

ii) $\phi : (-\varepsilon, \varepsilon) \times U \to M$ is smooth,

iii) for each $p \in U$ the map $c_p : (-\varepsilon, \varepsilon) \to M$ given by $c_p(t) = \phi(t, p)$ is an integral curve of X starting from p,

iv) if $t \in (-\varepsilon, \varepsilon)$ the map $\phi_t : U \to \phi_t(U)$ defined by $\phi_t(p) = \phi(t, p)$ is a smooth diffeomorphism between the open submanifolds U and $\phi_t(U)$ of M.

Such a flow box is unique in the sense that if $(U', \varepsilon', \phi')$ is another triple satisfying (i)-(iv) above then ϕ and ϕ' agree on the intersection of their domains. Also, if s, t and $s + t \in (-\varepsilon, \varepsilon)$ then $\phi_s \circ \phi_t = \phi_t \circ \phi_s = \phi_{s+t}$ wherever these compositions are defined and ϕ_0 is the identity map on U.

Finally if $U_t \equiv \phi_t(U)$ and if $U \cap U_t \neq \emptyset$, ϕ_t and ϕ_{-t} restrict to inverse maps between $U_t \cap U$ and $U_{-t} \cap U$.

If X is complete ϕ is a smooth map $\mathbb{R} \times M \to M$ and is referred to as the *flow* of X and the set $\{\phi_t : t \in \mathbb{R}\}$ as a *1-parameter group of diffeomorphisms*. If X is not complete each map ϕ_t, $t \in (-\varepsilon, \varepsilon)$, is called a *local diffeomorphism* or a *local flow (associated with* X*)*. [It is remarked here for occasional use later that if the vector field X is C^k $(k \geq 1)$ the associated local flows ϕ_t are C^k local diffeomorphisms.]

The way in which a flow box handles the case when X is not complete suggests the following formal definition of a local group action due to Palais [39]. Let G be a connected Lie group with identity e and M a smooth Hausdorff manifold. *A local G-transformation group acting on* M is a smooth map $\phi : D \to M$ where

i) D is an open subset of $G \times M$ such that for each $p \in M$ the subset $D_p \equiv \{g \in G : (g, p) \in D\}$ of G is a connected open neighbourhood of e.
ii) for each $p \in M$, $\phi(e, p) = p$.
iii) if $(h, p) \in D$, $(g, \phi(h, p)) \in D$ and $(gh, p) \in D$ then $\phi(gh, p) = \phi(g, \phi(h, p))$.

It should first be noted that if $D = G \times M$ then G acts on M as a Lie transformation group and the discussion of (5.7) applies. The local nature of the action is expressed in the facts that for $p \in M$ only members of D_p "act" on p and a particular $g \in G$ may only "act" on the open subset $D_g \equiv \{p \in M : (g, p) \in D\}$ of M and which it does so through the smooth map $\phi_g : D_g \to M$ given by $\phi_g(p) = \phi(g, p)$.

Let G be a connected Lie group as above. An *infinitesimal G-transformation group acting on* M is a Lie algebra preserving linear map from RG to the Lie algebra of smooth vector fields on M. The link between local and infinitesimal G-transformation groups is provided by another theorem due to Palais but a further definition is required before this can be done. Returning to the local G-transformation group action on M, define a smooth map $\phi_p : D_p \to M$ by $\phi_p(g) = \phi(g, p)$ and a linear map f from RG to the Lie algebra of smooth vector fields on M which associates with $\mathbf{v} \in T_eG$ the vector field given by $p \to \phi_{p*}\mathbf{v}$. This vector field is smooth and the map f, called the *infinitesimal generator* of ϕ, is Lie algebra preserving [39].

Theorem 5.13 *If G is a connected Lie group and ϕ a local G-*

transformation group acting on a Hausdorff manifold M then the associated infinitesimal generator f is an infinitesimal G-transformation group acting on M (and necessarily $\phi_{p*}X(g) = f(X(e))_{\phi(g,p)}$ for each right-invariant vector field X on G and $(g,p) \in D$). Conversely every infinitesimal G-transformation group acting on M is the infinitesimal generator of some local G-transformation group acting on M.

Since any finite-dimensional Lie algebra is the Lie algebra of some Lie group (Ado's theorem - see section 5.4) the general thrust of this theorem is that *finite-dimensional* subalgebras of the Lie algebra of global smooth vector fields on M arise as the result of local G-transformation group actions on M for various connected Lie groups G, and vice-versa (whereas the global actions (section (5.9)) arise as a result of finite-dimensional Lie algebras of global *complete* vector fields on M and vice-versa). As an example let M be a Hausdorff manifold and X a smooth vector field on M. Then, from theorem 5.13, X gives rise to a local $\mathbb{R}$-transformation group on M (cf. theorem 5.12) and, if X is complete, to a Lie transformation group on M ($G = \mathbb{R}$ in theorem 5.10) from section 5.7 example (vi).

In studying symmetry in general relativity one is, perhaps, tempted to appeal (in a not very precise way) to a certain family of local diffeomorphisms on the space-time M satisfying the particular symmetry under consideration. Even with a precise definition of such a family it may be rather difficult to work with in practice. In later chapters of this book when such symmetries are considered, a lead will be taken from this section by assuming them to arise as the local flows (again with the appropriate symmetry requirement) of a certain family S of global smooth vector fields on M. These local flows (symmetries) satisfy any reasonable requirement of such a symmetry and the "generating" vector fields in S are easier to work with. Usually (but not always - see chapter 13) S turns out to be a finite-dimensional Lie algebra of vector fields under the bracket operation and such Lie algebras are considered in the next section.

5.12 Lie Algebras of Vector Fields

This section is, in a sense, a completion of the work of section 4.13 and, in particular, of theorem 4.13. Let S be a non-trivial real Lie algebra of smooth vector fields on a smooth connected paracompact manifold M. The binary operation on S is the usual Lie bracket of vector fields and, as before, $S_m = \{X(m) : X \in S\} \subseteq T_m M$. Each $X \in S$ gives rise to a family

ϕ_t of local diffeomorphisms of M for appropriate t and S gives rise to a generalised distribution $m \to S_m$ of the type introduced in section 4.13.

Let k be a positive integer, let $X_1, \ldots, X_k \in S$ and let $\phi_t^1, \ldots, \phi_t^k$ be the local diffeomorphisms associated with $X_1, \ldots, X_k$ respectively. *The local group G of local diffeomorphisms of M generated by S* is the set of all maps (where they are defined)

$$m \to \phi_{t_1}^1\left(\phi_{t_2}^2(\ldots\phi_{t_k}^k(m)\ldots)\right) \qquad (m \in M) \tag{5.12}$$

for all choices of k, $X_1, \ldots, X_k$ and $(t_1, \ldots, t_k) \in \mathbb{R}^k$ under the usual formal rules of composition and inverses. The right hand side of (5.12) is defined in an open neighbourhood of m and for $(t_1, \ldots, t_k)$ in some open neighbourhood of the origin in $\mathbb{R}^k$. (The actual algebraic properties of G need not concern us here and the term "local group" will be taken to mean that which is described here and nothing more—see [30]).

The sets S and G lead to a structure on M which can now be discussed. Define an equivalence relation $\sim$ on M by $m_1 \sim m_2$ if and only if there exists $a \in G$ such that $a(m_1) = m_2$. The resulting equivalence classes are the *orbits* in M of (associated with) G (or S) and, from theorem 5.9, coincide (where applicable) to those defined in section 5.8. It would obviously be convenient if the orbits could be given some structure within M and if they could be related to the map $m \to S_m$. This is provided by the following extension of theorem 4.13 due to Hermann [28], Sussmann [30] and Stefan [27].

Theorem 5.14 *Let M be a smooth connected paracompact manifold and let S be a non-trivial Lie algebra of global smooth vector fields on M.*

i) Each orbit of S can be given the structure of a smooth leaf of M.
ii) If S is locally finitely generated, then there exists a unique maximal integral manifold of S through each $m \in M$ and each orbit O' of S is a maximal integral manifold of S with $\dim O' = \dim S_m$ for any $m \in O'$ and is also a leaf of M. Further, S is invariant in the sense that for each $m \in M$ and $f \in G$, $f_* S_m \subseteq S_{f(m)}$.

Again it is remarked for completeness that the conditions and conclusions of (ii) hold in the particular case when S is *finite-dimensional*. The orbits of S are not necessarily regular submanifolds of M. The relationship between the work of this section and that of sections 5.7–5.9, when the members of S are *complete* vector fields, is clear. It is noted that an alternative proof of theorem 5.7 is now available, in the case when G is connected, from theorems 5.9 and 5.14 and the discussion in section 4.11 since the orbits are then leaves of M.

5.13 The Lie Derivative

Let M be a smooth Hausdorff manifold and let T and X be a global smooth tensor field and a global smooth vector field on M, respectively. Let ϕ_t for appropriate t represent the local diffeomorphisms associated with X. A measure of the *change* in T along the integral curves of X in the vicinity of $m \in M$ can be obtained as a kind of Newton quotient involving the differences between $T(m)$ and the pullbacks $\phi_t^* T$ evaluated at m and the parameter t. This leads to a type of derivative of T along X called the *Lie Derivative*. It is a global smooth tensor field on M of the same type as T, denoted by $\mathcal{L}_X T$ and defined by

$$\mathcal{L}_X T(m) = \lim_{t \to 0} \frac{1}{t} [(\phi_t^* T)(m) - T(m)]$$

First it should be noted that this limit always exists [22]. The Lie derivative has the following properties, where f, Y and S are any global smooth real valued function, vector field and tensor field on M, respectively, and a, $b \in \mathbb{R}$ [20],[21],[22]. Here, smooth functions $M \to \mathbb{R}$ are regarded as $(0,0)$ tensors.

i) The operator $\mathcal{L}_X$ is always $\mathbb{R}$-linear and commutes with all contractions.
ii) $\mathcal{L}_X f = X(f)$.
iii) $\mathcal{L}_X Y = [X, Y]$.
iv) $\mathcal{L}_X (S \otimes T) = \mathcal{L}_X S \otimes T + S \otimes \mathcal{L}_X T$.
v) $\mathcal{L}_{aX+bY} T = a\mathcal{L}_X T + b\mathcal{L}_Y T$.
vi) $\mathcal{L}_{[X,Y]} T = \mathcal{L}_X (\mathcal{L}_Y T) - \mathcal{L}_Y (\mathcal{L}_X T)$.

If in some coordinate domain T and X have components $T^{a\ldots b}_{c\ldots d}$ and X^a, respectively, then the tensor $\mathcal{L}_X T$ has components given by

$$\begin{aligned}(\mathcal{L}_X T)^{a\ldots b}_{c\ldots d} =& T^{a\ldots b}_{c\ldots d,e} X^e - T^{e\ldots b}_{c\ldots d} X^a{}_{,e} \cdots - T^{a\ldots e}_{c\ldots d} X^b{}_{,e} \\ &+ T^{a\ldots b}_{e\ldots d} X^e{}_{,c} \cdots + T^{a\ldots b}_{c\ldots e} X^e{}_{,d}\end{aligned} \tag{5.13}$$

The left hand side of (5.13) will usually be written (with a traditional abuse of notation) as $\mathcal{L}_X T^{a\ldots b}_{c\ldots d}$. It is useful to note that if M admits a symmetric connection then the right hand side of (5.13) is unchanged if partial derivatives are replaced by covariant ones (*i.e.* the commas are replaced by semi-colons).

The following results relating the Lie derivative and the maps ϕ_t^* will be important in what is to follow (see *e.g.* [20],[40])

vii) $\phi_t^* T = T$ for all $\phi_t \Leftrightarrow \mathcal{L}_X T = 0$.

viii) $\phi_t^* T = e^{at} T$ for all ϕ_t and some fixed $a \in \mathbb{R} \Leftrightarrow \mathcal{L}_X T = aT$.

ix) $\phi_t^* T = e^{\chi(t)} T$ for all ϕ_t and where χ is a smooth real-valued function with appropriate domain and possibly dependent on the flow of X from which the maps ϕ_t arise $\Leftrightarrow \mathcal{L}_X T = \psi T$ for some smooth real-valued function ψ with appropriate domain.

It will be useful here to collect together some miscellaneous results regarding the members of the Lie algebra S of all global smooth vector fields on M. First let $X \in S$ with associated local diffeomorphisms ϕ_t and let $\phi : M \to M$ be a smooth diffeomorphism on M. Then the pushforward $\phi_* X \in S$ and its local diffeomorphisms are of the form $\phi \circ \phi_t \circ \phi^{-1}$. If $Y \in S$ with associated local diffeomorphisms ψ_t then $[X, Y] \in S$ and its local diffeomorphisms are of the form $\psi_t^{-1} \circ \phi_t^{-1} \circ \psi_t \circ \phi_t$ and from this result and those immediately above it one can show that the following statements are equivalent (where (iv), (v) and (vi) below are assumed to hold for all appropriate s and t) [21].

$$\begin{array}{l} \text{(i) } \mathcal{L}_X Y = 0 \quad \text{(ii) } \mathcal{L}_Y X = 0 \quad \text{(iii) } [X, Y] = 0 \\ \text{(iv) } \phi_{t*} Y = Y \ \text{(v) } \psi_{t*} X = X \ \text{(vi) } \phi_t \circ \psi_s = \psi_s \circ \phi_t \end{array} \tag{5.14}$$

In particular the *commuting condition* (vi) between the local diffeomorphisms associated with X and Y is equivalent to the vanishing of $[X, Y]$. Finally, if M_1 and M_2 are manifolds, $\phi : M_1 \to M_2$ a (smooth) map, X a vector field on M_1, Y a vector field on M_2 with $\phi_* X = Y$ and T a type $(0, s)$ tensor field on M_2 then [21]

$$\phi^*(\mathcal{L}_Y T) = \mathcal{L}_X(\phi^* T). \tag{5.15}$$

Chapter 6

The Lorentz Group

6.1 Minkowski Space

In this section a brief description of *Minkowski space* will be given. This space can be presented in two different ways each based on the set $\mathbb{R}^4$. First one may regard $\mathbb{R}^4$ as a 4-dimensional real vector space together with the Lorentz inner product η and defined by $\eta(\mathbf{u}, \mathbf{v}) = \eta_{ab}u^a v^b$ where $\mathbf{u}$, $\mathbf{v}$ are members of $\mathbb{R}^4$ with components u^a and v^a, respectively, and η_{ab} is the matrix $\text{diag}(-1, 1, 1, 1)$. In this form it will simply be referred to as *Minkowski space*. Second, one could regard $\mathbb{R}^4$ as a 4-dimensional smooth manifold admitting the global smooth Lorentz metric η whose components in the natural global chart on $\mathbb{R}^4$ are η_{ab}. In this form it will be referred to as *Minkowski space-time* because it is thus that it is usually used in special relativity. There is a very natural link between them obtained by noting that the tangent space at any point p of Minkowski space-time is just Minkowski space when taken with the inner product $\eta(p)$. Then the completeness of the Levi-Civita connection associated with η may be used to show that the exponential map at any point of Minkowski space-time provides a smooth diffeomorphism between it and Minkowski space (with its natural manifold structure as a vector space). This identification will always be assumed made.

A non-zero member $\mathbf{v}$ of Minkowski space is called, respectively, *spacelike*, *timelike* or *null* according as $\eta(\mathbf{v}, \mathbf{v})$ is positive, negative or zero. Let S, T and N denote, respectively, the subsets of $\mathbb{R}^4$ consisting of all the spacelike, timelike and null members. Then one has a disjoint decomposition (classification) of Minkowski space in the form $\mathbb{R}^4 = S \cup T \cup N \cup \{0\}$ where 0 is the zero vector of $\mathbb{R}^4$. A 1-dimensional subspace (henceforth called a 1-*space* or a *direction*) of Minkowski space is then called, respectively, *spacelike*, *timelike* or *null* if it is spanned by a spacelike, timelike or null vector. The collections of such subspaces are labelled in an obvious notation by S_1, T_1 and N_1, and this classifies all 1-dimensional subspaces of Minkowski space.

A vector $\mathbf{v}$ in Minkowski space is called a *unit* vector if $|\eta(\mathbf{v},\mathbf{v})| = 1$. A basis $\{\mathbf{e}_a\}$ of Minkowski space consisting of four mutually orthogonal unit vectors is called an *orthonormal* basis (or tetrad) and necessarily one member is timelike and the others spacelike. A basis $\{\mathbf{f}_a\}$ of Minkowski space is called a (*real*) *null basis* (or *tetrad*) if the only non-vanishing inner products between them are $\eta(\mathbf{f}_3,\mathbf{f}_3) = \eta(\mathbf{f}_4,\mathbf{f}_4) = \eta(\mathbf{f}_1,\mathbf{f}_2) = 1$ (so that $\mathbf{f}_1$ and $\mathbf{f}_2$ are null and $\mathbf{f}_3$ and $\mathbf{f}_4$ unit spacelike).

Now consider the set of all 2-dimensional subspaces of Minkowski space (henceforth called 2-*spaces*). This set may be disjointly decomposed in the form $S_2 \cup T_2 \cup N_2$ where the members of S_2 (called *spacelike* 2-*spaces*) contain no null vectors, where the null vectors in any member of N_2 (called a *null* 2-*space*) are confined to a single null direction and where the null vectors in any member of T_2 (called a *timelike 2-space* are confined to two distinct null directions. That this classification is exhaustive follows quickly by assuming a certain 2-space contains three distinct null 1-spaces and obtaining a contradiction from the fact that if $\mathbf{u}$ and $\mathbf{v}$ are orthogonal null vectors then $\mathbf{u} = \lambda\mathbf{v}$ for some $\lambda \in \mathbb{R}$. It easily follows that all non-zero vectors in any member of S_2 are spacelike and that any member of T_2 contains spacelike, timelike and null members. If $V \in N_2$ then V contains a unique null direction, called the *principal null direction* of V, and all other members of V are spacelike and orthogonal to (any non-zero member of) this null direction. The two null directions in a member of T_2 are also referred to as its *principal null directions*. A spanning pair of vectors for any 2-space V may always be chosen orthogonal but if they are, and V is null, one of them must span the principal null direction and if V is timelike one must be spacelike and one timelike. The orthogonal complement of a spacelike 2-space is timelike and vice-versa (here the fact that the only vectors orthogonal to a given timelike vector are spacelike is used) and in this case a spanning pair from each 2-space together give a basis for Minkowski space. The orthogonal complement of a null 2-space V is also null and V and its orthogonal complement $V^\perp$ have the same principal null direction. In this case V and $V^\perp$ intersect in exactly their common principal null direction.

Now consider the set of all 3-dimensional subspaces (3-*spaces*) of Minkowski space. Such a 3-space is called, respectively, *spacelike*, *time-*

like or *null* if its orthogonal complement is a timelike, spacelike or a null 1-space. Alternatively one can characterise such 3-spaces as spacelike if they have no null members, null if their null members are confined to a single null direction (again called its *principal null direction*) and timelike if they contain two (and hence infinitely many) null directions. The collections of such types of subspaces are labelled in an obvious notation by S_3, T_3 and N_3 and their union exhausts all possible 3-spaces. Spanning sets for a spacelike (or timelike) 3-space and its orthogonal complement together give a basis for Minkowski space but the same is not true for a null 3-space V because then $V^\perp$ coincides with the principal null direction of V and is hence contained in it. If $V \in S_3$, all non-zero members of V are spacelike whereas if $V \in T_3$, V contains timelike, spacelike and null members. If $V \in N_3$ then all members of V are either in its principal null direction or else orthogonal to any member of this null direction and spacelike. One may always choose a spanning set of mutually orthogonal vectors for any 3-space V but if V is timelike one of these vectors must be timelike and the others spacelike and if V is null one of them must span its principal null direction.

It is convenient at this point to introduce the idea of a *wave surface* [41]. Let $\mathbf{l}$ be a null vector in Minkowski space. A *wave surface to* $\mathbf{l}$ is a spacelike 2-space each member of which is orthogonal to $\mathbf{l}$. The set of all wave surfaces to $\mathbf{l}$ is denoted by $W(\mathbf{l})$ and clearly $W(\mathbf{l}) \subseteq S_2$. Also a consideration of orthogonal complements shows that any member of S_2 is a wave surface for the non-zero members of exactly two null 1-spaces.

Clearly S and T are open submanifolds of $\mathbb{R}^4$. The subset N is neither open nor closed in $\mathbb{R}^4$ and can be given the structure of a regular 3-dimensional submanifold of $\mathbb{R}^4$ (section 4.11, example (v)). The subset $N \cup \{0\}$ is closed in $\mathbb{R}^4$ and is the (topological) boundary of both S and T. The subset S is connected but T and N are not and this latter remark leads to the concept of 'future and past pointing' and will be discussed in section 6.2. The set N has two components and a consideration of the smooth maps $S^2 \times \mathbb{R}^+ \to \mathbb{R}^4$ given by $(\mathbf{x}, t) \to t(\mathbf{x}, \pm 1)$, whose combined ranges is the regular submanifold N of $\mathbb{R}^4$, shows that each component is diffeomorphic to $S^2 \times \mathbb{R}^+$. The topological and manifold theoretic properties of the sets S_n, T_n, N_n ($n = 1, 2, 3$) and $W(\mathbf{l})$ will be discussed in section 6.3.

6.2 The Lorentz Group

Minkowski space has associated with it an important group of linear transformations which preserve its inner product structure.

Let $\mathcal{L}$ be the set of all linear maps $f : \mathbb{R}^4 \to \mathbb{R}^4$ satisfying

$$\eta(f(\mathbf{x}), f(\mathbf{y})) = \eta(\mathbf{x}, \mathbf{y}) \qquad \forall \mathbf{x}, \mathbf{y} \in \mathbb{R}^4 \tag{6.1}$$

Thus $\mathcal{L}$ is the set of all linear maps $\mathbb{R}^4 \to \mathbb{R}^4$ which preserve the inner product η. It follows from (6.1) that

$$\eta(f(\mathbf{x}), f(\mathbf{x})) = \eta(\mathbf{x}, \mathbf{x}) \qquad \forall \mathbf{x} \in \mathbb{R}^4 \tag{6.2}$$

and so, obviously, f preserves the 'size' associated with η. The condition (6.2) appears at first sight to be weaker than (6.1) but is, in fact, equivalent to it as a consequence of the linearity of the members of $\mathcal{L}$. To see this let $f : \mathbb{R}^4 \to \mathbb{R}^4$ be linear, let x, $y \in \mathbb{R}^4$ and suppose that f satisfies (6.2). Then $\eta(f(\mathbf{x}+\mathbf{y}), f(\mathbf{x}+\mathbf{y})) = \eta(\mathbf{x}+\mathbf{y}, \mathbf{x}+\mathbf{y})$ and on expanding this out one easily achieves (6.1). The set $\mathcal{L}$ forms a group under the usual composition, inversion and identity operations. These are easily checked except, of course, one must show that each $f \in \mathcal{L}$ is bijective and hence has an inverse. So let $f \in \mathcal{L}$ and $\mathbf{y} \in \mathbb{R}^4$ and suppose that $f(\mathbf{y}) = 0$. Then for any $\mathbf{x} \in \mathbb{R}^4$

$$\eta(\mathbf{x}, \mathbf{y}) = \eta(f(\mathbf{x}), f(\mathbf{y})) = 0 \tag{6.3}$$

So $\eta(\mathbf{x}, \mathbf{y}) = 0 \ \forall \mathbf{x} \in \mathbb{R}^4$ and since η is non-degenerate, one finds $\mathbf{y} = 0$. It follows that f is injective and, since it is linear, it is also surjective and hence a bijection $\mathbb{R}^4 \to \mathbb{R}^4$. The corresponding inverse transformation f^{-1} is also a member of $\mathcal{L}$ because it is linear and because

$$\eta(f^{-1}(\mathbf{x}), f^{-1}(\mathbf{y})) = \eta(f(f^{-1}(\mathbf{x})), f(f^{-1}(\mathbf{y}))) = \eta(\mathbf{x}, \mathbf{y}) \tag{6.4}$$

$\forall \mathbf{x}, \mathbf{y} \in \mathbb{R}^4$. The set $\mathcal{L}$ with the binary operation of composition is thus a group called the *Lorentz group* and its members are called *Lorentz transformations*. Each member of $\mathcal{L}$ is an isomorphism of $\mathbb{R}^4$ and can be represented by a non-singular 4×4 real matrix. First choose the standard basis $\{\mathbf{e}_a\}$ of $\mathbb{R}^4$ as orthonormal with respect to η, that is, $\eta(\mathbf{e}_a, \mathbf{e}_b) = \eta_{ab}$. Suppose $f \in \mathcal{L}$ is represented with respect to this basis by the matrix $A = (a_{ab})$ so that (recalling the summation convention) $f(\mathbf{e}_a) = a_{ab}\mathbf{e}_b$. Then if $\mathbf{x}, \mathbf{y} \in \mathbb{R}^4$ with $\mathbf{x} = x_a\mathbf{e}_a$ and $\mathbf{y} = y_a\mathbf{e}_a$ one finds $f(\mathbf{x}) = a_{ab}x_a\mathbf{e}_b$ and $f(\mathbf{y}) = a_{ab}y_a\mathbf{e}_b$ and (6.1) is equivalent to

$$a_{ba}x_b a_{cd}y_c\eta_{ad} = x_b y_c \eta_{bc} \qquad \forall x, y \in \mathbb{R}^4 \tag{6.5}$$

This is equivalent to the matrix relation

$$A\eta A^{\mathrm{T}} = \eta. \tag{6.6}$$

So for a fixed orthonormal basis the transformations in $\mathcal{L}$ are characterised by their representative matrix A satisfying (6.6). The subgroup of members A of $GL(4,\mathbb{R})$ satisfying (6.6) is then isomorphic to the Lorentz group $\mathcal{L}$ according to the map which associates with $f \in \mathcal{L}$ the matrix A (and if $g \in \mathcal{L}$ is represented by the matrix B then $g \circ f$ is represented by AB – but see the remark at the end of section 2.2). If a different basis $\{\tilde{\mathbf{e}}_a\}$ is chosen for $\mathbb{R}^4$ with $\tilde{\mathbf{e}}_a = \Sigma t_{ab}\mathbf{e}_b$ ($T = (t_{ab}) \in GL(4,\mathbb{R})$) and $h_{ab} = \eta(\tilde{\mathbf{e}}_a, \tilde{\mathbf{e}}_b)$ then if the above $f \in \mathcal{L}$ has matrix $\tilde{A}$ with respect to the basis $\{\tilde{\mathbf{e}}_a\}$ then (6.1) is equivalent to

$$\tilde{A}h\tilde{A}^{\mathrm{T}} = h$$

and A and $\tilde{A}$ are related by $\tilde{A} = TAT^{-1}$.

These equations describe the Lorentz group in a different basis to the standard basis of $\mathbb{R}^4$. It can be regarded as a different representation of $\mathcal{L}$ (*i.e.* an isomorphic copy of $\mathcal{L}$) within $GL(4,\mathbb{R})$ under the isomorphism $A \to \tilde{A}$ determined by T.

Minkowski space has been viewed here as a real (Lorentz) inner product space. It is sometimes convenient to extend the inner product η to the complex vector space $\mathbb{C}^4$. If $\mathbf{z}, \mathbf{w} \in \mathbb{C}^4$ have components z_a and w_a, respectively, in the usual basis for $\mathbb{C}^4$ define $\eta(\mathbf{z}, \mathbf{w}) = \eta_{ab}z_a w_b$. This will be considered as a formal device for dealing with situations where the complex field unavoidably arises (such as in the next theorem). The terms *null* and *orthogonal* will still be used for complex vectors in an obvious way, using η.

The following theorem collects together some algebraic information about $\mathcal{L}$. In it, the term invariant 2-space will always refer to a *real* 2-space (*i.e.* a subspace of $\mathbb{R}^4$). The usual (fixed) basis is chosen for $\mathbb{R}^4$ (and hence $\mathbb{C}^4$). A Lorentz transformation is then identified with its matrix, and a vector with its components, in this basis. In compiling this theorem the papers [42]-[44] were useful.

Theorem 6.1

i) $A \in \mathcal{L} \Rightarrow \det A = \pm 1$

ii) $A \in \mathcal{L} \Rightarrow A^T \in \mathcal{L}$

iii) *If* $\alpha \in \mathbb{C}$ *is an eigenvalue of* $A \in \mathcal{L}$ *then* $\alpha \neq 0$ *and* α^{-1} *is an eigenvalue of* A^{-1} *and also of* A^T *and* A.

iv) If $\alpha \in \mathbb{C}$ is an eigenvalue of $A \in \mathcal{L}$ with a corresponding non-null eigenvector then $|\alpha| = 1$. The result is false if the non-null condition is dropped.

v) If $\mathbf{r}$ and $\mathbf{s}$ are eigenvectors of $A \in \mathcal{L}$ with eigenvalues α and β, respectively, then either $\eta(\mathbf{r}, \mathbf{s}) = 0$ or $\beta = \alpha^{-1}$.

vi) If $\alpha \in \mathbb{C} \backslash \mathbb{R}$ is an eigenvalue of $A \in \mathcal{L}$ then $|\alpha| = 1$ and the corresponding complex eigenvector is null with its real and imaginary parts orthogonal and spacelike and spanning a spacelike invariant 2-space of A.

vii) A $(\in \mathcal{L})$ always admits an orthogonal pair of invariant 2-spaces. A timelike invariant 2-space of A always contains two distinct real eigendirections. If A admits a spacelike invariant 2-space then either it contains two distinct real eigendirections or else A admits two (conjugate) complex null eigendirections whose real and imaginary parts span this 2-space. A null invariant 2-space of A always contains a real null eigendirection and possibly another (necessarily real spacelike) eigendirection. If f admits no non-null invariant 2-spaces then all its invariant 2-spaces are null with the same principal null direction and (from part (vi)) all its eigenvalues are real.

Proof. Part (i) is clear from (6.6). To prove (ii) take the inverse of each side of (6.6) to see that $(A^{-1})^{\mathrm{T}} \in \mathcal{L}$ and hence, by the group property, that $A^{\mathrm{T}} \in \mathcal{L}$. The first statement in (iii) is true for all non-singular matrices. The next part follows because if $\mathbf{x} \in \mathbb{C}^4$ is an eigenvector of A with eigenvalue $\alpha \in \mathbb{C}$ then $\mathbf{x}A = \alpha\mathbf{x}$ and multiplying on the right by A^{-1} gives $\mathbf{x}A^{-1} = \alpha^{-1}\mathbf{x}$ and so $\mathbf{x}$ is an eigenvector of A^{-1} with eigenvalue α^{-1}. Next, multiplying (6.6) on the left by $\mathbf{x}$ gives $(\mathbf{x}\eta)A^{\mathrm{T}} = \alpha^{-1}(\mathbf{x}\eta)$ and so $\mathbf{x}\eta$ is an eigenvector of A^{T} with eigenvalue α^{-1}. That α^{-1} is an eigenvalue of A follows because any matrix is similar to its transpose. To establish the first part of (iv) one has (using the notation of the proof of (iii)) $\eta(\mathbf{x}A, \mathbf{x}A) = \eta(\mathbf{x}, \mathbf{x})$ and on expanding this one finds $\eta(\mathbf{x}, \mathbf{x}) = \alpha^2\eta(\mathbf{x}, \mathbf{x})$ and the result follows. A counter example for the second part will be seen later. For (v) one has $\mathbf{r}A = \alpha\mathbf{r}$ and $\mathbf{s}A = \beta\mathbf{s}$ and so $\eta(\mathbf{r}A, \mathbf{s}A) = \alpha\beta\eta(\mathbf{r}, \mathbf{s})$. This leads to $\eta(\mathbf{r}, \mathbf{s}) = \alpha\beta\eta(\mathbf{r}, \mathbf{s})$ and the result follows. To establish (vi) let $\mathbf{x}$ be an eigenvector of A with eigenvalue $\alpha \in \mathbb{C} \backslash \mathbb{R}$. Then one has $\mathbf{x}A = \alpha\mathbf{x}$, $\bar{\mathbf{x}}A = \bar{\alpha}\bar{\mathbf{x}}$ and, as in the previous two proofs, $\eta(\mathbf{x}, \mathbf{x}) = \alpha^2\eta(\mathbf{x}, \mathbf{x})$ and $\eta(\mathbf{x}, \bar{\mathbf{x}}) = |\alpha|^2\eta(\mathbf{x}, \bar{\mathbf{x}})$. If $\mathbf{x}$ is not null then the first of these gives $\alpha^2 = 1$ and the contradiction that $\alpha \in \mathbb{R}$. So $\mathbf{x}$ is null. If, in addition, $\eta(\mathbf{x}, \bar{\mathbf{x}}) = 0$ then writing $\mathbf{x}$ in terms of its real and imaginary parts, $\mathbf{x} = \mathbf{a} + i\mathbf{b}$ for $\mathbf{a}, \mathbf{b} \in \mathbb{R}^4$, one finds that $\eta(\mathbf{a}, \mathbf{a}) = \eta(\mathbf{b}, \mathbf{b}) = \eta(\mathbf{a}, \mathbf{b}) = 0$ and so $\mathbf{x}$ is a (complex) multiple of a real null vector. This again implies that $\alpha \in \mathbb{R}$. Thus $\eta(\mathbf{x}, \bar{\mathbf{x}}) \neq 0$ and hence $|\alpha| = 1$. For the final part of (vi) one has a null eigenvector $\mathbf{x} = \mathbf{a} + i\mathbf{b}$ as above and so $\eta(\mathbf{x}, \mathbf{x}) = 0 \Rightarrow \eta(\mathbf{a}, \mathbf{a}) = \eta(\mathbf{b}, \mathbf{b})$ and $\eta(\mathbf{a}, \mathbf{b}) = 0$. Also the fact that the corresponding eigenvalue $\alpha \in \mathbb{C}\backslash\mathbb{R}$ means that neither $\mathbf{a}$ nor $\mathbf{b}$ is null. It follows that they are spacelike (and orthogonal). Finally, writing $\alpha = \gamma + i\delta$ (γ, $\delta \in \mathbb{R}$), the real and imaginary parts of the equation $(\mathbf{a} + i\mathbf{b})A = (\gamma + i\delta)(\mathbf{a} + i\mathbf{b})$ show that $\mathbf{a}$ and $\mathbf{b}$ span an invariant 2-space of A (which the previous sentence shows is spacelike). To prove (vii) one notes that an invariant 2-space always exists (section 2.6) and that *its orthogonal complement is also invariant* follows from the definition of the Lorentz group. So let U be an invariant timelike 2-space. Then U contains exactly two null 1-spaces spanned say by $\mathbf{l}$ and $\mathbf{n}$ which are chosen such that $\eta(\mathbf{l}, \mathbf{n}) = 1$. By the definition of the Lorentz group either these 1-spaces are the required eigendirections or $\mathbf{l}A = \lambda\mathbf{n}$, $\mathbf{n}A = \lambda^{-1}\mathbf{l}$ ($0 \neq \lambda \in \mathbb{R}$). In this case $\mathbf{l} \pm \lambda\mathbf{n}$ are real, non-null, orthogonal eigenvectors of A contained in U (with eigenvalues ± 1). If U is a spacelike invariant 2-space the restriction of a Lorentz transformation to U is a linear orthogonal transformation on a copy of 2-dimensional Euclidean space and the result follows. If U is a null invariant 2-space the definition of a Lorentz transformation and the uniqueness of the principal null direction of U shows that this latter null direction is a real null eigendirection of A. Later it will be seen that other (necessarily spacelike) eigendirections may, but need not, exist in U. Finally if f admits null invariant 2-spaces with distinct principal null directions any two of these latter directions give rise to an invariant timelike 2-space. This contradiction completes the proof.□

Let $\mathbf{a}$, $\mathbf{b}$ and $\mathbf{c}$ be, respectively, a timelike, a spacelike and a null vector in Minkowski space. Then it is straightforward to show that there exist f_1, f_2 and $f_3 \in \mathcal{L}$ such that $f_1(\mathbf{a}) = (\mu, 0, 0, 0)$, $f_2(\mathbf{b}) = (0, 0, 0, \nu)$ and $f_3(\mathbf{c}) = (1, 0, 0, 1)$ for appropriate positive real numbers μ and ν. By using this result it is easy to show that if V is any of the sets S_k, T_k or N_k ($k = 1, 2, 3$) or the set of all wave surfaces or the set of orthonormal bases or the set of all null vectors and if $f \in \mathcal{L}$ then, in an obvious notation, $f(V) = V$. Further $\mathcal{L}$ is *transitive* on V in the sense that if $\mathbf{x}, \mathbf{y} \in V$ there exists $f \in \mathcal{L}$ such that $f(\mathbf{x}) = \mathbf{y}$. The same is also true for the set of all timelike or spacelike vectors $\mathbf{v}$ of a fixed "size" $\eta(\mathbf{v}, \mathbf{v})$.

There is a decomposition of the Lorentz group which can now be described. Consider the set T of all timelike vectors in Minkowski space. Then it is easily shown by considering continuous curves in the open submanifold

T of $\mathbb{R}^4$ that T has two path components and hence two components T' and T'' and that in some (any) orthonormal frame these components are distinguished by the sign of the first component of the member of T. Now define, with the obvious topologies, a continuous map $f : T \times T \to \mathbb{R}\backslash\{0\}$ by $f(\mathbf{u}, \mathbf{v}) = \eta(\mathbf{u}, \mathbf{v})$. Clearly f takes positive and negative values. Now $T \times T$ has four components and it follows since $\mathbf{v} \in T \Rightarrow \eta(\mathbf{v}, \mathbf{v}) < 0$ and since $\mathbf{v} \in T' \Rightarrow -\mathbf{v} \in T''$ (and vice versa) that f maps the connected subsets $T' \times T'$ and $T'' \times T''$ to the negative reals and the connected subsets $T' \times T''$ and $T'' \times T'$ to the positive reals. Thus the relation $\sim$ on T defined for $\mathbf{u}, \mathbf{v} \in T$ by $\mathbf{u} \sim \mathbf{v} \Leftrightarrow \eta(\mathbf{u}, \mathbf{v}) < 0$ is an equivalence relation on T with two equivalence classes T' and T''. A similar argument then shows that the set N of null vectors in Minkowski space also has two components and a continuity argument shows that it satisfies the condition that these components may be labelled N' and N'' in such a way that $k \in N'$ (respectively $k \in N''$) $\Leftrightarrow \eta(\mathbf{k}, \mathbf{u}) < 0\ \forall \mathbf{u} \in T'$ (respectively $\forall \mathbf{u} \in T''$). Then the set C of all timelike and null vectors may be partitioned as $C = C_1 \cup C_2$ where $C_1 = T' \cup N'$ and $C_2 = T'' \cup N''$. The members of one of these subsets, say C_1, are called *future pointing* and those of the other *past pointing*.

Now let $f \in \mathcal{L}$. From the previous paragraph it is clear that f maps each of the partitions C_1 and C_2 onto itself or maps each one onto the other. In the former case f is called *future preserving* and in the latter case, *future reversing*. The future preserving members of $\mathcal{L}$ form a subgroup of $\mathcal{L}$ but the future reversing members do not. Now return to theorem 6.1 noting that each $A \in \mathcal{L}$ (in the usual representation) satisfies $det\ A = \pm 1$, the sign being independent of the orthonormal base chosen. Those with $det\ A = 1$ also form a subgroup of $\mathcal{L}$ whilst those with $det\ A = -1$ do not. The group $\mathcal{L}$ can now be decomposed into four disjoint subsets $\mathcal{L} = \mathcal{L}_+^\uparrow \cup \mathcal{L}_+^\downarrow \cup \mathcal{L}_-^\uparrow \cup \mathcal{L}_-^\downarrow$ where the superscripts $\uparrow$ and $\downarrow$ refer, respectively, to future preserving and future reversing and the subscripts $+$ and $-$ to the sign of the determinant. In this decomposition only $\mathcal{L}_+^\uparrow$ is a subgroup of $\mathcal{L}$ and it is particularly important. It is described by various adjectives in the literature ("proper", "orthochronous") and will be given the symbol $\mathcal{L}_0$ here. Its significance will become more apparent later when $\mathcal{L}$ is considered as a Lie group. The subgroup $\mathcal{L}_0$ turns out to be a normal subgroup of $\mathcal{L}$ and $\mathcal{L}_+^\downarrow$, $\mathcal{L}_-^\uparrow$ and $\mathcal{L}_-^\downarrow$ are the cosets of $\mathcal{L}_0$ in $\mathcal{L}$ which, in the matrix representation, can be obtained from $\mathcal{L}_0$ by multiplying on the left (or right), respectively, by the matrices

$$\begin{pmatrix} -1 & 0 & 0 & 0 \\ 0 & 1 & 0 & 0 \\ 0 & 0 & 1 & 0 \\ 0 & 0 & 0 & -1 \end{pmatrix} \in \mathcal{L}_+^{\downarrow} \quad \begin{pmatrix} 1 & 0 & 0 & 0 \\ 0 & 1 & 0 & 0 \\ 0 & 0 & 1 & 0 \\ 0 & 0 & 0 & -1 \end{pmatrix} \in \mathcal{L}_-^{\uparrow} \quad \begin{pmatrix} -1 & 0 & 0 & 0 \\ 0 & 1 & 0 & 0 \\ 0 & 0 & 1 & 0 \\ 0 & 0 & 0 & 1 \end{pmatrix} \in \mathcal{L}_-^{\downarrow} \tag{6.7}$$

The quotient group $\mathcal{L}/\mathcal{L}_0$ is isomorphic to the non-cyclic group (the Klein 4-group - see section 2.2) of order four.

These remarks together with theorem 6.1 can be used to give the following algebraic classification of members of $\mathcal{L}$ and $\mathcal{L}_0$. Again the liberty will be taken of extending the inner product η from $\mathbb{R}^4$ to $\mathbb{C}^4$ and regarding members of $\mathcal{L}$ as linear maps $\mathbb{C}^4 \to \mathbb{C}^4$ to facilitate the inclusion of complex eigenvectors and eigenvalues.

Theorem 6.2 *Let $f \in \mathcal{L}$. Then either*

i) f admits four independent (complex) eigendirections of which either exactly two can be chosen real or all four are real, or
ii) f admits exactly two independent eigendirections, in which case they are both real and span a null invariant 2-space of f and so f admits a unique null eigendirection. The corresponding (real) eigenvalues are either 1 or -1 and may be equal.

In (i) f is diagonable over $\mathbb{C}$ in the first case with Segre type $\{11z\bar{z}\}$ or $\{(11)z\bar{z}\}$ whilst in the second case f is diagonable over $\mathbb{R}$ with Segre type $\{1111\}$ or one of its degeneracies and at least two of the eigenvalues are each equal to ± 1. In (ii) the Segre type of f is $\{31\}$ or its degeneracy with the non-simple elementary divisor corresponding to the null eigendirection and the simple elementary divisor corresponding to a spacelike eigendirection. If the condition that $f \in \mathcal{L}_0$ is imposed the eigenvalues in (i) are of the general form e^{λ}, $e^{-\lambda}$, $e^{i\theta}$, $e^{-i\theta}$ where λ, $\theta \in \mathbb{R}$ and in (ii) the eigenvalues are both equal to 1. The allowed Segre types are then $\{11z\bar{z}\}$, $\{(11)z\bar{z}\}$, $\{11(11)\}$, $\{(11)(11)\}$ or $\{(1111)\}$ in (i) and $\{(31)\}$ in (ii).

Proof. Suppose f admits a non-null (and hence an orthogonal pair of non-null) invariant 2-spaces. Then the previous theorem shows that f is diagonalisable over $\mathbb{C}$ or $\mathbb{R}$. In the former case exactly two of the eigenvalues must be real and in the latter case all four are real and at least two of them must have unit modulus. If two eigenvalues are not of unit modulus they are inverses of each other (theorem 6.1). If f admits no non-null invariant 2-spaces then it admits only null ones with a common principal null direction (spanned, say, by $\mathbf{l}$ which is then an eigendirection of f) and all its eigenvalues are real (theorem 6.1). It is clear that f is not diagonalisable (over $\mathbb{R}$) because it would then necessarily admit a non-null invariant 2-space spanned by an appropriate pair of eigenvectors. If f admits *three* independent (real) eigenvectors (one of which may be taken as $\mathbf{l}$) then a non-null invariant 2-space of f necessarily results and hence a contradiction. If f admits a *single* eigendirection (necessarily spanned by $\mathbf{l}$) then its Segre type is $\{4\}$. But then f admits (at least) an orthogonal pair of null invariant 2-spaces spanned say by $\mathbf{l}$ and $\mathbf{x}$ and by $\mathbf{l}$ and $\mathbf{y}$ with $\mathbf{x}$ and $\mathbf{y}$ unit orthogonal spacelike vectors each orthogonal to $\mathbf{l}$. The properties of $\mathcal{L}$ together with part (iii) of the previous theorem show that one then has

$$f(\mathbf{l}) = \epsilon_1 \mathbf{l}, \qquad f(\mathbf{x}) = \epsilon_2 \mathbf{x} + b\mathbf{l}, \qquad f(\mathbf{y}) = \epsilon_3 \mathbf{y} + d\mathbf{l}$$

where ϵ_1, ϵ_2 and $\epsilon_3 = \pm 1$ and b, d are non-zero real numbers. Two possibilities arise: either ϵ_1 differs from at least one of ϵ_2 and ϵ_3 or $\epsilon_1 = \epsilon_2 = \epsilon_3$. In the former case suppose, say, that $\epsilon_1 = -\epsilon_2$. Then $\epsilon_1 \mathbf{x} - \frac{1}{2} b\mathbf{l}$ is an eigenvector of f (with eigenvalue $-\epsilon_1$) contradicting the uniqueness of the eigendirection spanned by $\mathbf{l}$. In the latter case, $\epsilon_1 = \epsilon_2 = \epsilon_3 = \epsilon$ say, and one sees that $d\mathbf{x} - b\mathbf{y}$ is an eigenvector of f with eigenvalue ϵ and again a contradiction is achieved. The conclusion is that f admits exactly two (necessarily real) eigendirections and hence that its Segre type is either $\{31\}$ or $\{22\}$. But the Jordan matrix associated with type $\{22\}$ easily reveals two invariant 2-spaces of f which intersect only in the zero vector and so this type is impossible since, now, no non-null invariant 2-spaces of f exist. Hence the Segre type is $\{31\}$ or its degeneracy and by theorem 6.1 (iii) and (iv) (since now there exists a non-null eigenvector) the eigenvalues may only be ± 1. If these two eigenvalues are distinct one can take the eigenvectors as a null vector $\mathbf{l}$ and a unit spacelike vector $\mathbf{x}$ which are orthogonal and have eigenvalues $+1$ and -1 (or vice versa). By extending $\mathbf{l}$ and $\mathbf{x}$ to a null tetrad a calculation shows that either an extra (forbidden) eigenvector arises in the null 2-space orthogonal to that spanned by $\mathbf{l}$ and $\mathbf{x}$ or else the determinant of the matrix representing f is -1. Hence if $f \in \mathcal{L}_0$ the Segre type is $\{(31)\}$ and the corresponding eigenvalue is unity since f is future preserving. A consideration of the Jordan canonical form for the matrix representing f together with the fact that all invariant 2-spaces of f are null and contain $\mathbf{l}$ then shows that $\mathbf{l}$ corresponds to the non-simple elementary divisor and $\mathbf{x}$ to the simple one. □

The above possibilities for $f \in \mathcal{L}_0$ ($f \neq$ identity) can be conveniently (and conventionally) represented as follows (*cf.*[42]) by piecing together the results of the previous two theorems.

$$f(\mathbf{l}) = e^{\lambda}\mathbf{l} \qquad f(\mathbf{n}) = e^{-\lambda}\mathbf{n} \qquad (0 \neq \lambda \in \mathbb{R}) \tag{6.8a}$$
$$f(\mathbf{x}) = \mathbf{x} \qquad f(\mathbf{y}) = \mathbf{y}$$

$$f(\mathbf{l}) = \mathbf{l} \qquad f(\mathbf{n}) = \mathbf{n} \tag{6.8b}$$
$$f(\mathbf{m}) = e^{i\theta}\mathbf{m} \qquad f(\bar{\mathbf{m}}) = e^{-i\theta}\bar{\mathbf{m}} \qquad (0 < \theta < 2\pi)$$

$$f(\mathbf{l}) = e^{\lambda}\mathbf{l} \qquad f(\mathbf{n}) = e^{-\lambda}\mathbf{n} \qquad (0 \neq \lambda \in \mathbb{R}) \tag{6.8c}$$
$$f(\mathbf{m}) = e^{i\theta}\mathbf{m} \qquad f(\bar{\mathbf{m}}) = e^{-i\theta}\bar{\mathbf{m}} \qquad (0 < \theta < 2\pi)$$

$$f(\mathbf{l}) = \mathbf{l} \qquad f(\mathbf{n}) = \mathbf{n} - \gamma\mathbf{y} - \frac{1}{2}\gamma^2\mathbf{l} \qquad (0 \neq \gamma \in \mathbb{R}) \tag{6.8d}$$
$$f(\mathbf{x}) = \mathbf{x} \qquad f(\mathbf{y}) = \mathbf{y} + \gamma\mathbf{l}$$

In these equations $(\mathbf{l}, \mathbf{n}, \mathbf{x}, \mathbf{y})$ is an appropriate null tetrad with $\mathbf{l}$ and $\mathbf{n}$ null vectors and the complex null vector $\mathbf{m}$ is given by $\sqrt{2}\mathbf{m} = \mathbf{x} + i\mathbf{y}$ (so that $\eta(\mathbf{m}, \bar{\mathbf{m}}) = 1$). In (6.8a) the Segre type is $\{11(11)\}$ with eigenvectors $\mathbf{l}$, $\mathbf{n}$, $\mathbf{x}$, $\mathbf{y}$ and corresponding eigenvalues e^{λ}, $e^{-\lambda}$, 1,1. In (6.8b) the Segre type is $\{(11)z\bar{z}\}$ with eigenvectors $\mathbf{l}$, $\mathbf{n}$, $\mathbf{m}$ and $\bar{\mathbf{m}}$ and corresponding eigenvalues 1, 1, $e^{i\theta}$, $e^{-i\theta}$ $(0 < \theta < 2\pi)$ except when $\theta = \pi$ when the Segre type is $\{(11)(11)\}$. In (6.8c) the Segre type is $\{11z\bar{z}\}$ with eigenvectors l, n, m, $\bar{m}$ and corresponding eigenvalues e^{λ}, $e^{-\lambda}$, $e^{i\theta}$, $e^{-i\theta}$ except when $\theta = \pi$ when the Segre type is $\{11(11)\}$. In (6.8d) the Segre type is $\{(31)\}$ with eigenvectors $\mathbf{l}$ and $\mathbf{x}$ and corresponding eigenvalues each equal to 1 and $\gamma \in \mathbb{R}$, $\gamma \neq 0$. The null invariant 2-spaces spanned by the pairs $\mathbf{l}, \mathbf{y}$ and $\mathbf{l}, \mathbf{x}$ contain one independent eigenvector and two independent eigenvectors, respectively. The identity member of $\mathcal{L}_0$ has Segre type $\{(1111)\}$. The transformations (6.8a) are called *boosts* (in the $\mathbf{ln}$ plane) whilst those in (6.8b) are called *(spatial) rotations* (in the $\mathbf{xy}$ plane) with angle θ. A transformation of type (6.8c) is called a *screw motion* whilst one of type (6.8d) is called a *null rotation* (about $\mathbf{l}$).

From the above results one can see that every (non-identity) member of $\mathcal{L}_0$ "fixes" at least one and at most two real null directions (*i.e.* admits at least one and at most two real null eigendirections). Both these results are false if $\mathcal{L}_0$ is replaced by $\mathcal{L}$ (simply consider the transformations $f(\mathbf{l}) = \mathbf{n}$, $f(\mathbf{n}) = \mathbf{l}$, $f(\mathbf{m}) = \mathbf{m}$ and $f(\mathbf{l}) = \mathbf{l}$, $f(\mathbf{n}) = \mathbf{n}$, $f(\mathbf{x}) = \mathbf{x}$, $f(\mathbf{y}) = -\mathbf{y}$ where in each case $f \in \mathcal{L}\backslash\mathcal{L}_0$). Also if $f \in \mathcal{L}_0$ ($f \neq$ identity) admits an orthogonal pair of invariant *non-null* 2-spaces then this pair is unique. Again this result is false if $\mathcal{L}_0$ is replaced by $\mathcal{L}$ (as the second of the above examples shows). Note also that the member $f \in \mathcal{L}\backslash\mathcal{L}_0$ given by

$$f(\mathbf{l}) = -\mathbf{l} \qquad f(\mathbf{n}) = -\mathbf{n} + \frac{1}{2}d^2\mathbf{l} + d\mathbf{y} \tag{6.8e}$$
$$f(\mathbf{x}) = \mathbf{x} \qquad f(\mathbf{y}) = -\mathbf{y} - d\mathbf{l}$$

where $d \in \mathbb{R}$, $d \neq 0$, has Segre type $\{31\}$ with eigenvalues ± 1. Finally the null rotation (6.8d) can, by changing to the real null tetrad $(\mathbf{l}', \mathbf{n}', \mathbf{x}', \mathbf{y}')$ where $\mathbf{l}' = \gamma\mathbf{l}$, $\mathbf{n}' = \gamma^{-1}\mathbf{n}$, $\mathbf{x}' = \mathbf{x}$, $\mathbf{y}' = \mathbf{y}$, be written in the simple form (*cf.*[42])

$$f(\mathbf{l}') = \mathbf{l}' \qquad f(\mathbf{n}') = \mathbf{n}' - \mathbf{y}' - \frac{1}{2}\mathbf{l}' \tag{6.8f}$$
$$f(\mathbf{x}') = \mathbf{x}' \qquad f(\mathbf{y}') = \mathbf{y}' + \mathbf{l}'$$

6.3 The Lorentz Group as a Lie Group

In this section $\mathcal{L}$ will be regarded as the subgroup of $GL(4,\mathbb{R})$ whose members A satisfy (6.6). Now $GL(4,\mathbb{R})$ is a Lie group of dimension 16 (section 5.2) and the map $g : GL(4,\mathbb{R}) \to M_4\mathbb{R}$ given by $A \to A\eta A^T - \eta$ is a smooth map such that $g^{-1}\{O\} = \mathcal{L}$ where O is the zero matrix in $M_4\mathbb{R}$. Hence $\mathcal{L}$ *is a closed subgroup* of $GL(4,\mathbb{R})$ and, since it is not discrete, it follows from theorem 5.1 that $\mathcal{L}$ *can be given a unique structure as a regular submanifold of* $GL(4,\mathbb{R})$ *and is then a Lie subgroup of* $GL(4,\mathbb{R})$.

There is another construction which gives more detail about $\mathcal{L}$, in particular, its dimension and its Lie algebra. Let $S(4,\mathbb{R})$ denote the set of symmetric 4×4 real matrices with its standard 10-dimensional manifold structure (section 4.3) and let ψ be the smooth map $\psi : GL(4,\mathbb{R}) \to S(4,\mathbb{R})$ given by $A \to A\eta A^T$. If x and y are the standard charts on $GL(4,\mathbb{R})$ and $S(4,\mathbb{R})$ with coordinate functions x_{ab} and y_{ab}, respectively, then $y_{ab} \circ \psi = x_{ac}\eta_{cd}x_{bd}$. Now compute ψ_{*e}, where e is the identity of $GL(4,\mathbb{R})$, by computing its action on a general member of $T_eGL(4,\mathbb{R})$ represented by

$\alpha_{ef}(\frac{\partial}{\partial x_{ef}})_e$ $(\alpha_{ef} \in M_4\mathbb{R})$

$$\begin{aligned}\psi_{*e}(\alpha_{ef}(\frac{\partial}{\partial x_{ef}})_e) &= \alpha_{ef}\psi_{*e}(\frac{\partial}{\partial x_{ef}})_e \\ &= \sum_{a\leq b}\alpha_{ef}(\frac{\partial}{\partial x_{ef}}(y_{ab}\circ\psi))_e(\frac{\partial}{\partial y_{ab}})_{\psi(e)} \\ &= \sum_{a\leq b}\alpha_{ef}(\eta_{ae}\delta_{bf}+\eta_{eb}\delta_{af})(\frac{\partial}{\partial y_{ab}})_{\psi(e)} \\ &= \sum_{a\leq b}(\alpha_{be}\eta_{ae}+\alpha_{ae}\eta_{eb})(\frac{\partial}{\partial y_{ab}})_{\psi(e)}\end{aligned} \tag{6.9}$$

From this result two important conclusions follow. Firstly, it is easy to check from the final step in (6.9) that ψ_{*e} is *surjective* and hence has rank 10 ($= \dim S(4,\mathbb{R})$). Also it follows from the definition of ψ that $\psi^{-1}(\psi(e)) = \mathcal{L}$ and that if $A \in GL(4,\mathbb{R})$ and $\tilde{A} \in \mathcal{L}$ then $\psi(A\tilde{A}) = \psi(A)$. Hence from theorem 5.2 it follows once again that $\mathcal{L}$ can be given the structure of a Lie subgroup of $GL(4,\mathbb{R})$ with an underlying *regular* submanifold structure. Further, the dimension of $\mathcal{L}$ equals $\dim GL(4,\mathbb{R}) - \dim S(4,\mathbb{R}) = 6$. The uniqueness remark at the end of the previous paragraph shows that the Lie group structures for $\mathcal{L}$ obtained here and in that paragraph are the same (*cf.*section 4.11 example (viii)) and this structure is then the conventional 6-dimensional Lie group structure for $\mathcal{L}$. Secondly, it also follows from section 5.4 that the *Lie algebra* L of $\mathcal{L}$ can be identified with the subalgebra $\psi_{*e}^{-1}(O)$ of the Lie algebra of $GL(4,\mathbb{R})$ and so from (6.9)

$$L = \{\alpha \in M_4\mathbb{R} : \alpha\eta + (\alpha\eta)^T = 0\} \tag{6.10}$$

with the Lie product being the usual matrix commutation as for $GL(4,\mathbb{R})$. Thus the members of L are those members of $M_4\mathbb{R}$ which are skew self-adjoint with respect to η or, in more familiar language, if one thinks of a member of $M_4\mathbb{R}$ as the components of a $(1,1)$ 'tensor' then L is the set of such tensors which, when the upper index is lowered using η, becomes a skew symmetric 'tensor.' L is called the *Lorentz algebra*.

With the knowledge that $\mathcal{L}$ is a Lie group consider the decomposition $\mathcal{L} = \mathcal{L}_0 \cup \mathcal{L}_-^{\uparrow} \cup \mathcal{L}_+^{\downarrow} \cup \mathcal{L}_-^{\downarrow}$ discussed earlier. Now equations (6.8a)-(6.8d) (and regarding $\mathcal{L}$ as a group of matrices as described earlier) show that any member of $\mathcal{L}_0$ can be joined by a continuous curve (in the topological space $GL(4,\mathbb{R})$ and lying in $\mathcal{L}_0$) to the identity e of $\mathcal{L}_0$. Thus since $\mathcal{L}$ is regular, $\mathcal{L}_0$ is a path connected (and hence a connected) subspace and subgroup of $\mathcal{L}$. Now let $\mathbf{v}$ be a fixed future-pointing timelike vector and consider the smooth maps $f : \mathcal{L} \to \mathbb{R}^4$ and $g : \mathcal{L} \to \mathbb{R}$ given by $f : A \to \mathbf{v}A$ and $g : A \to \det A$. Now let $U \subseteq \mathbb{R}^4$ be the subset of future-pointing timelike vectors. It follows that $f^{-1}U$ and $g^{-1}\mathbb{R}^+$ are open in $\mathcal{L}$ and that $\mathcal{L}_0 = f^{-1}U \cap g^{-1}\mathbb{R}^+$ is open in $\mathcal{L}$. Thus from section 5.2, since $\mathcal{L}_0$ is open, connected and contains the identity of $\mathcal{L}$, it follows that *$\mathcal{L}_0$ is the identity component of $\mathcal{L}$* and that $\mathcal{L}_0$ is diffeomorphic to its cosets $\mathcal{L}_-^\uparrow$, $\mathcal{L}_+^\downarrow$ and $\mathcal{L}_-^\downarrow$. Thus *$\mathcal{L}_0$ is a connected Lie subgroup of $\mathcal{L}$ and of $GL(4, \mathbb{R})$ with Lie algebra L.*

It was pointed out in section 5.7 that $GL(4, \mathbb{R})$ acts on $\mathbb{R}^4$ on the right as a Lie transformation group. Since $\mathcal{L}$ and $\mathcal{L}_0$ are Lie subgroups of $GL(4, \mathbb{R})$ these Lie groups also similarly act on $\mathbb{R}^4$ as Lie transformation groups. So let $\phi : \mathbb{R}^4 \times \mathcal{L}_0 \to \mathbb{R}^4$, where $\phi(\mathbf{v}, A) = \mathbf{v}A$, denote the action of the connected 6-dimensional Lie group $\mathcal{L}_0$ on Minkowski space $\mathbb{R}^4$. A consideration of this action is rather useful in determining the topological properties of certain features of Minkowski space. First one notes, from the linearity of the members of $\mathcal{L}_0$, that each $f \in \mathcal{L}_0$ induces a natural map (also denoted by f) $f : Z \to Z$ where Z is any of the sets S_n, T_n or N_n $(n = 1, 2, 3)$ of Minkowski space described earlier. Also one has transitivity in each case for if $\mathbf{a}, \mathbf{b} \in Z$, there exists $f \in \mathcal{L}_0$ such that $f(\mathbf{a}) = \mathbf{b}$. These maps can be explored more carefully by noting that the original action of $\mathcal{L}_0$ on $\mathbb{R}^4$ leads in a natural way to right actions of $\mathcal{L}_0$ on $\mathbb{R}^4 \times \mathbb{R}^4$ and on $\mathbb{R}^4 \times \mathbb{R}^4 \times \mathbb{R}^4$ as a Lie transformation group where the former is given for $\mathbf{u}, \mathbf{v} \in \mathbb{R}^4$ and $A \in \mathcal{L}_0$ by $\psi : (\mathbb{R}^4 \times \mathbb{R}^4) \times \mathcal{L}_0 \to \mathbb{R}^4 \times \mathbb{R}^4$ where $\psi((\mathbf{u}, \mathbf{v}), A) = (\mathbf{u}A, \mathbf{v}A)$ and the latter similarly. Now if $V(n, \mathbb{R}^4)$ is the Steifel manifold of n-frames in $\mathbb{R}^4$ $(n = 1, 2, 3)$ described in section 4.17 one easily sees that $V(n, \mathbb{R}^4)$ is diffeomorphic to an open and hence *regular* submanifold of the manifold which is the product of $\mathbb{R}^4$ with itself n times. It follows that $\mathcal{L}_0$ acts on each $V(n, \mathbb{R}^4)$ on the right as a Lie transformation group. But each Grassmann manifold $G(n, \mathbb{R}^4)$ is diffeomorphic to a quotient manifold of $V(n, \mathbb{R}^4)$ (section 4.17) and the associated equivalence relation is easily seen to be preserved by the action of $\mathcal{L}_0$. Hence $\mathcal{L}_0$ acts on the right as a Lie transformation group on each $G(n, \mathbb{R}^4)$ by theorem 5.5. This action is denoted by χ_n so that

$$\chi_n : G(n, \mathbb{R}^4) \times \mathcal{L}_0 \to G(n, \mathbb{R}^4)$$

It is clear then that each S_n, T_n and N_n is an orbit under χ_n and also, from the transitivity remark earlier in this paragraph, that the isotropy group for each action of $\mathcal{L}_0$ is not an open subgroup of $\mathcal{L}_0$ (since $\mathcal{L}_0$ is the only open subgroup of $\mathcal{L}_0$). Since $G(n,\mathbb{R}^4)$ is Hausdorff it follows that the orbits S_n, T_n and N_n may each be given the structure of a submanifold of $G(n,\mathbb{R}^4)$ which is diffeomorphic to a quotient manifold of the connected manifold $\mathcal{L}_0$ and hence *connected* (theorem 5.6). In fact, for $n = 1,2,3$, if $a \in G(n,\mathbb{R}^4)$ and I_a^n denotes the isotropy group at a (under $\mathcal{L}_0$) then the quotient manifold $R(\mathcal{L}_0, I_a^n)$ is diffeomorphic to the orbit O_a^n of a (under $\mathcal{L}_0$) and $\dim O_a^n = \dim \mathcal{L}_0 - \dim I_a^n$ (since I_a^n is never discrete). Also, it is clear that $I_a^1 = I_{a^\perp}^3$ and $I_a^2 = I_{a^\perp}^2$ where $a^\perp$ is the orthogonal complement of a (sections 2.5 and 6.1). In the next section it will be shown that

$$\begin{aligned}
&\dim I_t^1 = \dim I_s^3 = 3 &\qquad& \dim I_t^2 = 2 \\
&\dim I_s^1 = \dim I_t^3 = 3 && \dim I_s^2 = 2 \\
&\dim I_n^1 = \dim I_n^3 = 4 && \dim I_n^2 = 3
\end{aligned} \tag{6.11}$$

and so

$$\begin{aligned}
\dim T_1\ (= \dim O_t^1) = \dim O_s^3\ (= \dim S_3) = 3 &\quad \dim S_2\ (= \dim O_s^2) = 4 \\
\dim S_1\ (= \dim O_s^1) = \dim O_t^3\ (= \dim T_3) = 3 &\quad \dim T_2\ (= \dim O_t^2) = 4 \\
\dim N_1\ (= \dim O_n^1) = \dim O_n^3\ (= \dim N_3) = 2 &\quad \dim N_2\ (= \dim O_n^2) = 3
\end{aligned} \tag{6.12}$$

where s, t and n indicate, respectively, any spacelike, timelike or null member of $G(n,\mathbb{R}^4)$. Now $\dim G(1,\mathbb{R}^4) = \dim G(3,\mathbb{R}^4) = 3$ and $\dim G(2,\mathbb{R}^4) = 4$ (section 4.17) and so S_n and T_n $(n = 1,2,3)$, having the same dimension as the Grassmann manifold to which they belong, are *open* (and hence regular) submanifolds of $G(n,\mathbb{R}^4)$. The submanifolds N_n of $G(n,\mathbb{R}^4)$ are *closed* submanifolds. It follows by using orthogonal complementation (*cf.* section 4.17) that S_2 and T_2 are diffeomorphic submanifolds as are S_1 and T_3, S_3 and T_1 and N_1 and N_3. Now for each $n = 1,2,3$ consider the decomposition $G(n,\mathbb{R}^4) = S_n \cup T_n \cup N_n$ recalling that S_n and T_n are open and N_n closed (but not open since $G(n,\mathbb{R}^4)$ is connected). It follows that N_n is the topological boundary of S_n and of T_n. To see this suppose there exists $\mathbf{x} \in N_n$ and an open subset U of $G(n,R^4)$ such that $\mathbf{x} \in U \subseteq N_n$. Then by the transitivity remark earlier such an open subset would exist for each $\mathbf{x} \in N_n$ and this would give the contradiction that N_n was open in $G(n,\mathbb{R}^4)$. Hence any open subset of $G(n,R^4)$ containing $\mathbf{x} \in N_n$ must intersect $S_n \cup T_n$ non-trivially. If for some $\mathbf{x} \in N_n$ an open set U containing $\mathbf{x}$ were to intersect, say, S_n only, then the transitivity result would show that $S_n \cup N_n$ was open in $G(n,R^4)$ and hence that T_n was closed (and open) in $G(n,R^4)$ contradicting the connectedness of $G(n,R^4)$. Hence each such

U intersects both S_n and T_n non-trivially for each $\mathbf{x} \in N_n$ and so the result follows.

Now let $\mathbf{l}$ be a null vector in Minkowski space $\mathbb{R}^4$ and consider the set $W(\mathbf{l})$ of all *wave surfaces* to $\mathbf{l}$. Let $\mathbf{x}$ and $\mathbf{y}$ be unit orthogonal spacelike vectors each orthogonal to $\mathbf{l}$ so that they span a wave surface of $\mathbf{l}$. By extending $\mathbf{l}$, $\mathbf{x}$, $\mathbf{y}$ to a null tetrad it is straightforward to show that any other wave surface of $\mathbf{l}$ can be spanned by the spacelike vectors $\alpha\mathbf{x}+\beta\mathbf{y}+\gamma\mathbf{l}$ and $\mu\mathbf{x}+\nu\mathbf{y}+\rho\mathbf{l}$ ($\alpha, \beta, \gamma, \mu, \nu, \rho \in \mathbb{R}$) which, when chosen unit and orthogonal, and by taking appropriate linear combinations may be replaced by $\mathbf{x}+a\mathbf{l}$ and $\mathbf{y}+b\mathbf{l}$ and clearly each wave surface determines and is determined by the pair $(a,b) \in \mathbb{R}^2$. Hence $W(\mathbf{l})$ may be identified with $\mathbb{R}^2$. Now return to the original wave surface spanned by x and y. If one chooses a basis for $\mathbb{R}^4$ such that $\mathbf{x}$, $\mathbf{y}$ and $\mathbf{l}$ have components given, respectively, by $x^a = \delta^a_1$, $y^a = \delta^a_2$, $l^a = \delta^a_3$ then in the usual chart for $G(2,\mathbb{R}^4)$ with this original wave surface as origin (section 4.17) the members of $W(\mathbf{l})$ are represented by the subset with coordinates $(a,0,b,0)$. Thus the inclusion map $W(\mathbf{l}) \to G(2,\mathbb{R}^4)$ has, in the above coordinates, the representation $(a,b) \to (a,0,b,0)$ and is hence an immersion. Thus $W(\mathbf{l})$ admits the structure of a 2-dimensional (obviously regular) connected submanifold of $G(2,\mathbb{R}^4)$ (and of S^2 since S^2 is open in $G(2,\mathbb{R}^4)$) diffeomorphic to $\mathbb{R}^2$. It is not a closed submanifold of $G(2,\mathbb{R}^4)$ because each member of N_2 containing $\mathbf{l}$ is a limit point of $W(\mathbf{l})$ (section 3.3).

The following theorem summarises these results.

Theorem 6.3 *The connected Lie group $\mathcal{L}_0$ acts on the right on $\mathbb{R}^4$ and on $G(n,\mathbb{R}^4)$ $(n = 1,2,3)$ as a Lie transformation group. For the latter action, S_n, T_n and N_n are orbits and admit the structure of connected submanifolds of $G(n,\mathbb{R}^4)$ and have dimensions given by (6.12). The submanifolds S_n and T_n are open and the N_n are closed submanifolds of $G(n,\mathbb{R}^4)$ and N_n is the boundary of both S_n and T_n. The set of wave surfaces $W(\mathbf{l})$ to a null vector $\mathbf{l}$ is a 2-dimensional, regular non-closed connected submanifold of $G(2,\mathbb{R}^4)$ diffeomorphic to $\mathbb{R}^2$. Regarding the action of $\mathcal{L}$ on Minkowski space the orbits are $\{0\}$, the subsets of S and T of a particular fixed magnitude (with respect to η)and the subset N. The orbits of $\mathcal{L}_0$ are $\{0\}$, the future (respectively past) pointing members of T of a particular fixed magnitude, the future (respectively past) pointing members of N and the members of S of a particular fixed magnitude.*

6.4 The Connected Lie Subgroups of the Lorentz Group

To describe the connected Lie subgroups of $\mathcal{L}$ (which are just the connected Lie subgroups of $\mathcal{L}_0$) it follows from theorem 5.3 that it is sufficient to find all the subalgebras of L. The solution to this latter problem is well known and can now be described. The 6-dimensional Lie algebra L has been identified with the vector space of skew-symmetric 4×4 matrices which, in turn, were constructed from the actual members of L by an operation (index lowering) involving η. It follows that one can regard the members of L as being constructed from skewed products of pairs of vectors in ($\mathbb{R}^4$, η). If $(\mathbf{l}, \mathbf{n}, \mathbf{x}, \mathbf{y})$ is a null tetrad in Minkowski space, the vector space L can be spanned by $\mathbf{l} \wedge \mathbf{n}$ (which, in components, represents $l_a n^b - n_a l^b$), $\mathbf{l} \wedge \mathbf{x}$, $\mathbf{l} \wedge \mathbf{y}$, $\mathbf{x} \wedge \mathbf{y}$, $\mathbf{n} \wedge \mathbf{x}$ and $\mathbf{n} \wedge \mathbf{y}$ and this basis for L is particularly convenient for describing its subalgebras (although for the particular subalgebra type R_{13} an orthonormal basis $\mathbf{u}, \mathbf{x}, \mathbf{y}, \mathbf{z}$ (with $\mathbf{u}$ timelike) is more convenient).

Table 6.1 Subalgebras of L

	Basis	Dim		Basis	Dim
R_1	0	0	R_8	$\mathbf{l} \wedge \mathbf{x}$, $\mathbf{l} \wedge \mathbf{y}$	2
R_2	$\mathbf{l} \wedge \mathbf{n}$	1	R_9	$\mathbf{l} \wedge \mathbf{x}$, $\mathbf{l} \wedge \mathbf{y}$, $\mathbf{l} \wedge \mathbf{n}$	3
R_3	$\mathbf{l} \wedge \mathbf{x}$	1	R_{10}	$\mathbf{l} \wedge \mathbf{x}$, $\mathbf{n} \wedge \mathbf{x}$, $\mathbf{l} \wedge \mathbf{n}$	3
R_4	$\mathbf{x} \wedge \mathbf{y}$	1	R_{11}	$\mathbf{l} \wedge \mathbf{x}$, $\mathbf{l} \wedge \mathbf{y}$, $\mathbf{x} \wedge \mathbf{y}$	3
R_5	$\mathbf{l} \wedge \mathbf{n} + \rho \mathbf{x} \wedge \mathbf{y}$	1	R_{12}	$\mathbf{l} \wedge \mathbf{x}$, $\mathbf{l} \wedge \mathbf{y}$, $\mathbf{l} \wedge \mathbf{n} + \rho \mathbf{x} \wedge \mathbf{y}$	3
R_6	$\mathbf{l} \wedge \mathbf{n}$, $\mathbf{l} \wedge \mathbf{x}$	2	R_{13}	$\mathbf{x} \wedge \mathbf{y}$, $\mathbf{x} \wedge \mathbf{z}$, $\mathbf{y} \wedge \mathbf{z}$	3
R_7	$\mathbf{l} \wedge \mathbf{n}$, $\mathbf{x} \wedge \mathbf{y}$	2	R_{14}	$\mathbf{l} \wedge \mathbf{x}$, $\mathbf{l} \wedge \mathbf{y}$, $\mathbf{l} \wedge \mathbf{n}$, $\mathbf{x} \wedge \mathbf{y}$	4
			R_{15}	(=L)	6

Theorem 6.4 *The subalgebras of L can be classified into fifteen types which are represented by prescribing a basis for each subalgebra as shown in table 6.1 (see e.g.[44]). They are labelled $R_1 - R_{15}$ following a scheme in [45] where R_1 is the trivial subalgebra and $R_{15} = L$ and the symbols used to describe the bases are as defined in the previous paragraph. In the table, ρ is a non-zero real number and, henceforth, the symbols $R_1 - R_{15}$ will be used to represent either the algebra or the corresponding connected Lie subgroup of $\mathcal{L}_0$.*

A proof of the theorem can be found, for example, in [44]. All the connected Lie subgroups of $\mathcal{L}_0$ can now be found by the method of exponentiation described in section 5.5. In this respect the group $\mathcal{L}_0$ itself is particularly well behaved.

Theorem 6.5 *The Lie group $\mathcal{L}_0$ is an exponential Lie group (that is every $f \in \mathcal{L}_0$ is the exponential of some $F \in L$).*

Proof. (*cf* [44]) Let F be a member of the Lie algebra L of $\mathcal{L}_0$. Then from section 5.5

$$\exp(tF) = I_4 + \sum_{s=1}^{\infty} \frac{t^s}{s!} F^s \qquad (t \in \mathbb{R}) \tag{6.13}$$

Suppose $(\mathbf{l}, \mathbf{n}, \mathbf{x}, \mathbf{y})$ is a null tetrad. If F is the member of L corresponding to the bivector $\mathbf{l} \wedge \mathbf{n}$ then $\mathbf{l}F = -\mathbf{l}$, $\mathbf{n}F = \mathbf{n}$, $\mathbf{x}F = \mathbf{y}F = 0$ and so from section 5.5

$$\mathbf{l}[\exp(tF)] = e^{-t}\mathbf{l} \qquad \mathbf{x}[\exp(tF)] = \mathbf{x}$$
$$\mathbf{n}[\exp(tF)] = e^{t}\mathbf{n} \qquad \mathbf{y}[\exp(tF)] = \mathbf{y}$$

Thus $\exp(tF)$ is a Lorentz transformation of the type (6.8a). Similarly, if $F \in L$ corresponds to $\mathbf{x} \wedge \mathbf{y}$ then $\mathbf{l}F = \mathbf{n}F = 0$, $\mathbf{x}F = \mathbf{y}$, $\mathbf{y}F = -\mathbf{x}$ and, with $\sqrt{2}\mathbf{m} = \mathbf{x} + i\mathbf{y}$, $\mathbf{m}F = -i\mathbf{m}$,

$$\mathbf{l}[\exp(tF)] = \mathbf{l} \qquad \mathbf{m}[\exp(tF)] = e^{-it}\mathbf{m}$$
$$\mathbf{n}[\exp(tF)] = \mathbf{n} \qquad \bar{\mathbf{m}}[\exp(tF)] = e^{it}\bar{\mathbf{m}}$$

which is the Lorentz transformation (6.8b). If $F \in L$ corresponds to the bivector $\mathbf{l} \wedge \mathbf{y}$ then $\mathbf{l}F = \mathbf{x}F = 0$, $\mathbf{y}F = -\mathbf{l}$, $\mathbf{n}F = \mathbf{y}$ and so (since $F^k = 0$ for $k \geq 3$)

$$\mathbf{l}[\exp(tF)] = \mathbf{l} \qquad \mathbf{y}[\exp(tF)] = \mathbf{y} - t\mathbf{l}$$
$$\mathbf{x}[\exp(tF)] = \mathbf{x} \qquad \mathbf{n}[\exp(tF)] = \mathbf{n} + t\mathbf{y} - \frac{1}{2}t^2\mathbf{l}$$

which is the Lorentz transformation (6.8d). Similarly if $F \in L$ corresponds to $\mathbf{l} \wedge \mathbf{n} + \rho(\mathbf{x} \wedge \mathbf{y})$ $(0 \neq \rho \in \mathbb{R})$ then $\exp(tF)$ is the screw motion (6.8c). Since any $f \in \mathcal{L}_0$ $f \neq$ identity is of one of the types (6.8a)–(6.8d) and $\exp(O)$ is the identity the proof is complete. □

Although $\mathcal{L}_0$ is exponential not all of its subgroups are. However any 1-parameter subgroup of $\mathcal{L}_0$ is necessarily exponential (section 5.5) and any compact connected Lie group is exponential [37]. Thus the connected subgroups R_1, R_2, R_3, R_4, R_5 and R_{13} are exponential ($R_2 - R_5$ are 1-dimensional and hence 1-parameter subgroups from theorem 5.4 and R_{13} is Lie isomorphic to $SO(3)$ and is hence a compact Lie subgroup of $\mathcal{L}_0$). The subgroup R_{10} is Lie isomorphic to the identity component of the Lorentz group in three dimensions and is isomorphic to the identity component of $SL(2, \mathbb{R})$ and is then *not* exponential. For all the subgroups $R_1 - R_{15}$ one has the following result which follows from theorem 5.4.

Theorem 6.6 *If H is any connected Lie subgroup of $\mathcal{L}_0$ the members of H can be written as finite products of exponentials of members of the Lie subalgebra of L corresponding to H.*

Thus any Lorentz transformation can be regarded as a finite product of boosts, rotations, screw motions and null rotations. The next theorem translates certain algebraic properties of the subalgebras of L into geometrical properties of the associated connected Lie subgroups of $\mathcal{L}_0$.

Theorem 6.7 *Let H be any of the connected subgroups R_2–R_{15} of $\mathcal{L}_0$. A member $\mathbf{v} \in \mathbb{R}^4$ spans a fixed direction of H (i.e. $h(\mathbf{v}) \propto \mathbf{v}$ for each $h \in H$) if and only if $\mathbf{v}$ is an eigenvector of each matrix in the subalgebra L' of L corresponding to H. Further, $\mathbf{v}$ is a fixed point of H (i.e. $h(\mathbf{v}) = \mathbf{v}$ for each $h \in H$) if and only if $\mathbf{v}$ is an eigenvector of each member of L' with zero eigenvalue.*

Proof. The proof follows from results (iv) and (vii) in section 5.5 and theorem 6.6 (noting that the range of the exponential curve for each member of L' lies in H as was mentioned in section 5.5). □

The main geometrical features of the non-trivial connected subgroups $R_2 - R_{15}$ of $\mathcal{L}_0$ are listed in table 6.2. In this table entries in the 'fixed points' column like $\mathbf{a}$ or $< \mathbf{a}, \mathbf{b} >$ refer to all vectors in the 1-space spanned by $\mathbf{a}$ or in the 2-space spanned by $\mathbf{a}$ and $\mathbf{b}$ and entries in the 'fixed direction' column give a vector or vectors spanning such a direction or directions and is taken to imply that these directions contain no fixed points.

In table 6.2 the notation is as in table 6.1. Each fixed direction or 1-space of fixed points leads to an invariant 3-space of the corresponding subgroup namely its orthogonal complement. Also a pair of independent fixed points or directions spans an invariant 2-space of the group. The one-parameter subgroups R_2, R_3, R_4 and R_5 are, respectively, the boosts (6.8a), the null rotations (6.8d), the spatial rotations (6.8b) and the screw motions (6.8c) with the identity included in each case and where, in (6.8c), $\theta = \rho\lambda$ (from result (viii) in section 5.5). They are Lie isomorphic, respectively, to

Table 6.2

R	Fixed Points	Fixed Directions	R	Fixed Points	Fixed Directions
R_2	$<\mathbf{x},\mathbf{y}>$	$\mathbf{l}$, $\mathbf{n}$	R_9	0	$\mathbf{l}$
R_3	$<\mathbf{l},\mathbf{y}>$	-	R_{10}	$\mathbf{y}$	-
R_4	$<\mathbf{l},\mathbf{n}>$	-	R_{11}	$\mathbf{l}$	-
R_5	0	$\mathbf{l}$, $\mathbf{n}$	R_{12}	0	$\mathbf{l}$
R_6	$\mathbf{y}$	$\mathbf{l}$	R_{13}	$\mathbf{u}$	-
R_7	0	$\mathbf{l}$, $\mathbf{n}$	R_{14}	0	$\mathbf{l}$
R_8	$\mathbf{l}$	-	R_{15}	0	-

the Lie groups $\mathbb{R}$, $\mathbb{R}$, $SO(2)$ and $\mathbb{R}$. The group R_{10} is Lie isomorphic to the (identity component of the) Lorentz group in three dimensions. The group R_{13} is Lie isomorphic to the compact Lie group $SO(3)$. The groups $R_6 - R_9$, R_{11}, R_{12} and R_{14} are, perhaps, best described as subgroups of R_{14} and this will be done in the final paragraphs.

The subset of members of $\mathcal{L}_0$ which fix a particular direction of Minkowski space $\mathbb{R}^4$ is clearly a non-discrete subgroup of $\mathcal{L}_0$ and can be seen to be closed by considering isotropy groups of the associated action of $\mathcal{L}_0$ on the set $G(1, \mathbb{R}^4)$ of 1-spaces and noting that $G(1, \mathbb{R}^4)$ is Hausdorff (section 5.8). So let $\mathbf{l}$ be a null vector in $\mathbb{R}^4$ and let $N'(\mathbf{l})$ be the closed non-discrete subgroup of $\mathcal{L}_0$ which fixes the direction spanned by $\mathbf{l}$. Then, from theorem 5.1, $N'(\mathbf{l})$ may be given the structure of a regular submanifold of $\mathcal{L}_0$ and is then a Lie subgroup of $\mathcal{L}_0$. With this structure, the identity component $N(\mathbf{l})$ of $N'(\mathbf{l})$ is an open submanifold of $N'(\mathbf{l})$ and hence a regular submanifold of $\mathcal{L}_0$ and thus a Lie subgroup of $\mathcal{L}_0$. Now R_{14} (assumed adapted to the above null vector $\mathbf{l}$) is a connected Lie subgroup of $\mathcal{L}_0$ (contained in $N'(\mathbf{l})$) and hence a connected submanifold of $\mathcal{L}_0$. Thus R_{14} is a connected subset of $\mathcal{L}_0$ and hence also of $N'(\mathbf{l})$ and so one has $R_{14} \subseteq N(\mathbf{l})$. Since $N(\mathbf{l})$ is a regular submanifold of $\mathcal{L}_0$ it follows from example (ix), section 4.11, that R_{14} is a submanifold of $N(\mathbf{l})$ and so dim $N(\mathbf{l}) \geq$ dim $R_{14} = 4$. Now there are no 5-dimensional Lie subgroups of $\mathcal{L}_0$ and if dim $N(\mathbf{l}) = 6$ one would have that $N(\mathbf{l})$ is an open subgroup of $\mathcal{L}_0$ and hence the contradiction that $N(\mathbf{l}) = \mathcal{L}_0$ (section 5.2). So dim $N(\mathbf{l}) = 4$ and R_{14} is an open subgroup of $N(\mathbf{l})$. It follows from section 5.2 that R_{14} and $N(\mathbf{l})$ are identical as Lie subgroups of $\mathcal{L}_0$.

The members of the subgroup R_{14} can thus be calculated by finding all members of $\mathcal{L}_0$ that map a chosen null tetrad (l, n, x, y) to a null tetrad $(\mathbf{l}', \mathbf{n}', \mathbf{x}', \mathbf{y}')$ and preserve the null direction l (so that $\mathbf{l}' = A\mathbf{l}$ with $0 < A \in$

$\mathbb{R}$ by the future preserving condition). One finds for the members of R_{14}

$$\begin{aligned} \mathbf{l} \to e^{\lambda}\mathbf{l}, \qquad \mathbf{m} \to e^{i\theta}(\mathbf{m} - e^{\lambda}\bar{B}\mathbf{l}) \\ \mathbf{n} \to e^{-\lambda}\mathbf{n} + B\mathbf{m} + \bar{B}\bar{\mathbf{m}} - e^{\lambda}|B|^2\mathbf{l} \end{aligned} \tag{6.14}$$

where λ, $\theta \in R$, $0 \leq \theta < 2\pi$ and $B \in \mathbb{C}$. The subgroups $R_6 - R_9$, R_{11} and R_{12} of R_{14} can then be described by judicious choices of the parameters λ, θ and B. For example, a similar argument to that above shows that R_{11} consists of those members of $\mathcal{L}_0$ which fix a null *vector*, say $\mathbf{l}$ (from table 6.2). Thus R_{11} is described by the transformations (6.14) with $\lambda = 0$.

The group R_{14} is the *group of null rotations* (about $\mathbf{l}$). A member of R_{14} fixes at least one null direction and at most two (unless it is the identity). Also if $\mathbf{n}$ and $\mathbf{n}'$ are future pointing and span null directions neither of which is that spanned by $\mathbf{l}$ then they can be mapped one into the other by some member of R_{14}.

Finally, consider the isotropy groups I_a^n introduced in the last section. Let H be any of these isotropy groups so that (section 5.8) H is a Lie subgroup and can be given the structure of a regular submanifold of $\mathcal{L}_0$. Let H_0 be the identity component of H. Since H_0 is an open (hence regular) submanifold of H it is a regular submanifold of $\mathcal{L}_0$ (section 4.11 example (vii)). Thus H_0 is a connected Lie subgroup of $\mathcal{L}_0$. It follows that H_0 is one of the connected subgroups R_2 - R_{14}. It is now straightforward to check that, up to Lie isomorphism, and with (I_a^n) denoting the identity component of I_a^n (so that $\dim I_a^n = \dim(I_a^n)$)

$$\begin{aligned} (I_s^1) = (I_t^3) = R_{10}, \quad (I_t^1) = (I_s^3) = R_{13}, \quad (I_s^2) = (I_t^2) = R_7, \\ (I_n^1) = (I_n^3) = R_{14}, \quad (I_n^2) = R_9. \end{aligned} \tag{6.15}$$

The dimensions stated in (6.11) now follow from theorem 6.7 and table 6.1.

There are many discussions of the Lorentz group in the literature and the lectures of Trautman and Pirani in [46] and the paper by Sachs [47] are particularly convenient.

Chapter 7

Space-Times and Algebraic Classification

7.1 Space-Times

It is not the purpose of this book to enter into a logical or historical development of Einstein's field equations of general relativity theory. More detailed discussion of these equations can be readily found elsewhere (see, *e.g.* [48]-[51]). This section will thus content itself with the mathematics of the space-time as interpreted in classical general relativity and some elementary knowledge of general relativity will necessarily be assumed.

A *space-time* is a pair (M, g) where M is a 4-dimensional, smooth, connected, Hausdorff manifold and g a smooth Lorentz metric on M of signature $(-1, 1, 1, 1)$. The members of M are referred to as *points* or *events*. Any metrical property of a tensor *etc.* at $m \in M$ (*e.g.* "null", "non-null") will always be assumed to refer to the Lorentz metric $g(m)$. A global condition such as a "timelike vector field" will be a vector field which is timelike at each point where it is defined. The definition of a space-time requires the following remarks.

i) The usual agreement in general relativity is that "information" is transmitted along piecewise differentiable curves whose tangent vector, where defined, is timelike or null. Thus if M consisted of more than one path component the events in one such component could have no influence on the events in the other and in this sense, such components would have independent physics. Since, for a manifold with its natural manifold topology, path connectedness and connectedness are equivalent statements, then in order to have a single physics for M one assumes M is *connected*.

ii) The assumptions that M is 4-dimensional and Hausdorff are usually (and perhaps uncomfortably) attributed to "experience". The existence of the Lorentz metric g then implies that M is *paracompact* [32].

iii) The assumption that M is smooth is taken here mainly for convenience. It is mitigated by the fact (see section 4.5) that every complete C^r atlas ($r \geq 1$) of M contains a smooth atlas. The choice of a smooth metric g for M is also one of convenience and, of course, would make no sense unless M was smooth.

iv) The symmetric Levi-Civita connection arising from g, and its associated curvature tensor are usually referred to as *the connection and curvature (Riemann tensor) of M*. In this sense one calls a space-time *flat* if the curvature is identically zero on M (and so M is *not flat* if the curvature is not identically zero on M!) and *non-flat* if the curvature does not vanish over a non-empty open subset of M (*i.e.* the curvature is non-zero at each point of an open dense subset of M).

v) Suppose there exists a smooth diffeomorphism $f : M \to M$. Then f^*g and g are both Lorentz metrics for M. Although g and f^*g need not be the same metric for M, the space-time (M, g) and (M, f^*g) are *isometric* under f and, for the present purposes, identical. Formally one should, perhaps, identify such space-times and work within an equivalence class of such an equivalence relation. For the present purposes, however, a space-time will continue to be M together with a particular choice of Lorentz metric g on M.

One problem which concerns the global topology of a space-time manifold M is the question of whether or not it is *simply connected*. Since one may always choose a coordinate domain to be simply connected, local problems are indifferent to this (global) property of M. In fact simply connectedness will not be part of the definition of a space-time but will be added as an extra assumption from time to time in order to be able to deal with certain global mathematical constructions. It is remarked here that it is not hard to show from the work of section 4.14 that any space-time is isometric to the quotient manifold of a simply connected space-time. For if $\tilde{M}$ is the universal covering manifold of M with covering map $p : \tilde{M} \to M$ let $f : M \to \tilde{M}/\sim$ and $i : \tilde{M} \to \tilde{M}/\sim$ be the maps described in section 4.14 with $i = f \circ p$. Then $g_1 \equiv p^*g$ and $g_2 \equiv (f^{-1})^*g$ are metrics on $\tilde{M}$ and $\tilde{M}/\sim$, respectively, (and then (M, g) and $(\tilde{M}/\sim, g_2)$ are isometric) and $g_1 = p^*g = (f^{-1} \circ i)^*g = i^*g_2$. One result of interest (especially in the thermodynamical aspects of general relativity) is that if *M is a simply connected space-time then M admits a global smooth nowhere zero timelike vector field* and hence a consistent choice of the future direction of time on

M. This follows [52] from the existence of a 1-dimensional distribution on M guaranteed by theorem 4.15.

It is remarked that one normally chooses the space-time manifold M to be *not compact* (otherwise one has closed timelike curves and the associated philosophical problems [53]). Given that M is also paracompact one then has the existence on M of a global, nowhere zero, smooth vector field X [54] which, together with the positive definite metric on M guaranteed by theorem 4.14, leads to a Lorentz metric on M constructed from the 1-dimensional distribution determined by X and for which X is timelike everywhere.

The gravitational field is interpreted through the metric g and so field equations restricting g are required. These are the Einstein field equations given in each coordinate system of M by (see, *e.g.* [48])

$$R_{ab} - \frac{1}{2}Rg_{ab} + \Lambda g_{ab} = 8\pi T_{ab} \tag{7.1}$$

where (section 4.16) the R_{ab} are the Ricci tensor components, $R(\equiv R_{ab}g^{ab})$ is the Ricci scalar, Λ is a constant called the *cosmological constant* and the tensor T is the *energy-momentum tensor* with components $T_{ab}(= T_{ba})$ about which more will be said later. There are good reasons for supposing $|\Lambda|$ very small and in this book *the assumption* $\Lambda = 0$ *will be made unless otherwise specified.* The field equations (7.1) thus reduce to

$$R_{ab} - \frac{1}{2}Rg_{ab} \equiv G_{ab} = 8\pi T_{ab} \tag{7.2}$$

where G is the Einstein tensor. Using a semi-colon to denote a covariant derivative with respect to the space-time connection, the Bianchi identities $R_{ab[cd;e]} \equiv 0$ (see (4.36)) are easily shown to contract to the equivalent statements

$$G_a{}^b{}_{;b} \equiv 0, \qquad T_a{}^b{}_{;b} = 0 \tag{7.3}$$

The second of these is usually referred to as the *conservation law* for T.

The energy-momentum tensor is usually taken as formally representing the sources of the gravitational field. The more commonly used energy-momentum tensors will now be briefly discussed.

7.1.1 *Electromagnetic fields*

An electromagnetic field in space-time is described by the current density vector field j with components j^a, representing the electric charge and current distribution and the resulting electric and magnetic fields which are combined into the *Maxwell tensor* (or *Maxwell bivector* - see section 7.2) F with components $F_{ab}(=-F_{ba})$. Maxwell's equations are then

$$F^{ab}{}_{;b} = j^a, \qquad F_{[ab;c]} = 0. \tag{7.4}$$

The contribution of such a Maxwell field to the energy-momentum tensor is then given by

$$T_{ab} = \tfrac{1}{4\pi}(F_{ac}F_b{}^c - \tfrac{1}{4}F_{cd}F^{cd}g_{ab}). \tag{7.5}$$

If (7.5) is the total energy-momentum tensor, one usually refers to the gravitational field as an *Einstein-Maxwell* (or an *electrovac*) field.

7.1.2 *Fluid Space-Times*

Suppose the source of the gravitational field is a viscous fluid. The latter is described by its coefficients of *dynamic viscosity* η and *bulk viscosity* ξ, a unit timelike vector field u representing the *fluid flow*, its *energy density* ρ with respect to u, its *isotropic pressure* p, its *shear tensor* σ, its *expansion* θ and its *heat flow* vector field q. These quantities satisfy $u^a u_a = -1$, $u^a q_a = 0$, $\sigma_{ab} = \sigma_{ba}$, $\sigma_a{}^a = 0$ and $\sigma_{ab}u^b = 0$ and the resulting energy-momentum tensor is

$$T_{ab} = (p - \xi\theta + \rho)u_a u_b + (p - \xi\theta)g_{ab} - 2\eta\sigma_{ab} + 2u_{(a}q_{b)}. \tag{7.6}$$

There are several special cases to consider here and which result when some of the above physical quantities associated with the fluid take special values. Perhaps the most important is the case of a *perfect fluid*. This arises when $\xi = \eta \equiv 0$ and $q \equiv 0$ and T reduces to

$$T_{ab} = (p + \rho)u_a u_b + p g_{ab} \tag{7.7}$$

7.1.3 *The Vacuum Case*

This is a rather important case for general relativity and occurs when $T \equiv 0$ on M. Then M is called a *vacuum space-time* and Einstein's field equations reduce to either of the following two equivalent statements

$$R_{ab} = 0, \qquad G_{ab} = 0 \tag{7.8}$$

If $T \equiv 0$ over some non-empty open subset U of M then U with its open submanifold structure and induced metric is called a *vacuum region* of M.

Further details regarding the postulates of general relativity and details of the field equations can be found in [48] whilst a comprehensive list of exact solutions of (7.1), (7.2) and (7.8) is contained in [55]. A full discussion of the algebraic structure of the general energy-momentum tensor and of the particular examples above will be given later in this chapter.

The energy-momentum tensor is usually subject to the so-called *energy conditions* that at each $p \in M$ and for each timelike vector $u \in T_pM$

$$\text{(i)} \quad T_{ab}u^a u^b \geq 0, \qquad \text{(ii)} \quad T^a{}_b u^b \quad \text{is not spacelike} \tag{7.9}$$

By continuity, each condition is still true if u is null. The first condition expresses the non-negativity of the local energy-density of the field for any observer (represented by u) and the second the non-spacelike nature of the local energy flow vector. Condition (i) alone is sometimes referred to as the *weak* energy condition and (i) and (ii) together as the *dominant* energy conditions [48]. The restrictions they impose on T will be discussed later.

There is a useful decomposition of the curvature tensor of M given at $p \in M$ by [56],[57]

$$R_{abcd} = C_{abcd} + E_{abcd} + \tfrac{1}{6} R\, G_{abcd} \tag{7.10}$$

where C is the Weyl tensor and

$$E_{abcd} = \tilde{R}_{a[c}g_{d]b} + \tilde{R}_{b[d}g_{c]a} \qquad G_{abcd} = g_{a[c}g_{d]b} \tag{7.11}$$

and where $\tilde{R}_{ab} \equiv R_{ab} - \frac{1}{4}Rg_{ab}$ is the *tracefree* (part of the) *Ricci tensor* ($\tilde{R}_a{}^a = 0$). For a fixed g the expression (7.11) for E at $p \in M$ is a linear one-to-one relation between E and $\tilde{R}$ (since $E^c{}_{acb} = \tilde{R}_{ab}$) and, in particular, $E = 0 \Leftrightarrow \tilde{R} = 0 \Leftrightarrow R_{ab} = \frac{1}{4}Rg_{ab}$ which is equivalent to M satisfying the Einstein space condition (section 4.16) at p. The decomposition (7.10) will be further explored later.

7.2 Bivectors and their Classification

Let (M, g) be a space-time and $p \in M$. A second order skew-symmetric tensor F (at p) with components $F_{ab}(= -F_{ba})$ is called a *bivector* (at p). The set $B(p)$ of all bivectors at p is thus a 6-dimensional real vector space. If $F \in B(p)$ one can define the *dual of F*, denoted by ${}^*F$, by

$${}^*F_{ab} = \frac{1}{2}\epsilon_{abcd}F^{cd} \tag{7.12}$$

where ϵ_{abcd} is the usual pseudotensor (see, *e.g.* [51]) and $*$ is the duality operator. Thus ϵ_{abcd} is completely skew symmetric and, with g denoting $\det(g_{ab})$ (this should cause no confusion with the metric itself), ϵ^{abcd} is obtained from ϵ_{abcd} by the usual index raising and $\epsilon^{1234} = (-g)^{\frac{1}{2}} = -\epsilon_{1234}$. [It is remarked that unless M has the extra property of being *orientable* (which will not be assumed in this text) there are problems with a global definition of the dual. However, such a definition is satisfactory in an appropriately chosen coordinate domain and where such a domain is required it will be assumed so chosen without further comment.] These quantities and hence $^*F$ are only tensors under coordinate transformations where the transformation matrix has positive determinant. $^*F^{ab}$ is obtained from $^*F_{ab}$ by the usual raising of indices with g. If F and H are bivectors then standard formula [51] associated with the pseudotensor give

$$^{**}F_{ab} = -F_{ab} \tag{7.13}$$

$$F_{ab}F^{ab} = -{}^*F_{ab}{}^*F^{ab} \tag{7.14}$$

$$H^{ac}F_{bc} - {}^*F^{ac*}H_{bc} = \frac{1}{2}F^{de}H_{de}\delta^a_b \tag{7.15}$$

Since F_{ab} is skew symmetric it is a standard result in algebra that the rank of the matrix F_{ab} is even and hence, if $F \neq 0$, equal to two or four. If F has rank 2 it is called *simple* and if it has rank 4 it is called *non-simple*. Because the rank of a non-zero bivector F can never be three the existence of $k \in T_pM$, $k \neq 0$, satisfying $F_{ab}k^b = 0$ is then equivalent to F being simple. If F is simple it follows that the components F_{ab} may be chosen with two rows (and hence two columns) consisting entirely of zeros. From this it follows by inspection that there exist two independent covectors r and s at p such that $F_{ab} = r_a s_b - s_a r_b = 2r_{[a}s_{b]}$. The 2-space at p spanned by the vectors r^a and s^a is then easily checked (by considering orthogonal complements) to be well defined independently of the choice of r and s and is called the *blade* of F. If k is any vector at p orthogonal to this blade (*i.e.* orthogonal to every member of this blade) then $F_{ab}k^b = 0$ and conversely. Also, it follows from this that if F is simple and written in the above form then $^*F_{ab}r^b = {}^*F_{ab}s^b = 0$ and so $^*F$ is simple and can be written as $^*F = 2r'_{[a}s'_{b]}$ for independent vectors r'^a, s'^a orthogonal to the blade of F. Thus *F is simple if and only if $^*F$ is simple and then their blades are orthogonal complements of each other*. One now has the following theorem

Theorem 7.1 *The following statements are equivalent for a non-zero*

bivector F at p.

i) *F is simple.*
ii) *There exists $k \in T_pM$, $k \neq 0$, such that $F_{ab}k^b = 0$.*
iii) *There exist independent members $r, s \in T_p^*M$ such that $F_{ab} = 2r_{[a}s_{b]}$.*
iv) *${}^*F$ is simple.*
v) ${}^*F_{ab}F^{bc} = 0$.
vi) ${}^*F_{ab}F^{ab} = 0$.
vii) $F_{[ab}F_{c]d} = 0$.

Proof. The equivalence of conditions (i), (ii), (iii) and (iv) was established above. That these conditions together imply (v), (vi) and (vii) is then straightforward as are the implications (v) $\Leftrightarrow$ (vi) from (7.15). Now since $F \neq 0$, there exist $k \in T_p^*M$, $k \neq 0$, such that $F^{bc}k_c \neq 0$ and so immediately one has (v) $\Leftrightarrow$ (iv). Finally if (vii) holds choose $q \in T_pM$, $q \neq 0$, such that $F_{cd}q^d \equiv q'_c \neq 0$ and contract the condition (vii) with q^d to obtain $F_{[ab}q'_{c]} = 0$. A contraction with $t \in T_pM$ satisfying $t^a q'_a \neq 0$ then shows that (iii) holds. This completes the proof. □

The set of simple bivectors at p can be further classified into three types, *spacelike*, *timelike* and *null*, according as the corresponding blade of the bivector is a spacelike, timelike or null 2-space at p, respectively. It follows that the dual of a timelike bivector is spacelike and vice-versa and that the dual of a null bivector is null (section 6.1). It is clear that for any simple bivector, when written in the form (iii) of theorem 7.1 above, r and s may be chosen to be orthogonal. Also, a simple bivector uniquely determines its blade and, conversely, if a 2-space at p is the blade of the bivectors F_1 and F_2 then $F_2 = \lambda F_1$ ($\lambda \in \mathbb{R}$). It is also useful to note that if a bivector $F \neq 0$ at p satisfies $F_{[ab}k_{c]} = 0$ for $k \in T_p^*M$, $k \neq 0$, then F is simple and its blade contains k. This follows by contracting this last equation with ω^c for $\omega \in T_pM$ and $\omega^a k_a \neq 0$. Another remark is that it is easily checked that $\epsilon_{abcd;e} = 0$ and so the covariant derivative and dual operations on bivectors are interchangeable.

Theorem 7.2 *The following statements are equivalent for a non-zero bivector F at p.*

i) *F is null.*
ii) *There exists $k \in T_pM$, $k \neq 0$, such that $F_{ab}k^b = {}^*F_{ab}k^b = 0$.*
iii) $F_{ab}F^{ab} = {}^*F_{ab}F^{ab} = 0$.

Proof. If (i) holds one can check from the previous discussion of null 2-spaces in section 6.1 that (ii) holds with k spanning the principal null direction of the blade of F (and ${}^*F$). Conversely (ii) $\Rightarrow$ (i) because (ii) shows that the blades of F and ${}^*F$, although orthogonal complements, do not span T_pM and hence that these blades must be null. Further, if (i) holds then, again consulting the previous discussion of null 2-spaces, one may write (assuming an appropriate orientation of x and y) $F_{ab} = 2l_{[a}x_{b]}$, ${}^*F_{ab} = -2l_{[a}y_{b]}$ with l null, x and y spacelike and $l^a x_a = l^a y_a = x^a y_a = 0$. The condition (iii) is then immediate. Finally if (iii) holds the second condition in (iii) and the previous theorem show that F is simple. So write $F_{ab} = 2r_{[a}s_{b]}$ and ${}^*F_{ab} = 2r'_{[a}s'_{b]}$ with the pairs (r, s) and (r', s') chosen orthogonal. Then the first condition in (iii) and (7.14) give

$$(r_a r^a)(s_a s^a) = (r'_a r'^a)(s'_a s'^a) = 0 \tag{7.16}$$

so one of r and s, say r and one of r' and s', say r', is null. But $r_a r'^a = 0$ and so r and r' are proportional. It follows that F is null. $\square$

It should be noted that the condition (ii) in the last theorem implies that k is null and unique up to a scaling. In fact k spans the (necessarily null) direction in which the orthogonal null blades of F and ${}^*F$ intersect and this is called the (*repeated*) *principal null direction* of F (and ${}^*F$).

Let x and y be unit orthogonal spacelike vectors at $p \in M$ and suppose they are extended at p to a null tetrad l, n, x, y and also to a orthonormal tetrad x, y, z, t with l and n null, z unit spacelike, t unit timelike and $\sqrt{2}z = l + n$, $\sqrt{2}t = l - n$. Then two bases for the 6-dimensional vector space $B(P)$ of bivectors at p may be conveniently constructed from these tetrads. As a notational convenience the general simple bivector $2r_{[a}s_{b]}$ will be denoted by the symbol $r \wedge s$. These bases are

$$\begin{array}{lll} F^1 = l \wedge x & F^2 = l \wedge y & F^3 = x \wedge y \\ F^4 = n \wedge x & F^5 = n \wedge y & F^6 = l \wedge n \end{array} \tag{7.17}$$

and

$$\begin{array}{lll} G^1 = x \wedge y & G^2 = x \wedge z & G^3 = y \wedge z \\ G^4 = t \wedge x & G^5 = t \wedge y & G^6 = t \wedge z \end{array} \tag{7.18}$$

It should be noted that, up to signs, (F^1, F^2), (F^3, F^6), (F^4, F^5), (G^1, G^6), (G^2, G^5) and (G^3, G^4) are dual pairs. The bivectors F^1 and F^2 are null with principal null direction (spanned by) l, F^4 and F^5 are null with principal null direction n, F^3, G^1, G^2 and G^3 are spacelike and F^6, G^4, G^5 and G^6 are timelike. In fact, the way the null and spacelike bivectors are written above is typical for their type. For a timelike bivector there are two typical ways of writing it, namely as F^6 or as G^4 (G^5 or G^6). A timelike bivector has a (timelike) blade containing exactly two distinct null directions. They are called the *principal null directions* of the bivector.

So far only real bivectors have been discussed. Although no formal provision has been made for defining *complex* second order skew-symmetric tensors (*complex bivectors*) on a manifold this will informally be assumed. So a complex bivector will be regarded as an object at p whose real and imaginary parts are real bivectors in the sense defined above. The set of complex bivectors at p will be denoted by $CB(p)$ and is clearly a 6-dimensional complex vector space (the complexification of $B(p)$–see section 2.3). This vector space has two important subspaces denoted by $S^+(p)$ and $S^-(p)$ which, with the duality operation defined for members of $CB(p)$ by (7.12), are defined by

$$H \in S^+(p) \Leftrightarrow {}^*H = -iH, \quad H \in S^-(p) \Leftrightarrow {}^*H = iH \tag{7.19}$$

If one writes $H = A + iB$, $A, B \in B(p)$ then it follows from (7.19) that $H \in S^+(p) \Leftrightarrow B = {}^*A$ and that $H \in S^-(p) \Leftrightarrow B = -{}^*A$. Thus the members of $S^+(p)$ are precisely those complex bivectors of the form $F + i{}^*F$ and those in $S^-(p)$ precisely those of the form $F - i{}^*F$ for a real bivector F. Thus the members of $S^+(p)$ and $S^-(p)$ are each uniquely associated with a real bivector in a natural way and so denote members of $S^+(p)$ as $\overset{+}{F}$ and members of $S^-(p)$ as $\tilde{F}$ where (*cf.*[47])

$$\overset{+}{F} = F + i^*F \qquad \tilde{F} = F - i^*F \tag{7.20}$$

for a *real* bivector F. Clearly $\overset{+}{F}$ and $\tilde{F}$ are conjugates. Members of $S^+(p)$ are called *self-dual* and members of $S^-(p)$ *anti self-dual*.

Any complex bivector H at P can be written as the sum of a member of $S^+(p)$ and a member of $S^-(p)$ in exactly one way

$$H = \frac{1}{2}(H + i^*H) + \frac{1}{2}(H - i^*H) \tag{7.21}$$

with $H + i^*H \in S^+(p)$ and $H - i^*H \in S^-(p)$. It then follows easily that $CB(p)$ is the *direct vector space sum* (section 2.3)

$$CB(p) = S^+(p) \oplus S^-(p) \tag{7.22}$$

Now any $F \in B(p)$ can, from (7.17), be written as $F = \sum \alpha_k F^k$ $(1 \leq k \leq 6, \alpha_k \in \mathbb{R})$ and so any $\overset{+}{F} \in S^+(p)$ can be written as $\overset{+}{F} = \sum \alpha_k (F^k + i^* F^k)$ and any $\tilde{F} \in S^-(p)$ as $\tilde{F} = \sum \alpha_k (F^k - i^* F^k)$. Recalling the dual pairing in the bases (7.17) and (7.18) one easily sees how to choose bases for $S^+(p)$ and $S^-(p)$ each consisting of three members and so $\dim S^+(p) = \dim S^-(p) = 3$ (as complex vector subspaces of $CB(p)$).

The maps $B(p) \to B(p)$ given by $d : F \to {}^*F$ and $d' : F \to -{}^*F$ are linear maps on and, in fact, isomorphisms of $B(p)$. They have the property that $d^2 = d'^2 = -1$ where 1 is here the identity map on $B(p)$. Hence each gives rise to a complex structure on $B(p)$ (section 2.3) and hence to the complex vector spaces $S^-(p)$ and $S^+(p)$. A traditional (and useful) way of writing the basis for $S^+(p)$ is, using (7.17)

$$V = 2^{-1/2}(F^1 + i^* F^1), \qquad U = 2^{-1/2}(F^4 + i^* F^4), \qquad M = (F^6 + i^* F^6) \tag{7.23}$$

Then with the null tetrad appropriately oriented one has [47]

$$V_{ab} = \sqrt{2}(l_{[a}x_{b]} - i l_{[a}y_{b]}) = 2 l_{[a}\bar{m}_{b]} \tag{7.24}$$

$$U_{ab} = \sqrt{2}(n_{[a}x_{b]} + i n_{[a}y_{b]}) = 2 n_{[a}m_{b]} \tag{7.25}$$

$$M_{ab} = 2 l_{[a}n_{b]} + 2i x_{[a}y_{b]} = 2 l_{[a}n_{b]} + 2\bar{m}_{[a}m_{b]} \tag{7.26}$$

where the *complex null tetrad* $(l, n, m, \bar{m})$ has been introduced with $m = 2^{-1/2}(x + iy)$ (so that $m^a m_a = \bar{m}^a \bar{m}_a = 0$, $m_a \bar{m}^a = 1$). A basis for $S^-(p)$ then consists of the conjugates $\bar{V}$, $\bar{U}$ and $\bar{M}$. One easily finds the following relations between V, U, M, $\bar{V}$, $\bar{U}$ and $\bar{M}$

$$U_{ab}V^{ab} = \bar{U}_{ab}\bar{V}^{ab} = 2 \quad M_{ab}M^{ab} = \bar{M}_{ab}\bar{M}^{ab} = -4 \tag{7.27}$$

with any other such contraction between any two of these six complex bivectors equal to zero.

The orthonormal and null tetrads (x, y, z, t) and (l, n, x, y) and the complex null tetrad $(l, n, m, \bar{m})$ each satisfy a *completeness relation* linking it with the metric $g(p)$ at p. They are

$$g_{ab} = x_a x_b + y_a y_b + z_a z_b - t_a t_b \tag{7.28}$$

$$g_{ab} = 2 l_{(a}n_{b)} + x_a x_b + y_a y_b \tag{7.29}$$

$$g_{ab} = 2 l_{(a}n_{b)} + 2 m_{(a}\bar{m}_{b)} \tag{7.30}$$

These are understood in the following sense; if any of these relations hold, the covectors appearing on the right hand side constitute a tetrad of the appropriate type, and conversely. For example, if (l, n, x, y) is a null tetrad let S_{ab} denote the right hand side of (7.29). Then, clearly $S_{ab}l^b = l_a$, $S_{ab}n^b = n_a$, $S_{ab}x^b = x_a$, $S_{ab}y^b = y_a$. Hence for any $k \in T_pM$ $S_{ab}k^b = k_a$ and so $S_{ab} = g_{ab}$ and (7.29) holds. Conversely if (7.29) holds then clearly l, n, x and y are a basis for T_pM because if $k \in T_pM$, a contraction of (7.29) with k^b shows that they span T_pM. Next, successive contractions of (7.29) with l^b, n^b, x^b and y^b show that $l^a n_a = x^a x_a = y^a y_a = 1$ and that all the other inner products between these basis vectors vanish. Hence (l, n, x, y) is a null tetrad at p. The other proofs are similar.

Now consider the tensor G at p with components G_{abcd} given by (7.11). This has the following properties for any bivectors $F, H \in B(p)$

$$G_{abcd}F^{cd} = F_{ab} \qquad G_{abcd}F^{ab}H^{cd} = F^{ab}H_{ab} \tag{7.31}$$

It is clear that G acts as a metric on the vector space $B(p)$ according to the (coordinate independent) rule $G(F, H) = G(H, F) \equiv F^{ab}H_{ab}$. Also the basis (7.18) is such that, if re-defined by $\tilde{G}^k = 2^{-1/2}G^k$ $(1 \leq k \leq 6)$, it satisfies

$$G(\tilde{G}^a, \tilde{G}^a) = 1 \ (1 \leq a \leq 3), \qquad G(\tilde{G}^a, \tilde{G}^a) = -1 \quad (4 \leq a \leq 6) \tag{7.32}$$

with all other inner products zero. Hence G is a metric on $B(p)$ with signature $(+++---)$ and Sylvester canonical matrix equal to diag$(111-1-1-1)$. In this notation and using theorem 7.1 a (real) bivector F is simple if and only if $G(F, {}^*F) = 0$ and, given it is simple, it is then spacelike, timelike or null according as $G(F, F)$ is positive, negative or zero. G is sometimes referred to as the *bivector metric*. There is a completeness relation for bivectors. First one notes that the complex bivectors U, V and M satisfy

$$G_{abcd} + \frac{1}{2}i\epsilon_{abcd} = V_{ab}U_{cd} + U_{ab}V_{cd} - \frac{1}{2}M_{ab}M_{cd} \tag{7.33}$$

the proof being essentially the same as that for the space-time metric completeness relations above. By taking the real part of (7.33) one obtains the desired relation in terms of G and the real and imaginary parts of U, V and M (i.e. the real bivectors $F^1, \ldots, F^6$). It is noted here that if $\overset{+}{F} = F + i^*F$ is in $S^+(p)$ with $F \in B(p)$ then, informally extending the metric G to complex bivectors, one has

$$G(\overset{+}{F},\overset{+}{F}) = \overset{+}{F}_{ab}\overset{+}{F}^{ab} = 2(F_{ab}F^{ab} + iF_{ab}{}^{*}F^{ab}) \tag{7.34}$$

and so, from theorems 7.1 and 7.2, *F is null if and only if* $G(\overset{+}{F},\overset{+}{F}) = 0$ and *F is simple if and only if* $G(\overset{+}{F},\overset{+}{F}) \in \mathbb{R}$. A complex self dual bivector $\overset{+}{F} = F + i^{*}F$ is called *null* if F (and hence also ${}^{*}F$) is null. The common principal null direction of F and ${}^{*}F$ is then called the *(repeated) principal null direction of* $\overset{+}{F}$.

So far, little has been said about real non-simple bivectors. To deal with these first let $(l, n, m, \bar{m})$ be a *complex* null tetrad and F any *real* bivector at p. Under a null rotation about l at p (section 6.2) the tetrad $(l, n, m, \bar{m})$ is changed to, say, $(l', n', m', \bar{m}')$ where, from (6.14)

$$\begin{gathered} l' = Al, \qquad m' = e^{i\theta}(m - A\bar{B}l) \\ n' = A^{-1}n + Bm + \bar{B}\bar{m} - AB\bar{B}l \\ (A, \theta \in \mathbb{R},\ A > 0,\ B \in \mathbb{C}) \end{gathered} \tag{7.35}$$

The corresponding members U', V' and M' of $CB(p)$ defined as in (7.24)–(7.26) (but in terms of the primed tetrad components) satisfy

$$\begin{gathered} V' = Ae^{-i\theta}V, \qquad M' = 2A\bar{B}V + M \\ U' = A\bar{B}^{2}e^{i\theta}V + \bar{B}e^{i\theta}M + A^{-1}e^{i\theta}U \end{gathered} \tag{7.36}$$

Construct the complex self-dual bivector $\overset{+}{F} = F + i^{*}F$ corresponding to the real bivector F and write

$$\overset{+}{F} = F_1'V' + F_2'M' + F_3'U' = F_1V + F_2M + F_3U \tag{7.37}$$

for $F_1', \ldots, F_3 \in \mathbb{C}$, the intention being to choose U, V, M so that the expression for $\overset{+}{F}$ in terms of them is, in some sense, especially simple. A substitution of (7.36) into (7.37) and then equating coefficients of U, V, M gives

$$F_1 = Ae^{-i\theta}F_1' + 2A\bar{B}F_2' + A\bar{B}^{2}e^{i\theta}F_3' \tag{7.38a}$$

$$F_2 = F_2' + \bar{B}e^{i\theta}F_3' \tag{7.38b}$$

$$F_3 = A^{-1}e^{i\theta}F_3' \tag{7.38c}$$

Suppose now that $\overset{+}{F}$ is *not* null (equivalently F is *not* null). Suppose $F_3' = 0$ in the above expression for $\overset{+}{F}$. Then $F_3 = 0$. If also $F_2' = 0$ then $F_2 = 0$ and the contradiction that $\overset{+}{F}(= F^1 V)$ is null is achieved. Hence, if $F_3' = 0$, $F_2' \neq 0$. Then (7.38a) shows that there exists $B \in \mathbb{C}$ such that $F_1 = 0$ and so $\overset{+}{F} = F_2 M$. If $F_3' \neq 0$ then (7.38a) shows that there exists a value of B for which $F_1 = 0$. Thus one achieves a complex null tetrad, say $(l, n, m, \bar{m})$, in which $F_1 = 0$. Then a null rotation, this time about the null vector n (and again remembering that $\overset{+}{F}$ is not null), yields a complex null tetrad in which $F_1 = F_3 = 0$ and again one has $\overset{+}{F} = F_2 M$. Taking real parts one gets for any *non-null real bivector* F the expression

$$F_{ab} = 2\alpha l_{[a} n_{b]} + 2\beta x_{[a} y_{b]} \quad (\alpha, \beta \in \mathbb{R}) \tag{7.39}$$

in terms of a real null tetrad (l, n, x, y). If F is spacelike $\alpha = 0$ and if F is timelike $\beta = 0$ and previous results are recovered. If F is non-simple $\alpha \neq 0$ and $\beta \neq 0$ and one achieves a convenient expression for F.

A real or complex vector v at p is an *eigenvector* of $F \in CB(p)$ with *eigenvalue* $\lambda \in \mathbb{C}$ if $v^a F_a{}^b = \lambda v^b$ and the 1-dimensional subspace of $\mathbb{C}^4$ spanned by v is called an *eigendirection* of F. Thus, for the real bivector F in (7.39), l and n span real null eigendirections of F with eigenvalues $-\alpha$ and $+\alpha$ respectively and m and $\bar{m}$ span complex null eigendirections of F with eigenvalues $-i\beta$ and $+i\beta$, respectively. It is then clear that, just as a simple (null or non-null) real bivector determines its blade, a non-simple real bivector determines an orthogonal pair of 2-spaces, one timelike (spanned by l and n and which are defined to span the *principal null directions of* F) and one spacelike (spanned by x and y) and this pair is unique. They are called the *canonical pair of blades* of F. The following theorem summarises the situation [58]-[60].

Theorem 7.3 *A real bivector $F \neq 0$ at p is null if and only if it has a unique real null eigendirection (and the eigenvalue is necessarily zero) and non-null if and only if it admits exactly two independent real null eigendirections (with eigenvalues either both zero if F is spacelike or else differing only in sign if F is timelike or non-simple).*

So far the discussion of the algebraic properties of bivectors has been studied using "canonical" tetrads. Now a discussion through Segre types will be given. It is noted that, on account of the skew-symmetry of F, $F_{ab} g^{ab} = F_a{}^a = 0$ and so *the sum of the eigenvalues of F is zero.* For the same reason *any real or complex non-null eigenvector of F must have zero eigenvalue* (because $k^a k_a \neq 0$, $k^a F_a{}^b = \lambda k^b$ $(\lambda \in \mathbb{C}) \Rightarrow \lambda k^a k_a = 0$ $\Rightarrow \lambda = 0$). Also if all eigenvalues of F are real and if $k \in T_pM$ is a real eigenvector of F with a corresponding non-simple elementary divisor then there exists $k' \in T_pM$, independent of k, and $\lambda \in \mathbb{C}$ such that (section 2.6)

$$k^a F_a{}^b = \lambda k^b \quad k'^a F_a{}^b = \lambda k'^b + k^b \tag{7.40}$$

A contraction of these equations with k_b and k'_b gives

$$\lambda k_a k^a = 0 \qquad \lambda k'^a k'_a + k_a k'^a = 0 \qquad 2\lambda k'_a k^a + k_a k^a = 0 \tag{7.41}$$

If $\lambda \neq 0$ one easily gets the contradiction that k and k' are null and orthogonal. Hence $\lambda = 0$ and (7.41) shows that k is null and orthogonal to k'. *Hence any real eigenvector associated with a non-simple elementary divisor is null and the corresponding eigenvalue is zero.*

Let F be a real bivector at $p \in M$ with $F(p) \neq 0$. If F is non-null then, in terms of a null tetrad (l, n, x, y) at p, F takes one of the standard forms $2\lambda l_{[a}n_{b]}$ (timelike), $2\mu x_{[a}y_{b]}$ (spacelike) or $2\alpha l_{[a}n_{b]} + 2\beta x_{[a}y_{b]}$ (non-simple) for λ, μ, α and β non-zero real numbers. If F is null then $F_{ab} = 2\nu l_{[a}x_{b]}$ $(0 \neq \nu \in \mathbb{R})$. One then has

i) *F timelike:* Here the real eigenvectors may be taken as l (with eigenvalue $-\lambda$), $n(\lambda)$, $x(0)$ and $y(0)$. Hence $F_a{}^b$ is diagonalisable over $\mathbb{R}$ with Segre type $\{11(11)\}$.
ii) *F spacelike:* Here the real eigenvectors may be taken as l (with eigenvalue zero), $n(0)$ and the complex eigenvectors as $\sqrt{2}m$ $(= x + iy)$ and $\sqrt{2}\bar{m}$ (with eigenvalues $-i\mu$ and $+i\mu$, respectively). Hence $F_a{}^b$ is diagonalisable over $\mathbb{C}$ with Segre type $\{(11)z\bar{z}\}$.
iii) *F non-simple:* Here the real eigenvectors may be taken as l (with eigenvalue $-\alpha$), $n(\alpha)$ and the complex ones as $m(-i\beta)$ and $\bar{m}(i\beta)$. The Segre type is $\{11z\bar{z}\}$.
iv) *F null:* In this case, elementary algebra reveals that the only eigenvalues of F are real and the eigenvectors may be taken as l and y, each with zero eigenvalue. The Segre type can thus only be either $\{(31)\}$ or $\{(22)\}$. However the latter, having two non-simple elementary divisors, would, according to a result above, require two independent null eigendirections. Hence the Segre type is $\{(31)\}$.

As well as the eigenvector structure of a real bivector one could study the structure of its invariant 2-spaces. These results will be given briefly here and are easily proved using tetrad techniques. Let F be a real bivector at p. A 2-space $W \subseteq T_pM$ is called an *invariant 2-space of* F if whenever r^a is in W so also is $r^a F_a{}^b$. Any such bivector admits an invariant 2-space and if W is an invariant 2-space of F so also is the orthogonal complement of W. If W is a *null* invariant 2-space of F the unique null direction in W is a real null eigendirection of F. If W is a timelike invariant 2-space then the two real null directions contained in W are real null eigendirections of F. It is a straightforward exercise to link the existence of invariant 2-spaces of F with the latter's Ségre type.

There is a map $f_\theta : B(p) \longrightarrow B(p)$ given for $F \in B(p)$ and $0 \leq \theta < 2\pi$ by

$$F \longrightarrow f_\theta(F) = \cos\theta F - \sin\theta \, {}^*F \tag{7.42}$$

and called a *duality rotation* (through the angle θ). It can also be written in terms of the complex self-dual bivectors $\overset{+}{F}$ and $\overset{+}{F}{}'$ corresponding to F and $f_\theta(F)$, respectively, as $\overset{+}{F} \longrightarrow \overset{+}{F}{}' = e^{i\theta}\overset{+}{F}$. Clearly if F is null then so is $f_\theta(F)$ for each θ but if F is spacelike or timelike, $f_\theta(F)$ is, in general, non-simple. If F is non-simple then there exists θ such that $f_\theta(F)$ is simple and choices of θ exist such that $f_\theta(F)$ is spacelike or timelike. The blades of these latter bivectors are the canonical pair of blades of F as in (7.39).

The set $B(p)$ of real bivectors at $p \in M$ can be regarded as the manifold and vector space $\mathbb{R}^6$ (section 4.3, example (iv)). The set $Q'(p)$ of simple members of $B(p)$ is a subset but not a subspace of $B(p)$. Let $B_1, B_2 \in B(p)$ be non-zero and consider the equivalence relation $\sim$ on $B(p)$ given by $B_1 \sim B_2 \Leftrightarrow B_2 = \lambda B_1, \lambda \in \mathbb{R}$. Thus one obtains the quotient manifold $P^5\mathbb{R} = G(1, \mathbb{R}^6)$ of "projective bivectors" (section 4.17). Since the equivalence class under $\sim$ of a simple bivector consists entirely of simple bivectors one can define the set of equivalence classes Q of Q' under $\sim$ ("projective simple bivectors"). Now each simple bivector determines its blade uniquely and each such blade determines its corresponding bivector up to a real multiple. Now the set of blades (2-spaces) at p is just the Grassmann manifold $G(2, \mathbb{R}^4)$. Hence one has a natural *injective* map $\phi : G(2, \mathbb{R}^4) \longrightarrow P^5\mathbb{R}$ (section 4.17) whose range is Q. To obtain a coordinate representative of this map choose a basis of T_pM such that a particular member of $G(2, \mathbb{R}^4)$ is spanned by $(1, 0, x, y)$ and $(0, 1, a, b)$ for $x, y, a, b \in \mathbb{R}$. In the associated Grassmann chart for $G(2, \mathbb{R}^4)$ and the obvious (Grassmann) chart in $P^5\mathbb{R} = G(1, \mathbb{R}^6)$ the map ϕ has coordinate representative $(x, y, a, b) \longrightarrow (a, b, -x, -y, xb - ya)$. Elementary differentiation reveals this map to be of rank four and hence an immersion and so Q, which is the range of ϕ, may be given the structure of a 4-dimensional submanifold of $P^5\mathbb{R}$ diffeomorphic to $G(2,\mathbb{R}^4)$ and is, with its topology, compact and connected and is a closed regular (theorem 4.8 (i)) submanifold of the (Hausdorff) manifold $P^5\mathbb{R}$.

7.3 The Petrov Classification

As a preamble to this and other algebraic classifications it is useful to consider at a point p in the space-time M those tensors W of type (0, 4) satisfying the algebraic symmetries

$$W_{abcd} = -W_{bacd} = -W_{abdc}, \qquad W_{abcd} = W_{cdab} \tag{7.43}$$

$$W_{abcd} + W_{acdb} + W_{adbc} = 0, \qquad (\Leftrightarrow W_{a[bcd]} = 0) \tag{7.44}$$

the bracketed equivalence in (7.44) following from (7.43). For such a tensor one can define the *left dual* $^*W$ and the *right dual* W^* by

$$^*W_{abcd} = \frac{1}{2}\epsilon_{abef}W^{ef}{}_{cd}, \quad W^*_{abcd} = \frac{1}{2}W_{ab}{}^{ef}\epsilon_{cdef} \tag{7.45}$$

An immediate consequence (*cf.* (7.13)) is that

$$^{**}W_{abcd} = -W_{abcd} \qquad W^{**}_{abcd} = -W_{abcd} \tag{7.46}$$

Such tensors satisfy the Ruse-Lanczos identity (see, *e.g.* [61])

$$^*W^*_{abcd} + W_{abcd} = 2g_{a[c}\tilde{W}_{d]b} + 2g_{b[d}\tilde{W}_{c]a} \tag{7.47}$$

where $W_{ab} \equiv W^c{}_{acb}$ ($= W_{ba}$ from (7.44)), $\tilde{W}_{ab} \equiv W_{ab} - \frac{1}{4}Wg_{ab}$, $W \equiv W^a{}_a$ (and hence $\tilde{W}^a{}_a = 0$). It follows from (7.47) that $^*W^*_{a[bcd]} = 0$ and a contraction of (7.47) with g^{bd} reveals the equivalent statements

$$^*W^*_{abcd} = -W_{abcd} \; (\Leftrightarrow {}^*W_{abcd} = W^*_{abcd}) \Leftrightarrow \tilde{W}_{ab} = 0 \tag{7.48}$$

Next, the following statements are equivalent for W (even if only (7.43) is assumed)

$$W_{a[bcd]} = 0, \qquad W_{abcd}\epsilon^{ebcd} = 0, \qquad W^{*a}{}_{bad} = 0 \tag{7.49}$$

Thus if one takes W as the curvature tensor of the space-time one has, apart from the usual symmetries, the relation $R^{*a}{}_{bad} = 0$

Now take W as the Weyl tensor C (see section 4.16). Then it follows from (7.47) and the result $C^a{}_{bad} = 0$ that

$$^*C^*_{abcd} = -C_{abcd} \quad \text{and} \quad ^*C_{abcd} = C^*_{abcd} \tag{7.50}$$

Further one has

$$^*C_{abcd} = C^*_{abcd} = \frac{1}{2}C_{abmn}\epsilon^{mn}{}_{cd} = \frac{1}{2}C_{mnab}\epsilon^{mn}{}_{cd} = {}^*C_{cdab} \tag{7.51}$$

and so $^*C$ (and hence C^*) satisfies (7.43). Then applying (7.49) to $W(= C^*)$ using (7.50) yields $C^*{}_{a[bcd]} = 0$ and to C yields $C^{*a}{}_{bad} = 0$. Thus if one defines the *complex self-dual Weyl tensor* $\overset{+}{C}$ by

$$\overset{+}{C}_{abcd} = C_{abcd} + i^*C_{abcd} \tag{7.52}$$

then $\overset{+}{C}$ satisfies the symmetries (7.43) and (7.44), the relation $\overset{+}{C}{}^a{}_{bad} = 0$ and the *self dual property* $^*\overset{+}{C} = -i\overset{+}{C}$ (the position of $*$ being irrelevant).

There is a convenient way of labelling the components of a tensor such as W above. Allowing capital letters $A, B, \ldots$ to take the values $1, 2, \ldots, 6$ one can label a skew-symmetric pair of tensor indices with such a capital letter. Adopting the convention $[23] \leftrightarrow 1$, $[31] \leftrightarrow 2$, $[12] \leftrightarrow 3$, $[10] \leftrightarrow 4$, $[20] \leftrightarrow 5$, $[30] \leftrightarrow 6$, W can now be written as W_{AB} and is a symmetric 6×6 matrix since $W_{AB} = W_{BA}$. This notation is referred to as the 6×6 *notation* and the indices $A, B, \ldots$ as *bivector indices*. It was, as far as the author is aware, first used by Kretschmann [3] and will here be applied to the Weyl tensor. The *rank* of W is defined to be the rank of the matrix W_{AB}.

The Weyl tensor C on space-time is an important geometrical object and a general description of it was given in section 4.16. In the component form $C^a{}_{bcd}$ it is conformally invariant. A very important advance in general relativity was made by Petrov [62],[63] when he developed an algebraic classification for the Weyl tensor (more precisely he originally devised the classification for the Riemann tensor of a space-time which was an Einstein space). Petrov's work was extended by others in the decade that followed it and contributions by Géhéniau [64], Bel [65], Debever [66], Pirani [67], Penrose [68], Sachs [47] and Ehlers and Kundt [41] are of particular importance. For vacuum space-times, the Riemann and Weyl tensor are equal and the Petrov classification then applies to the Riemann tensor.

It is remarked here that the tensors G_{abcd} and ϵ_{abcd} used in the previous section have the algebraic properties (7.43) and also from (7.47) that $G^*_{abcd} = {}^*G_{abcd}(= \frac{1}{2}\epsilon_{abcd})$. Further, $G_{a[bcd]} = 0$ (and so $\epsilon^a{}_{bad} = 0$),

$G^a{}_{bad} = \frac{3}{2}g_{bd} \neq 0$ (and so $\epsilon_{a[bcd]} \neq 0$) and $\epsilon^*_{abcd} = {}^*\epsilon_{abcd} = -2G_{abcd}$. The complex self-dual tensor $\overset{+}{G} = G + i^*G$ with components $G_{abcd} + \frac{i}{2}\epsilon_{abcd}$ will be found useful later.

A convenient (but certainly not the only) approach to the algebraic classification of the Weyl tensor starts with the idea of an *eigenbivector*. It should be pointed out that this is an *algebraic* classification of C and that the details below apply at a single point $p \in M$. A (real or complex) bivector $F \in CB(p)$ is called an *eigenbivector* of C (respectively, of $\overset{+}{C}$) if the first (respectively, the second) equation in (7.53) holds

$$C_{abcd}F^{cd} = \lambda F_{ab} \quad (\lambda \in \mathbb{C})$$
$$\overset{+}{C}_{abcd}F^{cd} = \mu F_{ab} \quad (\mu \in \mathbb{C}) \tag{7.53}$$

and then λ (respectively, μ) is the associated *eigenvalue*. The classification sought is an algebraic (eigenbivector-eigenvalue) classification of the (real) Weyl tensor C when the latter is regarded as a linear map $B(p) \to B(p)$ (*i.e.* $\mathbb{R}^6 \to \mathbb{R}^6$) given by $H_{ab} \to C_{abcd}H^{cd}$. It turns out, however, to be more convenient to change the problem to an equivalent one (and an obvious abbreviated notation in which, for example, for a bivector H, CH means $C_{abcd}H^{cd}$, will be useful for this purpose). Such an algebraic description of the linear transformation $\mathbb{R}^6 \to \mathbb{R}^6$ represented by C in terms of matrix similarity can be equivalently described by regarding C as a linear map $CB(p) \to CB(p)$ by extending the definition of the original map to complex bivectors. Thus C is now regarded as a linear map $\mathbb{C}^6 \to \mathbb{C}^6$, the algebraic equivalence, in the above sense, being mentioned in section 2.6. Now if $H \in CB(p)$, one has ${}^*(CH) = {}^*CH = (C^*H) = C({}^*H)$. It follows from (7.19) that if $H \in S^+(p)$ (respectively, $S^-(p)$) then $CH \in S^+(p)$ (respectively, $S^-(p)$). Thus, from (7.22), C is completely described by its obvious restrictions C_1 and C_2 which are then, respectively, linear maps $C_1 : S^+(p) \to S^+(p)$ and $C_2 : S^-(p) \to S^-(p)$. Then introducing the conjugation operator $k : CB(p) \to CB(p)$, given by $k(H) = \overline{H}$, one has $C_2 = k \circ C_1 \circ k^{-1}$. It follows that C_1 and C_2 have identical Jordan forms as linear maps $\mathbb{C}^3 \to \mathbb{C}^3$ (including degeneracies and with eigenvalues differing only by conjugation). Since $S^+(p)$ and $S^-(p)$ are invariant subspaces of C whose direct sum is $CB(p)$, the Jordan form of C is just the common Jordan form of C_1 and C_2 "repeated" in an obvious way. This common Jordan form will be taken as the algebraic "type" of C.

Now consider the complex self-dual Weyl tensor $\overset{+}{C}$ at p as a linear map

$CB(p) \to CB(p)$ in the same way as for C. Then

$$H \in S^+(p) \Rightarrow \overset{+}{C}H = (C + iC^*)H = CH + iC(^*H) = 2CH \in S^+(p)$$
$$H \in S^-(p) \Rightarrow \overset{+}{C}H = (C + iC^*)H = CH + iC(^*H) = 0 \tag{7.54}$$

Thus $\overset{+}{C}$ has a range contained in $S^+(p)$ and maps $S^-(p)$ to zero. Further, the first equation in (7.54) shows that the restriction of $\overset{+}{C}$ to $S^+(p)$ is simply twice the restriction C_1 of C to $S^+(p)$. Thus the Jordan form of $\overset{+}{C}$ is determined by the Jordan form of its restriction to $S^+(p)$ which is, apart from the factor 2, identical to that of C_1 and hence to the algebraic type of C defined above. Thus the problem of classifying C has been transformed to the equivalent problem of classifying (the restriction of) $\overset{+}{C}$ on the 3-dimensional complex vector space $S^+(p)$ of complex self-dual bivectors at p. One further transformation of the problem is instructive.

Recalling the 6×6 notation mentioned earlier one can, using the symmetry of the 6×6 Weyl tensor, write

$$C_{AB} = \begin{pmatrix} M & N^{\mathrm{T}} \\ N & Q \end{pmatrix} \tag{7.55}$$

where M, N and Q and 3×3 real matrices with M and Q symmetric. The symmetry (7.44) for C when written out using $(a, b, c, d) = (0, 1, 2, 3)$ and the 6×6 index convention shows that N *is tracefree*. At this point a frame (in fact an orthonormal frame with respect to $g(p)$) will be introduced in T_pM but the final outcome will be independent of it. In this frame one has $g_{ab}(p) = \mathrm{diag}(-1, 1, 1, 1) = \eta_{ab}$ and then the identity $g^{ae}C_{ebad} = 0$ is written out in terms of the components C_{AB} in this frame. A little algebra shows that these ten conditions are equivalent to the ten restrictions $N = N^T$, $Q = -M$ and trace $M = 0$ in (7.55). Thus (7.55) becomes

$$C_{AB} = \begin{pmatrix} M & N \\ N & -M \end{pmatrix} \tag{7.56}$$

with M and N symmetric tracefree members of $M_3\mathbb{R}$. With this orthonormal frame fixed at p one has, at p, $g_{ab} = \eta_{ab}$ and converting the components of the bivector metric G (equation (7.11)) to 6×6 notation, one has

$$G_{AB} = \frac{1}{2} \begin{pmatrix} I_3 & 0 \\ 0 & -I_3 \end{pmatrix} \tag{7.57}$$

where I_3 is the unit 3×3 matrix. If ϵ is the 6×6 matrix of the tensor ϵ_{abcd} then, since $\sqrt{-g} = 1$ here, one has

$$\epsilon_{AB} = \begin{pmatrix} 0 & I_3 \\ I_3 & 0 \end{pmatrix} \tag{7.58}$$

A bivector F_{ab} in $CB(p)$ may be expressed in 6×6 notation as a row

$$F_A = (F_1, F_2, F_3, F_4, F_5, F_6)$$

where $F_1 = F_{23}, \ldots, F_6 = F_{30} \in \mathbb{C}$ and then if ${}^*F_A$ represents its dual, one has ${}^*F_A = \epsilon_{AB}F^B$ (the factor $\frac{1}{2}$ no longer being necessary). Also $F^A = 2G^{AB}F_B$ (noting the factor 2) where G^{AB} is the 6×6 representation of G^{abcd} and is identical to the matrix in (7.57). Then

$${}^*F_A = 2\epsilon_{AB}G^{BC}F_C \tag{7.59}$$

and so ${}^*F_A = (-F_4, -F_5, -F_6, F_1, F_2, F_3)$. So if one represents a bivector F by two triples of complex numbers in the above ordering, say $F_A = (A, B)$ with $A, B \in \mathbb{C}^3$ then ${}^*F_A = (-B, A)$ and F is self dual if and only if $F_A = (A, iA)$ and anti self-dual if and only if $F_A = (A, -iA)$.

The above discussion has shown that C can be represented rather compactly by two tracefree symmetric members M and N of $M_3\mathbb{C}$ as in (7.56) and hence by the complex 3×3 symmetric tracefree matrix $Q = M - iN$. Now compute $\overset{+}{C}_{AB}$

$$\begin{aligned} \overset{+}{C}_{AB} &= C_{AB} + i^*C_{AB} = C_{AB} + 2i\epsilon_{AC}G^{CD}C_{BD} \\ &= \begin{pmatrix} M - iN & N + iM \\ N + iM & -M + iN \end{pmatrix} = \begin{pmatrix} Q & iQ \\ iQ & -Q \end{pmatrix} \end{aligned} \tag{7.60}$$

Then with $F_A = (A, B)$, $F^A = (A, -B)$

$$\begin{pmatrix} Q & iQ \\ iQ & -Q \end{pmatrix} \begin{pmatrix} A \\ -B \end{pmatrix} = \begin{pmatrix} D \\ iD \end{pmatrix}$$

$$D = Q(A - iB) \tag{7.61}$$

Thus one recovers the results above that $\overset{+}{C}$ has range contained in $S^+(p)$ and that it maps $S^-(p)$ to zero. Now F_{ab} being an eigenbivector of $\overset{+}{C}_{abcd}$ ($\overset{+}{C}_{abcd}F^{cd} = 2\lambda F_{ab}$) is equivalent to F_A being an eigenvector of $\overset{+}{C}_{AB}$ ($\overset{+}{C}_{AB}F^B = \lambda F_A$ - and noting the factor $\frac{1}{2}$). It then follows from (7.61) that any eigenbivector of $\overset{+}{C}$ with non-zero eigenvalue is necessarily self-dual and that, irrespective of the eigenvalue, F is a *self-dual* eigenbivector of $\overset{+}{C}$ if and only if the complex 3-vector A associated with F is an eigenvector of Q $(F_A = (A, iA),\ \overset{+}{C}_{AB}F^B = \lambda F_A \Leftrightarrow QA = \frac{\lambda}{2}A)$. Hence the original problem given by the first equation in (7.53) has now been transferred to the essentially equivalent and much simpler one of determining the eigenvector-eigenvalue structure of the symmetric tracefree member $Q \in M_3\mathbb{C}$. This is done using the Jordan canonical forms (section 2.6) and is simplified by the facts that Q is tracefree and $\mathbb{C}$ algebraically closed. The possibilities for Q at $p \in M$ can now be given in terms of their Segre type and their traditional Petrov type label. If the Segre type is $\{111\}$ with distinct eigenvalues $\lambda_1, \lambda_2, \lambda_3 \in \mathbb{C}$ satisfying $\lambda_1 + \lambda_2 + \lambda_3 = 0$ (from the tracefree condition) the Petrov type is labelled **I**. If the Segre type is $\{1(11)\}$ with eigenvalues 2λ, $-\lambda$, $-\lambda$ $(\lambda \in \mathbb{C})$ the Petrov type is **D**. If the Segre type is $\{21\}$ with eigenvalues $-\lambda$, 2λ $(\lambda \in \mathbb{C})$ the Petrov type is **II**. If the Segre type is $\{(21)\}$ with all eigenvalues necessarily zero the Petrov type is **N** and if the Segre type is $\{3\}$ with all eigenvalues necessarily zero the Petrov type is **III**. In the case that $C(p) = 0$ the type is labelled **O**. Sometimes the (Petrov) type **D** is referred to as (Petrov) type **I** *degenerate* and type **N** as type **II** *null*. These six types **I**, **D**, **II**, **N**, **III** and **O** are the only possibilities.

By reinstating an orthonormal basis at p and choosing it appropriately for each type one can find the *Petrov canonical forms* taken by C_{abcd} as the matrix C_{AB} [63],[67]. For type **I** one has, for $\alpha_i, \beta_i \in \mathbb{R}$, $\sum \alpha_i = \sum \beta_i = 0$ and $\alpha_1, ...\beta_3$ not all zero (and consistent with the Segre type)

$$C_{AB} = \begin{pmatrix} \alpha_1 & 0 & 0 & \beta_1 & 0 & 0 \\ 0 & \alpha_2 & 0 & 0 & \beta_2 & 0 \\ 0 & 0 & \alpha_3 & 0 & 0 & \beta_3 \\ \beta_1 & 0 & 0 & -\alpha_1 & 0 & 0 \\ 0 & \beta_2 & 0 & 0 & -\alpha_2 & 0 \\ 0 & 0 & \beta_3 & 0 & 0 & -\alpha_3 \end{pmatrix} \quad \text{Type } \mathbf{I} \qquad (7.62)$$

Here, for C_{AB}, the eigenbivectors and their eigenvalues are $F^A = \delta_1^A \pm i\delta_4^A$ (eigenvalue $\alpha_1 \pm i\beta_1$), $\delta_2^A \pm i\delta_5^A$ ($\alpha_2 \pm i\beta_2$), $\delta_3^A \pm i\delta_6^A$ ($\alpha_3 \pm i\beta_3$). For type

II and $\alpha, \beta \in \mathbb{R}$ not both zero

$$C_{AB} = \begin{pmatrix} -2\alpha & 0 & 0 & 2\beta & 0 & 0 \\ 0 & \alpha-1 & 0 & 0 & -\beta & 1 \\ 0 & 0 & \alpha+1 & 0 & 1 & -\beta \\ 2\beta & 0 & 0 & 2\alpha & 0 & 0 \\ 0 & -\beta & 1 & 0 & 1-\alpha & 0 \\ 0 & 1 & -\beta & 0 & 0 & -1-\alpha \end{pmatrix} \quad \text{Type } \mathbf{II} \qquad (7.63)$$

Here the eigenvalues are $-2(\alpha \pm i\beta)$ and $\alpha \pm i\beta$. The eigenbivectors can be easily calculated and the ones associated with the non-simple elementary divisors are null. For type **III**

$$C_{AB} = \begin{pmatrix} 0 & -1 & 0 & 0 & 0 & 1 \\ -1 & 0 & 0 & 0 & 0 & 0 \\ 0 & 0 & 0 & 1 & 0 & 0 \\ 0 & 0 & 1 & 0 & 1 & 0 \\ 0 & 0 & 0 & 1 & 0 & 0 \\ 1 & 0 & 0 & 0 & 0 & 0 \end{pmatrix} \quad \text{Type } \mathbf{III} \qquad (7.64)$$

Here the eigenbivectors are null and the eigenvalues zero. For type **D** one sets $\alpha_2 = \alpha_3$ ($= \alpha$ say) and so $\alpha_1 = -2\alpha$, and $\beta_2 = \beta_3$ ($= -\beta$ say) and so $\beta_1 = 2\beta$ in the type **I** matrix (7.62). For type **N** one sets $\alpha = \beta = 0$ in the type **II** matrix (7.63). The above matrix expressions are usually referred to as the *Petrov canonical forms* and the orthonormal bases in which they are realised as the *Petrov tetrads*. The eigenvalues listed above (or sometimes their real and imaginary parts) are variously referred to as the *Petrov invariants*, *Petrov scalars* or *Weyl invariants*. It is remarked for future reference that the rank of the 6×6 matrix C_{AB} for each Petrov type is as follows: for type **I** the rank is 4 or 6, for type **II** and **D** it is 6, for type **III** it is 4, for type **N** it is 2 and for type **O** it is 0. Also all eigenvalues of C vanish if and only if the type is **III**, **N** or **O**.

It is useful to know how unique the Petrov tetrads are and this can be established from the work above. In fact (and ignoring tetrad "reflections") the Petrov tetrad for the Petrov types **I**, **II** and **III** is unique. For type **D** the Petrov tetrad is determined up to a spacelike rotation in the x^2x^3 2-space and a boost in the x^0x^1 2-space. For type **N** the tetrad is determined up to two independent null rotations about a null direction in the x^0x^1 2-space. This will be considered in more detail at the end of the next section.

7.4 Alternative Approaches to the Petrov Classification

So far, the Petrov Classification of the Weyl tensor has been described in terms of the latter's eigenvector-eigenvalue structure. There are many other approaches and two of them, due to Bel [65] (and Debever [66]) and Sachs [47], can now be described together. These result in a rather elegant approach to the Petrov canonical forms and the very useful *Bel criteria*.

Recalling the self-dual property ($^*\overset{+}{C} = -i\overset{+}{C}$) of $\overset{+}{C}$, one can write $\overset{+}{C} = \frac{1}{2}(\overset{+}{C} + i^*\overset{+}{C})$ and so

$$\begin{aligned}\overset{+}{C}_{abcd} &= \frac{1}{2}(\overset{+}{C}_{ab}{}^{ef}G_{efcd} + \frac{1}{2}i\overset{+}{C}_{ab}{}^{ef}\epsilon_{efcd}) \\ &= \frac{1}{2}\overset{+}{C}_{ab}{}^{ef}(G_{efcd} + \frac{1}{2}i\epsilon_{efcd}) \\ &= \frac{1}{4}\overset{+}{C}{}^{mnef}(G_{mnab} + \frac{1}{2}i\epsilon_{mnab})(G_{efcd} + \frac{1}{2}i\epsilon_{efcd}) \end{aligned} \tag{7.65}$$

Thus one may introduce the complex self-dual bivectors U, V and M from (7.33) to get

$$\begin{aligned}\overset{+}{C}_{abcd} &= AV_{ab}V_{cd} + B(V_{ab}M_{cd} + M_{ab}V_{cd}) + CM_{ab}M_{cd} + DU_{ab}U_{cd} \\ &\quad + E(V_{ab}U_{cd} + U_{ab}V_{cd}) + F(U_{ab}M_{cd} + M_{ab}U_{cd})\end{aligned} \tag{7.66}$$

where $A, \ldots, F \in \mathbb{C}$. Now the condition $\overset{+}{C}{}^a{}_{bad} = 0$ is equivalent to $C = E$ above and so one achieves the elegant canonical decomposition of $\overset{+}{C}$ in terms of the complex self-dual bivector basis U, V, M and $C^1, \ldots, C^5 \in \mathbb{C}$ given by Sachs [47]

$$\begin{aligned}\overset{+}{C}_{abcd} &= C^1V_{ab}V_{cd} + C^2(V_{ab}M_{cd} + M_{ab}V_{cd}) \\ &\quad + C^3(M_{ab}M_{cd} + V_{ab}U_{cd} + U_{ab}V_{cd}) + C^4(U_{ab}M_{cd} + M_{ab}U_{cd}) \\ &\quad + C^5U_{ab}U_{cd}\end{aligned} \tag{7.67}$$

To see how this relates to the Petrov classification one proceeds in a way similar to that used previously for bivectors. One performs null rotations about l given by (7.35) to see if the corresponding bivector transformations (7.36) will yield a particularly simple "canonical" form for $\overset{+}{C}$ in (7.66). Thus one imagines (7.67), but with primes on all quantities on the right hand side and then substitutes in (7.36). Equating back with the original unprimed expression for $\overset{+}{C}$ then gives after a lengthy but straightforward calculation

$$\begin{aligned}
C^1 &= A^2 e^{-2i\theta} C'^1 + 4A^2 \bar{B} e^{-i\theta} C'^2 + 6A^2 \bar{B}^2 C'^3 \\
&\quad + 4A^2 \bar{B}^3 e^{i\theta} C'^4 + A^2 \bar{B}^4 e^{2i\theta} C'^5 \\
C^2 &= A e^{-i\theta} C'^2 + 3A\bar{B} C'^3 + 3A\bar{B}^2 e^{i\theta} C'^4 + A\bar{B}^3 e^{2i\theta} C'^5 \\
C^3 &= C'^3 + 2\bar{B} e^{i\theta} C'^4 + \bar{B}^2 e^{2i\theta} C'^5 \\
C^4 &= A^{-1} e^{i\theta} C'^4 + A^{-1} \bar{B} e^{2i\theta} C'^5 \\
C^5 &= A^{-2} e^{2i\theta} C'^5
\end{aligned} \tag{7.68}$$

The null rotations (7.35) fix the direction of l but may change the direction spanned by n to any other null direction except that of l. Now at $p \in M$ suppose $\overset{+}{C}(p) \neq 0$ and consider those *null* vectors $k \in T_pM$ satisfying

$$k_{[e} \overset{+}{C}_{a]bc[d} k_{f]} k^b k^c = 0 \tag{7.69}$$

Then k (strictly speaking, the direction spanned by k, but this will be understood) is called a *principal null direction* of $\overset{+}{C}$ (or C) and it is clear that k satisfies the same relation with $\overset{+}{C}$ replaced by C or ${}^*C$. Suppose (7.69) has a solution. Choose it to be the null vector l' in a null tetrad $(l'n'x'y')$. Then in the expression (7.67) for $\overset{+}{C}$, but in terms of U', V' and M', one finds that this is equivalent to $C'^5 = 0$. Any other solution to (7.69) will then be (a multiple of) the vector n in some appropriately null rotated (about l') null tetrad. These solutions will show up by the vanishing of C^1 in (7.67) since this is equivalent to (7.69) with $k = n$. Now since $C'^5 = 0$ one obtains from (7.68) a cubic in $\bar{B}$ and hence there are at most three more principal null directions of $\overset{+}{C}$. If one does not assume that (7.69) has a solution, so that $C'^5 \neq 0$, then $C^1 = 0$ is a quartic in $\bar{B}$ with at least one and at most four solutions. Hence *there is at least one and at most four principal null directions of* $\overset{+}{C}$.

Now consider those *null* vectors $k \in T_pM$ satisfying

$$k_{[e} \overset{+}{C}_{a]bcd} k^b k^c = 0 \quad (\Leftrightarrow \overset{+}{C}_{abcd} k^b k^d \propto k_a k_c) \tag{7.70}$$

Then k is called a *repeated principal null direction* of $\overset{+}{C}$ (or C) and, again, k satisfies the same relation with $\overset{+}{C}$ replaced by C or ${}^*C$. Clearly a repeated principal null direction is a principal null direction. As before, if (7.70) has a solution choose it to be l'. Then in the appropriate primed expression for $\overset{+}{C}$ this is equivalent to $C'^4 = C'^5 = 0$. Then to find any other such solution n one requires $C^1 = C^2 = 0$ and (7.68) shows there is at most one more solution of (7.70). Thus *there are at most two repeated principal null directions* of $\overset{+}{C}$ (and possibly none).

To link the existence of these preferred null directions with the Petrov types consider first the type N case and the canonical form (7.63) for C with $\alpha = \beta = 0$. In this Petrov tetrad construct a null tetrad by the component relations

$$l^a = 2^{-1/2}(1,1,0,0) \quad n^a = 2^{-1/2}(-1,1,0,0) \quad m^a = 2^{-1/2}(0,0,1,-i) \tag{7.71}$$

Then, in 6-dimensional notation, one finds for the bivectors V, U and M the expressions

$$\begin{aligned} V_A &= \tfrac{1}{2}(0,-i,1,0,1,i), \qquad U_A = \tfrac{1}{2}(0,i,1,0,-1,i), \\ M_A &= (-i,0,0,1,0,0) \end{aligned} \tag{7.72}$$

Then one finds by inspection from (7.63) with $\alpha = \beta = 0$ that

$$\overset{+}{C}_{abcd} = 4V_{ab}V_{cd} \tag{7.73}$$

which is (7.67) with $C^1 = 4$, $C^2 = C^3 = C^4 = C^5 = 0$. In (7.73) the independent self-dual eigenbivectors are M and V each with eigenvalue zero and the principal null direction l of V satisfies

$$\overset{+}{C}_{abcd}l^d = 0 \tag{7.74}$$

Hence l is a repeated principal null direction of $\overset{+}{C}$ (or C) from (7.70). Regarding this vector as l' in the primed canonical expression for $\overset{+}{C}$ with $C'^1 \neq 0$, $C'^2 = C'^3 = C'^4 = C'^5 = 0$, it is easily checked that there are no other principal null directions of $\overset{+}{C}$. This is because one must then establish a null tetrad in which $C^1 = 0$ and (7.68) shows this to be impossible. The fact that C^1 turned out equal to four is not in any sense special. In fact if $\overset{+}{C}_{abcd} = C^1 V_{ab}V_{cd}$ then by a null rotation about l and suitable choices of A and θ in the first equation of (7.68) one may always set $C^1 = 1$ by appropriate choice of null tetrad. This also clarifies a remark on the uniqueness of the Petrov tetrad made earlier, since C^1 is unchanged under null rotations with $A = 1$, $\theta = 0$ and B arbitrary.

For the type **III** case the choices (7.71) and (7.72) and inspection of (7.64) lead to

$$\overset{+}{C}_{abcd} = 2(V_{ab}M_{cd} + M_{ab}V_{cd}) \tag{7.75}$$

which is (7.67) with $C^1 = C^3 = C^4 = C^5 = 0$, $C^2 = 2$. The null bivector V is the unique (up to complex scaling) self-dual eigenbivector and its eigenvalue is zero. Its principal null direction l satisfies

$$\overset{+}{C}_{abcd}l^d = 2V_{ab}l_c \tag{7.76}$$

Hence l is a repeated principal null direction of $\overset{+}{C}$ (or C). It is easily checked that n satisfies (7.69) (but not (7.70)) and so is a (non-repeated) principal null direction of $\overset{+}{C}$ (or C). Regarding this vector l as the vector l' in the primed canonical expression for $\overset{+}{C}$ with $C'^1 = C'^3 = C'^4 = C'^5 = 0$ one easily finds from (7.68) that there are no other principal null directions. Further, under null rotations, one finds from (7.68) that $C'^2 \to Ae^{-i\theta}C'^2$ and so one can ensure that $C^2 = 1$. The uniqueness (up to reflections) of the Petrov tetrad is easily checked.

For Petrov type **II** similar arguments with (7.71), (7.72) and (7.63) lead to the canonical expression

$$\overset{+}{C}_{abcd} = 4V_{ab}V_{cd} + 2(\alpha + i\beta)(M_{ab}M_{cd} + V_{ab}U_{cd} + U_{ab}V_{cd}) \tag{7.77}$$

This is (7.67) with $C^2 = C^4 = C^5 = 0$. The independent self-dual eigenbivectors are M and V and the principal null direction l of V satisfies

$$\overset{+}{C}_{abcd}l^b l^d = 2(\alpha + i\beta)l_a l_c \tag{7.78}$$

and is thus a repeated principal null direction of $\overset{+}{C}$ (or C). It is then easily checked from (7.68) that there are no more repeated principal null directions but that there are exactly two more principal null directions. Also the null tetrad may be chosen so that $C^1 = 1$ (whilst retaining $C^2 = C^4 = C^5 = 0$) but C^3 is unchanged under such null rotations. Again the uniqueness (up to reflections) of the Petrov tetrad is easily demonstrated.

For Petrov type **D**, similar procedures lead to

$$\overset{+}{C}_{abcd} = 2(\alpha + i\beta)(M_{ab}M_{cd} + V_{ab}U_{cd} + U_{ab}V_{cd}) \tag{7.79}$$

which is (7.67) with $C^1 = C^2 = C^4 = C^5 = 0$. The independent self-dual eigenbivectors are M, U and V. Further l and n each satisfy (7.78) and so each is a repeated principal null direction of $\overset{+}{C}$ (or C). There are no other principal null directions. The coefficient C^3 is unchanged under null rotations about l and the Petrov tetrad is determined up to null rotations about l with $B = 0$.

For Petrov type **I** the situation is a little more complicated. The procedures adopted so far, using (7.62), yield a canonical form

$$\overset{+}{C}_{abcd} = C^1 V_{ab}V_{cd} + C^3(M_{ab}M_{cd} + V_{ab}U_{cd} + U_{ab}V_{cd}) + C^5 U_{ab}U_{cd} \quad (7.80)$$

with C^1 and C^5 *non-zero* complex numbers (otherwise $\overset{+}{C}$ would be type **II** or **D**) and which is (7.67) with $C^2 = C^4 = 0$. It follows from (7.68) that a null rotation about l may be employed so that in the new tetrad $C^1 = C^5$. An alternative form can be found by choosing l as a principal null direction (which necessarily exists) and hence $C^5 = 0$. Now another principal null direction must exist. (If $C^1 = 0$ it is spanned by n and if $C^1 \neq 0$ the lack of another such null direction would necessitate $C^2 = C^3 = C^4 = 0$ and imply that $\overset{+}{C}$ is of type **N**, as follows from (7.68).) Building a null tetrad containing these principal null directions leads to (7.67) with $C^1 = C^5 = 0$ and the alternative form is

$$\begin{aligned}\overset{+}{C}_{abcd} = {} & C^2(V_{ab}M_{cd} + M_{ab}V_{cd}) + C^3(M_{ab}M_{cd} + V_{ab}U_{cd} + U_{ab}V_{cd}) \\ & + C^4(U_{ab}M_{cd} + M_{ab}U_{cd})\end{aligned} \quad (7.81)$$

where C^2 and C^4 are *non-zero* complex numbers. A computation of the self-dual eigenbivectors from either (7.80) with $C^1 = C^5$ or (7.81) is straightforward. In the first of these expressions, for example, they are M and $U \pm V$.

Thus one achieves the canonical forms of Sachs [47] with those of Bel [65] and Debever [66] following by taking the real parts of (7.73), (7.75), (7.77) and (7.79). One can say a little more here. Suppose $\overset{+}{C}$ admits a *repeated* principal null direction l. Then in any null tetrad (l, n, x, y) equation (7.67) shows that $C^4 = C^5 = 0$. Equation (7.68) then shows that, by a suitable null rotation, one can achieve exactly one of the expressions (7.73), (7.75), (7.77) or (7.79) for types **N**, **III**, **II** or **D**, respectively. Thus *if C is of type* **I** *it admits no repeated principal null directions.* Then recalling the results established earlier, one sees that for Petrov type **N** there is a single (necessarily repeated) principal null direction, for type **III** there are exactly two principal null directions, one repeated and one not, for type **II** there are exactly three principal null directions, one repeated and two not, for type **D** there are exactly two principal null directions, both repeated and for type **I** there are exactly four principal null directions, none of which is repeated. These were established above except for the existence of *four* distinct principal null directions in the type **I** case. This requires some algebra but follows more easily in the spinor approach due to Penrose [68].

It is usual to call the Petrov types **O**, **N**, **III**, **II** and **D** *algebraically special* and the type **I** *algebraically general*. Thus $\overset{+}{C}(p)$ (or $C(P)$) *is of an algebraically special Petrov type if and only if it admits a repeated principal null direction.* The repeated principal null directions in the type **N** and **III** cases are referred to, respectively, as *quadruply* and *triply* repeated and those in the type **II** and **D** cases as *doubly* repeated. Treating non-repeated principal null directions as *single*, one achieves a notion of "counting properly" in which the "sum" of the principal null directions in each Petrov type is *four*. This can be given an algebraic meaning in the formalism developed here but is clearer in the spinor approach [68]. Repeated (respectively, non-repeated) principal null directions are often referred to as repeated (respectively, non-repeated) *Debever-Penrose* directions. This terminology, whilst appropriately recognising the work of Debever and Penrose, unfortunately fails to recall the work of Bel (see reference [65] and earlier papers cited there).

The existence of repeated principal null directions through (7.70) implies that the Petrov type is algebraically special. However, it does not distinguish between those principal null directions which are doubly, triply or quadruply repeated. There exists a means of distinguishing such degeneracies and which links it to the Petrov classification. These features, essentially due to Bel [65] are often referred to as the *Bel criteria*.

Theorem 7.4 *Let M be a space-time, $p \in M$ and $C(p) \neq 0$. Then the condition that $C(p)$ is of type N is equivalent to any of the following statements for some $k \in T_pM$, $k \neq 0$.*

$$(a)\quad \overset{+}{C}_{abcd}k^d = 0 \quad (b)\quad C_{abcd}k^d = 0 \quad (c)\quad {}^*C_{abcd}k^d = 0$$

$$(d)\quad \overset{+}{C}_{ab[cd}k_{e]} = 0 \quad (e)\quad C_{ab[cd}k_{e]} = 0 \quad (f)\quad {}^*C_{ab[cd}k_{e]} = 0$$

In the conditions (a)-(f) k is necessarily null and unique up to scaling and spans the repeated principal null direction of C.

Proof. If $C(p)$ is of type **N** then (7.74) shows that (a) holds. Also, dualling on the index pair ab shows that $(b) \Leftrightarrow (c)$ and clearly $(a) \Rightarrow (b)$ and (c) and then (b) and (c) together imply (a). Thus (a), (b) and (c) are equivalent. Now for any tensor $T_{a\ldots bcd}$ at p, the condition $T_{a\ldots bcd}\epsilon^{bcde} = 0$, upon contraction with ϵ_{epqr}, reveals that $T_{a\ldots[bcd]} = 0$. So choosing $T_{abcde} = C_{abcd}k_e$, condition (c) gives $C_{abmn}\epsilon^{mn}{}_{cd}k^d = 0$ which implies (e) and clearly $(e) \Leftrightarrow (f)$. Then (e) or $(f) \Rightarrow (d)$ and (d) gives ${}^*\overset{+}{C}_{ab[cd}k_{e]} = 0$ and hence (a) holds. So $(a), \ldots, (f)$ are equivalent and are implied by the type **N** condition. Conversely, if any (and hence all) of $(a), \ldots, (f)$ hold, a contraction of (e) with k^c and use of (b) shows that k is necessarily null. Essentially the same trick then shows that if $(a), \ldots, (f)$ hold with k replaced by k' then k' is null and orthogonal to k. Thus the direction spanned by k is unique. Then choosing $l = k$ in the expansion (7.67) for $\overset{+}{C}$ shows that only C^1 does not vanish and reveals a type **N** expression like (7.73). □

Theorem 7.5 *Let M be a space-time, $p \in M$ and $C(p)$ not of type* **O** *or* **N**. *Then the statement that $C(p)$ is type III is equivalent to any of the following conditions for some $k \in T_pM$ and some real bivector F at p with k and F non-zero.*

$$(a)\ \ \overset{+}{C}_{abcd}k^bk^d = 0 \quad (b)\ \ C_{abcd}k^bk^d = {}^*C_{abcd}k^bk^d = 0$$

$$(c)\ \ \overset{+}{C}_{abcd}k^d = \overset{+}{F}_{ab}k_c \quad (d)\ \ C_{abcd}k^d = F_{ab}k_c \quad (e)\ \ {}^*C_{abcd}k^d = {}^*F_{ab}k_c$$

The vector k and bivector F are null, unique up to scalings and k spans the repeated principal null directions of F and $C(p)$.

Proof. Clearly $(a) \Leftrightarrow (b)$ and $(c) \Leftrightarrow (d) \Leftrightarrow (e)$ and any of these five conditions is implied by the type **III** condition expressed in (7.75). Now the second condition in (b), together with a dualling argument similar to that used in the previous theorem, show that $k^aC_{ab[cd}k_{e]} = 0$. A contraction of this equation with k^e (noting that $k^aC_{abcd} \neq 0$, otherwise $C(p)$ would be type **N**) shows that k is null. Condition (a) and (7.67) with k replacing l then show that $C^2 \neq 0$, $C^3 = C^4 = C^5 = 0$ and then the use of a null rotation to get $C^1 = 0$ recovers the type **III** expression (7.75). The uniqueness of k up to a scaling then follows. Finally, suppose (c), (d) and (e) hold. Then a contraction of any of these conditions with k^c shows that k is null and the condition $C^a{}_{bad} = 0$ shows that $F_{ab}k^b = {}^*F_{ab}k^b = 0$. Theorem 7.2 then shows that F is null with principal null direction k.

The uniqueness of F up to a scaling is now clear and (a) and (b) follow, completing the proof. □

Theorem 7.6 *Let M be a space-time and let $p \in M$.*

i) $C(p)$ is type **II** *if and only if there exists a unique $k \in T_pM$, $k \neq 0$, such that*

$$\overset{+}{C}_{abcd}k^b k^d = \rho k_a k_c$$

(or equivalently $C_{abcd}k^b k^d = \alpha k_a k_c$ and ${}^*C_{abcd}k^b k^d = \beta k_a k_c$) where $\rho = \alpha + i\beta \neq 0$, $\alpha, \beta \in \mathbb{R}$. The vector k is necessary null and spans the repeated principal null direction of $C(p)$.

ii) $C(p)$ is type **D** *if and only if there exists two independent vectors k and $k' \in T_pM$ satisfying the equation in part (i) above for non-zero complex numbers ρ and ρ'. It necessarily follows that $\rho = \rho'$ and that k and k' are null and unique up to interchange and scalings and span the repeated principal null directions of $C(p)$.*

Proof. The proof of this theorem has essentially been given in the previous discussion. □

It is convenient at this point to return briefly to the conditions (7.69) and (7.70) defining, respectively, a *principal* and a *repeated principal* null direction of the Weyl tensor at $m \in M$. Provided it is given that k is null these conditions may each be replaced by the equivalent condition at m obtained by replacing $\overset{+}{C}$ by C or by C^*

$$k_{[e}C_{a]bc[d}k_{f]}k^b k^c = 0 \qquad k_{[e}C^*_{a]bc[d}k_{f]}k^b k^c = 0 \tag{7.82a}$$

$$k_{[e}C_{a]bcd}k^b k^c = 0 \qquad k_{[e}C^*_{a]bcd}k^b k^c = 0 \tag{7.82b}$$

Then for k null the two conditions (7.82a) are equivalent to each other and to (7.69) and to the statement that k is a principal null direction of C. Similarly for k null the two conditions (7.82b) are equivalent to each other and to (7.70) and to the statement that k is a repeated principal null direction of C. These results can be checked from (7.67). It is remarked here that the first equation in (7.82b) implies that $C_{abcd}k^b k^c = \alpha k_a k_d$, $\alpha \in \mathbb{R}$. If $\alpha \neq 0$ a contraction with k^a shows that k is necessary null. Similar remarks apply to the second condition in (7.82b). Also in (7.82a) the first equation implies that $C_{abcd}k^b k^c = k_a P_d + P_a k_d$ for a covector P at m with $P^a k_a = 0$. A contraction with k^a then reveals that if $P \neq 0$, k is necessary null. Similar remarks apply to the second condition in (7.82a). Thus in either (7.82a) or

(7.82b) the extra condition $C_{abcd}k^b k^d \neq 0$ or $C^*_{abcd}k^b k^d \neq 0$ forces k to be null and then *either* condition in (7.82a) (respectively, (7.82b)) forces k to span a principal (respectively, a repeated principal) null direction of $C(p)$.

The multiplicity properties of the principal null directions are conveniently and traditionally represented for each Petrov type by the following self-explanatory (Penrose) diagram.

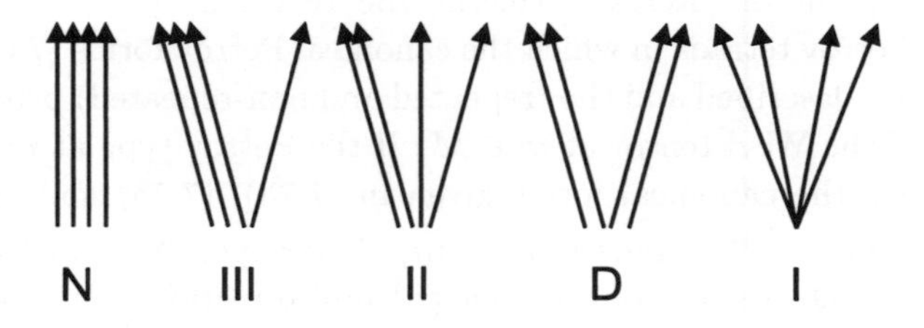

Fig. 7.1

It should be pointed out that the Petrov Classification is an algebraic statement about the Weyl tensor at $p \in M$ and the Petrov type may change if p changes. Some remarks about this will be made at the end of this chapter. If the Petrov type is the same at each $p \in M$ one refers to the *space-time* as being *of that Petrov type*.

Another important approach to the Petrov Classification is by using spinors. This approach, due to Penrose [68],[69], is rather elegant and is omitted here only because the tensor approach already discussed is more appropriate for the work in this book.

The canonical bivector decomposition given in this section allows the problem of the uniqueness of the Petrov tetrads to be reconsidered. Clearly the uniqueness or not of a frame where the Weyl tensor components assume a particular set of values can be established with any frame as starting point. The frames used in the above bivector decomposition are particularly convenient. For example, suppose $\overset{+}{C}(p)$ takes the type **N** form (7.73) with identical tetrad components in the null tetrads (l, n, x, y) and (l', n', x', y'). Then l' and l are proportional and these tetrads are thus related by a null rotation about l of the form (7.35). This rotation must be such that the coefficient C^1 of $\overset{+}{C}(p)$ is unchanged. Thus one finds from (7.68) that $C^1 = C^1 A^2 e^{-2i\theta}$ and so $A = 1$, $\theta = 0$. Hence the allowed null rotations are (7.35) with B arbitrary. For the Petrov type **D** a similar argument based on the canonical form (7.79) shows that any alternative real null tetrad must have its real null members spanning the two (necessarily repeated) principal null directions of $C(p)$ and that the coefficient C^3 must be unchanged. Hence from (7.35) and (7.68) one finds $B = 0$. Hence the allowed null rotations are (7.35) with $B = 0$ and A and θ arbitrary. Similarly arguments for the Petrov types **II**, **III** and **I** yield $A = 1$, $\theta = 0$, $B = 0$ and so in this sense and for these types the tetrad is uniquely determined.

Another point of interest concerns the relationship between the orthonormal Petrov tetrads in which the canonical Petrov forms (7.62), (7.63) and (7.64) are described and the (repeated and non-repeated) principal null directions of the Weyl tensor at $m \in M$. If the Petrov type at m is **N**, **III** or **D** one has the canonical forms given in (7.73), (7.75) and (7.79) and in these complex null tetrads $(l, n, m, \bar{m})$, l (for type **N**) and l and n for types **III** and **D** are the (only) principal null directions. In each case it is easily checked that (t, x^1, x^2, x^3) is a Petrov tetrad where $\sqrt{2}t = l - n$, $\sqrt{2}x^1 = l + n$ and $\sqrt{2}m = x^2 + ix^3$. However, the link between the Petrov tetrads and the principal null directions is less obvious for Petrov types **I** and **II**. The problem for the type **I** case was considered in [70] and here both types will be dealt with by a slightly different method. If C is of Petrov type **II** at m, the canonical form (7.77) applies and the Petrov tetrad is obtained from the complex null tetrad just as for the types **N**, **III** and **D** above. Now although l is the repeated principal null direction of C at m, the null vector n in the tetrad spans neither of the remaining principal null directions of C. Now perform two null rotations using (7.35) each on the above original complex null tetrad $(l, n, m, \bar{m})$ (the original tetrad chosen so that $C'^1 = 1$ in (7.67)) with $A = 1$, $\theta = 0$ and $\bar{B} = \pm F$, $F = i(6C'^3)^{-1/2}$. The null vector n is transformed, respectively, to n_1 and n_2 where

$$n_1 = n - |F|^2 l + p \quad n_2 = n - |F|^2 l - p \tag{7.83}$$

and where $p = Fm + \bar{F}\bar{m}$ is a real spacelike vector lying in the x^2x^3 2-space. It can then be checked that n_1 and n_2 span (non-repeated) principal null directions of C since in each of these new tetrads, $C^1 = 0$. Now put $\tau = n - |F|^2 l$ so that τ is timelike and orthogonal to p. Then $n_1 = \tau + p$, $n_2 = \tau - p$ ($\Rightarrow 2\tau = n_1 + n_2$, $2p = n_1 - n_2$) and so the direction of τ is fixed by the intersection of the timelike 2-spaces spanned by l and n and by n_1 and n_2. This enables figure 7.4 to be constructed and which shows the relation between the canonical null tetrad $(l, n, m, \bar{m})$ (or its associated Petrov tetrad (l, n, x^2, x^3) given above) and the principal null directions l, n_1 and n_2.

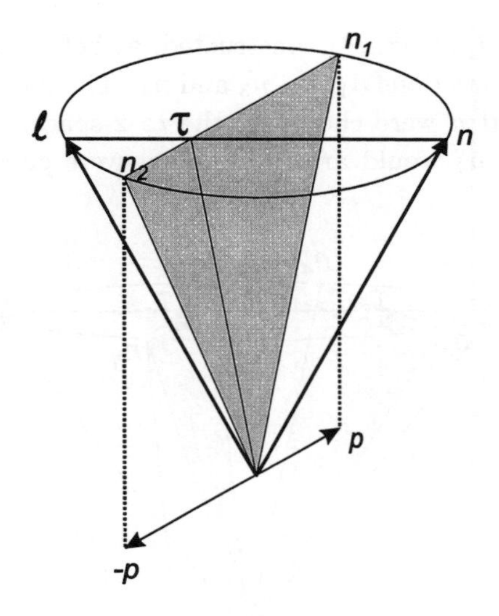

Fig. 7.2

If C is of Petrov type **I** at m it may be written in the form (7.80) with $C^1 = C^5$ and a Petrov tetrad (t, x^1, x^2, x^3) can be constructed just as before with $\sqrt{2}t = l - n$, $\sqrt{2}x^1 = l + n$ and $\sqrt{2}m = x^2 + ix^3$ and which is unique up to reflections. Here there are four distinct non-repeated principal null directions and none coincides with either l or n. Now perform four null rotations using (7.35) each on the original tetrad using, respectively, $A = 1$, $\theta = 0$ and the four values for $\bar{B}$ given by B_1, $-B_1$, B_1^{-1} and $-B_1^{-1}$ where

$$B_1 = \left[-3k + (9k^2 - 1)^{1/2}\right]^{1/2} \quad (k = \frac{C'^3}{C'^1}) \tag{7.84}$$

Then null vector n transforms to n_1, n_2, n_3 and n_4 where

$$n_1 = \tau_1 + p_1 \quad n_2 = \tau_1 - p_1 \quad n_3 = \tau_2 + p_2 \quad n_4 = \tau_2 - p_2 \tag{7.85}$$

where $\tau_1 = n - |B_1|^2 l$, $\tau_2 = n - |B_1|^{-2} l$, $p_1 = \bar{B}_1 m + B_1 \bar{m}$ and $p_2 = \bar{B}_1^{-1} m + B_1^{-1} \bar{m}$. Hence τ_1 and τ_2 are each timelike vectors in the ln 2-space and p_1 and p_2 are spacelike vectors in the $x^2 x^3$ 2-space. It can now be checked that the new C^1 coefficient in (7.67) is in each case zero and hence that $n_1, \ldots, n_4$ are the four principal null directions of C. Also $2\tau_1 = n_1 + n_2$ and $2\tau_2 = n_3 + n_4$ and so the directions of τ_1 and τ_2 are fixed, respectively, by the intersection of the ln 2-space first with the $n_1 n_2$ 2-space and then with the $n_3 n_4$ 2-space. This leads to figure 7.4 which shows the link between the Petrov tetrad (t, x^1, x^2, x^3) constructed, as before, from $l, n, m, \bar{m}$ and the principal null directions n_1, n_2, n_3 and n_4. The pairings n_1n_2 and n_3n_4 arise because l and n were chosen in the tz 2-space. The other possible pairings of $n_1, \ldots, n_4$ would arise if l and n were chosen in the tx or ty 2-spaces.

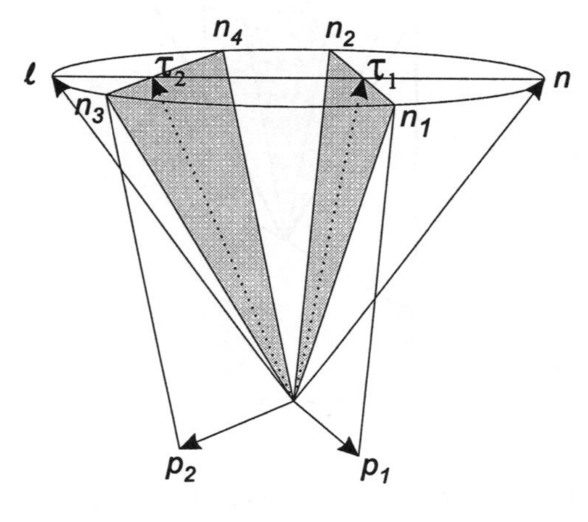

Fig. 7.3

7.5 The Classification of Second Order Symmetric Tensors

The classification of second order symmetric tensors at a point p in the space-time M has been given by several authors. Perhaps the first was due to Churchill [71] although the general ideas involved were known earlier from the work on the simultaneous reduction of the two quadratic forms, so important in algebraic geometry. This was followed by a comprehensive account by Plebanski [72] and, indeed, many other workers have contributed to this classification [48],[63],[69],[57], [73]-[84]. The approach followed here is largely based on [57],[78]. Although the classification to be described applies to any second order symmetric tensor, it is usually thought of as applying to the Ricci or the energy-momentum tensor.

Let S be a second order symmetric tensor at $p \in M$ with components $S_{ab}(= S_{ba})$. The eigenvector-eigenvalue problem for S consists in finding a (real or complex) eigenvector k and eigenvalue $\alpha \in \mathbb{C}$ such that $k^a S_a{}^b = \alpha k^b$, or equivalently, $k^a(S_{ab} - \alpha g_{ab}) = 0$. Because of the Lorentz signature of g, this formulation of the problem is not in the standard form usually discussed in the algebra textbooks (cf. chapter 2). To bring it into a standard form one rewrites it as $k^a(S_a{}^b - \alpha\delta_a^b) = 0$ so that S is regarded as a linear map $T_pM \to T_pM$. This reformulation has the disadvantages that $S_a{}^b$ is no longer symmetric and the Lorentz signature of $g(p)$ is "lost" in the Kronecker symbol. However, if one chooses this formulation, in order to proceed with the usual Jordan-Segre classification, the symmetry of S and the Lorentz signature of $g(p)$ can be reinstated afterwards (see, *e.g.* [57],[78]). Such a procedure will be described below. One can, however, proceed directly with the tensor S_{ab} in its symmetric form and such a method will also be described. A third procedure [57],[79] will also be described which, although originally inspired by the Churchill paper [71], will be approached in a different way and which shows many similarities to the techniques used in the Petrov classification.

Consider first the matrix $S_a{}^b$ at p and *assume that all eigenvalues are real* (the case of complex eigenvalues will be dealt with later). Then there are five possible Jordan canonical forms for S (assumed non-zero throughout) given by

$$(i)\begin{pmatrix}\rho_1 & 1 & 0 & 0\\ 0 & \rho_1 & 1 & 0\\ 0 & 0 & \rho_1 & 1\\ 0 & 0 & 0 & \rho_1\end{pmatrix} (ii)\begin{pmatrix}\rho_1 & 1 & 0 & 0\\ 0 & \rho_1 & 0 & 0\\ 0 & 0 & \rho_2 & 1\\ 0 & 0 & 0 & \rho_2\end{pmatrix} (iii)\begin{pmatrix}\rho_1 & 1 & 0 & 0\\ 0 & \rho_1 & 1 & 0\\ 0 & 0 & \rho_1 & 0\\ 0 & 0 & 0 & \rho_2\end{pmatrix} \quad (7.86)$$

$$(iv)\begin{pmatrix}\rho_1 & 1 & 0 & 0\\ 0 & \rho_1 & 0 & 0\\ 0 & 0 & \rho_2 & 0\\ 0 & 0 & 0 & \rho_3\end{pmatrix} \quad (v)\begin{pmatrix}\rho_1 & 0 & 0 & 0\\ 0 & \rho_2 & 0 & 0\\ 0 & 0 & \rho_3 & 0\\ 0 & 0 & 0 & \rho_4\end{pmatrix}$$

Here, the respective Segre types are {4}, {22}, {31}, {211} and {1111} together with their degeneracies indicated, as usual, with round brackets enclosing the appropriate entries and $\rho_1, \rho_2, \rho_3, \rho_4 \in \mathbb{R}$. The symmetric condition on S is reinstated by writing an arbitrary symmetric matrix for g_{ab} and imposing the condition $S_a{}^c g_{cb} = S_b{}^c g_{ca}$. It is a straightforward calculation to show that for each of the types {4} and {22} the symmetric relation implies that $\det g_{ab} \geqslant 0$ and hence, from the Lorentz signature, these Segre types can not occur. The other types can occur. For example if S has type {31} with eigenvalues ρ_1 and ρ_2 with $\rho_1 \neq \rho_2$, the above symmetric condition and the Lorentz signature of g require g to take the form g_{ab} with $g_{12} = g_{13} = g_{22} = g_{23} = g_{03} = 0$, $g_{11} = g_{02}$ and with $g_{02} > 0$ and $g_{33} > 0$ in the corresponding Jordan basis at p (the only other possibility $g_{02} < 0$ and $g_{33} < 0$ gives signature $(+,-,-,-)$). In this basis, eigenvectors may be chosen as $l^a \equiv (0,0,1,0)$ with eigenvalue ρ_1 and $y^a \equiv (0,0,0,1)$ with eigenvalue ρ_2. It then follows that $l^a l_a = l^a y_a = 0$ and $y^a y_a > 0$. A null tetrad can then be built around l and $y' = (y^a y_a)^{-1/2} y$ and a general expression for S constructed in this tetrad which may then be simplified by the use of the null rotations (7.35) about l. By this procedure one can construct canonical expressions for each of the remaining types (7.86) (further details may be found in [57]). However, one would still have to deal with the case when the eigenvalues are not all real and for this and other reasons a different approach will be used to obtain these canonical forms for S. This alternative approach deals directly with the symmetric tensor S_{ab} and has the Lorentz signature built into it from the outset. Also, the case when the eigenvalues are not all real poses no problems. This involves a brief digression into the invariant 2-space structure of S.

A 2-space $W \subseteq T_pM$ is called an *invariant 2-space of S at p* if, whenever u^a is in W, so also is $u^a S_a{}^b$. Now if S has all eigenvalues real at p it takes one of the Jordan types (7.86) and in such a basis the last two members of that basis span an invariant 2-space of S at p. If, on the other hand, S has a non-real eigenvector w at p, the real and imaginary parts of w span an invariant 2-space of S at p (section 2.6). *Hence S always admits an invariant 2-space at any $p \in M$* and the following theorem records some of their properties (*cf.*the comments on invariant 2-spaces of bivectors at the end of section 7.2).

Theorem 7.7 *Let M be a space-time, let $p \in M$ and let $S \neq 0$ be a second order symmetric tensor at p. Then at p*

i) if W is an invariant 2-space of S so also is its orthogonal complement,
ii) S admits a null invariant 2-space W if and only if S admits a real null eigenvector w (and w may be chosen in W),
iii) S admits a spacelike (equivalently, from (i), a timelike) invariant 2-space if and only if S admits an orthogonal pair of spacelike eigenvectors (which necessarily lie in the spacelike invariant 2-space).

Proof. It is useful here and elsewhere to note the decomposition of S

along a real null tetrad (l, n, x, y) at p. This is

$$S_{ab} = 2S^1 l_{(a}n_{b)} + S^2 l_a l_b + S^3 n_a n_b + 2S^4 l_{(a}x_{b)} + 2S^5 l_{(a}y_{b)} + 2S^6 n_{(a}x_{b)} \\ + 2S^7 n_{(a}y_{b)} + 2S^8 x_{(a}y_{b)} + S^9 x_a x_b + S^{10} y_a y_b \quad (7.87)$$

where $S^1, ..., S^{10} \in \mathbb{R}$. Such an expression for S_{ab} may be simplified by null rotations (say, about l) using the equations (7.35) written out in their real and imaginary parts so as to apply to the (real) null tetrad (l, n, x, y) with $\sqrt{2}m = x + iy$. The procedure is essentially the same as that used to obtain (7.68) for the Weyl tensor and will prove useful in what is to follow.
(i) Suppose that W is spacelike and taken to be spanned by x and y. Then it follows from (7.87) that $S^4 = S^5 = S^6 = S^7 = 0$ and then that l and n (which span the orthogonal complement of W) also span an invariant 2-space of S. If W is timelike one merely reverses the argument. If W is null and taken to be spanned by l and x then one finds $S^3 = S^6 = S^7 = S^8 = 0$ and so l and y span an invariant 2-space of S (and is again the orthogonal complement of W). It also follows that l is a null eigenvector of S.
(ii) If W is a null invariant 2-space of S then the last sentence of (i) above confirms the existence of a null eigenvector of S. Conversely if S admits a null eigenvector, taken as l, then (7.87) show that $S^3 = S^6 = S^7 = 0$. A null rotation (7.35) with $A = 1$ and $B = 0$ can then be used (if necessary) to form a new null tetrad in which $S^8 = 0$ also. Then, in this new tetrad, l and x span a null invariant 2-space of S.
(iii) If W is a spacelike invariant 2-space of S spanned, say, by x and y then (7.87) holds with $S^4 = S^5 = S^6 = S^7 = 0$. A null rotation (7.35) with $A = 1$ and $B = 0$ can again be used to set $S^8 = 0$ also and then x and y are eigenvectors of S. Conversely, if x and y are taken as the orthogonal pair of spacelike eigenvectors of S then (7.87) gives $S^4 = S^5 = S^6 = S^7 = S^8 = 0$ and the result is clear. □

The theorem which gives the complete algebraic description of S at p can now be given.

Theorem 7.8 *Let M be a space-time, let $p \in M$ and let $S \neq 0$ be a second order symmetric tensor at p. Then S takes one of the following* *canonical forms in terms of some (real) null tetrad* (l, n, x, y) *at* p.

$$S_{ab} = 2\rho_1 l_{(a}n_{b)} + \rho_2(l_a l_b + n_a n_b) + \rho_3 x_a x_b + \rho_4 y_a y_b \qquad (7.88a)$$

$$S_{ab} = 2\rho_1 l_{(a}n_{b)} + \lambda l_a l_b + \rho_2 x_a x_b + \rho_3 y_a y_b \qquad (7.88b)$$

$$S_{ab} = 2\rho_1 l_{(a}n_{b)} + 2l_{(a}x_{b)} + \rho_1 x_a x_b + \rho_2 y_a y_b \qquad (7.88c)$$

$$S_{ab} = 2\rho_1 l_{(a}n_{b)} + \rho_2(l_a l_b - n_a n_b) + \rho_3 x_a x_b + \rho_4 y_a y_b \qquad (7.88d)$$

where all $\rho_i \in R$ *and where* $0 \neq \lambda \in \mathbb{R}$ *in (7.88b) and* $\rho_2 \neq 0$ *in (7.88d). In (7.88a), the eigenvectors can be taken as* $t = 2^{-1/2}(l - n)$ *(with eigenvalue* $\rho_1 - \rho_2$*),* $z = 2^{-1/2}(l + n)$ $(\rho_1 + \rho_2)$*,* x (ρ_3) *and* y (ρ_4)*. The eigenvector* t *is timelike and* x, y *and* z *are spacelike and* (t, x, y, z) *is an orthonormal tetrad. The Segre type is* $\{1, 111\}$ *or one of its degeneracies. (In this type, a comma will be used to separate the eigenvalue arising from a timelike eigenvector from those arising from spacelike eigenvectors. This will be explained in more detail later.) Thus* S *is diagonalisable over* $\mathbb{R}$*. In (7.88b), the eigenvectors may be taken as* $l(\rho_1)$*,* $x(\rho_2)$ *and* $y(\rho_3)$ *and, without affecting the remainder of (7.88b), the null tetrad may be chosen so that* $\lambda = 1$ *(if* $\lambda > 0$*) and* $\lambda = -1$ *(if* $\lambda < 0$*). The Segre type is* $\{211\}$ *or one of its degeneracies. In (7.88c), the eigenvectors may be taken as* $l(\rho_1)$ *and* $y(\rho_2)$ *and the Segre type is* $\{31\}$ *or one of its degeneracies. The case (7.88d) is the only case where non-real eigenvectors occur and they may be taken as* $l \pm in(\rho_1 \pm i\rho_2)$*, the other eigenvectors being taken as* $x(\rho_3)$ *and* $y(\rho_4)$*. The Segre type is* $\{z\bar{z}11\}$ *or its degeneracy. An alternative from for (7.88a) is*

$$S_{ab} = (\rho_2 - \rho_1)t_a t_b + (\rho_1 + \rho_2)z_a z_b + \rho_3 x_a x_b + \rho_4 y_a y_b \qquad (7.88e)$$

Proof. First it should be pointed out that the description of the eigenvectors and eigenvalues in (7.88a) to (7.88d) can easily be checked. The statements regarding the Segre types of these canonical expressions for S then follow from the number of independent eigenvectors and the fact that only the above mentioned four Segre types are possible. Two general cases will be considered; when S admits a real null eigenvector and when it does not.

If S admits such a null eigenvector l, construct a null tetrad (l, n, x, y) so that S decomposes as in (7.87) with $S^3 = S^6 = S^7 = 0$. Then a null rotation (7.35) with $A = 1$ and $B = 0$ can be used to preserve these conditions and to achieve, in addition, $S^8 = 0$. Now four possibilities arise concerning the coefficients S^1, S^9 and S^{10}.

(a) $S^9 \neq S^1 \neq S^{10}$. A null rotation with $A = 1$ and $\theta = 0$ can be used to obtain a new tetrad in which $S^3 = S^4 = S^5 = S^6 = S^7 = S^8 = 0$ and hence S takes the form (7.88b) but with S^2 possibly zero. If $S^2 \neq 0$ the Segre type is thus $\{211\}$ or $\{2(11)\}$ (the latter occurring if $S^9 = S^{10}$). If $S^2 = 0$, equation (7.88a) can now be used to see that the Segre type is $\{(1,1)11\}$ or $\{(1,1)(11)\}$.

(b) $S^1 = S^9 \neq S^{10}$. If also $S^4 = 0$ then one has $S^3 = S^4 = S^6 = S^7 = 0$ and a null rotation with $A = 1$ and $\theta = 0$ can be used to obtain an expression like (7.88b) again and so the Segre type is $\{(21)1\}$ or $\{(1,11)1\}$ depending on whether $S^2 \neq 0$ or $S^2 = 0$. If, however, $S^4 \neq 0$ in the original tetrad a null rotation with $A = 1$ and $\theta = 0$ can be used to obtain a tetrad in which $S^2 = S^3 = S^5 = S^6 = S^7 = S^8 = 0$ and so an expression like (7.88c) results. A null rotation can then be used to set $R_4 = 1$. The Segre type is $\{31\}$.

(c) $S^1 = S^{10} \neq S^9$. This case is essentially the same as the last one.

(d) $S^1 = S^9 = S^{10}$. Here, the same techniques as above show that if $S^4 = S^5 = 0$ in the original tetrad, then one arrives at the Segre type $\{(211)\}$ or $\{(1,111)\}$. If, however S^4 and S^5 are not both zero a null rotation can be used to cast S into the Segre type $\{(31)\}$ canonical form.

Now suppose that S has no real null eigenvectors. Then from theorem 7.7(ii) S admits no null invariant 2-space. Since S must admit some invariant 2-space W, W must be spacelike or timelike and hence (theorem 7.7(iii)) S admits an orthogonal pair of (unit) spacelike eigenvectors, say x and y. On constructing a null tetrad (l, n, x, y) around x and y, it then follows that S takes the form (7.87) with $S^4 = S^5 = S^6 = S^7 = S^8 = 0$ and with $S^2 \neq 0 \neq S^3$ (otherwise real null eigenvectors would be admitted). A null rotation with $\theta = B = 0$ can then be used to arrange that $|S^2| = |S^3|$. The two possibilities $S^2 = S^3$ and $S^2 = -S^3$ lead, respectively, to (7.88a) with $\rho_2 \neq 0$ and $\rho_3 \neq \rho_1 - \rho_2 \neq \rho_4$ and to (7.88d) with $\rho_2 \neq 0$. These are the Segre types $\{1,111\}$, $\{1,1(11)\}$ or $\{1,(111)\}$ in the first case and $\{z\bar{z}11\}$ or $\{z\bar{z}(11)\}$ in the second.

The remark in the statement of the theorem about being able to choose $\lambda = \pm 1$ in (7.88b) follows by performing a null rotation with $\theta = B = 0$ and choosing A accordingly. □

There are some other simple algebraic properties of a second order symmetric tensor S that can now be collected together.

Theorem 7.9 *Let M be a space-time, let $p \in M$ and let $S \neq 0$ be a second order symmetric tensor at p.*

i) S admits at least two independent real eigenvectors, which may be cho-

sen orthogonal, and at least one of them is spacelike.

ii) S admits a (real) timelike eigenvector if and only if it is of Segre type $\{1,111\}$ or one of its degeneracies (equivalently S is diagonalisable over $\mathbb{R}$).

iii) S admits two (or more) independent (real) null eigenvectors if and only if it admits two (or more) independent (real) timelike eigenvectors (and if such is the case each of these null and timelike eigenvectors have the same eigenvalue). These conditions are equivalent to the statement that S has Segre type $\{(1,1)11\}$ or one of its degeneracies.

iv) S has all its eigenvalues real but possesses non-simple elementary divisors (i.e. it has Segre type $\{211\}$ or $\{31\}$ or one of their degeneracies) if and only if S has a unique (real) null eigendirection (which then corresponds to the non-simple elementary divisor).

Proof. The proofs of each part can essentially be obtained by a study of the eigenvectors which exist for each canonical form (7.88a)-(7.88e). □

It is useful here to note that if S admits eigenvectors u and v with distinct eigenvalues then *u and v must be orthogonal* as follows easily by contracting the equations $S_{ab}u^b = \rho_1 u_a$ and $S_{ab}v^b = \rho_2 v_a$ ($\rho_1 \neq \rho_2$) with v^a and u^a, respectively, and using the symmetry of S. The results in (ii) and (iii) above show that only in the Segre type $\{1,111\}$, or one of its degeneracies, can timelike eigenvectors occur and that a unique eigenvalue is associated with them. This "timelike eigenvalue" is indicated as the first digit in the Segre symbol and separated from the others by a comma (as mentioned, without justification, earlier).

7.6 The Anti–Self Dual Representation of Second Order Symmetric Tensors

A study of the algebraic structure of a second order symmetric tensor such as S above using its invariant 2-space structure was first considered by Churchill [71]. Here this idea will be taken up in a different way and which exhibits some similarities with the Petrov classification of the Weyl tensor [57],[79]. One notes here that the eigenvectors, Segre type (including degeneracies) and invariant 2-space structure (but, in general, not the eigenvalues) of S are identical to those of the *tracefree part of S*, denoted by $\tilde{S}$

and defined by

$$\tilde{S}_{ab} = S_{ab} - \frac{1}{4}(S^c{}_c)g_{ab} \quad (\Rightarrow \tilde{S}^c{}_c = 0) \tag{7.89}$$

To consider this approach to the classification of S, define a (0,4) tensor P at $p \in M$ by

$$P_{abcd} = \tilde{S}_{a[c}g_{d]b} + \tilde{S}_{b[d}g_{c]a} \quad (\Rightarrow P^c{}_{acb} = \tilde{S}_{ab}) \tag{7.90}$$

Then P satisfies the following symmetry relations (in which the bracketed result follows from (7.49))

$$P_{abcd} = -P_{bacd} = -P_{abdc} \quad P_{abcd} = P_{cdab} \tag{7.91}$$

$$P_{a[bcd]} = 0 \quad (\Leftrightarrow P^{a*}{}_{bad} = 0) \tag{7.92}$$

Thus P is the same type of tensor studied at the beginning of section 7.3 and so the Ruse-Lanczos identity (7.47) together with (7.90) give ${}^*P^* + P = 2P$ and hence

$${}^*P^*_{abcd} = P_{abcd} \quad (\Leftrightarrow {}^*P_{abcd} = -P^*_{abcd}) \tag{7.93}$$

From the definition (7.90) of P one finds

$$P_{abcd} = 0 \quad \Leftrightarrow \quad \tilde{S}_{ab} = 0 \quad \Leftrightarrow \quad S_{ab} = \frac{1}{4}(S^c{}_c)g_{ab} \tag{7.94}$$

Thus if one regards the definition of P as that of a linear map f from the 9-dimensional vector space of tracefree second order symmetric tensors at p into the vector space of type (0,4) tensor at p then (7.94) shows this map to be one-to-one and thus the type (0,4) tensors in the *range of* f (i.e of the form (7.90)) are, in this sense, "characteristic" of such second order symmetric tensors. Further, because of the conditions (7.93), P is called *anti self-dual*. Now one may define a complex tensor $\overset{+}{P}$ associated with P (and noting that, unlike the analogous situation with the Weyl tensor, the position of the dual symbol is important)

$$\overset{+}{P}_{abcd} = P_{abcd} + iP^*_{abcd} \tag{7.95}$$

which satisfies the first set of symmetry properties in (7.43) but not the second since now, using a similar argument to that given for Weyl tensor, one has $P^*_{abcd} = -P^*_{cdab}$ and hence $\overset{+}{P}_{abcd} = \overset{+}{\bar{P}}_{cdab}$. One also has the relations

$\overset{+}{P}{}^a{}_{bad} = \tilde{S}_{bd}$ (from (7.90) and (7.92)), $\overset{+}{P}{}^*_{abcd} = -i\overset{+}{P}_{abcd}$, ${}^*\overset{+}{P}_{abcd} = i\overset{+}{P}_{abcd}$ and the result that $\overset{+}{P} = 0 \Leftrightarrow \tilde{S} = 0$.

Now let $U \subseteq T_pM$ be an invariant 2-space of $\tilde{S}$ spanned, say, by r and s. Then

$$\tilde{S}_{ab}r^b = \alpha r_a + \beta s_a, \quad \tilde{S}_{ab}s^b = \gamma r_a + \delta s_a \quad (\alpha, \beta, \gamma, \delta \in \mathbb{R}) \tag{7.96}$$

Now the set of all simple bivectors whose blade is U is just the set of non-zero real multiples of the (real) bivector $F_{ab} = 2r_{[a}s_{b]}$ and this statement is independent of the representative bivector F chosen. Then it follows that (7.96) is equivalent to

$$S_{b[a}F^b{}_{c]} = \mu F_{ac} \quad (\mu \in \mathbb{R}) \tag{7.97}$$

where μ is independent of the representative bivector F. This condition, with the help of (7.90) and (7.95), can be written in the following three equivalent ways

$$\text{(i) } P_{abcd}F^{cd} = 2\mu F_{ab} \quad \text{(ii) } P_{abcd}{}^*F^{cd} = -2\mu {}^*F_{ab} \quad \text{(iii) } \overset{+}{P}_{abcd}\overset{+}{F}{}^{cd} = 4\mu \tilde{F}_{ab} \tag{7.98}$$

where $\overset{+}{F}$ and $\tilde{F}$ are as in (7.20). Thus an invariant 2-space U of $\tilde{S}$ (equivalently, of S) determines up to a real multiple a real simple eigenbivector F of P and, conversely, a real simple eigenbivector F of P determines a unique invariant 2-space of $\tilde{S}$ (or S) and, in fact, is the blade of F. Also (7.98) shows that such eigenbivectors of P come in dual pairs, reflecting the fact that invariant 2-spaces of $\tilde{S}$ (or S) come in orthogonal pairs (theorem 7.7) and, of course, the blades of a simple bivector and its dual are orthogonal. Thus the invariant 2-space problem for S become one of finding simple (real) eigenbivectors of P. But if F is *any* (real) eigenbivector of P satisfying (7.98(i)) then a contraction with ${}^*F^{ab}$ gives, using (7.93)

$$\begin{aligned} 2\mu F_{ab}({}^*F^{ab}) &= P_{abcd}({}^*F^{ab}F^{cd}) = {}^*P_{abcd}F^{ab}F^{cd} = -P^*_{abcd}F^{ab}F^{cd} \\ &= -P_{abcd}F^{ab*}F^{cd} = -P_{cdab}F^{ab*}F^{cd} = -2\mu F_{ab}({}^*F^{ab}) \end{aligned} \tag{7.99}$$

and so, if $\mu \neq 0$, $F_{ab}{}^*F^{ab} = 0$, which, from theorem 7.1, is equivalent to F being simple. If $\mu = 0$ then F and ${}^*F$ are eigenbivectors of P with zero eigenvalue and then some duality rotation of F (section 7.2) yields a real *simple* eigenbivector of P with zero eigenvalue. In this sense, the invariant 2-space problem for $\tilde{S}$ (or S) becomes that of finding (real) *simple* eigenbivectors of P. But now suppose (7.98)(iii) holds *for* $\mu \in \mathbb{C}$ and some complex self-dual bivector $\overset{+}{F} = F + i^*F$. Then a duality rotation $\overset{+}{F} \to \overset{+}{H} = e^{i\theta}\overset{+}{F}$ reveals a complex self-dual bivector $\overset{+}{H}$ satisfying (7.98(iii)) with $\mu \in \mathbb{R}$ for appropriate $\theta \in \mathbb{R}$. This follows since $\overset{+}{P}_{abcd}\overset{+}{H}^{cd} = 4\mu e^{2i\theta}\tilde{H}_{ab}$. It is now easily checked that (7.98)(i) and (ii) hold for the real and imaginary parts H and ${}^*H$ of $\overset{+}{H}$ (and $\mu \in \mathbb{R}$). Thus the invariant 2-space problem for $\tilde{S}$ (or S) becomes that of solving (7.98(iii)) for $\mu \in \mathbb{R}$.

The solution of this last problem proceeds on similar lines to that for the tensor $\overset{+}{C}$. One first chooses a *complex* null tetrad $(l, n, m, \bar{m})$, then the complex self-dual bivectors U, V and M as in (7.24)-(7.26) and then decomposes $\overset{+}{P}$ as follows [57],[79] (*cf.*(7.65)).

$$\begin{aligned}\overset{+}{P}_{abcd} &= \frac{1}{2}(\overset{+}{P}_{abcd} + i\overset{+}{P}{}^{*}_{abcd}) = \frac{1}{2}\overset{+}{P}_{ab}{}^{mn}(G_{mncd} + \frac{1}{2}i\epsilon_{mncd}) \qquad (7.100)\\ &= \frac{1}{4}(\overset{+}{P}_{ab}{}^{mn} - i^*\overset{+}{P}_{ab}{}^{mn})(G_{mncd} + \frac{1}{2}i\epsilon_{mncd})\\ &= \frac{1}{4}\overset{+}{P}{}^{pqmn}(G_{pqab} - \frac{1}{2}i\epsilon_{pqab})(G_{mncd} + \frac{1}{2}i\epsilon_{mncd})\end{aligned}$$

Then by using (7.33) and its conjugate one has

$$\begin{aligned}\overset{+}{P}_{abcd} &= P_1\bar{U}_{ab}U_{cd} + P_2\bar{V}_{ab}V_{cd} + P_3\bar{M}_{ab}M_{cd} + P_4\bar{U}_{ab}V_{cd} \qquad (7.101)\\ &\quad + P_5\bar{V}_{ab}U_{cd} + P_6\bar{U}_{ab}M_{cd} + P_7\bar{M}_{ab}U_{cd} + P_8\bar{V}_{ab}M_{cd} + P_9\bar{M}_{ab}V_{cd}\end{aligned}$$

where $P_1, ..., P_9 \in \mathbb{C}$ and where the condition $\overset{+}{P}_{abcd} = \overline{\overset{+}{P}}_{cdab}$ reveals that $P_1, P_2, P_3 \in \mathbb{R}$ and that $P_5 = \bar{P}_4$, $P_7 = \bar{P}_6$ and $P_9 = \bar{P}_8$. The classification is now based on the number of independent complex self-dual solutions $\overset{+}{F}$ of (7.98(iii)) and their type (null or non-null). From the resulting canonical expressions, $\tilde{S}$ and hence S, are recovered from the contraction $\tilde{S}_{bd} = \overset{+}{P}{}^{a}{}_{bad}$.

It is remarked here that this problem is similar to that encountered in the Petrov classification. Using the abbreviated notation employed in section 7.3, the tensors P and $\overset{+}{P}$ give rise to linear maps $CB(p) \to CB(p)$ which, using ideas very similar to those for the Weyl tensor, have the properties that P maps $S^+(p) \to S^-(p)$ and $S^-(p) \to S^+(p)$ and that $\overset{+}{P}$ maps $S^+(p) \to S^-(p)$ and $S^-(p) \to 0$ (and that for $F \in S^+(p)$, $\overset{+}{P}F = 2P\overset{+}{F}$).

Since there is always an invariant 2-space of $\tilde{S}$ there is always a non-trivial $\overset{+}{F}$ satisfying (7.98(iii)). Hence this latter equation has either one, two or three independent solutions in the complex 3-dimensional vector space of complex self-dual bivectors. Now if there are two independent *null* solutions then it is easy to see that a complex null tetrad $(l, n, m, \bar{m})$ may be chosen so that these solutions may be taken as U and V in (7.98(iii)). But (7.101) then reveals that $P_1 = P_2 = P_6 = P_7 = P_8 = P_9 = 0$ and hence that M is a (non-null) solution. On the other hand if *no* null independent solutions of (7.98(iii)) exist then there exists a non-null solution, which may be taken as M, and so (7.101) gives $P_6 = P_7 = P_8 = P_9 = 0$. But one must also have $P_1 \neq 0$ and $P_2 \neq 0$ (to avoid U or V becoming null solutions). A null rotation (7.35) with $B = 0$ can then be used to preserve the restrictions achieved so far and to obtain, in addition, $|P_1| = |P_2|$ and $P_4 \in \mathbb{R}$. If $P_1 = P_2$ then $U \pm V$ become extra independent non-null solutions of (7.98(iii)) whilst if $P_1 = -P_2$ it can be checked that M is the only solution. The separate cases can now be discussed and the techniques required to obtain the canonical forms are very similar to those above and to those used in the Petrov classification. Hence the details are omitted but can be found in [57],[79] (*cf.*[72]).

Case(i) Suppose there is one independent solution to (7.98(iii)) and it is *non-null*. Choosing it as M one may arrange the tetrad $(l, n, m, \bar{m})$ so that

$$\overset{+}{P}_{abcd} = P_1(\bar{U}_{ab}U_{cd} - \bar{V}_{ab}V_{cd}) + P_3\bar{M}_{ab}M_{cd} + P_4(\bar{U}_{ab}V_{cd} + \bar{V}_{ab}U_{cd}) \quad (7.102)$$

where $P_1, P_3, P_4 \in \mathbb{R}$ and $P_1 \neq 0$. A contraction of this yields the canonical form (7.88d) for $\tilde{S}_{ab}$ and hence the Segre type $\{z\bar{z}11\}$ or its degeneracy.

Case(ii) Suppose there is one independent solution to (7.98(iii)) and it is *null*. Choosing it as V one may arrange the tetrad so that

$$\overset{+}{P}_{abcd} = P_3(\bar{M}_{ab}M_{cd} - 2\bar{U}_{ab}V_{cd} - 2\bar{V}_{ab}U_{cd}) + P_8(\bar{V}_{ab}M_{cd} + \bar{M}_{ab}V_{cd}) \quad (7.103)$$

where $P_3, P_8 \in \mathbb{R}$ and $P_8 \neq 0$. A contraction of this leads to (7.88c) and the Segre type $\{31\}$ or its degeneracy.

Case(iii) Suppose there are exactly two independent solutions of (7.98(iii)). Then the remarks above show that they may be taken as a *null* solution and a *non-null* solution and, in a suitable tetrad, as V and M, respectively. Then

$$\overset{+}{P}_{abcd} = P_2\bar{V}_{ab}V_{cd} + P_3\bar{M}_{ab}M_{cd} + P_4(\bar{U}_{ab}V_{cd} + \bar{V}_{ab}U_{cd}) \quad (7.104)$$

with $P_2, P_3, P_4 \in \mathbb{R}$ and $P_2 \neq 0$. A contraction then leads to (7.88b) and the Segre type $\{211\}$ or one of its degeneracies.

Case(iv) Suppose there are three independent solutions to (7.98(iii)). If there are *no* null solutions the tetrad may be chosen so that

$$\overset{+}{P}_{abcd} = P_1(\bar{U}_{ab}U_{cd} + \bar{V}_{ab}V_{cd}) + P_3\bar{M}_{ab}M_{cd} + P_4(\bar{U}_{ab}V_{cd} + \bar{V}_{ab}U_{cd}) \quad (7.105)$$

with $P_1, P_3, P_4 \in \mathbb{R}$ (together with certain further restrictions on these coefficients to prevent null solutions). A contraction then reveals (7.88a) and the Segre type $\{1,111\}$ together with the degeneracies permitted by these coefficient restrictions and which are $\rho_2 \neq 0$, $\rho_3 \neq \rho_1 - \rho_2 \neq \rho_4$. The only other possibility is that there are at least two independent null solutions of (7.98(iii)). Then the tetrad may be chosen so that

$$\overset{+}{P}_{abcd} = P_3\bar{M}_{ab}M_{cd} + P_4(\bar{U}_{ab}V_{cd} + \bar{V}_{ab}U_{cd}) \quad (7.106)$$

with $P_3, P_4 \in \mathbb{R}$ and the usual contraction leads to (7.88a) with $\rho_2 = 0$ and hence to the Segre type $\{(1,1)11\}$ or one of its degeneracies. This completes the classification.

The similarity of the classification of the second order symmetric tensor S (though the tensor P) with the Petrov classification can be extended by linking the algebraic types of S with certain null directions which play a similar role with respect to P (and $\overset{+}{P}$) that the null directions of Bel, Debever and Penrose play with respect to the Weyl tensor C (and $\overset{+}{C}$). The next theorem, whose proof will be omitted since it is similar to proofs given in the Weyl tensor case, summarises the situation.

Theorem 7.10

Let M be a space-time, $p \in M$ and $S \neq 0$ a second order symmetric tensor at p with associated (0,4) tensor P.

i) The statement that S has Segre type $\{(211)\}$ is equivalent to any of the following statements for some $k \in T_pM$, $k \neq 0$

$$(a)\ \overset{+}{P}_{abcd}k^d = 0 \quad (b)\ P_{abcd}k^d = 0 \quad (c)\ P^*_{abcd}k^d = 0$$

$$(d)\ \overset{+}{P}_{ab[cd}k_{e]} = 0 \quad (e)\ P_{ab[cd}k_{e]} = 0 \quad (f)\ P^*_{ab[cd}k_{e]} = 0$$

In the conditions (a)-(f), k is necessarily null and unique up to a scaling and spans the (unique) real null eigendirection of S. Each of these *conditions is equivalent to* $\overset{+}{P}$ *taking the form (7.101) with* $l = k$ *and* P_2 *the only non-vanishing coefficient, and also to* $\tilde{S}_{ab} = P_2 k_a k_b$.

ii) The statement that S *has Segre type* $\{(31)\}$ *is equivalent to any of the following statements for some* $k \in T_pM$, $k \neq 0$ *and some real null bivector* $F \neq 0$.

$$(a)\ k^a \overset{+}{P}_{abcd} = k_b \overset{+}{F}_{cd}, \quad (b)\ k^a P_{abcd} = k_b F_{cd}, \quad (c)\ k^a P^*_{abcd} = k_b {}^*F_{cd}$$

In these conditions k *is necessarily null, and unique up to a real scaling and spans the (unique) real null eigendirection of* S *and the (unique) principal null direction of* F. *Each of these conditions is equivalent to* $\overset{+}{P}$ *taking the form (7.101) with* $l = k$ *and* P_2, P_8 *and* $\bar{P}_8$ *the only non-vanishing coefficients (and then* P_2 *can be set to zero by means of a null rotation) and also to* $\tilde{S}_{ab}$ *taking the form (7.88c) with* $\rho_1 = \rho_2 = 0$.

iii) The statement that a null vector $k \in T_pM$ *is an eigenvector of* S *with eigenvalue* $-\rho$ *is equivalent to any of the following (equivalent) conditions (with* $\rho \in \mathbb{R}$*)*

$$(a)\ k^a k^c \overset{+}{P}_{abcd} = \rho k_b k_d \quad (b)\ k^a k^c P_{abcd} = \rho k_b k_d$$

$$(c)\ k^a k^c P^*_{abcd} = 0.$$

iv) The statement that a null vector $k \in T_pM$ *satisfies* $S_{ab} k^a k^b = 0$ *(equivalently* $\tilde{S}_{ab} k^a k^b = 0$*) is equivalent to any of the following (equivalent) conditions.*

$$(a)\ k_{[e} \overset{+}{P}_{a]bc[d} k_{f]} k^b k^c = 0 \quad (b)\ k_{[e} P_{a]bc[d} k_{f]} k^b k^c = 0$$

$$(c)\ k_{[e} P^*_{a]bc[d} k_{f]} k^b k^c = 0.$$

Part (iii) of this theorem shows how, what might be termed a "repeated principal null direction" of P (or $\overset{+}{P}$), coincides with a null eigendirection of S and vice-versa. Part (iv) indicates the relevance to S of a "principal null direction" of P (or $\overset{+}{P}$). Theorem 7.10 describes what might be termed *generalised Bel criteria* for P (or $\overset{+}{P}$) or S. However, there are some differences due to the anti self-dual nature of P. For example, the number of distinct real null directions satisfying the conditions in either (iii) or (iv) of the above theorem could be none, finitely many or infinitely many. (*cf.*the case for the Weyl tensor where the respective numbers are 0, 1 or 2 and 1,2,3 or 4) [85].

There are several other approaches to the classification of second order symmetric tensors. Of particular interest is the approach due to Penrose [76] who classifies such a tensor by the type of curve on a complex quadric surface to which it corresponds. The relation between Penrose's approach and the present one has been given [81],[85]. A spinor approach is also available [72],[80] and is closely related to the approach using the tensor $\overset{+}{P}$ given above. A direct 6×6 classification of the tensor P can also be used and which, because of the symmetries of P, is quite straightforward [81].

7.7 Examples and Applications

The Ricci and energy-momentum tensor of space-time are second order symmetric tensors at any $p \in M$ and are hence subject to the procedures of the last two sections. The fourth order tensor arising from the Ricci tensor according to (7.90) is easily seen from (7.11) to be the tensor E. Then (7.10) can be viewed as a decomposition of the curvature at $p \in M$ into its self-dual and anti self-dual parts because

$$^*C = C^* \quad ^*G = G^* \quad ^*E = -E^* \tag{7.107}$$

The energy-momentum tensor T has also been subjected to the dominant energy conditions (7.9) that for any $p \in M$ and any timelike or null $u \in T_pM$ (i) $T_{ab}u^a u^b \geq 0$ and (ii) $T^a{}_b u^b$ is not spacelike. These conditions place restrictions on the Segre type of T at each $p \in M$ (see *e.g.* [72],[48],[57])

Theorem 7.11 *Let M be a space-time, $p \in M$ and let T be the energy-momentum tensor at p.*

i) If the energy conditions are satisfied T cannot be of Segre type $\{z\bar{z}11\}$ or $\{31\}$ or any of their degeneracies. (In fact, if T is of either of these types it cannot satisfy any of the conditions (i) and (ii) above.)

ii) If T has Segre type $\{211\}$ or one of its degeneracies (equation (7.88b)) the energy conditions are satisfied if and only if $\lambda > 0$, $\rho_1 \leq 0$, $\rho_1 \leq \rho_2 \leq -\rho_1$ and $\rho_1 \leq \rho_3 \leq -\rho_1$.

iii) If T has Segre type $\{1,111\}$ or one of its degeneracies (equation (7.88a)) the energy conditions are satisfied if and only if $\rho_1 \le 0$, $\rho_2 \ge 0$, $\rho_1-\rho_2 \le \rho_3 \le \rho_2-\rho_1$ and $\rho_1-\rho_2 \le \rho_4 \le \rho_2-\rho_1$.

Proof. For (i) it is simply noted that, for Segre type $\{z\bar{z}11\}$ or its degeneracy, the expression (7.88d) for T shows that $T_{ab}l^a l^b = -T_{ab}n^a n^b = -\rho_2 \neq 0$ whereas for Segre type $\{31\}$ or its degeneracy, (7.88c), the null vectors $v = l - n + \sqrt{2}x$ and $u = l - n - \sqrt{2}x$ satisfy $T_{ab}v^a v^b = -T_{ab}u^a u^b = -2\sqrt{2}$. Thus the energy condition (i) fails in each case. Further, for the $\{z\bar{z}11\}$ type, one of $T_{ab}l^b$ $(= \rho_1 l_a - \rho_2 n_a)$ and $T_{ab}n^b$ $(= \rho_1 n_a + \rho_2 l_a)$ is spacelike if $\rho_1 \neq 0$ whilst if $\rho_1 = 0$, $T_{ab}(l^b - n^b) = -\rho_2(l_a + n_a)$ is spacelike since $\rho_2 \neq 0$. For the $\{31\}$ type, $T_{ab}n^b = \rho_1 n_a + x_a$ is spacelike. The proofs of (ii) and (iii) are straightforward but a little lengthy. □

A second order symmetric tensor S at p in the space-time M may also be classified according to its *minimal polynomial equation* (section 2.6). If the minimal polynomial for a matrix $A \in M_n\mathbb{R}$ is of the form

$$(x-\lambda_1)^{s_1}\cdots(x-\lambda_p)^{s_p}$$

where $\lambda_1,\ldots,\lambda_p \in \mathbb{C}$ are the eigenvalues of A then denote it by $[s_1 \cdots s_p]$. Then A satisfies the equation

$$(A-\lambda_1 I_n)^{s_1}\cdots(A-\lambda_p I_n)^{s_p} = 0$$

and one can label the Segre types (including their degeneracies) with their minimal polynomial type. The classification by minimal polynomial is not so fine as by Segre type since certain minimal polynomial types become associated with more than one Segre type (see table 7.1) [57],[79].

Table 7.1

Minimal Polynomial Type	Segre Type
[1 1 1 1]	$\{1,111\}, \{z\bar{z}11\}$
[1 1 1]	$\{(1,1)11\}, \{1,1(11)\}, \{z\bar{z}(11)\}$
[1 1]	$\{1,(111)\}, \{(1,11)1\}, \{(1,1)(11)\}$
[1]	$\{(1,111)\}$
[2 1 1]	$\{211\}$
[2 1]	$\{2(11)\}, \{(21)1\}$
[2]	$\{(211)\}$
[3 1]	$\{31\}$
[3]	$\{(31)\}$

The minimal polynomial conditions give contracted relationships on S. For example if S has Segre type $\{(211)\}$ with zero eigenvalue this relationship is $S^a{}_b S^b{}_c = 0$ whereas if S has Segre type $\{(1,1)(11)\}$ with eigenvalues α and $-\alpha$ one finds $S^a{}_b S^b{}_c = \frac{1}{4}(S_{bd}S^{bd})\delta^a_c$. These minimal polynomial relations are just the *generalised algebraic Rainich conditions* [86], and apply to every Segre type (but not necessarily in a one-to-one way). The above two examples, together with the energy conditions of the last theorem (and the extra condition $S^a{}_a = 0$ in the second example to enable the Segre type $\{(1,1)(11)\}$ to be distinguished) are the original *algebraic Rainich Conditions* for Einstein-Maxwell (null and non-null) fields (see, *e.g.* [55]) as will be seen in the examples later.

Some applications of this classification as applied to well-known energy-momentum tensors can now be given.

(a) Electromagnetic Fields.

The Maxwell tensor F and energy-momentum tensor T for an Einstein-Maxwell field are related by (7.5). The Maxwell field is called *non-null* (respectively, null) at $p \in M$ if F is non-null (respectively, null) at p. If F is *non-null* at p, the expression (7.39) for F on substitution into (7.5) yields

$$T_{ab} = -\frac{1}{8\pi}(\alpha^2 + \beta^2)(2l_{(a}n_{b)} - x_a x_b - y_a y_b) \tag{7.108}$$

It is clear that T takes the form (7.88a) and has Segre type $\{(1,1)(11)\}$ with eigenvalues μ, μ, $-\mu$, $-\mu$, where $8\pi\mu = -(\alpha^2 + \beta^2)$, and so $\mu < 0$ if $F \neq 0$. The energy conditions are then clearly satisfied by theorem 7.11. If the Maxwell field is *null* at p with principal null direction spanned by l then, in the notation of section 7.2, $F_{ab} = 2\nu l_{[a}x_{b]}$ ($l^a l_a = l^a x_a = 0$, $x^a x_a = 1$) and T takes the form

$$T_{ab} = \frac{\nu^2}{4\pi} l_a l_b \tag{7.109}$$

and so, from (7.88b), has Segre type $\{(211)\}$ with all eigenvalues zero. Again, from theorem 7.11, the energy conditions are satisfied.

(b) Fluids.

The general expression for the energy-momentum tensor of a fluid is given by (7.6). In this general form, without appeal to any physical assumptions such as the energy conditions, there is no restriction on the Segre type of T. To see this let S be any second order symmetric tensor at any $p \in M$ and u be any timelike member of T_pM. Now decompose S with respect to u and use the completeness relation (7.28) for g in the form $g_{ab} = h_{ab} - u_a u_b$ where $h_{ab} = x_a x_b + y_a y_b + z_a z_b$ can be interpreted as the induced positive definite metric on the subspace of T_pM orthogonal to u. Then (see, *e.g.* [87],[88])

$$\begin{aligned} S_{ab} &= S^{cd} g_{ac} g_{bd} = S^{cd}(h_{ac} - u_a u_c)(h_{bd} - u_b u_d) \\ &= (S^{cd} u_c u_d + \frac{1}{3} S^{cd} h_c{}^e h_{ed}) u_a u_b + (\frac{1}{3} S^{cd} h_c{}^e h_{ed}) g_{ab} \\ &+ [S^{cd} h_{ac} h_{bd} - (\frac{1}{3} S^{cd} h_c{}^e h_{ed}) h_{ab}] - 2 S^{cd} u_d h_{c(a} u_{b)} \end{aligned} \tag{7.110}$$

The expression for S has the same general form as T in (7.6) when one identifies $q_a \equiv -2S^{cd} u_d h_{ca}$ and $-2\eta\sigma_{ab}$ with the term in square brackets in (7.110) ($\Rightarrow q^a u_a = 0$, $\sigma_{ab} u^b = 0$, $\sigma^a{}_a = 0$). In performing the calculation for (7.110) it is noted that the three terms carrying the factor $\frac{1}{3}$ cancel.

As a special case of (7.6) consider the situation when the *heat flow vector q in (7.6) is zero* [87],[88]. Then u is a timelike eigenvector of T and so the Segre type is $\{1, 111\}$ or one of its degeneracies (theorem 7.9(ii)). The other three independent eigenvectors of T may then be identified in an obvious way with three independent eigenvectors of σ_{ab} (when, because of the condition $\sigma_{ab} u^b = 0$, σ is identified, roughly speaking, with a real 3×3 symmetric matrix in the subspace of T_pM orthogonal to u). In fact it easily follows that a general second order symmetric tensor S can be projected as in (7.110) to a tensor of the form T in (7.6) with $q = 0$ if and only if S admits a timelike eigenvector (which when normalised provides the "projecting" vector u). Another special case arises when the fluid *has heat flow* but is *non-viscous* ($\xi = \eta = 0$). In this case it is clear that if x and y are independent members of T_pM orthogonal to the 2-space spanned by u and q then they are eigenvectors of T with the same eigenvalue p. Hence the possible Segre types are $\{1, 1(11)\}$, $\{2(11)\}$ and $\{z\bar{z}(11)\}$ or their degeneracies (theorem 7.8). They are distinguished by the number of independent eigenvectors in the 2-space spanned by u and q, and whether they are real or not, and which necessarily take the form $q + \lambda u$ ($\lambda \in \mathbb{C}$). A simple calculation then shows that the above listed Segre types (including their degeneracies) occur, respectively, when $(p + \rho)^2 > 4q^a q_a$, $(p + \rho)^2 = 4q^a q_a$ and $(p + \rho)^2 < 4q^a q_a$. By theorem 7.11 the imposition of the energy conditions would eliminate the last of these and ensure that $\rho > 0, p+\rho > 0$ and that no further degeneracies can occur in the two remaining possible Segre types. The final special case to consider is the *perfect fluid* case ($\xi = \eta = 0$, $q = 0$) given by (7.7). Here one has Segre type $\{1,(111)\}$ or its degeneracy and the timelike eigendirection spanned by u (with eigenvalue $-\rho$) and other eigenvalue p. The energy conditions give $\rho \geq 0$ and $p+\rho \geq 0$.

(c) Combination of Fields.

One may also consider certain combinations of non-interacting matter fields (individually satisfying the energy conditions) where, it is assumed, the energy-momentum tensor is obtained by adding together the energy-momentum tensors of the individual fields. Some of these have been discussed before [73],[89] but the present classification scheme allows a more convenient approach of which more details may be found in [87],[88].

For the combination of a perfect fluid and a null electromagnetic field one finds from (7.7) and (7.109), and using the same notation (but assuming, in addition, that $p + \rho > 0$ which means the perfect fluid eigenvalues $-\rho$ and p are distinct)

$$T_{ab} = (p+\rho)u_a u_b + p g_{ab} + \frac{\nu^2}{4\pi} l_a l_b \tag{7.111}$$

Here u and l span well-defined timelike and null directions at each $m \in M$. In the timelike 2-space defined by u and l one introduces another null direction n and then constructs a real null tetrad l, n, x, y at m satisfying the usual orthogonality relations. The expression (7.111) can then be written out, using a completeness relation for g, in terms of this tetrad. Thus one easily obtains (7.88a) with $\rho_3 = \rho_4$ and the Segre type is $\{1, 1(11)\}$ with no further degeneracies. The eigenvalues in terms of p, ρ and ν are easily calculated, as are the corresponding eigenvectors.

For the combination of a perfect fluid and a non-null electromagnetic field one has from (7.108) and (7.7) (where, clearly, $\mu < 0$ and one again chooses $p + \rho > 0$)

$$T_{ab} = (p+\rho)u_a u_b + p g_{ab} + \mu(2l_{(a}n_{b)} - x_a x_b - y_a y_b) \tag{7.112}$$

Here one must distinguish between the cases when the three well-defined directions spanned by l, n and u at some $m \in M$ are dependent (coplanar) or independent. In the former case the situation is similar to that of the previous paragraph and the Segre type is $\{1,1(11)\}$ with no further degeneracies. In the latter case one starts with l and n above and uses the freedom of choice of x and y in the xy 2-space to arrange that x is orthogonal to l, n and u. Next, one introduces null vectors k and k' (satisfying $k^a k'_a = 1$) in the timelike 2-space spanned by u and y and builds the real null tetrad k, k', x, z where the spacelike vector z is determined up to a sign by the tetrad orthogonality relations. Writing out (7.112) in this final tetrad at m leads to (7.88a) and Segre type $\{1,111\}$ with no further degeneracies. Again eigenvalues and eigenvectors can be computed.

Similar techniques show that the combination of two perfect fluids with different flow directions and each having non-negative pressure, or two different fields with energy-momentum tensors of the form (7.109) with different null directions each have Segre type $\{1,1(11)\}$ with no further degeneracies. Other possibilities and further details may be found in [87],[88].

It has been shown how the classification of a second order symmetric tensor S at a point p of a space-time M can be accomplished by considering an associated fourth order tensor P. In the event that S is the Ricci tensor, (7.90) shows that the associated fourth order tensor is the tensor E of (7.11). This remark leads to the following theorem which, although rather trivial, can be useful in calculations. It provides a link between the Segre type of the Ricci tensor and the Petrov type at p through the properties of the curvature at p [90].

Theorem 7.12 *Let M be a space-time and let $m \in M$. If l is a null vector at m, then any two of the following imply the third.*

$$(i)\, l^a l^c R_{abc[d} l_{e]} = 0 \quad (ii)\, l^a l^c C_{abc[d} l_{e]} = 0 \quad (iii)\, l^a l^c E_{abc[d} l_{e]} = 0$$

Thus if (i) holds, l is a repeated principal null direction of the Weyl tensor if and only if l is a (null) Ricci eigenvector (since condition (iii) is equivalent to l being a null Ricci eigenvector from theorem 7.10(iii)).

Proof. The first part follows immediately from (7.10) after noting that $l^a l^c G_{abc[d} l_{e]} = 0$. The second part follows from (7.82b) and theorem 7.10(iii).

□

7.8 The Local and Global Nature of Algebraic Classifications

It was pointed out earlier that the algebraic classification of the tensors considered here applies at a particular point in the space-time M. Two immediate questions spring to mind regarding the classification of the (smooth) Weyl and energy-momentum tensors. First, is there a convenient topological decomposition of M into subsets of M in each of which the algebraic type of the tensor in question is the same at each point? Second, if the tensor in question has the same type at each point of some open subset U of M (and so can be written in the same canonical form at each $p \in U$ as given in the previous sections) can it be written in this canonical form throughout U with the scalars, vectors, bivectors, *etc.* appearing now being smooth on U? The second question is perhaps the most important from the practical viewpoint of calculation as well as theoretically and is rarely, if ever, considered (but the answer is always assumed to be "yes"). There is a precise sense in which this answer is correct but it is not completely obvious. An attempt to answer some of these questions will be presented in this section. The next two theorems are taken from [31].

Theorem 7.13 *Let S be a smooth global second order symmetric tensor on M whose algebraic Segre type is the same (including any degeneracies) at each $m \in M$. Then the eigenvectors and eigenvalues of S may be chosen locally smooth in the sense that, given $p \in M$, there exists an open neighbourhood U of p over which the canonical form (7.88a)-(7.88e) appropriate to the type of S holds and for which the various functions ρ_i and λ are smooth functions and the l, n, x, y smooth vector fields on U. One may set $\lambda = \pm 1$ in (7.88b) over U without affecting any of these results. Similar local smoothness can also achieved for the bivector decomposition of the fourth order tensors P and $\overset{+}{P}$ appropriate to S as in section 7.6. Further the real eigenvalues of S may be regarded as global smooth functions on M. If, in addition, M is simply connected and S has only real non-degenerate eigenvalues (i.e. types $\{1,111\}$, $\{211\}$ or $\{31\}$ with no further degeneracies), then there exist global smooth vector fields on M which span the eigendirections of S at each $m \in M$ (i.e there exist global smooth eigenvector fields of S on M). In particular, if S has Segre type $\{1,111\}$ at each $m \in M$ and M is simply connected then M is* parallelisable *(section 4.8). In fact, under the same conditions, but with S of Segre type $\{211\}$, M is still parallelisable.*

Theorem 7.14 *Let C be the smooth Weyl tensor on M and suppose C has the same Petrov type at each $m \in M$. Then the (complex) Weyl eigenvalues and associated eigenbivectors of C (or $\overset{+}{C}$) can be chosen locally smoothly (i.e. smooth over some open neighbourhood U of $m \in M$) and the canonical forms (7.73), (7.75), (7.77) and (7.79)-(7.81) may be regarded as consisting of smooth functions and bivectors over U. Also, there exist, for some choice of U, four smooth vector fields on U which "span" a Petrov tetrad at each $m \in U$ and k (=1,2,3 or 4) smooth null vector fields on U which are, everywhere on U, the repeated or the non-repeated principal null directions of C. If, in addition, M is simply connected, M admits k global smooth null vector fields which are at each $m \in M$ the repeated or the non-repeated principal null directions of C. In particular if C is of Petrov type **I** at each $m \in M$ and M is simply connected then there exist four smooth null vector fields on M spanning the Debever-Penrose directions at each point and four mutually orthogonal global smooth vector fields on M, one timelike and three spacelike which "span" a Petrov tetrad at each point. As a consequence, M is parallelisable in this case.*

Similar remarks apply in an obvious way to the algebraic classification of smooth bivectors on M as described in section 7.2 [31].

The above theorems required the algebraic (Petrov and Segre) type to be the same at each $m \in M$ (including degeneracies) and it appears that without this, or some similar restriction, the theorems are in doubt. Such an assumption on both the Weyl and energy-momentum tensors is, in fact, usually made, especially in work on finding exact solutions of Einstein's equations. In this respect (and recalling the first question asked at the start of this section) the following result is relevant for a general space-time M and where, for the purposes of its statement only, the subset of M consisting of those points where the Petrov type is **I** (respectively **II**, **III**, **D**, **N** or **O**) will be denoted by the Petrov symbol for that type. Thus $\mathbf{I} \subset M$ is the subset of points of M where the Petrov type is **I**, *etc*.

Theorem 7.15 *Any space-time M can be written as a disjoint decomposition*

$$M = \mathbf{I} \cup \operatorname{int} \mathbf{II} \cup \operatorname{int} \mathbf{III} \cup \operatorname{int} \mathbf{D} \cup \operatorname{int} \mathbf{N} \cup \operatorname{int} \mathbf{O} \cup X$$

where the disjointness of the decomposition defines the (necessarily closed) subset X and where X has no interior, $\operatorname{int} X = \emptyset$.

Proof. The proof of this theorem is given in [91] but this paper contains several errors and a better account can be found in [92]. It can be briefly described here. First one notes that the complex 3×3 matrix Q representing the Weyl tensor (section 7.3) has, for the Petrov types **I**, **II**, **D**, **III**, **N** and **O**, the respective ranks 2 or 3, 3, 3, 2, 1 and 0. Also the points of $\mathbf{I} \subseteq M$ are characterised by Q having three distinct eigenvalues and this can be used to show that **I** is open in M. Next, the above remarks on the rank of Q and the rank theorem (section 3.11) show that the subsets $\mathbf{I} \cup \mathbf{II} \cup \mathbf{D}$, $\mathbf{I} \cup \mathbf{II} \cup \mathbf{D} \cup \mathbf{III}$ and $\mathbf{I} \cup \mathbf{II} \cup \mathbf{D} \cup \mathbf{III} \cup \mathbf{N}$ are open in M. The final part of the argument is to prove that $\text{int}\, X = \emptyset$ (which is the only step required to prove the theorem). This follows from a topological argument using the minimal polynomial of P. □

It is pointed out that since **I** is an *open* subset of M it is equal to its interior. It also follows, since M is connected, that $X = \emptyset$ if and only if the Petrov type is constant on M. Also $M \setminus X$ *is an open dense subset of* M *with the property that every point in it lies in an open subset of* M *in which the Petrov type is constant*. Thus the "constant Petrov type" property is available in an open neighbourhood of "almost every" point of M.

Following the proof of theorem 7.15 a similar result was given for the energy-momentum tensor T of space-time. This result, first given by a direct proof from T [93] and later [94] using the tensor E (section 7.6) states that M may be decomposed disjointly into a family of *open* subsets, in each one of which, the Segre type of T (including degeneracies) is constant, together with a closed subset X' which has empty interior. In particular, the subsets of M on which the Segre type of T is $\{1,111\}$ and $\{z\bar{z}11\}$ (no degeneracies in either case) are each open in M. This result may be combined with the previous theorem to yield the following general decomposition of M with regard to the Weyl and energy-momentum tensors. The proof requires nothing more than taking intersections between the open subsets in the two decompositions (the union of these intersections being $M\setminus(X \cup X'))$ and then showing that $\text{int}(X \cup X') = \emptyset$ (this latter result following since X and X' are each closed and have empty interior) [92].

Theorem 7.16 *Any space-time* M *can be written as a disjoint decomposition*

$$M = A_1 \cup \cdots \cup A_k \cup Y$$

for $k \in \mathbb{Z}^+$ *where each* A_i *is an open subset of* M *at each point of which the Petrov type of* C *and the Segre type (including degeneracies) of* T *are* *constant, and where Y is closed and* $\operatorname{int} Y = \emptyset$.

Thus any point of the open dense subset $M \setminus Y$ of M (*i.e.* "almost everywhere" in M) lies in an open neighbourhood in which the Petrov type and Segre type of T (including degeneracies) are constant.

Another point of theoretical interest concerns the *generic* nature of the Petrov, Segre (and other) classification systems. This question asks whether, in a given classification scheme, certain of the types arising are to be regarded as, in some sense, "general" and others as "special" (and this is already suggested intuitively by the nomenclature of the Petrov scheme). A rigorous approach to such a problem would require some precise way of saying when a set of space-times possessing a certain property was "large" or "small" within the set of all space-times and possible approaches are suggested by measure theory and topology. A topological approach to the restricted problem of performing such a task on the set of all smooth Lorentz metrics on a *fixed* space-time manifold has been given and makes use of the Whitney topology. The topological concept of a subset being open and dense is now taken as "large" whilst closed and nowhere dense is taken as "small". The following theorem gives information about the Petrov and energy-momentum tensor classifications and also information on the curvature tensor which will be useful later (for details see [95]).

Theorem 7.17 *Let M be a space-time and let Λ be the topological space of all smooth Lorentz metrics on M together with the Whitney C^∞ topology.*

i) There exists an open dense subset W_1 of Λ such that if $g \in W_1$ its associated Weyl tensor is of Petrov type **I** *at all points of M except possibly on a 2-dimensional regular submanifold of M where it is type* **II** *and on a set of isolated points where it is of type* **III** *or* **D**.

ii) There exists an open dense subset W_2 of Λ such that if $g \in W_2$ its associated energy-momentum tensor has Segre type $\{1, 111\}$ or $\{z\bar{z}11\}$ (no degeneracies) on an open dense subseet of M, and if its associated energy-momentum tensor satisfies the weak energy condition then its Segre type is $\{1, 111\}$ (no degeneracies) at all points of M except possibly on some 2-dimensional regular submanifold of M where it is $\{1, 1(11)\}$.

iii) There exists an open dense subset W_3 of Λ such that if $g \in W_3$ its associated curvature tensor, in the 6×6 formalism, has rank 6 everywhere (section 7.3) except possibly on some 3-dimensional regular submanifold of M where its rank is 5 and a 1-dimensional regular submanifold *of M where its rank is 4.*

Now if M' is a regular (but not an open) submanifold of M and A is a set of isolated points of M then $M \setminus M'$, $M \setminus A$ and $M \setminus (A \cup M')$ are dense subsets of M. The first of these follows from theorem (4.8)(vi) and the second by supposing $\emptyset \neq U \subseteq A$ for some open subset U of M. Then for $p \in U$ let V be an open subset of M such that $p \in V$ and $V \cap A = \{p\}$. Then the contradiction that $U \cap V = \{p\}$ is open in M follows. For the third let $\emptyset \neq U \subseteq A \cup M'$ with U open in M. The previous two results show that $U \nsubseteq A$ and $U \nsubseteq M'$ and so choose $p \in A \cap U$ and an open set V in M such that $V \cap A = \{p\}$. Then $W \equiv U \cap V$ is non-empty and open in M and then $W \cap (M \setminus \{p\})$ is non-empty, open in M and contained in M'. This contradiction completes the proof and leads to the following result.

Theorem 7.18 *Let M be a space-time and let Λ be the topological space of all smooth Lorentz metrics on M together with the Whitney C^∞ topology. Then there exist open dense subsets W_1, W_2, W_3 and W_4 of Λ such that*

i) if $g \in W_1$ its associated Weyl tensor is of Petrov type **I** *on an open dense subset of M,*

ii) if $g \in W_2$ its associated energy-momentum tensor has Segre type $\{1,111\}$ or $\{z\bar{z}11\}$ (no degeneracies in either case) on an open dense subset of M and if g satisfies the weak energy condition its Segre type is $\{1,111\}$ (no degeneracies) on an open dense subset of M,

iii) if $g \in W_3$ its associated curvature tensor has rank 6 on an open dense subset of M,

iv) if $g \in W_4$ then its associated Weyl, energy-momentum and curvature tensors are of Petrov type **I**, *Segre type $\{1,111\}$ or $\{z\bar{z}11\}$ with no degeneracies in either case (and of the former type if g satisfies the weak energy condition) and rank 6, respectively, on some open dense subset of M.*

Proof. Parts (i) and (ii) follow from the preceeding discussion, having noted from earlier remarks that those subsets of M on which the Petrov type is **I** and those on which the Segre type of T is $\{1,111\}$ and $\{z\bar{z}11\}$ (no degeneracies in either case) are open. For part (iii) let M_4 and M_5 be the *regular* submanifolds where the curvature rank is 4 and 5, respectively (see the previous theorem). The rank theorem then shows that M_4 is closed and $M \setminus (M_4 \cup M_5)$ is open in M. But this latter subset is also dense in M for if $\emptyset \neq U \subseteq M_4 \cup M_5$ with U open in M then $U \nsubseteq M_4$ and $U \nsubseteq M_5$ and so $U \cap (M \setminus M_4)$ is a non-empty open subset of M contained in M_5, which is a contradiction. Part (iv) follows by taking the appropriate intersections and recalling (section 3.3) that the intersection of finitely many open and dense subsets of M is open and dense in M. □

This theorem can be summarised by saying that "in general" a space-time metric on M has a type **I** Weyl tensor, a rank 6 curvature tensor and an energy-momentum tensor of Segre type $\{1, 111\}$ or $\{z\bar{z}11\}$ (no degeneracies in either case) "almost everywhere" on M (and the $\{z\bar{z}11\}$ possibility is ruled out if the weak energy condition is satisfied). In this sense, it confirms what might be called the *generic* types for these fundamental tensors on space-times.

Chapter 8

Holonomy Groups and General Relativity

8.1 Introduction

In Chapter 7 a classification of space-times was presented by means of an algebraic study of the Weyl and energy-momentum tensors. It was pointed out there that such classifications were pointwise and the algebraic types discussed would generally be expected to be different at different points of the space-time. In this chapter an alternative classification will be discussed which makes use of holonomy groups. This classification has the advantage of applying to the space-time itself and is not pointwise. However it has the disadvantage that, in a sense, its general case is too general and its special cases too special. Nevertheless it has a number of uses in the study of symmetries in general relativity and is also useful for a better understanding of the connection and curvature structure of space-times.

8.2 Holonomy Groups

Let M be an n-dimensional smooth paracompact (hence Hausdorff) connected manifold admitting a smooth *symmetric* (*linear*) connection Γ and let $m \in M$. The study of the holonomy group of M involves parallel transport (from Γ) around curves in M beginning and ending at m. Such a study is, perhaps, best done using the frame bundle of M but a less technical approach will be used here. An excellent text for holonomy theory is the book by Kobayashi and Nomizu [20] and this section is largely based on their work.

For an integer k with $1 \leqslant k \leqslant \infty$ let $C_k(m)$ denote the set of all piecewise C^k (*closed*) curves starting and ending at m. If $c \in C_k(m)$, then there is an associated vector space isomorphism τ_c on T_mM obtained by parallely transporting each member of T_mM around c. (The concept of parallel transport introduced in section 4.16 along smooth curves is easily extended to piecewise C^k curves [20].) Using a standard notation for combining and inverting curves (section 3.10) one has $\tau_{c^{-1}} = \tau_c^{-1}$ and $\tau_{c_1 \cdot c_2} = \tau_{c_1} \circ \tau_{c_2}$ for $c, c_1, c_2 \in C_k(m)$. It follows that the set $\{\tau_c : c \in C_k(m)\}$ of all isomorphisms of T_mM arising in this way from curves in $C_k(m)$ is a subgroup of the group $GL(T_mM)$ of all isomorphisms of T_mM. This subgroup will be called *the k-holonomy group of M at m.* More precisely one should add *with respect to the connection* Γ but this will be omitted when the connection referred to is clear. If m' is another point of M then, since M is connected (and hence path connected from section 3.9), there is a piecewise C^k curve γ from m to m' (section 4.14). Thus to every $c \in C_k(m)$ one can associate $c' \in C_k(m')$ by $c' = \gamma \cdot c \cdot \gamma^{-1}$. It is then easily checked that the k-holonomy groups of M at m and m' are isomorphic. Thus one refers to the *k-holonomy group* of M and drops the reference to the point m. It can also be shown that *the k-holonomy groups are independent of the differentiability index k* and so one arrives at the *holonomy group* Φ *of* M.

One could repeat the above operations but now using only those curves in $C_k(m)$ which are homotopic to zero. The above independence of m or k still holds and one arrives at the *restricted holonomy group* Φ° *of* M.

Now let $m \in M$ and let U be a connected open neighbourhood of m. Considering U as an open submanifold of M with the connection Γ of M restricted to it one can define the *restricted holonomy group* $\Phi^\circ(U)$ of U. Then one defines the *local holonomy group* Φ_m^* *of* M *at* m by the intersection of all the restricted holonomy groups $\Phi^\circ(U)$ for all connected open neighbourhoods U of m. In fact it turns out that there exists a connected open neighbourhood U^0 of m such that $\Phi_m^* = \Phi^\circ(U^0)$, that is, the local holonomy group of M at m is the restricted holonomy group of U^0.

The various types of holonomy groups defined above are clearly subgroups of $G \equiv GL(T_mM) = GL(n, \mathbb{R})$ for any $m \in M$. It is possible that Φ° and Φ_m^* could be trivial. If none of these is the case then the following holds [20].

Theorem 8.1 *Let M be a smooth connected paracompact manifold. Then with the above notation the following hold.*

i) The groups Φ, Φ° *and* Φ_m^* *are, for any* $m \in M$, *Lie subgroups of G with* Φ° *and* Φ_m^* *connected.*
ii) Φ° *is the identity component of* Φ *and hence is a normal Lie subgroup*

of Φ of the same dimension. The quotient group Φ/Φ° is countable.

iii) For each integer q the set $\{m \in M : \dim \Phi_m^* \leqslant q\}$ is an open subset of M.

iv) If $\dim \Phi_m^*$ is constant on M then for each $m \in M$ the groups Φ_m^* are equal and, in fact, equal to Φ°.

v) If M is simply connected, $\Phi = \Phi^{\circ}$ (and then Φ is connected).

The Lie algebras of Φ and Φ° are equal (by (ii) above - see section 5.4) and "equal" (*i.e.* isomorphic to) a subalgebra of the Lie algebra of G. They are denoted by ϕ and called the *holonomy algebra*.

Now let $m \in M$ and let U be a coordinate domain of m with coordinates x^a. Consider the curvature tensor arising from Γ and with components $R^a{}_{bcd}$. In the basis $\left(\frac{\partial}{\partial x^a}\right)_m$ of T_mM one can compute and represent Φ and ϕ in terms of matrices (section 5.4). Now with a semi-colon denoting a covariant derivative with respect to Γ one can compute, at m, the following matrices

$$R^a{}_{bcd}X^cY^d, \quad R^a{}_{bcd;e}X^cY^dZ^e, \quad \ldots \tag{8.1}$$

where $X, Y, Z, \ldots \in T_mM$. It turns out that the set (8.1) spans a *subalgebra* of the Lie algebra ϕ (and hence only a finite number of terms are required in (8.1).) [Since any bivector F^{ab} $(= -F^{ba})$ may be written as a linear combination of simple bivectors the span of the matrices in (8.1) equals the span of the matrices $R^a{}_{bcd}F^{cd}, R^a{}_{bcd;e}H^{cde}, \ldots$ where F is any bivector at m and H any tensor at m satisfying $H^{abc} = -H^{bac}$.] This subalgebra is denoted by ϕ'_m and called the *infinitesimal holonomy algebra* and the (unique connected) Lie subgroup of G that it then gives rise to is denoted by Φ'_m and called the *infinitesimal holonomy group* (of M) *at* m. In fact, ϕ'_m is a subalgebra of ϕ. If ϕ'_m is generated entirely by terms of the form $R^a_{bcd}X^cY^d$ $(X, Y \in T_mM)$ then ϕ'_m (and Φ'_m) is called *perfect*. Of course, Φ'_m could be trivial. If it is not, then the following holds [20]

Theorem 8.2 *Let M be a smooth connected paracompact manifold.*

i) For each $m \in M$ the infinitesimal holonomy group Φ'_m at m is a connected Lie subgroup of the local holonomy group Φ_m^* at m.

ii) For each integer q the set $\{m \in M : \dim \Phi'_m \geqslant q\}$ is an open subset of M.

iii) If $\dim \Phi'_m$ is constant on M then $\Phi'_m = \Phi_m^* = \Phi^{\circ}$ for each $m \in M$.

The infinitesimal holonomy group is defined without appealing to parallel transport and seems to be of a different nature to the holonomy (including the restricted and local holonomy) group. A link between them is contained in theorem 8.7. The infinitesimal holonomy group is often useful in calculations since it works locally with the curvature tensor. It is noted here that the local and infinitesimal holonomy groups Φ^*_m and Φ'_m depend on the point $m \in M$ whereas Φ and Φ° are properties of M itself. If, for example, M has a *flat region* (an open subset U of M on which the curvature tensor vanishes) then the local and infinitesimal holonomy groups are trivial at each $m \in U$ but Φ° and Φ may not be since they will "detect" the curvature outside U. It is, however, remarked here that if M and Γ are analytic then Φ^*_m and Φ'_m coincide with Φ° for each $m \in M$ (and also with Φ if M is simply connected) [20].

Now let $m \in M$ and suppose that V is a subspace of T_mM. Then V is called *holonomy invariant* if V is carried onto itself by parallel transport of its members around any differentiable (ie. C^k, $k \geqslant 1$) closed curve at m. It follows that the intersection of two holonomy invariant subspaces is holonomy invariant. Then it is straightforward to see that the parallel transport of a holonomy invariant subspace V from m to any other point m' of M along a differentiable curve from m to m' gives a subspace of $T_{m'}M$ of the same dimension as V which is independent of the curve from m to m' chosen and which is itself holonomy invariant. Thus each such V gives rise to a distribution (see section 4.13) which can be shown to be *smooth* (since M and Γ are) and *integrable* and is called the *holonomy invariant distribution generated by* V or by some basis of V. If such a subspace V exists, the holonomy group Φ is called *reducible* (otherwise *irreducible*). If a subspace of T_mM is such that it contains no non-trivial proper holonomy invariant subspaces it is called *irreducible* (otherwise *reducible*). The group Φ (or Φ° or Φ^*_m) is said to *act trivially* on $v \in T_mM$ (respectively on a subspace $U \subseteq T_mM$) if v (respectively any member of U) is unchanged upon parallel transport around any appropriate closed differentiable curve at m. The subset of all members of T_mM on which Φ acts trivially is a holonomy invariant subspace of T_mM.

Now suppose that the connection Γ on M is a *metric connection* so that M admits a smooth metric g compatible with Γ. Then parallel transport preserves inner products using g and the holonomy structure is enriched. For example, if V is a holonomy invariant subspace of T_mM as above then so is the orthogonal complement of V with respect to $g(m)$. If g is *positive definite* one has the important decomposition theorem of de Rham (see *e.g.* [20]).

Theorem 8.3 *Let M be a smooth connected manifold admitting a smooth positive definite metric g and corresponding Levi-Civita connection Γ. Let $m \in M$ and let V_0 be the maximal subspace of T_mM upon which the associated holonomy group acts trivially. Then the orthogonal complement of V_0 is holonomy invariant and may be written as a direct sum $V_1 \oplus \cdots \oplus V_k$ of subspaces of T_mM which are mutually orthogonal, holonomy invariant and irreducible. Further*

i) If M_i are the maximal integral manifolds through m of the holonomy invariant distributions on M corresponding to the subspaces V_i ($0 \leqslant i \leqslant k$) there exists an open neighbourhood U of m which, as an open submanifold of M, is the manifold product $U = U_0 \times \cdots \times U_k$ where each U_i is an open submanifold of M_i and where the metric on U (restricted from the metric g on M) is the product of the metrics on each U_i which are the restrictions of those on M_i (which, in turn, are induced from the original metric g on M). If M_0 is non-trivial, the induced metric on M_0 is locally Euclidean.

ii) If M is also simply connected and geodesically complete (see section 4.16) then M_0, if non-trivial, is Euclidean and each M_i ($1 \leqslant i \leqslant k$) is, with its induced metric from M, connected, simply connected and geodesically complete and M is isometric to the metric product $M_0 \times \cdots \times M_k$.

The conclusion in the first sentence in (i) above, with the condition that at least one of the subspaces $V_0, \ldots, V_k$ of T_mM is non-trivial and proper, is the definition of (M, g) being *locally decomposable* (i.e. that each $m \in M$ has a neighbourhood which is a non-trivial metric product of submanifolds of M together with their induced metrics from the metric on M). The conclusion in the first sentence of (ii) above, again with the condition that at least one of $V_0, \ldots, V_k$ is non-trivial and proper, is the definition of (M, g) being (*globally*) *decomposable* (i.e. M is a non-trivial metric product of submanifolds of M together with their metrics induced from the metric on M).

Some further results on holonomy can now be presented. The first links holonomy with flatness, the second describes a means of identifying a metric connection from its holonomy group and the third gives a useful result linking the various holonomy groups with their Lie algebras.

Theorem 8.4 *Let M be a smooth connected paracompact manifold admitting a smooth symmetric connection Γ. Then the following conditions* *are equivalent.*

i) Γ *is flat (i.e. the associated curvature tensor vanishes on* M*).*
ii) The holonomy algebra ϕ *is trivial.*
iii) Φ° *is trivial.*
iv) Φ^*_m *is trivial for each* $m \in M$*.*
v) Φ'_m *is trivial for each* $m \in M$*.*

Proof. Theorem 8.7 will show that (i) $\Rightarrow$ (ii). The implications (ii) $\Rightarrow$ (iii) and (iii) $\Rightarrow$ (iv) are immediate, theorem 8.2 shows that (iv) $\Rightarrow$ (v) and the definition contained in (8.1) shows that (v) $\Rightarrow$ (i). This completes the proof. □

Let f be an inner product on T_mM. The holonomy group Φ is said to *preserve* f if, for any closed differentiable curve c at m, $f(u,v) = f(\tau_c u, \tau_c v)$ for each $u, v \in T_mM$. One now has the following theorem to identify metric connections [96],[18].

Theorem 8.5 *Let* M *be a smooth connected paracompact manifold with a smooth symmetric connection* Γ*. Then* Γ *is metric if and only if the holonomy group* Φ *preserves an inner product on* T_mM *for some* $m \in M$ *of signature* (p,q) *(and then* Γ *is compatible with a smooth metric* g *on* M *of signature* (p,q)*.)*

Now let Φ be the holonomy group of M regarded as a Lie subgroup of G ($\equiv GL(T_mM)$) for $m \in M$. Denote the Lie algebra of G by A. Then the Lie algebra ϕ of Φ is a subalgebra of A. If Φ is a connected Lie group (or if M is simply connected and so $\Phi = \Phi^\circ$ is necessarily connected) then Φ is uniquely determined by the subalgebra ϕ of A (theorem 5.3). One can realise G as $GL(n,\mathbb{R})$ where $n = \dim M$ and A as the Lie algebra $M_n\mathbb{R}$ under matrix commutation and use can now be made of the exponential map between ϕ and Φ. To set up the situation let $m \in M$ and choose a chart domain U containing m with coordinates x^a so that $\left(\frac{\partial}{\partial x^a}\right)_m$ is a basis for T_mM. Then Φ is realised as a Lie subgroup of $GL(n,\mathbb{R})$ arising by parallel transport of members of T_mM as described earlier, and ϕ as the corresponding subalgebra of $M_n\mathbb{R}$. Members of T_mM are represented by their components in the basis $\left(\frac{\partial}{\partial x^a}\right)_m$. The idea of a "complex distribution" will be used in the following theorem which, although not formally defined, should be clear.

Theorem 8.6 *Let* M *be a smooth connected paracompact manifold with smooth connection* Γ *and let* $m \in M$*. Suppose, in addition, that* Φ *is a* *connected Lie group (as would necessarily be true if one assumed M to be simply connected).*

i) Let $\alpha \in \phi$ so that the corresponding 1-parameter subgroup of Φ is identified with the Lie subgroup of Φ given by $\{\exp t\alpha : t \in \mathbb{R}\}$. Then if v $(\neq 0)$ is a real or complex eigenvector of α with eigenvalue $\lambda \in \mathbb{C}$, v is an eigenvector of $\exp t\alpha$ with eigenvalue $e^{\lambda t}$ for each t, and conversely if v is an eigenvector of $\exp t\alpha$ with eigenvalue $\lambda(t)$ for each $t \in \mathbb{R}$ then $\lambda(t) = e^{\lambda t}$ $(\lambda \in \mathbb{C})$ and v is an eigenvector of α with eigenvalue λ. If v is an eigenvector of α for each $\alpha \in \phi$ (but with eigenvalue possibly depending on α) then M admits a holonomy invariant 1-dimensional distribution determined by v at m (in the above representation and in the obvious way if v is complex) and conversely (and then a real or complex recurrent smooth vector field exists in some open neighbourhood of m and whose value at m is v).

ii) In the notation of (i) and if in matrix notation $v\alpha = 0$ then $v(\exp t\alpha) = v$ for each t, and conversely. If $v\alpha = 0$ for each $\alpha \in \phi$, M admits a covariantly constant smooth vector field whose value at m is v (in the above representation), and conversely.

Proof. The first part of (i) was given in section 5.5. For the second part if v is an eigenvector of each $\alpha \in \phi$, theorem 5.4 shows that, since Φ is connected, each $f \in \Phi$ may be written as $f = (\exp \alpha_1) \ldots (\exp \alpha_k)$ for some positive integer k and $\alpha_1, \ldots, \alpha_k \in \phi$. Then k applications of the first part of (i) gives $f(v) = \rho v$ $(\rho \in \mathbb{C})$ and so v spans a holonomy invariant subspace of T_mM and also a holonomy invariant 1-dimensional distribution on M. Conversely if such a holonomy invariant distribution exists and is spanned by $v \in T_mM$ and if $\alpha \in \phi$ then a consideration of the 1-parameter subgroup associated with α shows that v is an eigenvector of α. With these conditions holding choose a connected open coordinate neighbourhood U of m, and a smooth real or complex vector field X on U which spans the distribution at each point of U and satisfies $X(m) = v$. To see that X is recurrent let $m' \in U$ and $u' \in T^*_{m'}M$ such that, at m', $X^a u'_a = 0$. Now let c be a curve in U passing through m' with parameter t. Define covectors $u(t)$ along c by parallel translation from u' (i.e. $u(0) = u'$). Since the direction of X is preserved by parallel translation one has $X^a u_a(t) = 0$ on c and a differentiation then shows that, at m', $X^a{}_{;b} u'_a k^b = 0$ where k is the tangent vector to c at m'. The arbitrariness of c (i.e. of k) and u' then shows that $X^a{}_{;b} w_a = 0$ at m' for each covector w satisfying $X^a w_a = 0$ at m'. It follows that $X^a{}_{;b} = X^a q_b$ holds at m' for some $q \in T^*_mM$ and hence that

X is recurrent on U.

In (ii) the first part follows from (i) and then the appropriate covariantly constant vector field may be defined unambiguously on M by parallel transport of $v \in T_mM$ along any piecewise differentiable curve from m to any other point of M. Conversely if such a covariantly constant vector field exists the definition of parallel transport shows that $f(v) = v$ for each $f \in \Phi$. □

An interesting and useful result which provides a link between the infinitesimal holonomy group, the curvature tensor and the holonomy group is the Ambrose-Singer theorem ([97] - see *e.g.* [20]). Let $m, m' \in M$, let c be a piecewise differentiable curve from m to m' and let τ_c be the linear map: $T_mM \to T_{m'}M$ defined by parallel transport along c.

Theorem 8.7 *Let M be a smooth connected paracompact manifold with a smooth symmetric connection Γ and associated curvature structure $\tilde{R}$ (section 4.16) and let $m \in M$. For any other $m' \in M$, any piecewise differentiable curve c as above and any $X, Y, Z \in T_mM$ define a linear map f from T_mM to itself by*

$$f(Z) = \tau_c^{-1}\left[\tilde{R}(\tau_c(X), \tau_c(Y))\tau_c(Z)\right]. \tag{8.2}$$

Then the set of all such linear maps for all choices of m', c, X and Y, when represented in matrix form with respect to some basis of T_mM equals the holonomy algebra ϕ in matrix representation when the holonomy group Φ of M is described as a matrix Lie subgroup of $G = GL(n, \mathbb{R})$ with respect to this basis of T_mM.

In words equation (8.2) says fix m and choose m' and c. To calculate the holonomy algebra ϕ compute all tensors of the form $R^a{}_{bcd}X'^cY'^d$ at m' ($X', Y' \in T_{m'}M$) and then parallel transport them to m along c and do this for all such points m' and curves c to get a matrix representation of ϕ.

8.3 The Holonomy Group of a Space-Time

Now let M be a smooth space-time with smooth Lorentz metric g and associated Levi-Civita connection Γ. The conclusions of the decomposition theorem 8.3 arose from the fact that one considered a positive-definite metric. For a Lorentz metric the situation is a little more complicated and arises from the existence of null vectors (or more precisely from the fact that if V is a subspace of T_mM then V and its orthogonal complement $V^{\perp}$ are not necessarily *complementary* (i.e. $V \cup V^{\perp}$ does not necessarily span T_mM). In fact V and $V^{\perp}$ are complementary if and only if V is non-null (and which is equivalent to $V^{\perp}$ being non-null). One must therefore be a little more discriminating about the reducibility of the holonomy group associated with Γ on M. To find a theorem similar to the previous one for space-times one must appeal to the decomposition theorem of Wu [98].

Let $m \in M$ and call a subspace V of T_mM *weakly irreducible* if it is either $\{0\}$, T_mM itself or non-null and if, in addition, it contains no non-trivial proper non-null holonomy invariant subspaces. Note that V is not prohibited from containing *null* holonomy invariant subspaces (and this can occur as will be clear later). Hence if V is irreducible it is weakly irreducible but not conversely. The concept of weak irreducibility plays the role for Lorentz metrics that irreducibility played for positive definite metrics in the de Rham theorem. The holonomy group of M is said to be *non-degenerately reducible* if for some (any) $m \in M$ some non-trivial proper *non-null* subspace of T_mM is holonomy invariant.

The analogue of the de Rham theorem for non-positive definite metrics is due to Wu [98] (see also [99],[100]). It is given here for simply connected space-times.

Theorem 8.8 *Let M be a simply connected space-time with metric g and let $m \in M$. Suppose the subspace V_0 of T_mM on which the holonomy group Φ (now equal to Φ°) of M acts trivially is either $\{0\}$ or proper and non-null (the trivial case $V_0 = T_mM$ is excluded - see theorem 8.4). Then the orthogonal complement of V_0 is holonomy invariant (and either equals T_mM or else is non-null) and may be written as a direct sum $V_1 \oplus \cdots \oplus V_k$ of subspaces of T_mM which are weakly irreducible, mutually orthogonal and holonomy invariant. Further, if M_i are the maximal integral manifolds through m of the holonomy invariant distributions on M corresponding to the subspaces V_i $(0 \leqslant i \leqslant k)$ each with their metric induced from g there exists an open neighbourhood U of m which, as an open submanifold of M with metric induced from g, is isometric to $U_0 \times U_1 \times \cdots \times U_k$ where each U_i is an open submanifold of M_i with metric induced from M_i. If V_0 is not trivial, M_0 is locally pseudo-Euclidean.*

If, in addition, M is geodesically complete then each M_i $(0 \leqslant i \leqslant k)$ is connected, simply connected and geodesically complete and M is isometric to the metric product $M_0 \times M_1 \times \cdots \times M_k$ with M_0 pseudo-Euclidean.

With the conditions and conclusions of the first paragraph of theorem

8.8 and the proviso that at least one of the subspaces $V_0, V_1, \ldots, V_k$ is non-trivial and proper, M is called *locally non-degenerately decomposable* and similarly for the second paragraph M is called (*globally*) *non-degenerately decomposable*.

For a space-time M parallel transfer preserves inner products. Thus Φ is a subgroup of the Lorentz group $\mathcal{L}$. But Φ is a Lie subgroup, hence a submanifold, of $GL(4,\mathbb{R})$ and $\mathcal{L}$ is a regular submanifold of $GL(4,\mathbb{R})$ (section 6.3). It follows from section 4.11, example (ix), that Φ is also a submanifold of $\mathcal{L}$ and hence that Φ is a Lie subgroup of $\mathcal{L}$. Hence Φ° (or Φ if M is simply connected) is a connected Lie subgroup of the identity component $\mathcal{L}_0$ of $\mathcal{L}$. Now the Lie algebra of $\mathcal{L}$ and $\mathcal{L}_0$ is L and the Lie algebra of Φ and Φ° is ϕ and so ϕ can be identified with a subalgebra of L. Since there is a one-to-one correspondence between the subalgebras of L and the connected Lie subgroups of $\mathcal{L}^0$ (theorem 5.3(iii)) the possibilities for the *restricted* holonomy group Φ° of a space-time are determined by the subalgebra structure of L. These were listed and discussed in section 6.4. In fact it was shown there that $\mathcal{L}_0$ is an *exponential group* (*i.e.* each $f \in \mathcal{L}_0$ is the exponential of some $v \in L$) and the other connected Lie subgroups of $\mathcal{L}_0$ (which are not necessarily exponential) were described in terms of the subalgebra types $R_1, \ldots, R_{15}$ of L. The label R_k ($1 \leqslant k \leqslant 15$) will be used not only for a subalgebra of L but also for its associated connected Lie subgroup of $\mathcal{L}_0$. The next theorem helps to clarify the situation for the holonomy groups of *simply connected* space-times and will be preceded by a brief description of the individual holonomy types using tables 6.1 and 6.2, theorem 8.6 and the bivector classification of section 7.2. Here the notation $<>$ will be used to denote the subspace spanned by the enclosed vector(s).

Type $\mathbf{R_1}$ is flat from theorem 8.4 and the **type $\mathbf{R_{15}}$** is irreducible (and will turn out to be the only irreducible type).

Type $\mathbf{R_2}$. This type has a Lie algebra spanned by a *timelike* bivector $\ell \wedge n$. At any $m \in M$ one can choose a real null tetrad ℓ, n, x, y and then $<\ell>$ and $<n>$ are holonomy invariant as also are $<x>$ and $<y>$ for any such x and y with $<\ell>$ and $<n>$ null and $<x>$ and $<y>$ non-null. This is because ℓ, n, x and y are each eigenvectors of $\ell \wedge n$ (see theorem 8.6). The timelike 2-space $<\ell, n>$ is also holonomy invariant and is weakly irreducible but not irreducible. The associated holonomy group is non-degenerately reducible and a reduction in the sense of theorem 8.8 is given by the subspaces $<\ell, n>$, $<x>$ and $<y>$. Thus M is *locally non-degenerately decomposable* and is sometimes loosely referred to, for obvious reasons, as "2 + 1 + 1 decomposable".

Type $\mathbf{R_3}$. Here the Lie algebra is spanned by a *null* bivector $\ell \wedge y$. In some tetrad ℓ, n, x, y, $<\ell>$ and $<x>$ are, respectively, null and spacelike holonomy invariant 1-spaces. The timelike 3-space $<\ell, n, y>$ is orthogonal to $<x>$ and so is holonomy invariant and is easily shown to be weakly irreducible but not irreducible. In fact the only holonomy invariant 1-space in $<\ell, n, y>$ is $<\ell>$ and if $V \subseteq <\ell, n, y>$ is a holonomy invariant 2-space then since $<x> \subseteq V^{\perp}$ one gets $\dim(V^{\perp} \cap <\ell, n, y>) = 1$. Hence $V^{\perp} \cap <\ell, n, y> = <\ell>$. Thus $<\ell> \subseteq V^{\perp}$ ($\Rightarrow V^{\perp} = <\ell, x>$) and so $V = <\ell, y>$ and is null. The associated holonomy group is non-degenerately reducible using the subspaces $<\ell, n, y>$ and $<x>$ and M is locally non-degenerately decomposable ("3 + 1 decomposable"). Strictly speaking the decomposition in theorem 8.8 does not apply to this type since V_0, although proper, is null. However, one still has the local reduction of the metric concluded in this theorem using the subspaces $\langle \ell, n, y \rangle$ and $\langle x \rangle$ and so this type will sometimes be referred to non-degenerately reducible (and M non-degenerately decomposable).

Type $\mathbf{R_4}$. The Lie algebra here is spanned by a spacelike bivector $x \wedge y$ and $<\ell>$ and $<n>$ are null holonomy invariant 1-spaces and $<x, y>$ is an irreducible spacelike holonomy invariant 2-space. Hence if $t = \ell - n$, $z = \ell + n$, $<t>$ and $<z>$ are holonomy invariant (and non-null) 1-spaces. Thus Φ is non-degenerately reducible using $<t>$, $<z>$ and $<x, y>$ and M is locally non-degenerately decomposable ("1 + 1 + 2 decomposable").

Type $\mathbf{R_5}$. This case will be seen to be impossible (theorem 8.9).

Type $\mathbf{R_6}$. The Lie algebra here is spanned by $\ell \wedge n$ and $\ell \wedge x$ and $<\ell>$ and $<y>$ are holonomy invariant as is $<\ell, n, x>$ the latter being timelike and although not irreducible is easily checked to be weakly irreducible. Then Φ is non-degenerately reducible using $<\ell, n, x>$ and $<y>$ and M is locally non-degenerately decomposable ("3 + 1 decomposable").

Type $\mathbf{R_7}$. The Lie algebra is spanned by $\ell \wedge n$ and $x \wedge y$ and $<\ell>$ and $<n>$ are the only holonomy invariant 1-spaces. Now the only holonomy invariant 2-spaces are the timelike 2-space $<\ell, n>$ and the spacelike 2-space $<x, y>$. To see this let V be a holonomy invariant 2-space. Now $W_1 = <\ell, x, y>$ and $W_2 = <n, x, y>$ are holonomy invariant being the orthogonal complements of $<\ell>$ and $<n>$, respectively, and hence so are $V \cap W_1$ and $V \cap W_2$. So either (a) $V \cap W_1 = <\ell>$ *or* (b) $V \subseteq W_1$ *and* either (c) $V \cap W_2 = <n>$ *or* (d) $V \subseteq W_2$. Now (a) and (c) together imply $V = <\ell, n>$ and (b) and (d) together imply $V = <x, y>$. The combinations of (a) and (d) and of (b) and (c) are impossible and the result holds. The 2-space $<x, y>$ is irreducible

and $<\ell, n>$ is weakly irreducible and so Φ is non-degenerately reducible with M locally non-degenerately decomposable ("$2+2$ decomposable").
Type $\mathbf{R_8}$. This type is spanned by $F = \ell \wedge x$ and $G = \ell \wedge y$ and the only holonomy invariant 1-space is $<\ell>$ and is null. Hence, by considering orthogonal complements, there are no non-null holonomy invariant 3-spaces. Let V be a non-null holonomy invariant 2-space. Then if W is the holonomy invariant null 3-space $<\ell, x, y>$ either $\dim V \cap W = 1$ (in which case $<\ell> \subseteq V$ and so V is timelike and $V^{\perp}$ is spacelike with $V^{\perp} \subseteq W$) or $V \subseteq W$ and so V is spacelike. So without loss of generality one may take V spacelike and $V \subseteq W$. Now if $f \in \Phi$ one may represent f as a finite product of the terms $\exp tF$ and $\exp t'G$ $(t, t' \in \mathbb{R})$ and since $\ell F = 0$, $\ell G = 0$, $xF = -\ell$, $xG = 0$, $yF = 0$, $yG = -\ell$ one finds $\ell(\exp tF) = \ell$, $x(\exp tF) = x - t\ell$, $y(\exp tF) = y$, $\ell(\exp t'G) = \ell$, $x(\exp t'G) = x$, $y(\exp t'G) = y - t'\ell$. It is now clear that $V \neq <x, y>$. Thus since V is spacelike, V is spanned by vectors of the form $\ell + x$ and $\ell + \alpha y$ $(\alpha \in \mathbb{R})$ after, if necessary, redefining x and y and taking linear combinations. Then $(\ell + x)\exp tF = (1-t)\ell + x$ and hence V cannot be holonomy invariant. It follows that for this type Φ is reducible but not non-degenerately reducible and that M is not locally non-degenerately decomposable.
Type $\mathbf{R_9}$. Here the Lie algebra generators are $\ell \wedge n$, $\ell \wedge x$ and $\ell \wedge y$ and again the only holonomy invariant 1-space is $<\ell>$ and an argument similar to that in the R_8 case shows that no non-null holonomy invariant 2- or 3-spaces exist. Thus Φ is reducible but not non-degenerately reducible and M is not locally non-degenerately decomposable.
Type $\mathbf{R_{10}}$. The Lie algebra generators are now $\ell \wedge n$, $\ell \wedge x$ and $n \wedge x$ and the only non-null holonomy invariant spaces of dimension 1, 2 or 3 are $<y>$ and its orthogonal complement and each is irreducible. Thus Φ is non-degenerately reducible and M is locally non-degenerately decomposable ("$3+1$ decomposable").
Type $\mathbf{R_{11}}$. The Lie algebra is spanned by $\ell \wedge x$, $\ell \wedge y$ and $x \wedge y$ and arguments similar to those for types R_8 and R_9 show that $<\ell>$ is the only holonomy invariant 1-space and that no non-null holonomy invariant subspaces of dimension 1, 2 or 3 exist. Hence Φ is reducible but not non-degenerately reducible and M is not locally non-degenerately decomposable.
Type $\mathbf{R_{12}}$. The Lie algebra is now spanned by $\ell \wedge x$, $\ell \wedge y$ and $\ell \wedge n + \rho(x \wedge y)$ and the result is as obtained for type R_{11}.
Type $\mathbf{R_{13}}$. Here the Lie algebra (in an orthonormal basis u, x, y, z) is spanned by $x \wedge y$, $x \wedge z$ and $y \wedge z$. The only holonomy invariant subspaces are $<u>$ and its orthogonal complement and each is non-null and irreducible.

Thus Φ is non-degenerately reducible and M is locally non-degenerately decomposable ("$1+3$ decomposable").

Type $\mathbf{R_{14}}$. The Lie algebra is spanned by $\ell \wedge n$, $x \wedge y$, $\ell \wedge x$, $\ell \wedge y$ and the situation is again as for R_{11} and R_{12}.

It should be remarked here that if M has holonomy type R_1 then M is flat and conversely (see theorem 8.4). This means that, in an obvious sense, M is locally Minkowski space-time. For a space-time M to actually be (*i.e.* be isometric to) Minkowski space-time it is necessary and sufficient that M be flat, simply connected and geodesically complete [18].

This completes the preliminary discussion of the possible holonomy types. It follows that each type except R_{15} is reducible and that types exist (R_8, R_9, R_{11}, R_{12} and R_{14}) which are reducible but not non-degenerately reducible. The other types R_1 (trivially) and $R_2, R_3, R_4, R_6, R_7, R_{10}$ and R_{13} are non-degenerately reducible. This latter group can be collected into "splitting" types as $R_1(1+1+1+1)$, $R_2(2+1+1)$, $R_4(1+1+2)$, $R_7(2+2)$, R_3, R_6 and $R_{10}(3+1)$ and $R_{13}(1+3)$. A remark arising from the above discussion is that each proper subgroup of $\mathcal{L}_0$ preserves some "direction" since each type except R_{15} possesses a holonomy invariant 1-space.

Theorem 8.9 *Let M be a simply connected space-time with (connected) holonomy group Φ ($= \Phi^\circ$) and associated holonomy algebra ϕ.*

i) Each of the types $R_1, \ldots, R_{15}$ with the exception of R_5 can occur as the type of the holonomy group Φ. Type R_5 can not occur as a space-time holonomy group or as a space-time local or infinitesimal holonomy group at any point of M.

ii) Generically, Φ is of type R_{15}, that is, Φ is the Lie group $\mathcal{L}_0$.

iii) The members of ϕ admit a common eigenvector $v \in T_mM$, $v \neq 0$, if and only if Φ is reducible (and which, in turn, is equivalent to M admitting a global smooth recurrent vector field whose value at m is v).

iv) The members of ϕ admit a common eigenvector $v \in T_mM$ with zero eigenvalue for each member of ϕ if and only if M admits a global smooth nowhere zero covariantly constant vector field whose value at m is v.

Proof. The existence part of (i) will be covered in sections 8.4 and 8.5. To show that Φ cannot be of type R_5 suppose it is. Then from theorem 8.4 M cannot be flat and so there exists $m \in M$ at which the curvature tensor is not zero. Since $\dim \Phi = 1$ the infinitesimal holonomy group Φ'_m must satisfy $\dim \Phi'_m = 1$ and so the single member F of L generating the holonomy and infinitesimal holonomy algebras (see (8.1)) must satisfy, in component form, $R_{abcd} = \alpha F_{ab}F_{cd}$ at m for $0 \neq \alpha \in \mathbb{R}$. But then $R_{a[bcd]} = 0 \Rightarrow F_{[ab}F_{c]d} = 0 \Rightarrow F$ is a simple bivector (theorem 7.1) and this contradicts the fact that the generator for the R_5 subalgebra is not simple [[101],[102] - see also [45],[103]]. This proof contains within it the impossibility of a type R_5 local or infinitesimal holonomy group at any $m \in M$.

In (ii) one uses the generic condition described in theorem 7.18 to see that, generically on the set of smooth Lorentz metrics on M, there exists $m \in M$ at which the curvature tensor has rank equal to 6 and hence that $\dim \Phi'_m = 6$. Hence, since $\dim \mathcal{L}_0 = 6$ one sees that Φ'_m and Φ° (and Φ) coincide with $\mathcal{L}_0$ and the holonomy type is R_{15} generically.

In (iii) if such a vector v exists then it follows from theorem 8.6 that Φ is reducible. Conversely if Φ is reducible then the remarks immediately before this theorem show that at $m \in M$ there exists a holonomy invariant 1-space spanned, say, by $v \in T_mM$ and hence, by theorem 8.6, v is an eigenvector of each member of the holonomy algebra ϕ. Theorem 8.6 then shows that for each $m \in M$ there is an open neighbourhood U of m and a smooth nowhere zero *recurrent* vector field X on U which belongs to the holonomy invariant distribution generated by v and thus this "direction" is well defined globally on M. Now M, being paracompact, admits a global smooth positive definite metric γ (theorem 4.14). The vector field X defined on U above is defined only to within a smooth nowhere zero scaling factor and can (and will) be replaced on U by the vector field $X' = [\gamma(X,X)]^{-1/2}X$ so that $\gamma(X',X') = 1$. (It is easily checked that if X is recurrent on U then so is fX for $f : U \to \mathbb{R} \setminus \{0\}$ and so X' is recurrent on U.) Now M can be covered by connected coordinate neighbourhoods such as U above and the vector fields X' arising on each such U agree up to a sign wherever such neighbourhoods intersect. The construction of a global recurrent vector field on M can now be briefly described. First consider the set $\tilde{M}$ which is the set theoretic union of all such subsets U each taken twice, once with each sign for X' (so that each point of $\tilde{M}$ is a point $m \in M$ and a choice of either $X'(m)$ or $-X'(m)$). Then $\tilde{M}$ is a 4-dimensional manifold with the above modified subsets as an atlas. [To see these coordinate neighbourhoods of $\tilde{M}$ satisfy the intersection property for manifolds let $p : \tilde{M} \to M$ be the obvious smooth projection map defined by preserving the point of $\tilde{M}$ in M but "forgetting" the sign for X'. Then if $\tilde{U}_1$ and $\tilde{U}_2$ are such coordinate neighbourhoods in $\tilde{M}$ with associated vector fields X_1 and X_2 and if $\tilde{m} \in \tilde{U}_1 \cap \tilde{U}_2$ then $X_1(m) = X_2(m)$ where $m = p(\tilde{m})$. Then if $U_1 = p(\tilde{U}_1)$, $U_2 = p(\tilde{U}_2)$, the definition of X' shows that there exists an open neighbourhood U_3 of m in M such that

$m \in U_3 \subseteq U_1 \cap U_2$ and which gives an open neighbourhood $\tilde{U}_3 \subseteq \tilde{U}_1 \cap \tilde{U}_2$ of $\tilde{m}$. The smoothness of the associated coordinate transformation in $\tilde{M}$ now follows from that in M.] Also, it follows by construction that $\tilde{M}$ is Hausdorff and that the local vector fields X' on M now give rise to a *global* smooth vector field on $\tilde{M}$. Now let $\tilde{U}_1$ be a (connected) coordinate neighbourhood in $\tilde{M}$ (as described above) and let $\tilde{M}_1$ be the component of $\tilde{M}$ containing $\tilde{U}_1$. Then $\tilde{M}_1$ is a connected open submanifold of $\tilde{M}$ and $p(\tilde{M}_1) = M$. To see this last result one assumes that $p(\tilde{M}_1) \neq M$ and notes that, since p is a diffeomorphism $\tilde{U}_1 \to U_1$ for each such U_1, p is an open map $\tilde{M} \to M$ (section 3.10). Hence $p(\tilde{M}_1)$ is open in M but not closed since M is connected. So let m be a limit point of $p(\tilde{M}_1)$ not contained in it. Then if U is one of the above special subsets in M containing m, $p^{-1}(U)$ intersects $\tilde{M}_1$ non-trivially and its appropriate component extends $\tilde{M}_1$ as a connected subset of $\tilde{M}$ contradicting the fact that $\tilde{M}_1$ is a component. Thus $p(\tilde{M}_1) = M$. Then for each such U in M, $p^{-1}(U)$ consists of a disjoint union of two connected subsets of $\tilde{M}$ each diffeomorphic to U and each is either contained in $\tilde{M}_1$ or disjoint from it (since $\tilde{M}_1$ is a component). It follows that the restriction p' of p to $\tilde{M}_1$ is a smooth covering $\tilde{M}_1 \to M$ (section 4.14). Since M is simply connected it follows that, for $m \in M$, $p'^{-1}(m)$ consists of a single point (section 3.10). Thus p' is a diffeomorphism $\tilde{M}_1 \to M$ and the global vector field on $\tilde{M}$ described earlier, when restricted to $\tilde{M}_1$ and then pushed down to M using p'_*, supplies the required *global* recurrent vector field on M. Conversely the existence of such a (global recurrent) vector field X' on M gives rise to a 1-dimension distribution on M. So if $m \in M$ and c is a curve from m to some other $m' \in M$ let m'' be any point in the image of c. Then for $u \in T^*_{m''}M$ one can define by parallel transport a smooth covector field $u(t)$ along c where t is the parameter of c. Now if one defines the function $z(t)$ along (the image of the) curve c by $z(t) = (X'^a u_a)(t)$ one finds $\frac{dz}{dt} = zf(t)$ for some smooth real valued function f. Now if u is chosen so that z is zero at m'' the solution of this differential equation is $z(t) = 0$ and so $X'^a u_a = 0$ in some neighbourhood of m'' on c for all such choices of $u \in T^*_{m''}M$. It now follows that the subset of points on c at which the *direction* of X' is preserved by parallel transport along c is an open subset of c (and clearly the subset of points of c at which it is not so preserved is also open). Since (the image of) c is connected it follows that the direction of X' is preserved by parallel transport from m to m'. That the distribution defined by X' is holonomy invariant now follows.

The proof of (iv) is similar to that of theorem 8.6 part (ii). □

So for a simply connected space-time M one can now summarise the situation (and remark in advance that *any recurrent or covariantly constant vector field on M is of the same type* (*i.e.* timelike, spacelike or null) *at each $m \in M$*). If the holonomy type is R_3, R_4, R_8 or R_{11}, M admits a covariantly constant global *null* vector field (and for the R_4 type two (independent) such vector fields are admitted). If the holonomy type is $R_2, R_6, R_7, R_9, R_{12}$ or R_{14}, M admits a global *recurrent null* vector field which is not covariantly constant (and for the R_2 and R_7 types two such independent vector fields are admitted). To see that the recurrent null vector field is not covariantly constant one merely considers the Lie algebra for each type and uses theorem 8.9. If the holonomy type is R_2, R_3, R_4, R_6 or R_{10} a global covariantly constant *spacelike* vector field is admitted by M (and for the R_2, R_3 and R_4 types two independent such vector fields are admitted). If the holonomy type is R_4 or R_{13} a global covariantly constant *timelike* vector field is admitted by M (and the R_4 type admits two independent such vector fields). The only non-degenerately reducible type not to admit a covariantly constant vector field is R_7. For type R_1, of course, there are four independent covariantly constant vector fields. In summary, types R_6, R_8, R_{10}, R_{11} and R_{13} admit exactly one independent covariantly constant vector field (as described above), types R_2, R_3 and R_4 admit exactly two independent covariantly constant vector fields which, respectively, span a spacelike, a null and a timelike 2-space at each $m \in M$ and type R_1 admits four independent covariantly constant vector fields. The types R_7, R_9, R_{12}, R_{14} and R_{15} admit no such vector fields.

Some further points are usefully noted here. First, as mentioned earlier, if X is a recurrent vector field on M and $\lambda : M \to \mathbb{R}$ is nowhere zero and smooth then λX is recurrent. It was seen above how M could admit a *properly recurrent null* vector field X, that is, X is recurrent but there does *not* exist a smooth nowhere zero function $\lambda : M \to \mathbb{R}$ such that λX is covariantly constant. Such a scaling function may, however, exist locally, that is, suppose one has a global recurrent vector field X on M which in some local coordinate domain U satisfies $X_{a;b} = X_a P_b$ for some covector field P on U. The existence of a nowhere zero function $\lambda : U \to \mathbb{R}$ such that λX is covariantly constant implies, by the Ricci identity, that $R_{abcd}X^d = 0$ on U. Conversely if $R_{abcd}X^d = 0$ on U, the recurrence condition and the Ricci identity give $P_{[a;b]} = 0$ and so in some open neighbourhood V of any $m \in U$ one has $P_a = \psi_{,a}$ for some function $\psi : V \to \mathbb{R}$ and then $e^{-\psi}X$ is covariantly constant on V. Thus properly recurrent vector fields may, over open subsets of M, give rise to covariantly constant null vector fields.

This argument is quite general. If M is simply connected, X recurrent and $R_{abcd}X^d = 0$ on M then the consequent condition $P_{[a;b]} = 0$ on M reveals a *global* function $\psi : M \to \mathbb{R}$ (section 4.16) such that $P_a = \psi_{,a}$ and so $e^{-\psi}X$ is a global covariantly constant vector field on M. Hence if M is simply connected, a recurrent vector field X on M can be globally scaled so as to be covariantly constant if and only if $R_{abcd}X^d = 0$ on M. The idea of *proper* recurrence is restricted to *null* vector fields in the sense that if X is a global recurrent *non-null* vector field on M then the global vector field $X' = |g(X,X)|^{-1/2}X$ is covariantly constant on M. To see this note that $X'^a X'_a$ is constant and non-zero and that X' is also recurrent (and so in any local coordinate system $X'_{a;b} = X'_a P_b$). A contraction with X'^a gives $P_b = 0$ in this (and any other) coordinate domain and the result follows. Bearing in mind theorem 8.9 this is essentially the result that a non-null eigenvector of a bivector necessarily has a zero eigenvalue. The types $R_2, R_6, R_7, R_9, R_{12}$ and R_{14} admit a properly recurrent null vector field (and R_2 and R_7 admit two independent such fields).

For those simply connected space-times which are locally non-degenerately decomposable (*i.e.* Φ ($= \Phi^\circ$) non-degenerately reducible) one has a convenient form for the metric and which arises from the fact that from theorem 8.8 each $m \in M$ has a neighbourhood which is locally a metric product of manifolds (and which explains the "splitting types" given earlier).

For holonomy type R_{13} (splitting type $1+3$) and for $m \in M$ one may decompose T_mM as $V_0 \oplus V_1$ with $\dim V_0 = 1$ and $\dim V_1 = 3$ (as in theorem 8.8). Then one may choose a product coordinate neighbourhood U of the type described in theorem 8.8 (with $U = I \times V$ where I is an open interval of $\mathbb{R}$ and V an open connected subset of $\mathbb{R}^3$) about m and coordinates t, x^α ($\alpha = 1,2,3$) such that, in U, the metric g takes the form

$$ds^2 = -dt^2 + g_{\alpha\beta}dx^\alpha dx^\beta \qquad (\beta = 1,2,3) \tag{8.3}$$

where the $g_{\alpha\beta}$ are independent of t. The vector field $\frac{\partial}{\partial t}$ on U agrees (on U) with the global covariantly constant timelike vector field admitted by M. The components $g_{\alpha\beta}$ give rise to a positive definite metric on the submanifolds $t =$ constant of U and two such submanifolds, say $t = t_0$ and $t = t_1$ are isometric under the diffeomorphism $(t_0, x^\alpha) \to (t_1, x^\alpha)$ on U ("following the integral curves $s \to (s,0,0,0)$ of $\frac{\partial}{\partial t}$"). For the metric (8.3) the only non-vanishing Christoffel symbols are those of the form $\Gamma^\alpha_{\beta\gamma}$ ($\alpha, \beta, \gamma = 1,2,3$) and these are easily checked to be the Christoffel symbols of the metric $g_{\alpha\beta}$. Similarly, the only non-vanishing curvature tensor components are of the form $R^{\alpha}{}_{\beta\gamma\delta}$ $(\delta = 1, 2, 3)$ and these are the curvature components of the Levi-Civita connection associated with $g_{\alpha\beta}$.

For holonomy types R_3, R_6 and R_{10} (splitting type $3 + 1$) one may similarly choose a product coordinate neighbourhood U about any $m \in M$ and coordinates x, x^{α} $(\alpha = 0, 2, 3)$ such that, in U, the metric takes the form

$$ds^2 = dx^2 + g_{\alpha\beta}dx^{\alpha}dx^{\beta} \qquad (\beta = 0, 2, 3) \tag{8.4}$$

where the $g_{\alpha\beta}$ are independent of x and give rise to a Lorentz metric on the submanifolds of constant x in U. The vector field $\frac{\partial}{\partial x}$ on U agrees (on U) with the unique independent global covariantly constant spacelike vector field admitted by M in the R_6 and R_{10} types and one of the family of such vector fields in the R_3 case. The submanifolds of constant x are isometric as in the previous case and the comments regarding the Christoffel symbols and curvature components made in the previous case (appropriately modified) also hold.

For the holonomy type R_7 (splitting type $2+2$) one may choose "product" coordinates $x^A, x^{A'}$ $(A = 0, 1, A' = 2, 3)$ about any $m \in M$ such that the metric in this coordinate system is

$$ds^2 = g_{AB}dx^A dx^B + g'_{A'B'}dx^{A'}dx^{B'} \tag{8.5}$$

where $B = 0, 1$, $B' = 2, 3$ and where the g_{AB} are independent of x^2 and x^3 and the $g'_{A'B'}$ are independent of x^0 and x^1. The components g_{AB} (respectively $g'_{A'B'}$) give rise to a Lorentz (respectively a positive definite) metric on the submanifolds of constant x^2 and x^3 (respectively of constant x^0 and x^1) in this coordinate domain. The link between the local geometry and the global holonomy structure is clear. It is remarked here that, as shown earlier, M admits two global null recurrent vector fields ℓ and n. Thus in local components $\ell_{a;b} = \ell_a p_b$, $n_{a;b} = n_a q_b$. If ℓ and n are scaled such that $\ell^a n_a = 1$ (and note that scaling does not affect recurrence) and this last equation covariantly differentiated one finds $q_a = -p_a$. It then follows that if F is the bivector $F_{ab} = 2\ell_{[a}n_{b]}$, F is covariantly constant, $F_{ab;c} = 0$, and so ${}^*F_{ab;c} = 0$. (Here it is assumed that the coordinate neighbourhood is chosen so that ${}^*F$ is defined - see the remark about this in section 7.2.) The Christoffel symbols Γ^a_{bc} for (8.5) can only be non-zero if each index a, b, c is either 0 or 1 or each is either 2 or 3. In this way these symbols split into two groups in an obvious way which are then the Christoffel symbols for g_{AB} and $g'_{A'B'}$. The same comments apply to the curvature components which then split into the curvature components associated with g_{AB} and $g'_{A'B'}$.

For holonomy type R_2 (splitting type $2+1+1$) the local "product" coordinates can be chosen so that

$$ds^2 = g_{AB}dx^A dx^B + dy^2 + dz^2 \qquad (A, B = 0, 1). \tag{8.6}$$

The g_{AB} are independent of y and z and give rise to a Lorentz metric on the submanifolds of constant y and z in the coordinate domain and in this same domain $\frac{\partial}{\partial y}$ and $\frac{\partial}{\partial z}$ are covariantly constant and spacelike. Again the link with the global holonomy is clear. The Christoffel symbols and curvature components can only be non-zero if each index is 0 or 1 and then they are the Christoffel symbols and curvature components for g_{AB}.

For holonomy type R_4 (splitting type $1+1+2$) one can choose the local product coordinates such that

$$ds^2 = -dt^2 + dx^2 + g'_{A'B'}dx^{A'}dx^{B'} \qquad (A', B' = 2, 3) \tag{8.7}$$

where the $g'_{A'B'}$ are independent of x and t and give rise to a positive definite metric on the submanifolds of constant x and t in this coordinate domain and in the same domain $\frac{\partial}{\partial x}$ and $\frac{\partial}{\partial t}$ are covariantly constant and spacelike and timelike, respectively. Again the link with the global holonomy is clear. The remarks about the Christoffel symbols and curvature components made in the previous example (appropriately modified) still hold.

The above charts will be referred to as *special* (product) *charts* for the holonomy concerned. In the cases where the holonomy is reducible but not non-degenerately reducible (types R_8, R_9, R_{11}, R_{12} and R_{14}) the local forms for the metric are less convenient (see, *e.g.* [55]).

The holonomy structure of space-time can also be described in terms of the covariant constancy of second order (symmetric and skew symmetric) tensors on M [102]. This will be discussed later in Chapter 9.

8.4 Vacuum Space-Times

For a vacuum space-time the vanishing of the Ricci tensor means that the curvature and Weyl tensors are equal everywhere. Thus the curvature tensor inherits all the properties of the Weyl tensor and, in particular, the property

$$R^*_{abcd} = {}^*R_{abcd} \qquad (\Leftrightarrow {}^*R^*_{abcd} = -R_{abcd}). \tag{8.8}$$

This enables a simple and useful property of the infinitesimal holonomy algebra and group to be established [104] (cf. [45]).

Lemma 8.1 *Let M be a vacuum space-time. Then at any $m \in M$ the infinitesimal holonomy group is of type R_1, R_8, R_{14} or R_{15} and all these types are possible (as will be clear from section 8.5).*

Proof. Thinking of the infinitesimal holonomy algebra in terms of the matrices in (8.1) one notes that because of (8.8)

$$\begin{aligned} V_{ab} &\equiv R_{abcd}X^cY^d = R_{abcd}F^{cd} \qquad (F^{cd} = \frac{1}{2}(X^cY^d - Y^cX^d) = -F^{dc}) \\ &\Rightarrow {}^*V_{ab} = {}^*R_{abcd}F^{cd} = R^*_{abcd}F^{cd} = R_{abcd}{}^*F^{cd} \\ W_{ab} &\equiv R_{abcd;e}X^cY^dZ^e = R_{abcd;e}F^{cd}Z^e \\ &\Rightarrow {}^*W_{ab} = R_{abcd;e}{}^*F^{cd}Z^e \end{aligned}$$

and so on. Thus a certain (skew-symmetric) matrix is a member of the infinitesimal holonomy algebra ϕ'_m if and only if its dual is. From this it follows (not completely trivially) that $\dim \phi'_m$ is even (or 0). To see this suppose $F \in \phi'_m$. Then ${}^*F \in \phi'_m$ and so, if there are no further independent members of ϕ'_m, $\dim \phi'_m = 2$. If there is a third independent member $G \in \phi'_m$ then ${}^*G \in \phi'_m$. Now $F, {}^*F, G$ and ${}^*G$ are independent members of ϕ'_m because if

$$\alpha F + \beta\, {}^*F + \gamma G + \delta\, {}^*G = 0 \qquad (\alpha, \beta, \gamma, \delta \in \mathbb{R} \text{ and not all zero})$$

then the independence of $F, {}^*F$ and G shows that $\delta \neq 0$ and a similar argument applied to the dual of the above relation shows that $\gamma \neq 0$. Then the above relation and its dual can be used to eliminate ${}^*G$. This shows that $(\gamma^2 + \delta^2)G$ is a linear combination of F and ${}^*F$ and this contradiction shows that F, ${}^*F$, G and ${}^*G$ are independent. If no other independent members of ϕ'_m exist then $\dim \phi'_m = 4$. If they do then $\dim \phi'_m = 6$ either by repeating the previous argument or by reference to the classification of the subalgebras of the Lorentz algebra since no such subalgebra has dimension 5. The facts that $\dim \phi'_m$ is even and that $F \in \phi'_m \Leftrightarrow {}^*F \in \phi'_m$ now shows that the only possibilities for the type of ϕ'_m are R_1, R_7, R_8, R_{14} and R_{15} (table 6.1).

If for some $m \in M$ the type of ϕ'_m is R_7 then a consideration of table 6.1 and equation (8.1) shows that if F represents the bivector $\ell \wedge n$ in the table then there exist $\alpha, \beta, \gamma, \in \mathbb{R}$ and tensors P, Q, S at m such that the curvature tensor and its covariant derivatives at m satisfy

$$\begin{aligned} R_{abcd} &= \alpha F_{ab}F_{cd} + \beta\, {}^*F_{ab}\, {}^*F_{cd} + \gamma({}^*F_{ab}F_{cd} + F_{ab}\, {}^*F_{cd}) \\ R_{abcd;e\ldots f} &= F_{ab}F_{cd}P_{e\ldots f} + {}^*F_{ab}\, {}^*F_{cd}Q_{e\ldots f} \\ &\quad + ({}^*F_{ab}F_{cd} + F_{ab}\, {}^*F_{cd})S_{e\ldots f} \end{aligned} \tag{8.9}$$

But then the vacuum condition and the symmetry requirement $R_{a[bcd]} = 0$ imply that $\alpha = \beta = 0$ and $P = Q = 0$ and that $\gamma = 0$ and $S = 0$. (To get the latter conditions note that from (8.9) with $\alpha = \beta = 0$, $\ell^a R_{a[bcd]} = 0 \Rightarrow \gamma \ell_{[b} \overset{*}{F}_{cd]} = 0 \Rightarrow \gamma = 0$ since ℓ is not in the blade of $\overset{*}{F}$ (section 7.2). The proof for S is similar. This contradicts the R_7 assumption and it follows that the admissible types for ϕ'_m are R_1, R_8, R_{14} and R_{15}. □

Theorem 8.10 *[104] Let M be a vacuum space-time. The holonomy algebra ϕ of M and hence the restricted holonomy group of M (equal to the holonomy group if M is simply connected) is of type R_1, R_8, R_{14} or R_{15} and all these types are possible (as will be clear from section 8.5).*

Proof. One can directly compute the holonomy algebra ϕ as a subalgebra of the Lie algebra $M_4\mathbb{R}$ under matrix commutation by choosing $m \in M$ and a basis for T_mM and then applying the Ambrose-Singer theorem 8.7 at m. This theorem shows how to generate linear transformations of T_mM into itself whose matrices in the chosen basis constitute ϕ. These transformations arise from a choice of $m' \in M$, a piecewise differentiable curve c from m to m' and $X', Y' \in T_{m'}M$. One then constructs the bivector F' at m' given by

$$F'^a{}_b = R^a{}_{bcd}X'^cY'^d \tag{8.10}$$

and parallel transports it to m along c to obtain the matrix $F^a{}_b$ at m which is then in ϕ. Since the pseudotensor ϵ_{abcd} is locally defined and satisfies $\epsilon_{abcd;e} = 0$, parallel transport along a particular curve "preserves duality" in the sense that the parallel transport of the dual of a bivector is, up to a sign, the dual of the parallel transport of the bivector. Since, from the proof of lemma 8.1, ${}^*F_{ab}{}'$ may be obtained at m' as F'_{ab} was in (8.10) it follows that ${}^*F_{ab}$ is obtained at m, as was F_{ab}, and hence ${}^*F^a{}_b$ is in ϕ. Hence ϕ satisfies the conditions derived previously for the infinitesimal holonomy algebra and so ϕ is of type R_1, R_7, R_8, R_{14} or R_{15}. It is pointed out that the argument ruling out the type R_7 case for the *infinitesimal holonomy* was specifically for that case. To show that ϕ cannot be of type R_7 suppose it is. Then for $m \in M$ the infinitesimal algebra ϕ'_m is a subalgebra of ϕ and, from the previous theorem, ϕ'_m must be of type R_1. Since m is arbitrary this forces M to be flat and hence the contradiction that ϕ is of type R_1. This completes the proof. □

If one makes the simplifying assumptions that the infinitesimal holonomy group is the same at all points of a (vacuum) space-time then the conclusions of theorem 8.10 follow from lemma 8.1 and theorem 8.2(iii) and the infinitesimal holonomy type at $m \in M$ can then be related to the Petrov type of the Weyl tensor (in this case equal to the curvature tensor) at m. In fact the canonical Petrov forms can be easily used to see that if the infinitesimal holonomy type at m is R_8 then the Petrov type at m is O or N whereas for type R_{14} it is O, N or III at m. Some of the early studies of the (not necessarily vacuum) space-time *infinitesimal* holonomy group were given in [[45],[103],[105]-[108]].

8.5 Examples

In this section some standard space-time metrics can be classified according to holonomy type. Some of these metrics will be discussed in more detail later.

Consider the *pp-wave metric* given in a *simply connected* coordinate domain with standard coordinates u, v, x, y by [41],[55]

$$ds^2 = H(x, y, u)du^2 + 2dudv + dx^2 + dy^2. \tag{8.11}$$

Before continuing, it is remarked that this differs from the definition in [55] in that this reference asks only that a global covariantly constant null vector field be admitted. As will be seen below this condition is also satisfied by (8.11) but (8.11) demands more. (In fact the space-time represented by (8.11) admits a global covariantly constant null bivector field.) However, it can be shown that if *either* the Petrov type is O or N everywhere *or* the Ricci tensor is either zero or Segre type $\{(211)\}$ with zero eigenvalue (and both these conditions hold for (8.11)) then the two definitions are locally equivalent [102],[55] (and, in particular, this holds for vacuum space-times).

The Ricci tensor satisfies

$$R_{ab} = \left(\frac{\partial^2 H}{\partial x^2} + \frac{\partial^2 H}{\partial y^2}\right)\ell_a \ell_b \qquad \ell_a = u_{,a}. \tag{8.12}$$

The vector field ℓ^a is null and covariantly constant and a dual pair of covariantly constant null bivectors $F_{ab} = 2\ell_{[a}x_{b]}$ and ${}^*F_{ab} = -2\ell_{[a}y_{b]}$ are admitted where $x_a = x_{,a}$ and $y_a = y_{,a}$. The Petrov type is N or O everywhere with ℓ as repeated principal null direction and the curvature tensor takes the form

$$R_{abcd} = \alpha F_{ab}F_{cd} + \beta\, {}^*F_{ab}{}^*F_{cd} + \gamma(F_{ab}{}^*F_{cd} + {}^*F_{ab}F_{cd}) \tag{8.13}$$

for real valued functions α, β and γ which are independent of v. If the vacuum field equations $R_{ab} = 0$ hold the metric (8.11) is called a *vacuum pp-wave* [41] and then for each u, H is the real part of a complex analytic function and $\beta = -\alpha$ in (8.13). In this case the infinitesimal holonomy group is, from (8.1), of the (perfect) R_8 type everywhere and hence, from theorem 8.2(iii), the holonomy group is of type R_8. Such metrics could, from (8.12), also represent null electromagnetic (null Einstein-Maxwell) fields and here a special case is possible where the infinitesimal holonomy group is everywhere of the R_3 type and hence so is the holonomy group [103],[108]. It is remarked here that any vacuum space-time with nowhere vanishing curvature tensor and admitting a nowhere zero covariantly constant vector field ℓ is locally isometric to a vacuum *pp*-wave [41]. The Ricci identity and theorem 7.4 show that ℓ is necessarily null.

Consider now the *Einstein static universe metric* given in a standard global coordinate system by [55]

$$ds^2 = -dt^2 + (1 - \Lambda r^2)^{-1}dr^2 + r^2(d\theta^2 + sin^2\theta d\phi^2) \tag{8.14}$$

where Λ is a positive constant. This metric has a non-zero cosmological constant and an energy-momentum tensor of the perfect fluid type. From (8.3) this metric, which admits a global non-zero covariantly constant timelike vector field $\frac{\partial}{\partial t}$, can be seen to be of the holonomy type R_{13} (the type R_4 possibility cannot arise because of theorem 8.11(iii)). Similarly, from (8.4) the *Gödel metric* given in a standard global coordinate system by [55]

$$ds^2 = dz^2 + dx^2 - dt^2 - \frac{1}{2}e^{2\sqrt{2}\omega x}dy^2 - 2e^{\sqrt{2}\omega x}dydt \tag{8.15}$$

where ω is a positive constant, admits a global non-zero covariantly constant spacelike vector field $\frac{\partial}{\partial z}$ and has holonomy type R_{10} (the types R_2, R_3 and R_6 are again eliminated by theorem 8.11(iii)) since it also has a non-zero cosmological constant and a perfect fluid type energy-momentum tensor).

The existence part of theorem 8.9(i) can now be attended to. Already the existence of space-times of holonomy type R_3, R_8, R_{10} and R_{13} has been exhibited. It is clear from (8.5)-(8.7) how one may construct examples of type R_7, R_2 and R_4. Minkowski space-time is of type R_1 and the (vacuum) Schwarzschild metric is of type R_{15}. This latter result is a consequence of the fact that for this metric the curvature tensor is easily checked to be everywhere of rank 6 (and so the infinitesimal holonomy type is R_{15} everywhere) and of theorem 8.2(iii). For the remaining types one notes that space-times which have these types as *infinitesimal* holonomy types everywhere have been constructed in [105] and again theorem 8.2(iii) completes the argument. The vacuum type R_{14} examples in [105],[109],[110] complete the final sentences in lemma 8.1 and theorem 8.10. A study of the *infinitesimal* holonomy structure of Einstein-Maxwell fields was given in [103] and a detailed study of *infinitesimal* holonomy for arbitrary space-times (with some errors!) can be found in [108].

An investigation of the *full* holonomy group of a space-time was undertaken in [111]. In this work attention was directed to space-times with the more commonly studied energy-momentum tensors such as perfect fluids and null and non-null Einstein-Maxwell fields, and also to conformally flat space-times. For such space-times, convenient expressions for the curvature tensor can be obtained from (7.10) and (7.11). Then applications of the general techniques of this chapter yield the following theorem [111].

Theorem 8.11 *Let M be a simply connected space-time on which Einstein's equations are assumed with zero cosmological constant. Then*

i) if M is null Einstein-Maxwell with nowhere zero Maxwell tensor F and nowhere zero associated null direction ℓ (section 7.7(a)), the holonomy group of M is of type R_3, R_8, R_{10}, R_{14} or R_{15},

ii) if M is non-null Einstein-Maxwell with nowhere zero associated energy-momentum tensor (section 7.7(a)), the holonomy group of M is of type R_7, R_{14} or R_{15},

iii) if M is a perfect fluid space-time with fluid flow vector field u and with isotropic pressure p and energy density ρ in (7.7) and where M is not an Einstein space, the holonomy group of M is of type R_{10}, R_{13} or R_{15},

iv) if M is conformally flat but not flat, the holonomy group of M is of type R_7, R_8, R_{10}, R_{13}, R_{14} or R_{15},

v) if M is a proper Einstein space (i.e. the Ricci scalar $R = constant \neq 0$), the holonomy group of M is of type R_7, R_{14} or R_{15}.

Proof. The general idea of the proof is to note that either the holonomy type is R_{15} or it is reducible (see the remarks just before theorem 8.9) and that in the latter case a global recurrent vector field is admitted from theorem 8.9. Proceeding with this latter case, with k the recurrent vector field, one uses the information on the energy-momentum tensor (or the Weyl tensor in part (iv)) and the Ambrose-Singer theorem 8.7 to achieve the results claimed. The perfect fluid case (iii) will be briefly established here and the proofs for the other cases can be found in [111].

In (iii) one has from (7.7)

$$R_{ab} = 8\pi \left[\frac{1}{2}(\rho - p)g_{ab} + (\rho + p)u_a u_b \right]. \tag{8.16}$$

Now let k be a recurrent vector field on M. If k is, or can be scaled to be, covariantly constant (and this is always the case if k is not null) then the Ricci identity yields $R_{abcd}k^d = 0$ and $R_{ab}k^b = 0$. If not, then k is null on M and $k_{a;b} = k_a p_b$ for some global 1-form p on M and the Ricci identity gives $k^a R_{abcd} = k_b F_{cd}$ where $F_{ab} = 2p_{[a;b]}$. The algebraic Bianchi identity then gives $k_{[a}F_{bc]} = 0$ and so (section 7.2) F is a simple bivector whose blade contains k. It then follows that $k^a R_{ab} = \lambda k_b$ for some function $\lambda : M \to \mathbb{R}$. Thus, whatever the recurrent vector field k is, it is a Ricci eigenvector, $R_{ab}k^b = \lambda k_a$. It then follows by contracting (8.16) with k^b that

$$\lambda k_a = 4\pi(\rho - p)k_a + 8\pi(\rho + p)(u_b k^b)u_a. \tag{8.17}$$

Since k and u are nowhere zero on M it follows that if k is null, $u_a k^a$ is nowhere zero on M and so $\rho + p \equiv 0$ on M. Thus, from (8.16) M is an Einstein space and the clause in theorem 8.11(iii) is contradicted. If k is not null then, on M, one can assume that $k_{a;b} = 0$ and $\lambda = 0$. In particular, if k is timelike, $u^a k_a$ is never zero on M and (8.17) shows that either $\rho - p = \rho + p = 0$ on M (which means that M is a vacuum space-time and hence an Einstein space) or that u is proportional to k on M. Thus u is recurrent and, since $u^a u_a = -1$, it is covariantly constant and one may assume $u = k$ on M. Then $u_{a;b} = 0 \Rightarrow R_{abcd}u^d = 0 \Rightarrow R_{ab}u^b = 0$ and, from (8.16), $3p + \rho = 0$ on M. Thus

$$R_{ab} = -16\pi p(g_{ab} + u_a u_b). \tag{8.18}$$

The identity $T_a{}^b{}_{;b} = 0$ contracted with u^a shows first that $p_{,a}u^a = 0$ and then that p and ρ are each constant on M. The relation $R_{abcd}u^d = 0$ together with (8.18) then show that if (u, x, y, z) is a pseudo-orthonormal tetrad at any $m \in M$, then at m

$$R_{abcd} = -8\pi p(X_{ab}X_{cd} + Y_{ab}Y_{cd} + Z_{ab}Z_{cd}) \tag{8.19}$$

where X, Y and Z are, respectively, the bivectors $x \wedge y$, $x \wedge z$ and $y \wedge z$. It follows from (8.19) that the infinitesimal holonomy group, and hence the holonomy group of M has dimension at least three and the existence of the covariantly constant timelike vector field k (or u) then shows the holonomy type is R_{13}. If k is spacelike, the procedure is similar. One again finds $\lambda = 0$ on M and then, from (8.17) (rejecting the vacuum case), that $u^a k_a$ and $\rho - p$ vanish on M. Then $R_{ab} = 16\pi\rho u_a u_b$ and the identity $T_a{}^b{}_{;b} = 0$ shows that p and ρ are non-zero constants on M. The curvature tensor is

$$R_{abcd} = 8\pi\rho(C_{ab}C_{cd} + D_{ab}D_{cd} + E_{ab}E_{cd}) \tag{8.20}$$

where C, D and E are, respectively, the bivectors $u\wedge y$, $u\wedge z$ and $y\wedge z$ in some orthonormal tetrad u, k, y, z at any $m \in M$. Again the holonomy group of M is at least 3-dimensional and so is of type R_{10}. It is remarked here that for the R_{13} type above, (8.18) and (8.19) show that M is conformally flat since the Weyl tensor at m is unchanged under $SO(3)$ rotations in the x, y, z 3-space at m and this is impossible unless it vanishes (see the end of section 7.3). In the R_{10} case, however, one only achieves such a Weyl symmetry in the y, z 2-space at m. Hence the Petrov type is D or O. Thus for a perfect fluid which is not a proper Einstein space (see part (v) of the theorem) one either has holonomy type R_{13} with p and ρ constants satisfying $3p + \rho = 0$, holonomy type R_{10} with $p = \rho =$ constant or holonomy type R_{15} (for which the F.R.W. models yield examples since, generically, they have curvature rank six). It is remarked that if the cosmological constant is reinstated in the perfect fluid case the same results are obtained except for the restrictions on p and ρ which now involve the cosmological constant. In this case the Einstein static universe gives an example of such a holonomy type R_{13} whilst the Gödel metric provides one for the R_{10} type. These last few remarks correct some errors in [111]. □

It is also remarked here that the various holonomy possibilities in each of the parts of theorem 8.11 actually exist. For parts (i) and (ii) they can be found in [103],[111], for part (iii) they were given above and for parts (iv) and (v) in [111][112].

In all holonomy types except R_{10}, R_{13} and R_{15} a recurrent or covariantly constant *null* vector field ℓ is admitted. It was shown in the proof of the last theorem that k is an eigenvector field of the Ricci tensor and from this argument one can easily deduce that $\ell^a \ell^c R_{abcd} = \beta \ell_b \ell_d$ holds on M for some function $\beta : M \to \mathbb{R}$. Thus, from theorem 7.12, ℓ is a repeated principal null direction of the Weyl tensor and the latter is thus algebraically special in the Petrov classification. If M has holonomy type R_{13} one has at each $m \in M$ a timelike vector u satisfying $R_{abcd}u^d = 0$. From this it follows that the Petrov type at m is O, I or D [113]. If M has holonomy type R_2, R_3 or R_4 then at each $m \in M$ the curvature tensor is either zero or of rank one and so (7.10) gives at m

$$E_{abcd} = \lambda F_{ab}F_{cd} - C_{abcd} - \frac{R}{6}G_{abcd} \tag{8.21}$$

where $\lambda \in \mathbb{R}$ (and may be zero) and F is a simple bivector. If F is timelike (respectively, spacelike) the principal null directions (spanned say by ℓ and n) of F (respectively $^*F$) satisfy the conditions of theorem 7.12(i) and $R_{ab}\ell^b = R_{ab}n^b = 0$. Hence, from the same theorem, ℓ and n are repeated principal null directions of the Weyl tensor at m and so the Petrov type is O or D at each $m \in M$. However, if the Weyl tensor vanishes at m, (8.21) shows that the tensor E_{abcd} admits five (or six if $\lambda = 0$) real simple eigenbivectors (including a dual pair) with the same eigenvalue $-\frac{1}{6}R$. Equation (7.98) then shows that $R = 0$ and then (8.21) reveals that $E_{abcd} = 0$ and finally $\lambda = 0$. Thus if $\lambda \neq 0$ the Petrov type at m is D and if $\lambda = 0$ it is clearly zero at m. If F is null in (8.21) the principal null direction ℓ of F satisfies $R_{abcd}\ell^d = 0$, $R_{ab} \propto \ell_a \ell_b$ and hence $R = 0$. Thus the Petrov type at m is O or N and a similar argument to that given above shows that this type is O if and only if $\lambda = 0$ (*i.e.* $R_{abcd} = 0$) at m. One thus has the following theorem relating the holonomy and Petrov classifications.

Theorem 8.12 *Let M be a simply connected space-time. If the holonomy type of M is $R_2, R_3, R_4, R_6, R_7, R_8, R_9, R_{11}, R_{12}$ or R_{14} the Petrov type at any $m \in M$ is algebraically special. If the holonomy type is R_{13} the Petrov type is O, I or D at each $m \in M$. If the holonomy type is R_2 or R_4 the Petrov type is O or D at each $m \in M$ and if it is R_3 the Petrov type is O or N at each $m \in M$. In the R_2, R_3 and R_4 cases the type O possibility at m occurs if and only if the curvature tensor vanishes at m.*

As a final remark it is pointed out that some early important work which is very closely related to holonomy theory (and, in particular to theorem 8.6) was undertaken in [114]-[117] and further references quoted therein. In the holonomy classification given in [99] it seems that types R_9 and R_{11} are omitted and that typographical errors occur in the representation of types R_{12} and R_{14}.

thus the Petrov type at m is G_1 [illegible] D (or I). If M has holonomy type R_{10} or R_{11} then at each $m \in M$ the curvature tensor is either zero or of [illegible] one and so (8.10) gives [illegible]

$$[illegible] \qquad (8.24)$$

where λ [illegible] and may be zero, and [illegible] is a [illegible] simple bivector [illegible] is the [illegible] (respectively spacelike) [illegible] null eigenvectors spanned [illegible] and x_2 of F respectively [illegible] satisfy the conditions of theorem 7.12(?) and [illegible] $R_{ab}{}^{cd}$ [illegible] $= 0$. Hence from the same theorem [illegible] and [illegible] repeated principal null directions of [illegible] sense of the Petrov type [illegible] O or D at each $m \in M$. [illegible] the Weyl tensor [illegible] (8.24) shows that [illegible] (or $\lambda = 0$) [illegible] simple [illegible] bivectors (including [illegible]) [illegible] with the same eigenvalue [illegible]. Equation (7.24)(?) then shows that [illegible] and then (8.24) reveals that [illegible] $= 0$ and finally [illegible] $= 0$. Thus if [illegible] the Petrov type [illegible] D and if [illegible] $= 0$ [illegible] clearly zero at m. If [illegible] is of type R_{11} [illegible] the principal null directions [illegible] of F [illegible] Hence [illegible] the Petrov type [illegible] is O or D and a similar argument to that given above shows that the type is O [illegible] and only if [illegible] $= 0$ [illegible] this [illegible] that the following theorem relating the holonomy and Petrov type [illegible]

Theorem 8.12 *Let M be a space-time [illegible]. If the holonomy [illegible] of M is R_7, R_{10} or R_{11} [illegible] then the Petrov type of [illegible] If the holonomy [illegible] is R_{10} or R_{11} [illegible] Petrov type is O or D [illegible] Petrov type is O [illegible] and R_{10} [illegible] the type is O [illegible]*

As a final remark [illegible] some [illegible] holonomy [illegible] is [illegible] related to [illegible] the [illegible] of the [illegible] in [illegible] and [illegible] [illegible] occur in the [illegible] of type [illegible]

Chapter 9

The Connection and Curvature Structure of Space-Time

9.1 Introduction

Let M be a space-time with metric g, associated Levi-Civita connection Γ and curvature tensor $\tilde{R}$ (with components $R^a{}_{bcd}$). By definition g uniquely determines Γ and Γ uniquely determines $\tilde{R}$. One now asks the following reverse question. To what extent does the above Levi-Civita connection Γ determine the metric g and to what extent does the above curvature tensor $\tilde{R}$ determine Γ and g? Clearly some ambiguity in this determination exists since if $0 < \alpha \in \mathbb{R}$ then g and αg give rise to the same Levi-Civita connection and hence to the same curvature tensor. This chapter will examine this problem which is not only interesting in itself but will yield some results which will be useful in later chapters.

9.2 Metric and Connection

With the notation of the previous section the question of finding all alternative metrics g' for M with the same Levi-Civita connection Γ as the original metric g amounts, in the first instance, to finding all second order symmetric tensors on M which are covariantly constant with respect to Γ. Then, of course, one must consider other properties of such tensors such as non-degeneracy and signature.

Let S be the set of global smooth covariantly constant type $(0,2)$ symmetric tensors on M, so that $h \in S$ implies that $\nabla h = 0$ or, in components, $h_{ab;c} = 0$. Now S is a real vector space and since $h \in S$ is uniquely determined by its value at any $m \in M$, $\dim S \leq 10$. Further, since $g \in S$, $\dim S \geq 1$. If there exists $h \in S$ such that h is not a multiple of g at some (and hence any) point of M then h is called *non-trivial*. If $h \in S$ and h is non-trivial then this is equivalent to the existence of a Lorentz metric g' on M, other than a constant multiple of the original metric g, such that g and g' are each compatible with the original Levi-Civita connection Γ associated with g on M. To see this first choose a fixed basis of T_mM at m and let G_m be the set of all metrics at m (of any signature) each represented in this basis as a symmetric, real, non-singular matrix. The example at the end of section 5.8 applied to the non-singular members of $S(4, \mathbb{R})$ then shows that the subset U of G_m consisting of Lorentz metrics at m is an *open* subset of G_m. It then follows, since g is Lorentz, that there exists $\lambda \in \mathbb{R}$ such that $g(m) + \lambda h(m)$ is both Lorentz and not a multiple of $g(m)$. So set $g' = g + \lambda h \in S$ to obtain a global Lorentz metric on M which is not a constant multiple of g and which is compatible with Γ, $\nabla g' = 0$. The constancy of the Lorentz signature of g' follows from a remark after (4.39).

The set Γ_g of all metrics on M compatible with Γ and of the same signature as g satisfies $\Gamma_g \subseteq S$ but, because of the restrictions of non-degeneracy and signature, is not a subspace of S. However, the previous argument does show that if $h \in S$ is non-trivial, there exists $0 \neq \lambda \in \mathbb{R}$ such that $g + \lambda h \in \Gamma_g$ and so h is a linear combination of members of Γ_g. It follows that $S = \text{Sp}(\Gamma_g)$.

Now let Φ be the holonomy group of M (*i.e.* of Γ) and ϕ the holonomy algebra. Then the members of Φ, regarded as linear isomorphisms of T_mM for $m \in M$, are members of the Lorentz group associated with $g(m)$ and hence the members of ϕ may be represented by matrices F which are skew self-adjoint with respect to $g(m)$ (section 6.3). If an alternative metric such as g' above exists on M then the same is true for $g'(m)$. Thus for each $F \in \phi$ one has at m

$$g_{c(a}F^c{}_{b)} = 0, \qquad g'_{c(a}F^c{}_{b)} = 0. \tag{9.1}$$

This equation supplies an important relationship between g and g' at each $m \in M$ which depends on the members of ϕ. Also, it follows from the remarks above that (9.1) applies not only to $g' \in \Gamma_g$ but to every member of S. The members of ϕ depend on the holonomy type of M (table 6.1). This relationship can be determined from the following theorem [118] in which it is agreed that all raising and lowering of indices is done using the original metric g (so that for $F \in \phi$, $F_{ab} = F_a{}^c g_{cb}$).

Theorem 9.1

i) If (9.1) holds for a simple bivector F then each member v of the blade of F is an eigenvector of g' (with respect to g) with the same eigenvalue *i.e.* $g'_{ab}v^b = \alpha g_{ab}v^b$, $\alpha \in \mathbb{R}$, *for each* v *in the blade of* F.

ii) If (9.1) holds for a non-simple bivector F *then each of the pair of canonical blades for* F *(see (7.39)) consists of eigenvectors of* g' *(with respect to* g*) with the same eigenvalue (but with possibly different eigenvalues for the two blades).*

Proof. In (i) suppose F is spacelike with $F_{ab} = 2x_{[a}y_{b]}$ for unit orthogonal spacelike vectors x and y (with respect to g) at m. Then a substitution into the second equation of (9.1) and contracting firstly with x^a and secondly with y^a gives

$$(g'_{ac}x^a x^c)y_b - (g'_{ac}x^a y^c)x_b = g'_{bc}y^c \tag{9.2}$$

$$(g'_{ac}y^a x^c)y_b - (g'_{ac}y^a y^c)x_b = -g'_{bc}x^c \tag{9.3}$$

A contraction of (9.2) with x^b then gives $g'_{ab}x^a y^b = 0$ and then (9.2) and (9.3) confirm that x and y are eigenvectors of g'. The first of these eigenvector equations when contracted with x reveals that the eigenvalues are equal. If F is timelike or null one writes, respectively, $F_{ab} = 2\ell_{[a}n_{b]}$ and $F_{ab} = 2\ell_{[a}x_{b]}$ for null vectors ℓ and n and a spacelike vector x at m satisfying $\ell^a n_a = x^a x_a = 1$, $\ell^a x_a = 0$. In the timelike case the second equation in (9.1) is contracted, successively, with ℓ^a, n^a, $\ell^a\ell^b$ and $n^a n^b$ to reveal that ℓ and n are eigenvectors of g' with the same eigenvalue. In the null case the corresponding contractions are with ℓ^a, x^a, n^a and $n^a n^b$.

In (ii) for F non-simple one writes $F_{ab} = 2\ell_{[a}n_{b]} + 2\lambda x_{[a}y_{b]}$ $(0 \neq \lambda \in \mathbb{R})$ for a real null tetrad (ℓ, n, x, y) at m and substitutes into (9.1). Then one contracts successively with $\ell^a x^b$, $\ell^a y^b$, $n^a x^b$ and $n^a y^b$ to get $g'_{ab}\ell^a x^b = g'_{ab}\ell^a y^b = g'_{ab}n^a x^b = g'_{ab}n^a y^b = 0$, and with $\ell^a\ell^b$ and $n^a n^b$ to get $g'_{ab}\ell^a\ell^b = g'_{ab}n^a n^b = 0$. The remainder of the proof is straightforward. □

The information in theorem 9.1 together with the possibilities for the holonomy algebra ϕ from table 6.1 then lead to the following theorem [101],[102].

Theorem 9.2 *Let* M *be a simply connected space-time and let* V *denote the vector space of global covariantly constant vector fields on* M. *Then the following hold.*

i) If M *is flat,* $\dim V = 4$. *Otherwise,* $\dim V \leq 2$. *In particular, if the holonomy type is* R_2, R_3 *or* R_4, $\dim V = 2$, *if the holonomy type is* R_6, R_8, R_{10}, R_{11} *or* R_{13}, $\dim V = 1$ *and if the holonomy type is* R_7, R_9, R_{12}, R_{14} *or* R_{15}, $\dim V = 0$.

ii) If M is flat, $\dim S = 10$*. Otherwise,* $\dim S \leq 4$ *(and* $\neq 3$*). In particular, if the holonomy type is* R_2*,* R_3 *or* R_4*,* $\dim S = 4$*, if the holonomy type is* R_6*,* R_7*,* R_8*,* R_{10}*,* R_{11} *or* R_{13}*,* $\dim S = 2$ *and if the holonomy type is* R_9*,* R_{12}*,* R_{14} *or* R_{15}*,* $\dim S = 1$*.*

iii) If h is a non-trivial member of S, the Segre type (including degeneracies) of h is the same at each $p \in M$. The eigenvalues of h are global constant functions on M and for any non-degenerate eigenvalue the associated eigenvector field may be chosen as a member of V.

iv) The set Γ_g of alternative metrics on M compatible with Γ consists, for the holonomy types R_9, R_{12}, R_{14} and R_{15}, only of (positive) constant multiples of g and this is the generic situation. For the holonomy types R_1, R_2, R_3, R_4, R_6, R_7, R_8, R_{10}, R_{11} and R_{13}, however, Γ_g contains members which are not constant multiples of g. All these cases are listed in table 9.1.

Proof. Part (i) was proved in section 8.3. For part (ii) one uses the fact that each $h \in S$ satisfies (9.1) (with g' replaced by h) together with the first four columns of table 9.1 (which have now been justified). In particular, if the holonomy type is R_9, R_{12}, R_{14} or R_{15}, theorem 9.1 shows that, in each case, ℓ, n, x and y (and hence every member of T_pM) is an eigenvector of h (with respect to g) with the same eigenvalue, at each $p \in M$. Thus $h = \phi g$ on M for some function $\phi : M \to \mathbb{R}$ which is clearly smooth (since g and h are). The covariant constancy of h and g then forces ϕ to be constant. The other holonomy types are dealt with similarly. For example, if the holonomy type is R_{13}, the possibilities for F in (9.1) are, at each $p \in M$, $x \wedge y$, $y \wedge z$ and $x \wedge z$ where (u, x, y, z) is a pseudo-orthonormal tetrad at p and u is the global covariantly constant unit timelike vector field admitted by this type. Thus x, y and z are eigenvectors of h at p with equal eigenvalues and so in any chart of M from section 7.5 and (7.28)

$$h_{ab} = \phi g_{ab} + \alpha u_a u_b \tag{9.4}$$

where ϕ and α are global functions $M \to \mathbb{R}$ and $u_a = g_{ab}u^b$. To see that ϕ and α are smooth one contracts (9.4) first with g^{ab} and second with $u^a u^b$ and uses the smoothness of g, h and u to see that $4\phi - \alpha$ and $-\phi + \alpha$ are smooth. The smoothness of ϕ and α follows. Finally, a covariant differentiation of (9.4) and use of the covariant constancy of h, g and u shows that ϕ and α are constant. Thus the members of S for this holonomy type are as in (9.4) with ϕ and α independent constants (and so $\dim S = 2$). As another example, consider the holonomy type R_7. Then the choices for

F in (9.1) are $\ell \wedge n$ and $x \wedge y$ and so the pairs (ℓ, n) and (x, y) span 2-dimensional eigenspaces of h with not necessarily equal eigenvalues at each $p \in M$, where ℓ and n are global recurrent null vector fields on M (section 8.3). Thus the Segre type of h is $\{(1,1)(11)\}$ or its degeneracy everywhere, and from (7.29) and (7.88a) one has in any chart of M

$$h_{ab} = \phi g_{ab} + 2\alpha \ell_{(a} n_{b)} \tag{9.5}$$

where ϕ and α are global functions $M \to \mathbb{R}$, $\ell_a = g_{ab}\ell^b$, $n_a = g_{ab}n^b$ and ℓ and n are chosen so that $\ell_a n^a = 1$. This last choice ensures that the recurrence 1-forms of ℓ and n differ only in sign and hence that the global type $(0,2)$ tensor with local components $\ell_{(a}n_{b)}$ is covariantly constant. The smoothness of ϕ and α in (9.5) then follow after contractions with g^{ab} and $\ell^a n^b$ and then a covariant differentiation of (9.5) reveals the constancy of ϕ and α on M. The other holonomy types can be handled in a similar way and give the results listed in table 9.1. The results of parts (iii) and (iv) of the theorem can then be read off from the table. □

It is remarked that the restrictions that h be in Γ_g are, for (9.4), that $\alpha < \phi > 0$ and, for (9.5), that $\phi + \alpha \neq 0 < \phi$. It is also remarked that, from theorem 8.9, the situation when $\dim V = 0$ or $\dim S = 1$ or when Γ determines g up to a positive constant multiple is generic. Finally, it is remarked that the holonomy type R_7 is the only one for which $\dim S > 1$ (equivalently, Γ admits compatible metrics other than constant multiples of g), but $\dim V = 0$.

9.3 Metric, Connection and Curvature

Now return to the conditions described at the beginning of section 9.1. There it was stated that for $0 < \alpha \in \mathbb{R}$ the space-time metrics g and αg give rise to the *same curvature tensor* $\tilde{R}$. Now one asks the question whether this "ambiguity" in the metric which gives rise to the given curvature tensor is the only one possible or are there "less trivial" metric changes with the same property.

Let g be the original metric on the space-time M so that the curvature tensor components $R^a{}_{bcd}$ satisfy the skew symmetry property $g_{ae}R^e{}_{bcd} + g_{be}R^e{}_{acd} = 0$. If g' is another Lorentz metric on M with the same curvature tensor $\tilde{R}$ as g then at each $m \in M$ one has similarly

$$g'_{ae}R^e{}_{bcd} + g'_{be}R^e{}_{acd} = 0. \tag{9.6}$$

Table 9.1 The first column gives the usual holonomy labelling and the second gives a bivector basis for the holonomy algebra in terms of a real null tetrad ℓ, n, x, y or, in the case of R_{13}, a pseudo-orthonormal tetrad u, x, y, z. The third column gives a basis for the vector space of covariantly constant vector fields on M and the fourth gives a complete set of properly recurrent null vector fields, up to scaling, on M. The fifth column gives, in the appropriate tetrads, the most general form on M for the members of the vector space S of covariantly constant type $(0,2)$ tensors on M, where $\phi, \alpha, \beta, \gamma \in \mathbb{R}$. This same column also gives the most general form for the metrics on M compatible with the connection Γ, but now with the appropriate constraints on ϕ, α, β and γ to ensure non-degeneracy and preserve signature.

Label	Bivectors	Cov. Const.	Rec.	S
R_2	$\ell \wedge n$	$\langle x, y\rangle$	ℓ, n	$\phi g_{ab} + \alpha x_a x_b + \beta y_a y_b$ $+\gamma(x_a y_b + y_a x_b)$
R_3	$\ell \wedge y$	$\langle \ell, x\rangle$	—	$\phi g_{ab} + \alpha x_a x_b + \beta \ell_a \ell_b$ $+\gamma(\ell_a x_b + x_a \ell_b)$
R_4	$x \wedge y$	$\langle \ell, n\rangle$	—	$\phi g_{ab} + \alpha \ell_a \ell_b + \beta n_a n_b$ $+\gamma(\ell_a n_b + n_a \ell_b)$
R_6	$\ell \wedge n,\ \ell \wedge x$	$\langle y\rangle$	ℓ	$\phi g_{ab} + \alpha y_a y_b$
R_7	$\ell \wedge n,\ x \wedge y$	—	ℓ, n	$\phi g_{ab} + \alpha(\ell_a n_b + n_a \ell_b)$
R_8	$\ell \wedge x,\ \ell \wedge y$	$\langle \ell\rangle$	—	$\phi g_{ab} + \alpha \ell_a \ell_b$
R_9	$\ell \wedge n,\ \ell \wedge x$ $\ell \wedge y$	—	ℓ	ϕg_{ab}
R_{10}	$\ell \wedge n,\ \ell \wedge x$ $n \wedge x$	$\langle y\rangle$	—	$\phi g_{ab} + \alpha y_a y_b$
R_{11}	$\ell \wedge x,\ \ell \wedge y$ $x \wedge y$	$\langle \ell\rangle$	—	$\phi g_{ab} + \alpha \ell_a \ell_b$
R_{12}	$\ell \wedge x,\ \ell \wedge y$ $\ell \wedge n + \rho x \wedge y$	—	ℓ	ϕg_{ab}
R_{13}	$x \wedge y,\ y \wedge z$ $x \wedge z$	$\langle u\rangle$	—	$\phi g_{ab} + \alpha u_a u_b$
R_{14}	$\ell \wedge x,\ \ell \wedge y$ $\ell \wedge n,\ x \wedge y$	—	ℓ	ϕg_{ab}
R_{15}	L	—	—	ϕg_{ab}

So consider the set B_m of type $(1,1)$ tensors at m given by

$$B_m = \{R^a{}_{bcd}V^{cd} : V \text{ a type } (2,0) \text{ skew tensor at } m\}. \tag{9.7}$$

Clearly B_m is a real vector space and each member of B_m is skew-self adjoint with respect to $g(m)$ and $g'(m)$, that is, $F \in B_m$ implies that (9.1) holds. Thus the dimension of B_m, called the *rank* of the curvature tensor (or the *curvature rank*) at m (cf. the equivalent notion of rank used for the Weyl tensor in section 7.3) satisfies $\dim B_m \le 6$. It is thus clear that by interpreting (9.7) as a map f from skew-symmetric type $(2,0)$ tensors at m

to type $(1,1)$ tensors at m according to $f: V^{cd} \to R^a{}_{bcd}V^{cd}$ then f is linear and B_m is the range of f. Since (9.1) holds for each $F \in B_m$ the dimension and nature of B_m will, from theorem 9.1, impose algebraic constraints on any alternative metric g' which has $\tilde{R}$ as its curvature tensor.

At this point it is convenient to introduce an algebraic classification of the curvature tensor at a point $m \in M$ which is suited to the work of this section and which is described in terms of B_m [118],[119]. All index raising and lowering is done using the original metric g.

Class A This class covers all possibilities with $\tilde{R}(m) \neq 0$ not covered by B, C or D. In this case the curvature rank at m is either $2, 3, 4, 5$ or 6.

Class B Suppose the range of f has dimension two and is spanned by a g-orthogonal pair of simple bivectors one (g)-timelike and one (g)-spacelike (so that they form a dual pair). If (ℓ, n, x, y) is a real null tetrad at m with respect to g and these bivectors are $F_{ab} = 2\ell_{[a}n_{b]}$ and ${}^*F_{ab} = 2x_{[a}y_{b]}$ then at m (since $R_{a[bcd]} = 0$)

$$R_{abcd} = \alpha F_{ab}F_{cd} - \beta\, {}^*F_{ab}{}^*F_{cd} \tag{9.8}$$

for $\alpha, \beta \in \mathbb{R}$, $\alpha \neq 0 \neq \beta$.

Class C Suppose the range of f has dimension two or three and the members of this range space have a common eigenvector $\omega \in T_mM$ with zero eigenvalue. Then each member of B_m is simple (theorem 7.1) and ω is determined up to a multiplicative factor.

Class D Suppose the range of f is of dimension one and spanned, say, by a tensor $F^a{}_b$. Then, at m, $R_{abcd} = \alpha F_{ab}F_{cd}$ $(\alpha \in \mathbb{R})$ and the symmetry $R_{a[bcd]} = 0$ shows that F_{ab} is a simple bivector (theorem 7.1).

Class O Here the curvature tensor is zero.

In connection with this classification it is remarked that the equation at m given for $k \in T_mM$ by

$$R^a{}_{bcd}k^d = 0 \tag{9.9}$$

has no non-zero solutions (for k) in classes A and B, a 1-dimensional subspace of T_mM spanned by ω as its solutions in class C and a 2-dimensional subspace of T_mM which is the g-orthogonal complement of F as its solutions in class D.

The above classification is clearly pointwise and the particular class of the curvature tensor will be expected to vary from point to point subject, of course, to continuity requirements and the rank theorem (section 3.11). A space-time will be said to be of *class A* (respectively, B, C, D or O) if it is of that class at every point. In the general case a topological decomposition of the space-time into the various classes can be given. In fact, if one introduces the labels A, B, C, D and O to denote the subsets of points of M at which the curvature has that respective class then M can be disjointly decomposed as $M = A \cup \text{int}B \cup \text{int}C \cup \text{int}D \cup \text{int}O \cup Z$ where Z is closed and has empty interior. It can be shown that A is open in M ($A = \text{int}A$) and clearly if M is non-flat, int$O$ is empty.

If M is a space-time of class C the solutions of (9.9) give a map which associates with $m \in M$ a 1-dimensional subspace $\Delta(m)$ of T_mM. This is a smooth distribution on M but this requires a proof that for $m \in M$ there is an open neighbourhood U of m and a smooth vector field on U which spans $\Delta(m)$ at each $m \in U$. To see this note that since $\dim \Delta(m) = 1$ at each $m \in M$, one may choose three equations of the form $\sum_{j=0}^{3} \alpha_{ij} k^j = 0$ ($i = 1, 2, 3$) from the set (9.9) such that, say, the matrix α_{ij} ($i, j = 1, 2, 3$), which consists of smooth curvature components, is non-singular at, and in some neighbourhood U of, any $m \in M$. Then the equations $\sum_{j=1}^{3} \alpha_{ij} k^j = -\alpha_{i0}$ ($i = 1, 2, 3$) have a unique solution for k^1, k^2, k^3 which is smooth on U. This solution together with the choice $k^0 = 1$ completes the proof and so a "locally smooth solution" of (9.9) exists.

If M is a space-time of class B then (9.8) holds at each point. If at $m \in M$ $\alpha \neq \beta$, then α and β will differ in some open neighbourhood of m. In this case it follows [31] that (9.8) holds in some neighbourhood U of m with α, β, F and $\overset{*}{F}$ chosen smoothly and, further, ℓ, n, x and y may be chosen smoothly in U such that at each point of U they constitute a null tetrad with ℓ and n spanning the blade of F and x and y the blade of $\overset{*}{F}$. Hence two smooth distributions are determined in U by the blades of F and $\overset{*}{F}$. The same is true if $\alpha = \beta$ in some neighbourhood of m. If, however, $\alpha = \beta$ at m but not in any neighbourhood of m the situation is not clear. In this case the above smoothness will be assumed. With this smoothness one can now substitute (9.8) into the Bianchi identity $R_{ab[cd;e]} = 0$ and contract the resulting equation with $\ell^a x^b \ell^c n^d x^e$. A short calculation then reveals that, on U, $\ell_{a;b} x^a x^b = 0$. Similar contractions with $\ell^a x^b \ell^c n^d y^e$, $\ell^a x^b x^c y^d \ell^e$ and $\ell^a x^b x^c y^d n^e$ reveal, respectively on U, the results $\ell_{a;b} x^a y^b = 0$, $\ell_{a;b} y^a \ell^b = 0$ and $\ell_{a;b} y^a n^b = 0$. Use of the interchange symmetries $\ell \leftrightarrow n$, $x \leftrightarrow y$, obvious in (9.8), then yields $\ell_{a;b} y^a y^b = \ell_{a;b} x^a \ell^b = \ell_{a;b} x^a n^b = 0$ together with similar such relationships on the covariant derivative of n. It follows from these that $\ell_{a;b} x^a = \ell_{a;b} y^a = 0$ (and, since ℓ is null, $\ell_{a;b} \ell^a = 0$) and so at each $m \in U$ and for each $s \in T_mM$, $\ell_{a;b} s^b$ is proportional to ℓ.

From this it easily follows that ℓ is recurrent on U, $\ell_{a;b} = \ell_a p_b$, for some smooth covector field p on U. Similar comments apply to n so that n is also recurrent on U. Since $\ell^a n_a = 1$ on U one has $n_{a;b} = -n_a p_b$ on U. It is then easily checked that, in (9.8), $F_{ab;c} = {}^*F_{ab;c} = 0$. This enables the covariant derivatives of the curvature tensor to be computed easily and then (8.1) shows that the *infinitesimal* holonomy group is 2-dimensional at each point of M. Theorem 8.2 and table 9.1 then show that Φ^0 is of type R_7. Section 8.3 then shows that, if M is simply connected, ℓ and n may be regarded as global null recurrent vector fields on M and the *holonomy type of M is R_7*. It also follows that in the local product coordinates used in (8.5) the functions α and β in (9.8) satisfy $\alpha = \alpha(x^0, x^1)$ and $\beta = \beta(x^2, x^3)$.

If M is a space-time of class D then the solutions of (9.9) give rise to a map which associates with $m \in M$ a 2-dimensional subspace $\Delta(m)$ of T_mM. Again this distribution is smooth. To see this note that the curvature tensor (given above) is $R_{abcd} = \alpha F_{ab} F_{cd}$ at each $m \in M$. It will be shown that the function $\alpha : M \to \mathbb{R}$ and bivector F can be chosen smoothly. Let B be a smooth bivector chosen such that $R_{abcd}B^{cd}$ is nowhere zero in some neighbourhood U of m. Then $R_{abcd}B^{cd}$ is smooth and thus F_{ab} is proportional to a *smooth* bivector G, $F_{ab} = \lambda G_{ab}$, with $\lambda : M \to \mathbb{R}$ not necessarily smooth. But then $R_{abcd} = \beta G_{ab} G_{cd}$ and β is clearly smooth since the curvature is. Thus the above claim is justified with β and G chosen for α and F, respectively.

The following algebraic statement about (9.6) can now be made [119].

Theorem 9.3 *Let M be a space-time with Lorentz metric g and let g' be another metric on M with the same curvature tensor as the original metric g. Then with all index raising and lowering and metric statements regarding orthogonality etc. made with respect to g, one has the following at $m \in M$.*

i) If the curvature is of class D and $u, v \in T_mM$ span the 2-space at m orthogonal to the blade of F, there exist $\phi, u, v, \lambda \in \mathbb{R}$ such that

$$g'_{ab} = \phi g_{ab} + \mu u_a u_b + 2\nu u_{(a} v_{b)} + \lambda v_a v_b. \tag{9.10}$$

ii) If the curvature is of class C, there exist $\phi, \gamma \in \mathbb{R}$ such that

$$g'_{ab} = \phi g_{ab} + \lambda w_a w_b. \tag{9.11}$$

iii) If the curvature is of class B, there exist $\phi, \lambda \in \mathbb{R}$ such that

$$g'_{ab} = \phi g_{ab} + 2\lambda \ell_{(a} n_{b)} = (\phi + \lambda) g_{ab} - \lambda (x_a x_b + y_a y_b). \tag{9.12}$$

iv) If the curvature is of class A, there exists $\phi \in \mathbb{R}$ such that

$$g'_{ab} = \phi g_{ab}. \tag{9.13}$$

Proof. In part (i), B_m consists of a single independent simple bivector F and theorem 9.1 then shows that each member of its blade is an eigenvector of g' with the same eigenvalue. If F is spacelike, say $F_{ab} = 2x_{[a}y_{b]}$ for *orthogonal* unit spacelike vectors x and y, then, from (7.88a)-(7.88d), the Segre type of g' is either $\{1,1(11)\}$, $\{2(11)\}$ or $\{z\bar{z}(11)\}$ or their degeneracies and one can write in terms of a real null tetrad (ℓ, n, x, y) and for $a, b, c, d \in \mathbb{R}$

$$g'_{ab} = a(x_a x_b + y_a y_b) + b\ell_a \ell_b + 2c\ell_{(a}n_{b)} + dn_a n_b. \tag{9.14}$$

This immediately converts into (9.10) on using the completeness relation (7.29). The proof when F is timelike is similar. If F is null, say $F_{ab} = 2\ell_{[a}x_{b]}$, with ℓ null, x unit spacelike and $\ell^a x_a = 0$, then g' has either Segre type $\{(1,11)1\}$ (the extra degeneracy occurring since in the diagonalisable case another null eigenvector of g' with the same eigenvalue as for ℓ and x must be present from theorem 7.9(iii)) or $\{(21)1\}$ or $\{(31)\}$ or their degeneracies. Choosing x appropriately, one obtains from the canonical forms (7.88a)-(7.88d), an expression for g' which readily converts to (9.10) after use of the completeness relation (7.29).

In part (ii) the set of all members of T_mM orthogonal to w span an eigenspace of g'. Thus if w is timelike (respectively, spacelike), g' has Segre type $\{1,(111)\}$ (respectively, $\{(1,11)1\}$) or its degeneracy whereas if w is null the Segre type is $\{(1,111)\}$ or $\{(211)\}$. The result now follows. The proof of (iii) is similar.

For (iv) one notes that the conditions are such that $\dim B_m \geq 2$, that if $\dim B_m = 2$, B_m cannot consist of the span of an orthogonal pair of non-null simple bivectors (class B) and that no member of T_mM is orthogonal to each member of B_m (class C). From this it follows that the members of B_m must give rise to a pair of 2-dimensional eigenspaces U and V of g' which are not orthogonal and from whose members a basis of T_mM may be chosen. Since U and V are not orthogonal, their associated eigenvalues are equal (see end of section 7.5), and so the Segre type of g' is $\{(1,111)\}$ and so (9.13) holds. □

A reference to theorem (7.17(iii)) reveals that, generically, the curvature tensor has rank ≥ 4 at each $m \in M$ and thus M is of class A. In this case a strong result can be established [119].

Theorem 9.4 *Let M be a space-time of class A. Then the curvature tensor associated with the space-time metric g determines g up to a constant conformal factor and hence determines the Levi-Civita connection uniquely. This situation is generic. If M is a vacuum space-time of constant curvature class the curvature tensor determines g up to a constant conformal factor (and the Levi-Civita connection uniquely) unless it is of Petrov type* **N**.

Proof. If g' is an alternative metric on M with the same curvature tensor as g the previous theorem (equation (9.13)) shows that, on M, $g' = \phi g$ where ϕ is a nowhere zero function $M \to \mathbb{R}$. Since g and g' are smooth, ϕ is smooth. Now with a semi-colon and a stroke denoting covariant derivatives with respect to the Levi-Civita connections of g and g', respectively, the Bianchi identity (4.36) contracted over the indices a and e give

$$R^a{}_{bcd;a} + 2R_{b[c;d]} = 0 \tag{9.15}$$

$$R^a{}_{bcd|a} + 2R_{b[c|d]} = 0 \tag{9.16}$$

Denoting the Christoffel symbols associated with g and g' by Γ^a_{bc} and Γ'^a_{bc}, respectively, one finds

$$P^a_{bc} \equiv \Gamma'^a_{bc} - \Gamma^a_{bc} = \frac{1}{2}\phi^{-1}(\phi_c\delta^a_b + \phi_b\delta^a_c - \phi^a g_{bc}) \tag{9.17}$$

where $\phi_a = \phi_{,a}$ and $\phi^a = g^{ab}\phi_b$. Now subtract (9.15) from (9.16) to get

$$\begin{aligned} R^e{}_{bcd}P^a_{ea} - R^a{}_{ecd}P^e_{ba} - R^a{}_{bed}P^e_{ca} - R^a{}_{bce}P^e_{da} \\ + R_{be}P^e_{dc} + R_{ed}P^e_{bc} - R_{ec}P^e_{bd} - R_{be}P^e_{cd} = 0. \end{aligned} \tag{9.18}$$

Now substitute (9.17) into (9.18), using (4.22), to get

$$-R_{cdbe}\phi^e + R_{ec}\phi^e g_{bd} - R_{ed}\phi^e g_{bc} = 0 \tag{9.19}$$

A contraction of (9.19) with g^{bd} gives the result

$$R_{ab}\phi^b = 0. \tag{9.20}$$

But then (9.19) confirms that $R_{abcd}\phi^d$ vanishes at m contradicting the fact that a class A space-time admits no solutions to (9.9) at any point unless $\phi_a(m) = 0$. Thus $\phi_a \equiv 0$ on M and so ϕ is a constant function on M. The generic condition follows from theorem 7.17(iii). The vacuum part of the result now follows since now the Weyl and curvature tensors are equal and only in the type **N** case can there be non-trivial solutions for k in (9.9).

That the type **N** clause is necessary follows from an example following the proof of theorem 9.6. □

Thus it is generic for the space-time curvature tensor to determine its metric up to a constant conformal factor. The techniques in the proof of this result are easily modified to lead to the following theorems, the first of which is due to Brinkmann [120] (see also [41],[119]) and the second gives a partial converse to the well-known result that the Weyl tensor C with components $C^a{}_{bcd}$ is unchanged under a conformal rescaling of the metric (see section 4.16). The statement of Brinkmann's theorem is adapted to the present needs and differs from the original. Brinkmann, in fact, proved more.

Theorem 9.5

Let M be a space-time manifold admitting space-time metrics g and g' each of which is non-flat and vacuum. Suppose that g and g' are conformally related on M, so that $g' = \phi g$ for a positive smooth function $\phi : M \to \mathbb{R}$. Let U be the (open) subset of M on which $d\phi$ is not zero and V the (open, dense) subset of M on which the curvature tensor of g (necessarily equal to that of g') is nowhere zero. Let $W = U \cap V$. Then there is a disjoint decomposition of M given by

$$M = W \cup \mathrm{int}(M \setminus U) \cup A$$

where the disjointness defines A. Further, one has

i) ϕ is constant on each component of $\mathrm{int}(M \setminus U)$,
ii) A is closed and nowhere dense (i.e.. $W \cup \mathrm{int}(M \setminus U)$ is open and dense in M),
iii) each point of W admits a (coordinate) neighbourhood on which g and g' are each pp-wave metrics.

Proof. Since g and g' are conformally related they have the same Weyl tensor C and since they are both vacuum they therefore have the same curvature tensor. This justifies the remark in the statement of the theorem regarding the set V. The work leading to (9.20) now shows that, since W is open, if $m \in W$ there is a coordinate neighbourhood D of m such that $D \subseteq W$ and $R^a{}_{bcd}\phi^d = 0$ on D where $\phi^a = g^{ab}\phi_{,b}$ is nowhere zero on D. Since the curvature tensor is nowhere zero on D, it follows from theorem 7.4 that ϕ^a is the unique solution to this equation up to a scaling, the Petrov type of g and g' is N and ϕ^a is null with respect to g and g' at each point of D. Using the covariant derivative notation of theorem 9.4 one then uses (9.17) to show that on D

$$\phi_{a;b} = \phi_{a|b} + \phi_c P^c_{ab} = \phi_{a|b} + \phi^{-1}\phi_a\phi_b. \tag{9.21}$$

Now on D one has $\phi^a\phi_{a;b} = \phi^a\phi_{a|b} = 0$ and $\phi_{[a;b]} = \phi_{[a|b]} = 0$. Thus a further covariant differentiation of (9.21) and use of (9.17) gives

$$\phi_{a;[bc]} = \phi_{a|[bc]} + \frac{1}{2}\phi^{-1}\phi_{a|[c}\phi_{b]}. \tag{9.22}$$

The Ricci identity (4.33) then reveals that $\phi_{a|[b}\phi_{c]} = 0$ and so it easily follows that $\phi_{a|b} = \alpha\phi_a\phi_b$ for some smooth function α. If one writes $\phi_{a|b} = \phi_a P_b$ $(P_a = \alpha\phi_a)$ another application of the Ricci identity then shows that $P_{[a|b]} = 0$ and hence that $P_a = \beta_{,a}$ for some smooth function β. It follows that $(e^{-\beta}\phi_a)_{|b} = 0$. Similar arguments show the existence of a smooth function γ such that $(e^{-\gamma}\phi_a)_{;b} = 0$. These latter arguments may require a shrinking of the original domain D to a coordinate neighbourhood D' which then admits nowhere zero covariantly constant null vector fields with respect to either g or g'. It now follows [41] that g and g' are *pp*-waves in some coordinate neighbourhood of m. This establishes (iii). Part (i) is clear since $d\phi$ vanishes on $M \setminus U$. For (ii) it is clear that A is closed so let B be a non-empty open subset of M contained in A. By definition of V and since V is dense there exists $p \in B$, and some open subset B' of M satisfying $p \in B' \subseteq B$, on which the curvature tensor never vanishes. Then, again by definition of A, $d\phi$ cannot vanish everywhere on B' (since this would give $B' \subseteq \text{int}(M \setminus U)$ and hence the contradiction $A \cap \text{int}(M \setminus U) \neq \emptyset$. Hence at some point of B', $d\phi$ and the curvature tensor are non-zero (*i.e.* $A \cap W \neq \emptyset$) and this contradicts the definition of A. Thus B is empty and A is nowhere dense. Taking complements shows finally that $W \cup \text{int}(M \setminus U)$ is open and dense in M. □

Brinkmann's theorem is sometimes loosely paraphrased as "if g and ϕg are vacuum metrics then either ϕ is constant or g and ϕg are *pp*-waves."

Theorem 9.6 *([121]) Let M be a space-time manifold and let g and g' be space-time metrics on M which have identical (type $(1,3)$) Weyl tensors on M. Then*

i) the Petrov type at any $m \in M$ is independent of whether it is taken with respect to g or g',

ii) if the subset of points M' of M at which the Petrov type is **O** *or* **N** *has no interior then g and g' are conformally related on M. (The clause here is necessary as a later example will show.)*

Proof. Just as for the curvature tensor, the Weyl tensor gives rise to a map from $(2,0)$ skew-symmetric tensors to type $(1,1)$ tensors at any $m \in M$ according to $V^{ab} \to C^a{}_{bcd}V^{cd}$. It can then be checked from the end of section 7.3 that the Weyl tensor rank as a 6×6 matrix equals the rank of this map and is 4 or 6 if the Petrov type at $m \in M$ is **I**, 6 if the Petrov type is **II** or **D**, 4 if the Petrov type is **III**, 2 if the Petrov type is **N** and zero if it is of type **O**. Now the classification of the curvature tensor given earlier in this section applies equally well to the Weyl tensor and, in fact, the classes D and B are impossible, the first on the grounds of rank (since the Weyl tensor, being self dual, has even rank) and the second because of the identity $C^a{}_{bad} = 0$. Class C applies only when the Weyl tensor has rank 2 and then, from theorem 7.4, ω is null with respect to the relevant metric and the Petrov type is **N**. If class A applies the Petrov type is **I**, **II**, **D** or **III**.

Now theorem 9.3 applies also to the Weyl tensor and then, from part (iv) of this theorem, if the Petrov type with respect to g is **I**, **II**, **D** or **III** at m then g and g' are conformally related at m. Since the Petrov type can be thought of as being determined by the tensors C_{abcd} and the bivector metric G_{abcd} in (7.11), it is clear that the Petrov type at m for these types is the same for the metric g'. If the Petrov type for g at m is **N** then g' satisfies (9.11) with ω null with respect to g (and hence g'). In this case the rank of the Weyl tensor is the same for g and g' (and equals 2) and so the Petrov type for g' is also **N**. The type **O** case is trivial. Thus (i) is established. Under the conditions of (ii), theorem 9.3(iv) shows that g and g' are conformally related on the subset $M \backslash M'$ of M. But the rank theorem (section 3.11) shows that this subset is *open* in M and it is also *dense* since M' has no interior. It easily follows that g and g' are conformally related on M. To see this let $g' = \phi g$ on M' with $\phi' : M' \to \mathbb{R}$ smooth. Then define a global smooth function $\psi : M \to \mathbb{R}$ by $\psi = \frac{1}{4}g^{ab}g'_{ab}$ and note that ψ and ϕ agree on $M \setminus M'$. Then the global smooth tensor field $g' - \psi g$ vanishes on $M \setminus M'$ and hence on M. It is remarked that the situation described in part (ii) here is, from theorem 7.17(i) the generic case. □

To see the importance of the clause in part (ii) consider the vacuum *pp*-wave metric g [41] in a single coordinate domain with coordinates v, u, x, y given in (8.11). The covector field with components $\ell_a = u_{,a}$ satisfies

$\ell_{a;b} = 0$ and the function H may be chosen so that the Petrov type is N and the vacuum condition holds everywhere. Now define a metric g' on this domain by

$$g'_{ab} = g_{ab} + \lambda(u)\ell_a\ell_b \tag{9.23}$$

where λ is an arbitrary smooth function on this domain. It is easily checked that g' is Lorentz with inverse $g'^{ab} = g^{ab} - \lambda\ell^a\ell^b$ ($\ell^a = g^{ab}\ell_b$) and, using a similar notation to that in theorem 9.4, one finds $g'_{ab;c} = \dot{\lambda}\ell_a\ell_b\ell_c$ ($\dot{\lambda} = \frac{d\lambda}{du}$). Then by using useful formulae in [122] one gets

$$\begin{aligned} P^a_{bc} = \Gamma'^a_{bc} - \Gamma^a_{bc} &= \tfrac{1}{2}g'^{ad}(g'_{db;c} + g'_{dc;b} - g'_{bc;d}) \\ &= \tfrac{1}{2}\dot{\lambda}\ell^a\ell_b\ell_c \end{aligned} \tag{9.24}$$

$$R'^a{}_{bcd} - R^a{}_{bcd} = 2P^a_{b[d;c]} + 2P^a_{f[c}P^f_{d]b} = 0 \tag{9.25}$$

where $R^a{}_{bcd}$ and $R'^a{}_{bcd}$ are the curvature tensor components associated with g and g', respectively. Then (9.25) shows that g' is a vacuum metric (since g is) and that g and g' have the same curvature tensor. It follows that g and g' are not conformally related (unless one selects $\lambda \equiv 0$) but that their Weyl tensors are equal everywhere. It is straightforward to check that g' is also a (type **N**) vacuum *pp*-wave metric with ℓ^a null with respect to g' and convariantly constant with respect to Γ'.

The techniques of this section lead to another result which can now be briefly sketched. Suppose that M is a space-time with metric g, type $(1,3)$ Weyl tensor C and energy-momentum tensor T (with local components T_{ab}). Which other metrics g' on M give the same tensors C and T on M? First make the additional assumptions that C is not zero or Petrov type **N** over any non-empty open subset of M and that there are no non-trivial solutions for k (and the curvature tensor of g) of (9.9) over any non-empty open subset of M. Then using a prime to denote quantities constructed from g' one easily finds that $g' = \phi g$ for some smooth nowhere zero function $\phi : M \to \mathbb{R}$ (theorem 9.6(ii)), that in local coordinates $g'^{ab} = \phi^{-1}g^{ab}$ and from the Einstein field equations $R = \phi R'$ and $R'_{ab} = R_{ab}$. Then, from (7.10) and (7.11) $\tilde{R}'_{ab} = \tilde{R}_{ab}$, $E'^a{}_{bcd} = E^a{}_{bcd}$ and $G'^a{}_{bcd} = \phi G^a{}_{bcd}$ and so one finally arrives at $R'^a{}_{bcd} = R^a{}_{bcd}$. Thus g and g' have identical curvature tensors. Then the proof of theorem 9.3 reveals that $k^a = g^{ab}\phi_{,b}$ satisfies (9.9) and hence that ϕ is constant on M. It is pointed out that the assumptions made to achieve this result do not stop the space-time being "generic" (see theorem 7.17). The following result has been proved [123] (see also [124]).

Theorem 9.7 *Let M be a space-time with metric g, Weyl tensor C and energy-momentum tensor T. Suppose C is not of Petrov type* **O** *or* **N** *over a non-empty open subset of M and that there are no non-trivial solutions for k of (9.9) for the curvature tensor of g over a non-empty open subset of M. Then any other metric on M with the same Weyl and energy-momentum tensor as g is a constant (conformal) multiple of g. Thus it is generically true that the Weyl and energy-momentum tensors of a space-time determine the metric up to a constant conformal factor.*

This theorem is a formal attempt to show that the sources of the gravitational field (the tensor T) and the "vacuum contribution" to this field (represented formally by the tensor C) determine the metric (up to the units of measurement). However, the fact that T is restricted to its tensor type $(0, 2)$ form means that this result must remain purely formal. It is remarked, however, that an alternative argument based on [125] yields somewhat similar, but less easily interpreted, results irrespective of the tensor type of the energy-momentum tensor.

9.4 Sectional Curvature

The idea of the Gauss curvature of a surface leads, for higher dimensional manifolds, to the concept of *sectional* (sometimes called *Riemannian*) *curvature*. Let M be a manifold of dimension n ($n \geq 2$) with a (smooth) metric g of any signature. At $p \in M$ let G_p denote the set of 2-dimensional subspaces (2-spaces) of T_pM so that G_p may be taken as the Grassmann manifold $G(2, \mathbb{R}^n)$ (section 4.17). As pointed out at the end of section 7.2 the set G_p may be identified with the set of projective simple bivectors at p (the equivalence classes of rank 2 bivectors at p where the equivalence relation is the usual equality up to a real scaling). Although these concepts were introduced earlier for $n = 4$, the generalised versions here are straightforward. Now define a real valued map σ_p on a certain subset of G_p (to be specified later) as follows: let $F \in G_p$ be a 2-space at p which in some basis of T_pM (or some coordinate system x^a about p where the basis of T_pM is taken as $(\frac{\partial}{\partial x^a})_p$) is the blade of the simple bivector F^{ab} (or any non-zero real multiple of F^{ab}) and take

$$\sigma_p(F) = \frac{R_{abcd}F^{ab}F^{cd}}{2G_{abcd}F^{ab}F^{cd}} \tag{9.26}$$

where G is defined in (7.11) and with the proviso that the denominator ($= 2F_{ab}F^{ab}$) in (9.26) does not vanish. This last condition fixes the subset of G_p upon which σ_p is defined. The discussion preceding (9.26) and the fact that the right hand side of (9.26) is unchanged if F^{ab} is replaced by λF^{ab} ($0 \neq \lambda \in \mathbb{R}$) shows that $\sigma_p(F)$ is well defined. The real number $\sigma_p(F)$ is called the *sectional curvature* of F and σ_p the *sectional curvature function at* p. Now let $G(M) = \bigcup_{p\in M} G_p$ be the *Grassmann bundle* of all 2-spaces at all points of M. Then the *sectional curvature function* σ on M is the map from a certain subset of $G(M)$ to $\mathbb{R}$ given by $\sigma(F) = \sigma_p(F)$ if $F \in G_p$ and $\sigma_p(F)$ is defined.

If F is a 2-space at p one can interpret $\sigma_p(F)$ in the following way. By use of normal coordinates (section 4.16) one can show the existence of an open coordinate neighbourhood U of p such that the geodesics in M starting from p and with initial tangent vector at p in F generate a 2-dimensional submanifold N of U. If N admits a metric induced from g then $\sigma_p(F)$ is defined and equals the Gauss curvature of N at p.

If g is positive definite then σ_p is clearly defined on the whole of G_p and hence σ on the whole of $G(M)$. For a space-time M with Lorentz metric g one may decompose G_p into its timelike, spacelike and null members as discussed in some detail in section 6.1. Here a change of notation is convenient and these subsets of G_p will be denoted, respectively, by T_p, S_p and N_p with the suffix p indicating the point of M to which they are attached. (Earlier they were referred to as T_2, S_2 and N_2, respectively, and were 2-dimensional subspaces of Minkowski space.) Thus $G_p = T_p \cup S_p \cup N_p$. Then, since if F^{ab} is a simple bivector ($\Leftrightarrow F_{ab}\overset{*}{F}{}^{ab} = 0$ from theorem 7.1) the condition $G_{abcd}F^{ab}F^{cd} = 0$ is equivalent to $F_{ab}F^{ab} = 0$ and hence to F^{ab} being *null* (theorem 7.2), and so σ_p is defined only on the subset $\overline{G}_p = T_p \cup S_p$ of G_p and is *smooth* (in fact, analytic). Also T_p and S_p, and hence $\overline{G}_p$, are 4-dimensional open submanifolds of the 4-dimensional Grassmann manifold G_p and $N_p = G_p \setminus \overline{G}_p$ is a 3-dimensional closed submanifold of G_p and is the boundary of T_p and of S_p (theorem 6.3). If $\ell \in T_pM$ is null the set of wave surfaces (sections 6.1 and 6.3) to ℓ at p will be denoted by $W_p(\ell)$ and is a 2-dimensional submanifold of G_p and of $\overline{G}_p$ (and S_p). Each of T_p, S_p, N_p and $W_p(\ell)$ is connected and T_p and S_p are diffeomorphic.

Clearly $\overline{G}_p$ is an open dense subset of G_p.

Again, if g is positive definite, σ_p is a globally defined smooth map $G_p \to \mathbb{R}$ and hence a (bounded) real-valued function on the compact connected space $G(2, \mathbb{R}^n)$. For a space-time, however, σ_p is now only defined on the subset $\overline{G}_p$ of G_p and since $\overline{G}_p$ is not a connected subspace of G_p or a compact subspace of G_p (otherwise $\overline{G}_p$ would be a closed subset of (the Hausdorff space) G_p contradicting the fact that N_p is not open) σ_p has a richer and more interesting structure. One might first ask about the possibility of extending the function σ_p to points of $G_p \setminus \overline{G}_p$, that is, to null 2-spaces at p. Clearly in the case when σ_p is a constant function on $\overline{G}_p$, say, $\sigma_p(F) = c \in \mathbb{R}$ for each $F \in \overline{G}_p$ then one can (smoothly) extend σ_p to G_p by $\sigma_p(F) = c$ for each $F \in G_p$. In this case one has at p

$$R_{abcd} = \tfrac{1}{6} R G_{abcd} = \tfrac{1}{12} R(g_{ac}g_{bd} - g_{ad}g_{bc}) \tag{9.27}$$

so that M has "constant curvature at p". It turns out (and will be shown later) that if σ_p can be continuously extended to *any* point of $G_p \setminus \overline{G}_p$ then σ_p is constant on $\overline{G}_p$ and (9.27) holds at p. This result was given in [126] following an earlier weaker result [127]. In the general case with M of arbitrary dimension at least three and admitting a metric g of arbitrary signature, the statement that σ_p (where defined) is a constant function is equivalent to (9.27) at p. Also, if these conditions (including that on dimension) hold at each $p \in M$, the Bianchi identities (4.36) show that R is constant on M and M has constant curvature (Schur's theorem - see [35]).

For any manifold with metric g of any signature it is clear that g uniquely determines the sectional curvature function σ. One now asks how tightly the function σ determines g. If g is positive definite it was shown in [128] that if σ_p is for no $p \in M$ a constant function then σ uniquely determines g. However for a space-time M with Lorentz metric g this tight fixing of g by σ is no longer true (although it is still rather tight). This corresponding result for a space-time was given in [126],[129],[130] (and note that [131] is in error). The situation for a space-time will now be described. This will be done through the following lemmas and theorem (which actually establish more than is required for the main result which follows it) [132].

Lemma 9.1 *Let (M, g) be a space-time.*

i) Let F be a real null bivector at $p \in M$. Suppose that, at p, $R_{abcd}H^{ab}H^{cd} = 0$ for all real null bivectors H where $\overset{+}{H}_{ab} = \alpha \overset{+}{F}_{ab}$ for *any $\alpha \in \mathbb{C}$ in some neighbourhood of $1 \in \mathbb{C}$. Then at p*

$$\overset{+}{C}_{abcd}\overset{+}{F}^{ab}\overset{+}{F}^{cd} = 0 \quad \text{and} \quad \overset{+}{E}_{abcd}\overline{\overset{+}{F}}^{ab}\overset{+}{F}^{cd} = 0 \tag{9.28}$$

and hence the principal null direction (spanned by) ℓ of F at p is then a principal null direction of the Weyl tensor C and also a principal null direction of the tensor E, or equivalently, $R_{ab}\ell^a\ell^b = 0$ (section 7.6).

ii) If $R_{abcd}H^{ab}H^{cd} = 0$ holds at $p \in M$ for each null bivector H in some non-empty open subset of the set of all null bivectors at p then $R_{abcd} = \frac{1}{6}RG_{abcd}$ at p.

Proof. For (i) note that the given condition on the curvature tensor holds for $H = F + a^*F$ with $a \in \mathbb{R}$ in some open interval $(-\varepsilon, \varepsilon)$, $\varepsilon > 0$. Hence it holds for $H = F$ and $H = {}^*F$ and one also finds $R_{abcd}F^{ab*}F^{cd} = 0$. A manipulation of the duals then gives

$${}^*R^*{}_{abcd}F^{ab}F^{cd} = {}^*R^*{}_{abcd}{}^*F^{ab}{}^*F^{cd} = {}^*R^*{}_{abcd}F^{ab}{}^*F^{cd} = 0 \tag{9.29}$$

and (9.28) now follows after a short calculation from the fact that F is null and from the identities

$$R_{abcd} + {}^*R^*{}_{abcd} = 2E_{abcd} \tag{9.30}$$

$$R_{abcd} - {}^*R^*{}_{abcd} = 2C_{abcd} + \tfrac{1}{3}RG_{abcd} \tag{9.31}$$

which themselves follow from (7.10), (7.50) and (7.93). Finally, by identifying the complex null bivector $\overset{+}{F}$ with V one sees from (7.67) and (7.69) that ℓ is a principal null directon of the Weyl tensor at p. A similar identification and consideration using (7.101) and theorem (7.10)(iv) with $P = E$ shows that ℓ is a principal null direction of the tensor E and that $R_{ab}\ell^a\ell^b = 0$ at p.

For (ii), and using the result in (i), it can be checked that infinitely many distinct null directions satisfy the principal null direction condition at p for C and E. Since only finitely many such directions for C are possible if $C(p) \neq 0$ one sees that $C(p) = 0$. Although infinitely many principal null directions are possible for E at p they would, in this case, have to include an *open* subset of the set of null directions at p. Thus the expression $R_{ab}\ell^a\ell^b$ would have to vanish over an *open* subset of null directions at p. This is only possible if $R_{ab} \propto g_{ab}$ at p and this is equivalent to the vanishing of E at p (equation (7.94)). The result now follows from (7.10). □

The next lemma will be stated without proof. The proof can be found in [132].

Lemma 9.2 *Let (M, g) be a space-time, let $p \in M$ and let $A \subseteq \overline{G}_p$. Suppose that $F \in N_p$ is a limit point of A and that for any bivector F_{ab} whose blade is F, $R_{abcd}F^{ab}F^{cd} \neq 0$. Then*

i) if U is any open neighbourhood of F in G_p, σ_p is unbounded on $A \cap U$,
ii) if F is a limit point of $A \cap S_p$ and $A \cap T_p$ then σ_p is neither bounded above nor below on A.

These lemmas lead to the following theorem.

Theorem 9.8 *Let (M, g) be a space-time with $p \in M$.*

i) If $F \in N_p$ and U is an open neighbourhood of F in G_p and if σ_p is bounded on $U \cap \overline{G}_p$ then σ_p is a constant function on $\overline{G}_p$ (and is hence continuously extendible to a constant function on G_p).
ii) If σ_p is continuously extendible from $\overline{G}_p$ to a single member of N_p then σ_p is a constant function on $\overline{G}_p$.
iii) If $\ell \in T_pM$ is null and σ_p is bounded on the set $W_p(\ell)$ of wave surfaces to ℓ at p then σ_p is constant on $W_p(\ell)$. Further, if $C_{abcd}(p) \neq 0$ (respectively $E_{abcd}(p) \neq 0$) ℓ is a principal null direction of the Weyl tensor (respectively the tensor E) at p.

Proof. For (i) one applies lemma 9.2(i) with $A = S_p \cup T_p$ to see that $R_{abcd}H^{ab}H^{cd} = 0$ for each null bivector H_{ab} representing some null 2-space in the open subset $U \cap N_p$ of N_p. It now follows from lemma 9.1(ii) that $R_{abcd} = \frac{1}{6}RG_{abcd}$ at p and hence that σ_p is a constant function on $\overline{G}_p$. Clearly σ_p may then be continuously extended to a constant function on G_p.

For (ii) if σ_p is continuously extendible to $F \in N_p$ then one can regard σ_p as a continuous map $\sigma_p : \overline{G}_p \cup \{F\} \to \mathbb{R}$. If $\sigma_p(F) = a \in \mathbb{R}$ and I is a bounded open interval of $\mathbb{R}$ containing a then clearly σ_p is bounded on the open subset $\tilde{U} \equiv \sigma_p^{-1}I$ of $\overline{G}_p \cup \{F\}$. But then $\tilde{U} = (\overline{G}_p \cup \{F\}) \cap U$ where U is open in G_p and also $\tilde{U} \cap \overline{G}_p = U \cap \overline{G}_p$. It follows that U is an open neighbourhood of F in G_p and that σ_p is bounded on $U \cap \overline{G}_p$. Hence, from part (i), σ_p is a constant function on $\overline{G}_p$ and is thus continuously extendible to G_p.

To prove (iii) one first notes that (theorem 6.3) $W_p(\ell)$ is a submanifold of G_p diffeomorphic to $\mathbb{R}^2$ and so the action of σ_p on $W_p(\ell)$ is that of a polynomial in two variables. The constancy of σ_p on $W_p(\ell)$ then follows from its boundedness. Thus the sectional curvatures of the wave surfaces to ℓ at p are equal. So choose a real null tetrad (ℓ, n, x, y) at p so that the wave surfaces to ℓ are spanned by a pair of vectors $x' = x + \alpha\ell$ and $y' = y + \beta\ell$ with $(\alpha, \beta) \in \mathbb{R}^2$. Then construct the simple bivector $F^{ab} = 2x'^{[a}y'^{b]}$ representing this wave surface and substitute in (9.26) noting that the resulting sectonal curvature must be independent of α and β. A straightforward calculation then yields, in the notation of (7.17), the result that at p

$$RF^1F^3 = RF^1F^1 = RF^2F^3 = RF^2F^2 = RF^1F^2 = 0 \tag{9.32}$$

where for bivectors P and Q, $RPQ = R_{abcd}P^{ab}Q^{cd}$. So if one defines a tensor $B_{ab} = R_{cadb}\ell^c\ell^d$ $(= B_{ba})$ at p, (9.32) gives

$$B_{ab}x^ax^b = B_{ab}y^ay^b = B_{ab}x^ay^b = 0, \qquad B_{ab}\ell^b = 0. \tag{9.33}$$

Hence from (7.87) there exist $\mu, \nu, \lambda \in \mathbb{R}$ such that

$$B_{ab} = \mu\ell_a\ell_b + 2\nu\ell_{(a}x_{b)} + 2\lambda\ell_{(a}y_{b)}. \tag{9.34}$$

From the definition of B,

$$\ell_{[e}R_{a]bc[d}\ell_{f]}\ell^b\ell^c = \ell_{[e}E_{a]bc[d}\ell_{f]}\ell^b\ell^c = 0 \tag{9.35}$$

where the second equation in (9.35) follows from the equation $R_{ab}\ell^a\ell^b = 0$ and theorem (7.10)(iv) with $P \equiv E$. Thus if $E(p) \neq 0$, ℓ is a principal null direction of the tensor E at p and from (7.10) and (7.69), if $C(P) \neq 0$, ℓ is a principal null direction of the Weyl tensor at p. This completes the proof.

□

It is also possible to establish that if σ_p is bounded above or below on $\overline{G}_p$ or if σ_p is bounded on either S_p or T_p then, again, σ_p is a constant function on $\overline{G}_p$. The proof of this and other similar results (including part (i) of the last theorem) can be found in [132]. Part (ii) of theorem 9.8 was first established by a different proof in [133]. Part (iii) was given in [134],[132].

Now let (M, g) be a space-time, let $p \in M$ and consider the set N_p of null 2-spaces at p. Let $F, F' \in N_p$ with representative bivectors F_{ab} and F'_{ab}, respectively. If F and F' have the same principal null direction then one may write $F_{ab} = 2\ell_{[a}r_{b]}$, $F'_{ab} = 2\ell_{[a}s_{b]}$ where ℓ is null and r and s are spacelike and orthogonal to ℓ. It then follows that for $\lambda \in \mathbb{R}$, $\ell^a(r_a + \lambda s_a) = 0$ and also, if F and F' are distinct, F_{ab} and F'_{ab} are not proportional. Thus ℓ, r and s are independent and hence $r + \lambda s$ is spacelike for each $\lambda \in \mathbb{R}$. It follows that the 2-spaces with representative bivectors $F_{ab} + \lambda F'_{ab}$ are also members of N_p for each $\lambda \in \mathbb{R}$. Conversely, if $F, F' \in N_p$ and are distinct but intersect in the direction spanned by k^a then one may choose $e, f \in T_p^*M$ so that $F_{ab} = 2k_{[a}e_{b]}$ and $F'_{ab} = 2k_{[a}f_{b]}$ are representative bivectors where, from the properties of null 2-spaces, $k^a k_a \geqslant 0$, $e^a e_a \geqslant 0$, $f^a f_a \geqslant 0$, $k^a e_a = k^a f_a = 0$. Now suppose that $F_{ab} + \alpha F'_{ab}$ also represents a member of N_p for each $\alpha \in \mathbb{R}$. Then (theorem 7.2)

$$(F_{ab} + \alpha F'_{ab})(F^{ab} + \alpha F'^{ab}) = 0 \Longrightarrow F_{ab}F'^{ab} = 0$$

and so $(k^a k_a)(e^a f_a) = 0$. If $k^a k_a \neq 0$ then, since k and e span a null 2-space *and are orthogonal*, e must be null (section 6.1). Similarly f is null and one achieves the contradiction that $e^a f_a = 0$. Thus k is null. It has therefore been shown that if $F, F' \in N_p$ and if, for each $\alpha \in \mathbb{R}$, $F_{ab} + \alpha F'_{ab}$ represents a member of N_p then $F \cap F'$ is the common principal null direction of F and F', and conversely. It now follows that if one is given the set N_p but not the (Lorentz) metric from which it arises at p then those pairs of members F, F' such that the bivector $F^{ab} + \alpha F'^{ab}$ also represents a member of N_p for each $\alpha \in \mathbb{R}$ determine a direction $F \cap F'$ which is null with respect to the metric at p. Hence the set of all null directions at p is determined by N_p and hence the metric at p is determined up to a conformal factor.

Now suppose g and g' are Lorentz metrics for a space-time M which give rise to the same sectonal curvature function σ on M such that σ_p is not a constant function on G_p for any $p \in M$. Then by theorem (9.8)(ii) the domain of σ_p determines, by complementation, the common set of null 2-spaces at p for g and g'. It follows from the previous paragraph that g and g' are conformally related at each $p \in M$ and hence that $g' = \phi g$ on M for some smooth nowhere zero function $\phi : M \to \mathbb{R}$. The following algebraic preliminary to the main theorem can now be established [126].

Theorem 9.9 *Let g and g' be Lorentz metrics for a space-time manifold M which have the same sectonal curvature function σ and where σ_p is not a constant function for any $p \in M$. Then at any $p \in M$ (and where primed quantitites refer to the metric g')*

(i) $g'_{ab} = \lambda g_{ab}$ *(ii)* $R'_{abcd} = \lambda^2 R_{abcd}$

(iii) $R'^a{}_{bcd} = \lambda R^a{}_{bcd}$ *(iv)* $R'_{ab} = \lambda R_{ab}$

(v) $R'_{ab}g'^{ab} = R_{ab}g^{ab}$ *(vi)* $C'^a{}_{bcd} = \lambda C^a{}_{bcd}$

where $\lambda = \phi(p) \in \mathbb{R}$.

Proof. Since $g' = \phi g$ on M, (i) holds and the conditions of the theorem and (9.26) show that at p

$$R'_{abcd}F^{ab}F^{cd} = \lambda^2 R_{abcd}F^{ab}F^{cd} \tag{9.36}$$

for every non-null simple bivector F^{ab} at p (since $\overline{G}_p$ is the same for the metrics g and g'.) Since $\overline{G}_p$ is an open dense subset of G_p a simple continuity argument shows that (9.36) is true for all simple bivectors F^{ab} at p. Then by choosing a pseudo-orthonormal basis (t, x, y, z) for T_pM (with respect to g, say) and substituting $F = G_1, \ldots, G_6$ (from (7.18)) together with certain of their (simple) linear combinations into (9.36) and using the symmetries of the curvature tensor one can, using the 6×6 formalism of chapter 7, find that (ii) of theorem (9.9) holds. From this result, since $g'^{ab} = \lambda^{-1}g^{ab}$, (iii) (iv) and (v) follow immediately. Then (7.10) and (7.11) show that (vi) is true. □

Now let $U \subseteq M$ be the open subset of points at which the Weyl tensor is not zero. Since $g' = \phi g$ on M, it follows from theorem 9.9(vi) that $\phi \equiv 1$ on U. Now let $V \subseteq M$ be the open subset of points at which $d\phi \neq 0$. Then $d\phi = 0$ on $M \setminus V$ and so ϕ is constant on each component of $\text{int}(M \setminus V)$. Thus $R'^a{}_{bcd} = R^a{}_{bcd}$ on $\text{int}(M \setminus V)$ and so from theorem (9.9)(iii) and the fact that neither of these curvature tensors can vanish at any point of M (because σ_p is for no $p \in M$ a constant function) it follows that $\phi \equiv 1$ on $\text{int}(M \setminus V)$. Now let $W \subseteq M$ be the closed subset of points at which $\phi = 1$. The above arguments show that $U \subseteq W$ and $\text{int}(M \setminus V) \subseteq W$. Then consider the *disjoint* decomposition of M given by $M = V \cup \text{int}\, W \cup A$ where the disjointness defines the closed subset A of M. The set A has no interior because any non-empty open subset of M contained in A would necessarily be contained in $M \setminus V$ (since A is) and hence in $\text{int}(M \setminus V)$ and hence in W and finally in $\text{int}\, W$, contradicting the disjointness of the decomposition. Thus, apart from the closed nowhere dense subset A, one may think of M as comprising the open subset $\text{int}\, W$, where $\phi = 1$ and hence $g' = g$, together with the open subset V on which $d\phi$ is never zero and which, since $V \cap U = \emptyset$, is conformally flat.

Now restricting attention to V and using similar techniques to those used in section 9.3 consider the (contracted) Bianchi identities for g and g', bearing in mind theorem 9.9, where a semi-colon and a stroke denote covariant derivatives with respect to the Levi-Civita connections of g and g', respectively. One finds

$$R^a_{bcd;a} + R_{bc;d} - R_{bd;c} = 0 \tag{9.37}$$

$$(\phi R^a{}_{bcd})_{|a} + (\phi R_{bc})_{|d} - (\phi R_{bd})_{|c} = 0. \tag{9.38}$$

Multiplying (9.37) by ϕ and subtracting from (9.38) gives

$$\begin{aligned}\phi(R^e{}_{bcd}P^a_{ea} - R^a{}_{ecd}P^e_{ba} - R^a{}_{bed}P^e_{ca} - R^a{}_{bce}P^e_{da} - R_{ec}P^e_{bd}\\ -R_{be}P^e_{cd} + R_{ed}P^e_{bc} + R_{be}P^e_{dc}) + \phi_a R^a{}_{bcd} + \phi_d R_{bc} - \phi_c R_{bd} = 0\end{aligned} \tag{9.39}$$

where $\phi_a = \phi_{,a}$, Γ^a_{bc} and Γ'^a_{bc} are the Christoffel symbols formed from g and g', respectively, and P^a_{bc} is given in (9.17). All indices are raised and lowered using the metric g. One now substitutes (9.17) into (9.39) and performs a straightforward algebraic computation using the identity $R_{a[bcd]} = 0$ to obtain

$$3R^e{}_{bcd}\phi_e + R_{ec}\phi^e g_{bd} - R_{ed}\phi^e g_{bc} + 2R_{bc}\phi_d - 2R_{bd}\phi_c = 0 \tag{9.40}$$

which on contraction with g^{bd} yields the equivalent statements

$$R_{ab}\phi^b = \tfrac{1}{4}R\phi_a \quad (\Leftrightarrow \tilde{R}_{ab}\phi^b = 0). \tag{9.41}$$

On substituting (9.41) back into (9.40) one finds

$$\begin{aligned}R^a{}_{bcd}\phi_a &= \tfrac{1}{12}R(g_{bc}\phi_d - g_{bd}\phi_c) + \tfrac{2}{3}(R_{bd}\phi_c - R_{bc}\phi_d)\\ &= \tfrac{2}{3}(\tilde{R}_{bd}\phi_c - \tilde{R}_{bc}\phi_d) + \tfrac{1}{12}R(g_{bd}\phi_c - g_{bc}\phi_d).\end{aligned} \tag{9.42}$$

So far the conformally flat condition has not been used. To introduce it one sets $C = 0$ in (7.10) to get

$$R^a{}_{bcd} = \tilde{R}^a{}_{[c}g_{d]b} + \tilde{R}_{b[d}\delta^a_{c]} + \tfrac{1}{6}R\delta^a{}_{[c}g_{d]b}. \tag{9.43}$$

A contraction of this with ϕ_a and use of (9.41) then gives

$$R^a{}_{bcd}\phi_a = \tfrac{1}{2}(\tilde{R}_{bd}\phi_c - \tilde{R}_{bc}\phi_d) + \tfrac{1}{12}R(g_{bd}\phi_c - g_{bc}\phi_d). \tag{9.44}$$

A comparison of this with (9.42) then yields the equivalent statements

$$\tilde{R}_{ab}\phi_c = \tilde{R}_{ac}\phi_b \quad (\Leftrightarrow \tilde{R}_{ab} = \psi\phi_a\phi_b) \tag{9.45}$$

for some real valued function ψ on V. Hence the Ricci tensor satisfies

$$R_{ab} = \psi\phi_a\phi_b + \tfrac{1}{4}Rg_{ab}. \tag{9.46}$$

A contraction with g^{ab} then reveals that $\psi\phi_a\phi^a = 0$ on V. Now if for $p \in V$, $\psi(p) = 0$, one has the Einstein space condition $\tilde{R}_{ab} = 0$ at p and hence, from (7.11), the tensor E vanishes at p. Since $C(p) = 0$ it follows from (7.11) that (9.27) holds at p making σ_p a constant function. Since this is forbidden one sees that, on V, ψ never vanishes and $\phi^a\phi_a = 0$ and so the important step of showing that ϕ^a is *null* (with respect to g and g') on V is achieved.

Now return to (9.17) and compare the two covariant derivatives of $d\phi$

$$\phi_{a;b} = \phi_{a|b} + \phi^{-1}\phi_a\phi_b. \tag{9.47}$$

A further covariant differentiation leads to

$$2\phi_{a;[bc]} = 2\phi_{a|[bc]} - \tfrac{1}{2}\phi^{-1}(\phi_{a;c}\phi_b - \phi_{a;b}\phi_c). \tag{9.48}$$

This allows the use of the Ricci identities for the connections associated with g and g' and with the curvature components related by theorem 9.9(iii). One finds

$$\phi R^d{}_{abc}\phi_d = R^d{}_{abc}\phi_d + \tfrac{1}{2}\phi^{-1}(\phi_{a;c}\phi_b - \phi_{a;b}\phi_c). \tag{9.49}$$

Then a contraction of (9.49) with ϕ^c and use of (9.44) gives

$$R^a{}_{bcd}\phi_a\phi^c = -\tfrac{1}{12}R\phi_b\phi_d. \tag{9.50}$$

If (9.49) is contracted with ϕ^b and use made of (9.50) one finds that on M

$$R(\phi - 1) = 0. \tag{9.51}$$

If, for some $p \in V$, $R(p) \neq 0$, then R will be non-zero over some open neighbourhood W of p and then ϕ will take the constant value 1 over W and so $d\phi$ will vanish on W contradicting the assumption that $d\phi$ vanishes nowhere on V. It follows that $R = 0$ and hence from (9.46) that $R_{ab} = \psi\phi_a\phi_b$ on V. Since ψ never vanishes on V the Ricci tensor has Segre type $\{(211)\}$ with eigenvalue zero at each point of V. One then immediately sees from (9.44) that $R_{abcd}\phi^d = 0$ and this relation when substituted into (9.49) and use made of (9.47) and the symmetry of the tensors $\phi_{a;b}$ and $\phi_{a|b}$ then gives

$$\phi_{a;b} = \alpha\phi_a\phi_b \qquad \phi_{a|b} = \beta\phi_a\phi_b \tag{9.52}$$

for real valued functions α and β on V. The Ricci identities for g and g' then show that $\alpha_{,a}$ and $\beta_{,a}$ are proportional to ϕ_a and hence that α and β are functions of ϕ. Thus in some neighbourhood of any $p \in V$ there exist functions σ and ρ of ϕ such that $\alpha\phi_a = \sigma_{,a}$ and $\beta\phi_a = \rho_{,a}$. Then one finds

$$(e^{-\sigma}\phi_a)_{;b} = 0 \qquad (e^{-\rho}\phi_a)_{|b} = 0. \tag{9.53}$$

Thus, locally in some neighbourhood U of any $m \in V$, g admits a covariantly constant null vector field, say ℓ ($\Rightarrow \ell^a{}_{;b} = 0$) and its Ricci tensor takes the form $R_{ab} = \gamma \ell_a \ell_b$ for some function $\gamma : U \to \mathbb{R}$. So, on U, one has $C = 0$, $R = 0$ and, from the twice contracted Bianchi identities, $R_a{}^b{}_{;b} = 0$. Thus $\tilde{R}_a{}^b{}_{;b} = 0$ and inserting this information into (7.10) and covariantly differentiating, one obtains $2R^a{}_{bcd;a} = \tilde{R}_{bd;c} - \tilde{R}_{bc;d} = R_{bd;c} - R_{bc;d}$. On comparing this with (9.37) one finds $R_{ab;c} = R_{ac;b}$ and so $\gamma_{,a}$ is proportional to ℓ_a. One can now set up a coordinate system u, v, x, y about m so that $\ell_a = u_{,a}$ and then the last result about γ shows that $\gamma = \gamma(u)$. It follows that this coordinate system may be chosen so that the metric g takes the form (8.11) with $H(x, y, u) = f(u)(x^2 + y^2)$ for some smooth function f. This means that g is a (conformally flat) *plane wave metric* [41],[55]. The same arguments hold for g' and so g' is also a plane wave metric. Thus one has the following theorem [126].

Theorem 9.10 *Let g and g' be Lorentz metrics for a space-time manifold M which have the same sectional curvature function σ and where σ_p is not a constant function for any $p \in M$. Then M may be disjointly decomposed as $M = V \cup \operatorname{int} W \cup A$ where A is closed and has empty interior, where $g' = g$ on $\operatorname{int} W$ and where V is an open subset, each point of which lies in a coordinate domain in which g' and g are conformally related, conformally flat plane waves.*

The conditions on σ_p in this theorem require the curvature tensor to be non-vanishing on M. However if the curvature tensor vanishes only over a (necessarily closed) subset B of M with no interior (the *non-flat* condition) then the above theorem still holds (with B included in the subset A).

Also, if one has the extra information that g is a non-flat vacuum metric then in the notation of the calculation leading to the previous theorem, U is open and dense in M and σ_p is necessarily nowhere a constant function on U. Hence $\phi = 1$ on U and so $\phi \equiv 1$ on M. The following theorem results without the need of the clauses required in the previous theorem [126].

Theorem 9.11 *Let M be a space-time and g a non-flat vacuum metric on M with associated sectional curvature function σ. If g' is any other Lorentz metric on M with the same sectional curvature function σ, then $g' \equiv g$.*

The previous two theorems (perhaps especially theorem 9.11) suggest that, because of the tightness of the link between the metric and the sectional curvature function of M, the latter be used as an alternative to the metric as the "field variable" of general relativity [135]. This idea will not be pursued any further here except for a few remarks on the description of a space-time using the function σ. First, the Petrov classification of space-times can be restated in sectional curvature terms by a theorem which links the Petrov type at $p \in M$ with the critical point structure of the function $\sigma_p : \overline{G}_p \to \mathbb{R}$. This was first discussed for vacuum space-times in [127] and refined and given for a general space-time in [132]. In fact, a straightforward initial result in this area is provided by theorem 9.8(iii). Second, certain space-time symmetries have been linked with symmetries of the sectional curvature function [135],[132],[136]. Thirdly, it is noted that if $F \in \overline{G}_p$ with representative bivector F^{ab} and if ${}^*F$ is the orthogonal complement of F with representative bivector ${}^*F^{ab}$ then $G_{abcd}F^{ab}F^{cd} = -G_{abcd}{}^*F^{ab}\,{}^*F^{cd}$ and so from (9.26) and (7.10)

$$\begin{aligned}\sigma_p(F) - \sigma_p({}^*F) &= \frac{R_{abcd}F^{ab}F^{cd}}{2G_{abcd}F^{ab}F^{cd}} - \frac{R_{abcd}{}^*F^{ab}\,{}^*F^{cd}}{2G_{abcd}{}^*F^{ab}\,{}^*F^{cd}} \\ &= \frac{(R_{abcd} + {}^*R^*{}_{abcd})F^{ab}F^{cd}}{2G_{abcd}F^{ab}F^{cd}} = \frac{E_{abcd}F^{ab}F^{cd}}{G_{abcd}F^{ab}F^{cd}}.\end{aligned} \tag{9.54}$$

Thus the tracefree part of the energy-momentum tensor (which is easily seen to be equal to the trace-free Ricci tensor) is conveniently related, through the tensor E (section 7.5), to the difference in the sectional curvatures of all orthogonal pairs of members of $\overline{G}_p$. A special case of this result, which is that $\sigma_p(F) = \sigma_p({}^*F)$ for each $F \in \overline{G}_p$ if and only if $E(p) = 0$ (which is equivalent to the Einstein space condition at p from section 7.5) was noted in [41]. Further mathematical properties of the sectional curvature function can be found in [135].

It was shown earlier that, *generically*, the Levi-Civita connection Γ on M determines g up to a constant conformal factor, that the curvature tensor (with components $R^a{}_{bcd}$) determines g up to a constant conformal factor and Γ uniquely, that the Weyl tensor (with components $C^a{}_{bcd}$) determines g up to a conformal factor and that this Weyl tensor and the energy-momentum tensor (with components T_{ab}) determine g up to a constant conformal factor. It is now seen in a similar way from theorems 9.10 and 7.17(i) that, *generically*, the sectional curvature function on M uniquely determines g.

To see that the sectional curvature function does not always determine the metric uniquely one must consider conformally flat plane wave space-times. In fact, it is easily checked from theorem 9.9 that if g and $g' = e^{2\sigma}g$ are conformally flat metrics on M then their sectional curvature functions agree on M if and only if the associated Ricci tensors satisfy $R'_{ab} = e^{2\sigma}R_{ab}$ on M. This follows from (7.10) and (7.11) by showing that such a condition on the Ricci tensors forces all the conditions (i)-(vi) of theorem 9.9 to hold. Now let (M, g) be the plane wave space-time (8.11) with $H = f(u)(x^2+y^2)$. A direct calculation of the Ricci tensor R'_{ab} for $g'_{ab} = e^{2\sigma(u)}g_{ab}$ in terms of the Ricci tensor R_{ab} for g_{ab} (and noting that $\sigma_{,a} = \dot{\sigma}\ell_a$ with $\dot{\sigma} \equiv \frac{d\sigma}{du}$ and $\ell_{a;b} = 0$) shows that the condition $R'_{ab} = e^{2\sigma}R_{ab}$ is equivalent to the condition [129],[130],[136]

$$\ddot{\sigma} - \dot{\sigma}^2 = f(u)(1 - e^{2\sigma}). \tag{9.55}$$

Thus for a given plane wave one can always, at least locally, find metrics like g'. An explicit solution of (9.55) is found [129],[130] by starting with (8.11) with H as above for $u > 0$ and $f(u) = \dfrac{e^{2u}}{e^{2u}-1}$ and choosing $\sigma(u) = -\frac{1}{2}\ell n(e^{2u} - 1)$.

9.5 Retrospect

The interrelations between the metric, its associated Levi-Civita connection and the corresponding curvature structure have attracted many workers both in the purely mathematical aspects as well as those applicable to theoretical physics. For the case of a positive definite metric the problem of recovering a metric from its curvature tensor was considered by Kowalski [137] whilst an early study of the same problem and which applied to general relativity theory was carried out by Ihrig [138],[139] who drew attention to the significance of equation (9.6) in this context. This work was followed up in [140] and led, through [118],[119], to the methods described in section 9.3. Further attention was given to this and related problems in [95] and [141]-[144]. The relationship between the metric and its associated connection studied in section 9.2 has been reconsidered in [145]-[148]. In the sectional curvature problem discussed in section 9.4 the contributing papers [126] and [129],[130] were independent efforts. The present author was unaware of the work in [129] when [126] was being written and is indebted to Professor Schimming in Greifswald for pointing it out to him and to Professor Voss in Zurich for sending him a copy of it. Some applications of the sectional curvature function together with a further bibliography can be found in [149],[150].

There is another rather interesting topic which is not unrelated in spirit to the work of this chapter. This is the so-called "equivalence problem" of Cartan [151] and others (and reviews of it can be found in [152],[153]). The application of this work to space-time manifolds has been actively pursued in the literature and involves significant use of computers. It will not be discussed any further here except to suggest the following references [154]-[157] for further reading.

Chapter 10

Affine Vector Fields on Space-Time

10.1 General Aspects of Symmetries

The remaining chapters of this book are concerned with symmetries in general relativity and this section will describe some general features of the mathematical study of such symmetries. The remainder of the chapter will concentrate on affine (including homothetic and Killing) symmetry. Throughout, M will be a space-time with (smooth) Lorentz metric g.

A symmetry of the space-time M, loosely speaking, is a smooth local diffeomorphism $f : U \to V$ where U and V are open subsets (*i.e.* open submanifolds) of M which preserves some geometrical feature of M. If one requires such local symmetry everywhere on M one would require many such triples (each triple consisting of f, U and the geometrical feature preserved). This is usually achieved, not by specifying such triples, but by assuming the existence of a smooth vector field on M whose associated local diffeomorphisms and domains (section 5.11) together with the geometrical feature preserved play the role of such triples. If these local diffeomorphisms each satisfy the required symmetry condition then the associated vector field is regarded as a symmetry vector field (of the appropriate type). This concept of a "symmetry vector field" on M will be used throughout the remainder of this text. The details of which geometrical feature of M is to be preserved by the local diffeomorphisms will be described when required.

The symmetries to be discussed will constitute a real vector space S of global smooth vector fields on M (and usually S will be a Lie algebra under the Lie bracket operation). The clause that each member of S is global is largely for convenience. If they are not global and if the intersections of the domains of the members of S is a non-empty open subset U of M, then one could concentrate attention on the (components of the) open submanifold

U, thus effectively reinstating the global nature of these vector fields.

The symmetries which will be discussed in this and subsequent chapters are represented by vector fields whose associated local diffeomorphisms either

i) preserve geodesics (projective vector fields) or
ii) preserve geodesics together with their affine parameters (affine vector fields) or
iii) preserve the metric up to a conformal factor (conformal vector fields) or
iv) preserve the metric up to a constant conformal factor (homothetic vector fields) or
v) preserve the metric (Killing vector fields) or
vi) preserve some other fundamental part of space-time geometry, *e.g.* the curvature tensor.

A more precise definition of these symmetries will be given when required. It is remarked here that a smooth vector field X is said to *preserve* a smooth tensor T on M if, for each smooth local diffeomorphism ϕ_t associated with X, the tensors T and $\phi_t{}^*T$ are equal on the domain U of ϕ_t. An equivalent statement is (section 5.13) that $\mathcal{L}_X T = 0$ on M. In fact if one chooses a coordinate domain $V \subseteq U$ with coordinate functions x^a then $\phi_t(V)$ is also a coordinate domain of M with coordinates $y^a = x^a \circ \phi_t^{-1}$. Then at $p \in V$

$$\phi_{t*}\left(\frac{\partial}{\partial x^a}\right)_p = \left(\frac{\partial}{\partial y^a}\right)_q, \qquad \phi_t^{-1*}(dx^a)_p = (dy^a)_q \tag{10.1}$$

where $q = \phi_t(p)$. Then from the definition of pullback and of tensor components (sections 4.7 and 4.9) one finds (writing T_q for $T(q)$, *etc*)

$$\begin{aligned}(\phi_t{}^*T_q)_p\left(\frac{\partial}{\partial x^a},\cdots,dx^b,\cdots\right) &= T_q\left(\phi_{t*}\left(\frac{\partial}{\partial x^a}\right)_p,\cdots,\phi_t^{-1*}(dx^b)_p,\cdots\right)\\ &= T_q\left(\left(\frac{\partial}{\partial y^a}\right)_q,\cdots,(dy^b)_q,\cdots\right)\end{aligned} \tag{10.2}$$

Thus, quite generally, the components of $\phi_t{}^*T$ at p in the coordinate system x^a equal the components of T at q in the coordinate system y^a. Then if, in addition, ϕ_t is a symmetry of T, that is, $\phi_t{}^*T = T$, then the components of T in V in the coordinates x^a are pointwise equal (under ϕ_t) to the components of T in $\phi_t(V)$ in the coordinates y^a. This coordinate calculation is the justification for the equation $\mathcal{L}_X T = 0$ defining a symmetry vector field of T (and, roughly speaking, demands that the components of T are unaltered if the coordinate system is 'carried along' by X (*i.e.* by the maps ϕ_t)).

10.2 Affine Vector Fields

A global smooth vector field X on a space-time M is called an *affine vector field* (or an *affine collineation*) if each of the smooth local diffeomorphisms ϕ_t associated with X is an *affine map*, that is, each map ϕ_t preserves the geodesics of M *and* their affine parameters. Thus, with appropriate domains and ranges, if I is an open interval of $\mathbb{R}$ and $c : I \to M$ is an affinely parametrised geodesic of M then $\phi_t \circ c$ is also an affinely parametrised geodesic of M. There are two important results associated with an affine vector field X and its local diffeomorphisms ϕ_t. First, the geodesic preserving property of ϕ_t shows that ϕ_t 'commutes' with the exponential map $\exp_m$ at $m \in M$ in the sense that if $m' = \phi_t(m)$ [20]

$$\phi_t \circ \exp_m = \exp_{m'} \circ \phi_{t*} \tag{10.3}$$

This result says that the action of ϕ_t is mirrored in the tangent space by ϕ_{t*} through the exponential maps at m and m'. It will be useful later. The second result involves the definition of the Lie derivative of the Levi-Civita connection ∇ on M associated with the space-time metric g with respect to a smooth vector field X on M [158],[159]. It is denoted by $\mathcal{L}_X\nabla$ and acts on two vector fields Y and Z on (some open subset U of) M, converting them to the vector field on U defined by

$$\mathcal{L}_X\nabla(Y, Z) = [X, \nabla_Y Z] - \nabla_{[X,Y]}Z - \nabla_Y[X, Z] \tag{10.4}$$

It is easily checked that if f and g are smooth maps : $U \to \mathbb{R}$ then $\mathcal{L}_X\nabla(fY, gZ) = fg\mathcal{L}_X\nabla(Y, Z)$. From this result, an argument similar to that given for the curvature tensor in section 4.16 shows that $\mathcal{L}_X\nabla$ leads to a type (1,2) tensor field on U with components in some coordinate system x^a denoted by C^a_{bc} and given by

$$\mathcal{L}_X\nabla\left(\frac{\partial}{\partial x^b}, \frac{\partial}{\partial x^c}\right) = C^a_{bc}\frac{\partial}{\partial x^a} \tag{10.5}$$

From the definitions of the Lie bracket and ∇ one finds

$$C^a_{bc} = X^a{}_{;bc} - R^a{}_{bcd}X^d \tag{10.6}$$

Now *the condition that* X *is affine can be shown equivalent [159] (see also [20], [158]) to the condition* $\mathcal{L}_X\nabla = 0$ and hence, from (10.6), to

$$X^a{}_{;bc} = R^a{}_{bcd}X^d \tag{10.7}$$

An alternative and useful way of rewriting (10.7) is first to decompose the covariant derivative of X into its self adjoint and skew-self adjoint parts. This leads to a symmetric tensor h and a skew-symmetric tensor F (called the *affine bivector* of X) satisfying in each coordinate domain

$$\begin{gathered} X_{a;b} = \tfrac{1}{2}h_{ab} + F_{ab} \\ h_{ab} = 2X_{(a;b)} = \mathcal{L}_X g_{ab} = h_{ba}, \quad F_{ab} = -F_{ba} \end{gathered} \tag{10.8}$$

On substituting this into (10.7) one finds

$$\tfrac{1}{2}h_{ab;c} + F_{ab;c} = R_{abcd}X^d \tag{10.9}$$

Symmetrising on the indices a and b then gives the conditions

$$(i)\ h_{ab;c} = 0 \qquad (ii)\ F_{ab;c} = R_{abcd}X^d \tag{10.10}$$

on h and F. Conversely suppose (10.10)(i) holds. Then

$$X_{a;bc} = F_{ab;c} \tag{10.11}$$

and then the Ricci identity (4.33) on X gives

$$F_{ab;c} - F_{ac;b} = X^d R_{dabc} \tag{10.12}$$

Permuting the indices a, b and c in (10.12) then gives

$$F_{ba;c} - F_{bc;a} = X^d R_{dbac} \tag{10.13}$$

$$F_{cb;a} - F_{ca;b} = X^d R_{dcba} \tag{10.14}$$

Adding the last three equations and using the curvature symmetry $R_{a[bcd]} = 0$ then easily leads to (10.10)(ii) and hence to (10.7). It follows that the condition (10.10)(i) is equivalent to the statement that X is an affine vector field and that the condition (10.10)(ii) on F follows automatically from (10.10)(i). From this condition on F it follows that $F_{ab;c}X^c = 0$ and so F is parallely transported along the integral curves of X. It now easily follows (see section 7.2) that *the nature (spacelike, timelike, null or non-simple) of F is constant along such integral curves.*

Regarding differentiability, so far an affine vector field has been assumed smooth. Suppose, however, X is C^2. Then (10.7) shows, since the curvature tensor is smooth, that X must be C^3 and then that X is C^4 and so on. *Thus the assumption that an affine vector field is smooth is no more restrictive than (in fact, equivalent to) assuming it C^2* as a consequence of the smoothness of the connection Γ. The smoothness of h and F is then immediate from (10.8).

Now let X be an affine vector field on M. Then one may characterise this fact by the equations

$$X_{a;b} = \frac{1}{2}h_{ab} + F_{ab} \qquad h_{ab;c} = 0 \qquad F_{ab;c} = R_{abcd}X^d \tag{10.15}$$

These equations are constructed out of first order partial derivatives of the components of the quantities X, h and F together with the terms supplied by the geometry of M (*i.e.* g, Γ and the curvature). This observation is, in many ways, more useful than the "second order in X" original characterisation given in (10.7). An immediate use will be made of this now.

Suppose X and Y are two affine vector fields on M with X satisfying (10.8) and Y satisfying

$$Y_{a;b} = \frac{1}{2}H_{ab} + G_{ab} \qquad (H_{ab} = H_{ba}, \quad G_{ab} = -G_{ba}) \tag{10.16}$$

Suppose also that at $m \in M$, $X(m) = Y(m)$, $h(m) = H(m)$ and $F(m) = G(m)$ (This is equivalent to X and Y and their first covariant derivatives agreeing at m, $X(m) = Y(m)$ and $\nabla X(m) = \nabla Y(m)$). Now let c be a smooth curve in M defined on an interval $(-\epsilon, \epsilon)$ $(0 < \epsilon \in \mathbb{R})$ and starting at m such that the tangent vector $T(t)$ to c is nowhere zero for $t \in (-\epsilon, \epsilon)$ and such that the range of c lies in a coordinate domain of M (containing m). Then a contraction of (10.8) with T^b and (10.10) with T^c reveals a system of first order differential equations of the form

$$\frac{dX_a}{dt} \left(= X_{a,b}T^b\right) = D_a, \qquad \frac{dh_{ab}}{dt} = C_{ab}, \qquad \frac{dF_{ab}}{dt} = D_{ab} \tag{10.17}$$

where D_a, C_{ab} and D_{ab} depend only on X, h and F and on t through the (given) geometry of M. Similar remarks apply to Y. Now the theory of first order differential equations shows that if the (ordered) triple of quantities (X, h, F) is equal to the triple (Y, H, G) at some $t_0 \in (-\epsilon, \epsilon)$ they will be equal on some open interval containing t_0. Also, if these triples differ at some $t_0 \in (-\epsilon, \epsilon)$ they will (by continuity) differ in some open interval containing t_0. Thus, since $(-\epsilon, \epsilon)$ is connected and since these triples are equal at $m = c(0)$, they agree for each $t \in (-\epsilon, \epsilon)$. It now follows by using the freedom in the choice of c that if these triples agree at some point of M they will agree in some open neighbourhood of that point (and clearly if they are not equal at some point of M they will differ in some open neighbourhood of that point). Thus, the points where these triples agree and where they disagree constitute disjoint open subsets of M, with the former not empty, and since M is connected it follows that $X = Y$ on M. *Thus an affine vector field X on M is uniquely determined by the values of X, h and F (or, alternatively, X and ∇X) at some (any) point m of M.* Another important result also follows from this argument. It is clear from (10.7) that *the set (denoted by $A(M)$) of affine vector fields on M forms a real vector space.* In fact $A(M)$ is also a *Lie algebra* under the usual bracket operation for vector fields. This can be shown in several ways (see [20]) and is clear from a consideration of the local diffeomorphisms associated with the bracket of two members of $A(M)$ (section 5.13). A direct proof in the present formalism can be obtained by letting X and Y be affine vector fields on M and let $Z = [X, Y]$. Then in local coordinates $Z^a = Y^a{}_{;b}X^b - X^a{}_{;b}Y^b$. First one computes $Z_{a;b}$ and finds its symmetric part $P_{ab} = Z_{(a;b)}$. The idea then is to show that $P_{ab;c} = 0$. This can be achieved after a little calculation during which it is useful to note that if M_{ab} is a bivector then $R_{abcd}M^{bd}$ is skew-symmetric in a and c and also that the Ricci identities on h and H yield

$$h_{ae}R^e{}_{bcd} + h_{be}R^e{}_{acd} = H_{ae}R^e{}_{bcd} + H_{be}R^e{}_{acd} = 0 \tag{10.18}$$

Finally, since the zero vector field on M is always affine and since any affine vector field X is uniquely determined by $X(m)$, $h(m)$ and $F(m)$ at any $m \in M$ it follows that if these three quantities vanish for any m then X is identically zero on M. Thus the taking of linear combinations and a simple counting procedure shows that $A(M)$ is finite dimensional and that $\dim A(M) \leqslant 4+10+6 = 20$. The set $A(M)$ will be referred to as the *affine algebra of M*.

The following theorem has thus been proved.

Theorem 10.1 *Let M be a space-time and let $A(M)$ be the set of affine vector fields on M.*

i) Each member X of $A(M)$ satisfies (10.7) and (10.10)(i) and either of these conditions is sufficient to ensure that X is affine.

ii) Each member X of $A(M)$ is uniquely determined by specifying the quantities $X(m)$, $h(m)$ and $F(m)$ or, alternatively, $X(m)$ and $\nabla X(m)$ at any $m \in M$ (and hence if an affine vector field vanishes over a non-empty open subset of M it vanishes everywhere on M)

iii) $A(M)$ is a finite-dimensional Lie algebra of smooth vector fields on M under the Lie bracket operation and $\dim A(M) \leqslant 20$. (For a connected n-dimensional manifold M with a linear connection $\dim A(M) \leqslant n(n+1)$ [20].)

The upper bound of 20 in part (iii) of theorem 10.1 is achieved in Minkowski space-time (and, as will be seen later, this is essentially the only way it is achieved). For in Minkowski space-time, (10.7) yields $X^a{}_{,bc} = 0$ for the components of an affine vector field in the usual global coordinate system. This is readily solved, the general solution being that each component X^a is a general linear function of the coordinates

$$X^a = P^a{}_b x^b + P^a \tag{10.19}$$

where $x^a = (t, x, y, z)$ are the usual Minkowski coordinates and the $P^a{}_b$ and P^a are 20 constants with the symmetric and skew-symmetric parts of $P_{ab} \equiv \eta_{ac} P^c{}_b$ playing the role of h_{ab} and F_{ab}, respectively.

10.3 Subalgebras of the Affine Algebra; Isometries and Homotheties

The affine algebra $A(M)$ of a space-time M has some important subalgebras. The two which are, perhaps, the most important in the present context are the *Killing* and *homothetic* subalgebras and these can now be discussed.

Let $X \in A(M)$ so that X is an affine vector field on M. Then X satisfies (10.8) and the affine condition is the condition that h is covariantly constant. If $h \equiv 0$ on M then X satisfies the equivalent conditions

$$X_{a;b} = F_{ab} \Leftrightarrow X_{a;b} + X_{b;a} = 0 \Leftrightarrow \mathcal{L}_X g = 0 \tag{10.20}$$

This equation is called *Killing's equation* and the vector field X is then called a *Killing vector field*. The associated bivector F will be called the

Killing bivector (of X). The subset of $A(M)$ consisting of all Killing vector fields on M will be called the *Killing algebra* and is denoted by $K(M)$.

Now suppose $X \in A(M)$ and in the resulting equation (10.8) $h = 2cg$ on M where c is a (constant) real number. Then X satisfies the equivalent conditions

$$X_{a;b} = cg_{ab} + F_{ab} \Leftrightarrow \mathcal{L}_X g = 2cg \tag{10.21}$$

and is called a *homothetic vector field*. The associated bivector F is then called the *homothetic bivector* and c the *homothetic constant* (of X). The subset of $A(M)$ consisting of all homothetic vector fields on M will be referred to as the *homothetic algebra* and denoted by $H(M)$. A consideration of the case $c = 0$ shows that $K(M) \subseteq H(M)$. If $X \in H(M)$ and the corresponding constant $c \neq 0$ then X is called *proper homothetic*. For later convenience it will be assumed that if X is proper homothetic then (by changing X to $-X$ if necessary) $c > 0$. If $X \in A(M)$ is neither Killing nor homothetic it is called *proper affine*. The subsets of proper affine and proper homothetic vector fields are *not* subspaces of $A(M)$.

The definitions of affine, Killing and homothetic vector fields given in (10.10), (10.20) and (10.21) apply quite generally for manifolds (where they make sense) and will, on occasions, be used for manifolds other than space-times together with the obvious appropriate notation.

If one refers to the collection of results regarding the Lie derivative given in section 5.13 it is clear that if X, Y and $Z = \alpha X + \beta Y \in A(M)$, $\alpha, \beta \in \mathbb{R}$, then $\mathcal{L}_X g = \mathcal{L}_Y g = 0 \Rightarrow \mathcal{L}_Z g = \mathcal{L}_{[X,Y]} g = 0$ and $\mathcal{L}_X g = 2c_1 g$, $\mathcal{L}_Y g = 2c_2 g \Rightarrow \mathcal{L}_Z g = 2c_3 g$ and $\mathcal{L}_{[X,Y]} g = 0$ where c_1, c_2 and c_3 ($= \alpha c_1 + \beta c_2$) are constants. It follows that $K(M)$ and $H(M)$ *are each Lie subalgebras of* $A(M)$ and that $K(M)$ is a Lie subalgebra of $H(M)$). Also any $X \in K(M)$ is uniquely determined by the values $X(m)$ and $F(m)$ at some (any) $m \in M$ (since, then $h(m) = 0$ is necessarily true) and any $X \in H(M)$ is uniquely determined by the values $X(m)$, $F(m)$ and the constant c (since, then $h(m)$ necessarily equals $cg(m)$). These results follow from theorem 10.1. One can now collect these results (and more) together in the following theorem.

Theorem 10.2 *Let M be a space-time with Killing algebra $K(M)$ and homothetic algebra $H(M)$.*

i) The members of $K(M)$ are characterised by their satisfying equation (10.20) and each $X \in K(M)$ is uniquely determined by the values $X(m)$

and $F(m)$ at any $m \in M$.

ii) The members of $H(M)$ are characterised by their satisfying the equation (10.21) and each $X \in H(M)$ is uniquely determined by the values $X(m), F(m)$ and c at any $m \in M$.

iii) The Lie algebras $K(M)$ and $H(M)$ are finite-dimensional Lie algebras of smooth vector fields on M. $K(M)$ and $H(M)$ are subalgebras of $A(M)$ and $K(M)$ is a subalgebra of $H(M)$. Further, $\dim K(M) \leqslant 10$, $\dim H(M) \leqslant 11$ *and* $\dim H(M) \leqslant \dim K(M) + 1$. *(In fact in the general case of an n-dimensional manifold M admitting a metric of any signature,* $\dim K(M) \leqslant \frac{1}{2}n(n+1)$ *and* $\dim H(M) \leqslant \frac{1}{2}n(n+1) + 1$.*)*

iv) If $X, Y \in H(M)$, $[X, Y] \in K(M)$.

Regarding the upper bounds on the dimensions in part (iii) of this theorem it is a classical result (see, *e.g.* [35] and the final paragraph of this chapter) that if $\dim K(M) = \frac{1}{2}n(n+1)$ ($= 10$ for a space-time) then M has constant curvature and that if M has constant curvature then M admits a local Killing algebra of dimension $\frac{1}{2}n(n+1)$ in the sense that each $m \in M$ admits a connected open neighbourhood U such that U with induced metric from g on M satisfies $\dim K(U) = \frac{1}{2}n(n+1)$. For example, in Minkowski space-time, one can solve Killing's equation by imposing (10.20) on the affine solution (10.19) for Minkowski space. The result is

$$X^a = P^a{}_b x^b + P^a \qquad P_{ab} + P_{ba} = 0 \qquad P_{ab} = \eta_{ac} P^c{}_b \tag{10.22}$$

and the $6 + 4 = 10$ independent constants in $P^a{}_b$ and P^a show the Killing algebra to have dimension 10. It will be shown later that if $\dim K(M) > 7$ then M has constant curvature and that $\dim K(M) \neq 9$ (see end of chapter). If $\dim H(M) = 11$ then, as will also be seen later, M is flat and a flat space-time always admits an 11-dimensional local homothety algebra (in the same sense as the above local Killing algebra), since Minkowski space-time does, namely the vector fields in (10.22) together with the proper homothetic vector field X with components $X^a = (t, x, y, z)$. Similar results apply in an obvious way to manifolds of any dimension and which admit a metric.

For the last inequality in part (iii) one notes that since $K(M)$ is a subspace of $H(M)$, $\dim K(M) = \dim H(M)$ if and only if $K(M) = H(M)$. If $K(M) \neq H(M)$ then there exists a proper homothetic vector field X

satisfying $\mathcal{L}_X g = 2cg$ with c a non-zero constant. If Y is any other homothetic vector field satisfying $\mathcal{L}_Y g = 2c'g$ for some non-zero constant c' then $Z = c'X - cY$ satisfies $\mathcal{L}_Z g = 0$ and so Z is Killing. Hence $Y = c'c^{-1}X - c^{-1}Z$ may be written as a linear combination of the original homothetic vector field X and a Killing vector field. It follows in this case that $\dim H(M) = \dim K(M) + 1$ because a basis for $K(M)$ augmented with X is a basis for $H(M)$. Part (iv) follows immediately from the remarks preceding the theorem.

The geometrical interpretation of an affine vector field arises from the fact that the corresponding local diffeomorphisms ϕ_t preserve the geodesics and their affine parameters. In fact they preserve the connection Γ as can be made precise in an elegant way [20]. This latter fact can also be demonstrated at an elementary level by considering one such local affine transformation ϕ_t whose domain U is a coordinate neighbourhood with coordinates x^a. Then ϕ_t is smooth (since X is) and so (section 10.1) $V \equiv \phi_t(U)$ can be considered a coordinate domain of M with coordinates $y^a = x^a \circ \phi_t^{-1}$. The geodesic property of ϕ_t can now easily be used to see that the connection coefficients Γ^a_{bc} in U in the x^a coordinates are equal, pointwise under ϕ_t, to the coefficients Γ'^a_{bc} in V in the y^a coordinates. This should be compared to a similar argument for symmetries of tensors given in section 10.1.

The geometrical interpretation for a Killing vector field X on M arises from the condition $\mathcal{L}_X g = 0$ and hence $\phi_t{}^* g = g$ (section 10.1 with $T = g$) for each local diffeomorphism ϕ_t associated with X. Hence each ϕ_t is a local symmetry of the metric g.

If X is a homothetic vector field on M one has $\mathcal{L}_X g = 2cg$ with c constant and so (section 5.13) each associated local diffeomorphism ϕ_t of X satisfies $\phi_t{}^* g = e^{2ct} g$. Thus each given ϕ_t is a dilatation of the metric g.

The existence of an affine, Killing or homothetic vector field on M leads to a family of special coordinate systems on M. Let X be a Killing vector field on M and suppose that $m \in M$ and $X(m) \neq 0$. Then from theorem 4.12 there exists a coordinate domain U containing m with coordinates x^a such that, in U, $X = \frac{\partial}{\partial x^1}$. Now Killing's equation (10.20) can be rewritten from (5.13) as

$$g_{ab,c}X^c + g_{ac}X^c{}_{,b} + g_{bc}X^c{}_{,a} = 0 \tag{10.23}$$

which in the above coordinates on U gives $\frac{\partial g_{ab}}{\partial x^1} = 0$ and so *the metric components are, in this coordinate system, independent of the coordinate* x^1. A similar argument with the above coordinate system when X $(=\frac{\partial}{\partial x^1})$ is a homothetic vector field satisfying (10.21) shows that $\frac{\partial g_{ab}}{\partial x^1} = 2cg_{ab}$ and so $g_{ab} = e^{2cx^1}q_{ab}$ *where the* q_{ab} *are independent of* x^1 (and q_{ab} is a Lorentz metric on U for which X is Killing). The geometrical interpretation of an affine vector field given above shows that when X is an affine vector field then, in the above coordinate system *the coefficients* Γ^a_{bc} *are independent of* x^1. Conversely, suppose that in some coordinate domain U with coordinates x^a the metric components are independent of x^1. Then from (10.20), $X = \frac{\partial}{\partial x^1}$ is a Killing vector field on U. Similarly if the metric components satisfy $g_{ab} = e^{2cx^1}q_{ab}$ for some constant c and where the q_{ab} are independent of x^1 then $X = \frac{\partial}{\partial x^1}$ is a homothetic vector field on U and, again, if the coefficients Γ^a_{bc} are independent of x^1 then $X = \frac{\partial}{\partial x^1}$ is an affine vector field on U. The extension of these results to several independent vector fields of the appropriate type is clear from theorem 4.12. Thus if $X_1, \cdots, X_N \in K(M)$ and none vanishes at $m \in M$ and $[X_i, X_j] = 0$ $(1 \leqslant i, j < N)$ then there is a coordinate domain about m in which $X_1 = \frac{\partial}{\partial x^1}, \cdots, X_N = \frac{\partial}{\partial x^N}$ and g_{ab} is independent of $x^1, \cdots, x^N$ and, conversely, if g_{ab} is independent of $x^1, \cdots, x^N$, $\frac{\partial}{\partial x^1}, \cdots, \frac{\partial}{\partial x^N}$ are Killing vector fields on the coordinate domain (and the Lie bracket of any two of them vanishes). Similar results hold for $H(M)$ and $A(M)$ (where, if $\frac{\partial}{\partial x^1}, \cdots, \frac{\partial}{\partial x^N}$ are homothetic with homothetic constants $c_1 \cdots, c_N$, $g_{ab} = e^{2c_1x^1 + \cdots + 2c_Nx^N}p_{ab}$ with p_{ab} independent of $x^1, \cdots, x^N$).

It is useful to note that if X is a homothetic vector field on M satisfying (10.21) and if $\sigma = g(X, X)$ (in components $\sigma = X^aX_a$) then it follows that

$$\sigma_{,a}X^a = 2c\sigma \tag{10.24}$$

Now if $m \in M$ and $X(m) = 0$ the integral curve of X through m is a constant map with range $\{m\}$. If, however, $X(m) \neq 0$ then (10.24) shows that along any integral curve of X through m with parameter t, $\frac{\partial \sigma}{\partial t} = 2c\sigma$. It follows that *the nature (spacelike, timelike or null) of* X *is constant along its integral curves and that along such curves,* σ *is constant if* $X \in K(M)$. The same result is not true for a proper affine vector field, a counterexample coming from (10.19) by considering the proper affine vector field $X^a = (t, 1, 0, 0)$ in Minkowski space. Also, if $X \in H(M)$ and $fX \in H(M)$ with $f : M \to \mathbb{R}$ then it is easily checked that f is necessarily constant. Again this result is not true for $A(M)$ as the affine vector fields $X^a = (1, 0, 0, 0)$ and tX^a in Minkowski space show.

The situation will occasionally arise when the idea of a *local symmetry* (vector field) is required. This is defined to be a vector field of a type discussed so far, or still to be introduced, but which is defined on an open subset U of M.

10.4 Fixed Point Structure

Let X be a non-trivial affine vector field on the space-time M and suppose $m \in M$. If $X(m) = 0$ then m is a *fixed point* or a *zero* of X. The term fixed point arises since, if ϕ_t is a local diffeomorphism associated with X, $\phi_t(m) = m$, and so m is a fixed point of any such ϕ_t. If U is a coordinate neighbourhood of such a zero m with coordinates y^a then the vector space isomorphism $\phi_{t*} : T_m M \to T_m M$ is represented in the basis $(\frac{\partial}{\partial y^a})_m$ by the transpose of the matrix [21] (see 5.9)

$$\exp tA = \exp t \left(\frac{\partial X^b}{\partial y^a}\right)_m \tag{10.25}$$

where A, with components $(\frac{\partial X^b}{\partial y^a})_m$, is the *linearisation* of X at m. The need to transpose this matrix is to conform to the fact that a different definition of the matrix of a linear transformation is taken here (chapter 2) from that in [21]. From (10.8) it follows that A has components $A^a{}_b = (\frac{1}{2}h^a{}_b + F^a{}_b)(m)$ and from (10.3) that if $\exp_m$ is the usual exponential diffeomorphism from some open neighbourhood of $0 \in T_p M$ onto some open neighbourhood V of m then

$$\phi_t \circ \exp_m = \exp_m \circ \phi_{t*} \tag{10.26}$$

In the normal coordinate system x^a with domain V resulting from the basis $\left(\frac{\partial}{\partial y^a}\right)_m \left(= \left(\frac{\partial}{\partial x^a}\right)_m\right)$ of $T_m M$ the integral curves of X satisfy $\frac{dx^a}{dt} = A^a{}_b x^b$. This follows because if $q \in V$, the integral curve $t \to \phi_t(q)$ of X through q is $t \to \phi_t \circ \exp_m(q') = \exp_m \circ \phi_{t*}(q') = \exp_m(q'(\exp tA)^T) = \exp_m((\exp tA)q')$ where $\exp_m q' = q$. Then use property (v) following theorem 5.4. From this it follows that in this coordinate system the components of X are linear functions of the coordinates, $X^a = A^a{}_b x^b$. A vector field X is called *linearisable about* m if, in some coordinate neighbourhood of m, the components X^a of X are *linear* functions of the coordinates. By theorem 4.12(i) this is true for any vector field X if $X(m) \neq 0$. The above shows that *any affine vector field on* M *is linearisable about any point of* M. This property which follows from the fact that the maps ϕ_t preserve geodesics and their affine parameters is useful. For example, the set of zeros of X lying in the above normal coordinate neighbourhood V have coordinates x^a satisfying $A^a{}_b x^b = 0$. Thus, if $\operatorname{rank} A < 4$ *the set of zeros of X which lie in V constitute a regular submanifold of V* and hence of M of dimension $4 - \operatorname{rank} A$ (section 4.11 (v) and (vii)). If $\operatorname{rank} A = 4$ then m is an *isolated zero* of X, that is, there exists an open neighbourhood of m in which m is the only zero of X.

A discussion of the zeros of Killing and homothetic vector fields was given in [160] and of the zeros of affine vector fields in [161]. Here, attention will be restricted to zeros of members of $K(M)$ and $H(M)$. So let $X \in K(M), X \not\equiv 0$, and let $m \in M$ with $X(m) = 0$. From (10.20) the linearisation A of X at m satisfies $A = F(m)$ where F is the Killing bivector of X. Since $X(m) = 0$, theorem 10.2(i) shows that $F(m) \neq 0$. Thus $A = F(m)$ is either spacelike, timelike, null or non-simple at m with respective ranks 2,2,2 and 4. *If $F(m)$ is non-simple m is an isolated zero of X*. In the other cases there exists an open neighbourhood V of m such that the zeros of X *in* V form a regular 2-dimensional submanifold of V and hence of M.

The existence of a zero m of a non-trivial Killing vector field X on M has consequences for certain tensors at m. First define a subset $\overset{*}{K}_m$ of $K(M)$ by

$$\overset{*}{K}_m = \{X \in K(M) : X(m) = 0\} \tag{10.27}$$

It is clear that $\overset{*}{K}_m$ is a vector subspace of $K(M)$ and, in fact, $\overset{*}{K}_m$ is a Lie subalgebra of $K(M)$ because $X \in \overset{*}{K}_m$, $Y \in \overset{*}{K}_m \Rightarrow [X,Y](m) = 0 \Rightarrow [X,Y] \in \overset{*}{K}_m$. $\overset{*}{K}_m$ is called the *isotropy (sub)algebra of K(M) at m*. Also, if $X \in \overset{*}{K}_m$ with associated local diffeomorphisms ϕ_t, then $\phi_t(m) = m$ and, by definition for $u, v \in T_m M$,

$$g(m)(u,v) = ({\phi_t}^* g)(m)(u,v) = g(m)(\phi_{t*}u, \phi_{t*}v) \tag{10.28}$$

Thus ϕ_{t*} is a member of the (identity component of the) Lorentz group with respect to the metric $g(m)$ (see the discussion following (6.6)). In a general coordinate system about M with coordinates x^a and basis $\left(\frac{\partial}{\partial x^a}\right)_m$ for $T_m M$, this representation of the Lorentz group has a Lie algebra L consisting of matrices $B^a{}_b$ which are skew self-adjoint with respect to the components g_{ab} of $g(m)$ in these coordinates, $g_{ac}B^c{}_b + g_{bc}B^c{}_a = 0$ (section 6.3). Thus the map f which associates $X \in \overset{*}{K}_m$ with its Killing bivector $F^a{}_b(m)$ at

m is a linear map from $\overset{*}{K}_m$ to L. Also, if $Y \in \overset{*}{K}_m$ with Killing bivector G then $[X, Y] \in \overset{*}{K}_m$ and its Killing bivector at m is easily calculated to be $(G^a{}_c F^c{}_b - F^a{}_c G^c{}_b)(m)$. *Thus f is a Lie algebra homomorphism from $\overset{*}{K}_m$ to L.* Furthermore, if $X \in \overset{*}{K}_m$ with $f(X) = 0$ then $X \equiv 0$ from theorem 10.2 and so f is injective and hence an isomorphism onto its range in L. Thus $\overset{*}{K}_m$ *is isomorphic to a subalgebra of the Lorentz group* and is thus isomorphic to one of the algebras described in table 6.1. Also, the linear map from $K(M)$ to the subspace $\{X(m) : X \in K(M)\}$ of $T_m M$ given by $X \to X(m)$ has kernel equal to $\overset{*}{K}_m$ and so from the theory of linear maps this subspace has dimension equal $\dim K(M) - \dim \overset{*}{K}_m$.

One can also define the *isotropy (sub)algebras $\overset{*}{H}_m$ and $\overset{*}{A}_m$ of $H(M)$ and $A(M)$*, respectively, at m by $\overset{*}{H}_m = \{X \in H(M) : X(m) = 0\}$ and $\overset{*}{A}_m = \{X \in A(M) : X(m) = 0\}$. Similar arguments to those given for $\overset{*}{K}_m$ show that they are isomorphic to the Lie algebra of matrices $\{X^a{}_{;b}(m) : X \in \overset{*}{A}_m\}$ and similarly for $\overset{*}{H}_m$. Clearly $\overset{*}{K}_m$ is a subalgebra of $\overset{*}{H}_m$ and $\overset{*}{H}_m$ a subalgebra of $\overset{*}{A}_m$. The above dimension relation, suitably modified, also applies to $H(M)$ and $A(M)$.

Now for $X \in K(M)$, $X \not\equiv 0$, the fact that the associated maps ϕ_t are isometries together with the discussion of section 10.1 with $T = g$ shows that, if P is any of the curvature, Weyl or Ricci tensors, then $\phi_t{}^* P = P$ and so

$$(i) \mathcal{L}_X C^a{}_{bcd} = 0 \quad (ii) \mathcal{L}_X R^a{}_{bcd} = 0 \quad (iii) \mathcal{L}_X R_{ab} = 0 \tag{10.29}$$

From the first of these one also has $\mathcal{L}_X C_{abcd} = 0$. Now suppose that $X(m) = 0$. Since $\phi_t(m) = m$, the tensor C' with components C_{abcd} and the tensors $\phi_t{}^* C'$ agree at m. This requires some consistency between the map $\phi_{t*} : T_m M \to T_m M$ and the type of ambiguity in the tetrads at m yielding the canonical Petrov forms for C' (chapter 7 - c.f. [41]). For example if the Killing bivector F of X is such that $F(m)$ is spacelike then there exist exactly two independent null vectors l, $n \in T_m M$ such that $F_{ab} l^b = F_{ab} n^b = 0$. Hence, from (10.25) and property (iv) following theorem 5.4, $\phi_{t*} l = l$, $\phi_{t*} n = n$. Now construct a real null tetrad (l, n, x, y) at m so that x and y span the blade of $F(m)$. Then the effect of ϕ_{t*} is, from the discussion of the previous paragraph and the equations (6.8b), that of a spatial rotation in the x, y plane. Since C' and $\phi_t{}^* C'$ agree at m

one finds that the components C_{abcd} are the same in the frames (l, n, x, y) and $(l, n, \phi_{t*}x, \phi_{t*}y)$. It also follows that since C, if non-zero at m, admits a *finite* number of principal null directions, the continuity of ϕ_{t*} means that *each principal null direction must be preserved by* ϕ_{t*}. Since ϕ_{t*} is a Lorentz transformation there are at most two principal null directions (either l, n or both). Thus (chapter 7) the Petrov type is either **N**, **III**, **D** or **O**. Reference to the discussion of Petrov tetrads at the end of section 7.3 yields a contradiction in the type **N** and **III** cases and so C must have type **D** or **O** at m with l and n spanning the (necessarily repeated) principal null directions of C in the type **D** case. A similar argument and conclusion follow if $F(m)$ is timelike or non-simple. If $F(m)$ is null one can find a null tetrad (l, n, x, y) such that $F_{ab}l^b = F_{ab}y^b = 0$ at m and so $\phi_{t*}l = l$, $\phi_{t*}y = y$ and with ϕ_{t*} yielding a null rotation in the l, x plane (equation (6.8d). Since l spans the unique null direction preserved by the ϕ_{t*}, at most one principal null direction of the Weyl tensor at m is now permitted. Thus the Petrov type at m is **N** or **O** with l spanning the (necessarily repeated) principal null direction in the type **N** case.

Now consider the Ricci tensor, Ricc, and the third equation in (10.29). One now has that Ricc and ${\phi_t}^*$ Ricc agree at m. Now the condition that k is an eigenvector of Ricc at m with eigenvalue $\mu \in \mathbb{C}$ is equivalent to the condition that $\text{Ricc}(k, k') = \mu g(k, k')$, $\forall k' \in T_m M$. So

$$\begin{aligned} \text{Ricc}(k, k') = \mu g(k, k') &\Rightarrow {\phi_t}^* \text{Ricc}(k, k') = \mu {\phi_t}^* g(k, k') \\ &\Rightarrow \text{Ricc}(\phi_{t*}k, \phi_{t*}k') = \mu g(\phi_{t*}k, \phi_{t*}k') \qquad (10.30) \end{aligned}$$

Since ϕ_{t*} is an isomorphism of $T_m M$ it follows that $\phi_{t*}k$ is also an eigenvector of Ricc with eigenvalue μ. This produces degeneracies in the algebraic structure of Ricc depending on the nature of ϕ_{t*}, that is, on $F(m)$. The individual details can be read off from the possible Segre types for Ricc in chapter 7 ([160], [162], [55]). They are given in table 10.1. An alternative proof here follows by noting that (10.29)(iii) applied at m is

$$R_{ac}F^c{}_b + R_{bc}F^c{}_a = 0 \qquad (10.31)$$

The same results can now be read off from theorem 9.1 replacing the tensor g' there by Ricc. The following theorem can now be stated.

Theorem 10.3 *Let X be a non-trivial Killing vector field on a space-time M which has a zero at $m \in M$ and let F be the Killing bivector of X. Then $F(m) \neq 0$ and the following hold.*

i) If $F(m)$ is non-simple the zero m of X is isolated.

ii) If $F(m)$ is simple there exists an open neighbourhood V of m such that the zeros of X in V constitute a 2-dimensional regular submanifold of V (and hence of M).

iii) If $F(m)$ is null the Weyl tensor at m is either zero or Petrov type **N** *and in the latter case the principal null direction of $F(m)$ coincides with the principal null direction of the Weyl tensor at m.*

iv) If $F(m)$ is non-null the Weyl tensor at m is either zero or Petrov type **D** *and in the latter case the timelike blade of $F(m)$ or $\overset{*}{F}(m)$ (whichever is timelike) if $F(m)$ is simple or the timelike member of the canonical pair of blades of $F(m)$ if $F(m)$ is non-simple contains the two principal null directions of the Weyl tensor at m.*

v) The algebraic type of the Ricci (and hence of the energy-momentum) tensor at m is determined by table 10.1 below which also incorporates the results (iii) and (iv) above. (In this table the middle column of Segre types should be read as those Segre types together with any of their degeneracies.)

Table 10.1

$F(m)$	Segre type of R_{ab} or T_{ab} at M	Petrov Type at m
Spacelike	$\{1,1,(11)\}$ or $\{z\bar{z}(11)\}$ or $\{2(11)\}$	**D** or **O**
Timelike	$\{(1,1)11\}$	**D** or **O**
Null	$\{(1,11)1\}$ or $\{(21)1\}$ or $\{(31)\}$	**N** or **O**
Non-simple	$\{(1,1)(11)\}$	**D** or **O**

[The author has recently discovered that parts (i) and (ii) of the above theorem were given in [163].]

In the general case, $\overset{*}{K}_m$ is one of the subalgebras listed in table 6.1. In table 10.2 a complete list is given for the Segre type of the Ricci (or energy-momentum) tensor and the Petrov type of the Weyl tensor for each possible subalgebra for $\overset{*}{K}_m$, where again, column two is intended to include, also, all degeneracies of the stated types. It is easily compiled from tables 6.1 and 10.1 and the following result.

Theorem 10.4 *Let M be a space-time and let $m \in M$, If* $\dim \overset{*}{K}_m \geqslant 3$

the Weyl tensor vanishes at m and if $\dim \overset{*}{K}_m \geqslant 4$ *the Ricci tensor is a multiple of the metric at m.*

Proof. For the first part suppose $\dim \overset{*}{K}_m \geqslant 3$ and $C(m) \neq 0$. Let $\tilde{B}$ be the vector space of Killing bivectors at m arising from members of $\overset{*}{K}_m$. If there exists a null member of $\tilde{B}$ then, from theorem 10.3, the Petrov type at m is **N** and all members of $\tilde{B}$ must be null with common principal null direction. A contradiction now follows by identifying $\overset{*}{K}_m$ with a subalgebra of L and consulting table 6.1. If there is a non-null member in $\tilde{B}$ then theorem 10.3 shows that they are all non-null, that the Petrov type at m is **D** and that the blade of any timelike member of $\tilde{B}$, the orthogonal complement of the blade of any spacelike member of $\tilde{B}$ and the timelike member of the pair of canonical blades of any non-simple member of $\tilde{B}$ must coincide, being determined by the two principal null directions of the Weyl tensor at m. Again table 6.1 reveals a contradiction and so $C(m) = 0$.

For the second part suppose that $\dim \overset{*}{K}_m \geqslant 4$. Since it is easy to check from table 6.1 that the union of the blades (including both members of a canonical pair of blades if a member of $\tilde{B}$ is non-simple) of all members of any subspace of $\tilde{B}$ of dimension four or more spans T_mM, it follows that the (real) eigenvectors of the Ricci tensor at m span T_mM. It is also straightforward to check geometrically that the associated eigenvalues are equal (theorem 9.1). The result follows. □

These results are best possible in the sense that if $\dim \overset{*}{K}_m \leqslant 2$ in the first case, and $\dim \overset{*}{K}_m \leqslant 3$ in the second, each result fails, counterexamples being provided by the type **N** plane waves and F.R.W. models, respectively (see sections 10.5 and 10.7).

If $X \in \overset{*}{K}_m$ with bivector F, so that $A = F(m)$ in (10.25), it was shown following (10.26) that the integral curves $x^a(t)$ of X in a normal coordinate domain U about m satisfy $\frac{dx^a}{dt} = A^a{}_b x^b$. This allows the components of X in U to be calculated in the following way. Suppose $F(m)$ is null and, say, $F^a{}_b(m) = A^a{}_b = l^a x_b - x^a l_b$ for $l, x \in T_mM$ with l null and x unit spacelike and orthogonal to l. If one extends l and x to a null tetrad (l, n, x, y) at m and uses this basis of T_mM to construct normal coordinates in some neighbourhood U of m the action of the maps ϕ_{t*} associated with X is given from (10.25) (and noting that, at m, the only non-zero metric

Table 10.2

$\overset{*}{K}_m$	Segre type of R_{ab} (T_{ab})	Petrov Type at m
R_2	$\{(1,1)11\}$	**D** or **O**
R_3	$\{(1,11)1\}$, $\{(21)1\}$, $\{(31)\}$	**N** or **O**
R_4	$\{1,1(11)\}$, $\{z\bar{z}(11)\}$, $\{2(11)\}$	**D** or **O**
R_5	$\{(1,1)(11)\}$	**D** or **O**
R_6	$\{(1,11)1\}$	**O**
R_7	$\{(1,1)(11)\}$	**D** or **O**
R_8	$\{(1,111)\}$, $\{(211)\}$	**N** or **O**
R_9	$\{(1,111)\}$	**O**
R_{10}	$\{(1,11)1\}$	**O**
R_{11}	$\{(1,111)\}$, $\{(211)\}$	**O**
R_{12}	$\{(1,111)\}$	**O**
R_{13}	$\{1,(111)\}$	**O**
R_{14}	$\{(1,111)\}$	**O**
R_{15}	$\{(1,111)\}$	**O**

components are $g_{01} = g_{10} = g_{22} = g_{33} = 1$) according to

$$(F^a{}_b)_m = A^a{}_b = \begin{pmatrix} 0 & 0 & 1 & 0 \\ 0 & 0 & 0 & 0 \\ 0 & -1 & 0 & 0 \\ 0 & 0 & 0 & 0 \end{pmatrix}, \quad \exp tA = \begin{pmatrix} 1 & -\frac{1}{2}t^2 & t & 0 \\ 0 & 1 & 0 & 0 \\ 0 & -t & 1 & 0 \\ 0 & 0 & 0 & 1 \end{pmatrix} \tag{10.32}$$

The integral curve of X starting from (a_0, a_1, a_2, a_3) in this coordinate system in U is then (labelling the coordinates as v, u, x, y with $l = \left(\frac{\partial}{\partial v}\right)_m$, $n = \left(\frac{\partial}{\partial u}\right)_m$, $x = \left(\frac{\partial}{\partial x}\right)_m$, $y = \left(\frac{\partial}{\partial y}\right)_m$)

$$t \to (a_0, a_1, a_2, a_3)(\exp tA)^\top = \left(a_0 - \frac{1}{2}t^2 a_1 + ta_2, a_1, -ta_1 + a_2, a_3\right) \tag{10.33}$$

Then the components of X in U are from (10.32) or (10.33)

$$X^a = \frac{dx^a}{dt} = (x, 0, -u, 0) \tag{10.34}$$

The cases when $F(m)$ is timelike, spacelike or non-simple can be taken together in an obvious way by writing $F_{ab}(m) = 2\alpha l_{[a}n_{b]} + 2\beta x_{[a}y_{b]}$ for some null tetrad (l, n, x, y) at m and $\alpha, \beta \in \mathbb{R}$. Similar arguments to those in the previous calculation give

$$\begin{aligned} t \to \left(a_0 e^{\alpha t}, a_1 e^{-\alpha t}, a_2 \cos\beta t - a_3 \sin\beta t, a_2 \sin\beta t + a_3 \cos\beta t\right) \\ X^a = (\alpha v, -\alpha u, -\beta y, \beta x) \end{aligned} \tag{10.35}$$

Now let $X \in H(M)$ with X *proper*, $m \in M$ and $X(m) = 0$. So X satisfies (10.21) with $c \neq 0$ and with homothetic bivector F. By replacing X by $-X$, if necessary, one can arrange that $c > 0$ and, with this assumed done, the linearisation A of X at m is then given by $A = cI + F(m)$ where I is the unit 4x4 matrix. Recalling the general remarks about the zeros of affine vector fields made earlier, one is led to study the equation $Av = 0$ at m for $v \in T_mM$. This equation is equivalent to the statement that $F_{ab}v^b = -cv_a$ at m. From this (see section 7.2) it follows that if $v \neq 0$ it is null and unique up to a scaling factor. Hence rank $A = 3$ or 4 and in the latter case, m is an isolated zero of X. In the former case the set of zeros of X constitute (part of) a null geodesic through m. To see this one simply notes that in a normal coordinate domain U about m the coordinates x^a of the zeros of X satisfy $Ax = 0$ and constitute a closed regular 1-dimensional submanifold of U (section 4.11, example (v)) and the only solutions to this equation are $x = sv$ for appropriate s. Thus the zeros of X in U are the points on the geodesic $s \to \exp_m(sv)$ which, since this geodesic starts from m with initial direction v, is null.

Now return to equation (10.29) These still hold if X is homothetic because the maps ϕ_t preserve the metric up to a constant conformal factor ($\phi_t{}^*g = e^{2ct}g$). One then appeals to (a slightly modified version of) section 10.1 with $T = g$. A similar argument to that given in (10.30) for Killing vectors now leads, for a Ricci eigenvector k at m with eigenvalue $\mu \in \mathbb{C}$, to

$$\mathrm{Ricc}\,(\phi_{t*}k, \phi_{t*}k') = \mu e^{2ct} g\,(\phi_{t*}k, \phi_{t*}k') \tag{10.36}$$

Since ϕ_{t*} is an isomorphism of T_mM it follows that $\phi_{t*}k$ is a Ricci eigenvector with eigenvalue μe^{2ct} for each admissible t. Since there are at most four distinct Ricci eigenvalues at m it follows that $\mu = 0$. Hence all Ricci eigenvalues vanish at m and so, from theorem 7.8, the corresponding Segre type at m is, if not trivial, either $\{(211)\}$ or $\{(31)\}$ each with zero eigenvalue [160]. An almost identical argument applied to the eigenbivector structure of the Weyl tensor given in section 7.3 shows that all Weyl eigenvalues vanish at m and so the Petrov type at m is either **III**, **N** or **O** [160].

Again from equation (10.29) one finds using (10.21), the definition of the Lie derivative and the condition $X(m) = 0$ that, at m,

$$2cR_{ab} + R_{ac}F^c{}_b + R_{cb}F^c{}_a = 0 \tag{10.37}$$

$$2cC^a{}_{bcd} + C^a{}_{bce}F^e{}_d + C^a{}_{bed}F^e{}_c + C^a{}_{ecd}F^e{}_b - C^e{}_{bcd}F^a{}_e = 0 \tag{10.38}$$

Now, $\phi_t{}^*$ preserves the Ricci and Weyl tensors. Thus if the Ricci tensor is not zero at m, the Segre types obtained in the previous paragraph and theorem 7.8 show that it admits a unique null eigendirection. This direction must be preserved by ϕ_{t*} at m, that is, if l spans this direction, $\phi_{t*}l \propto l$. It follows that l is a null eigenvector of the linearisation A and hence of $F(m)$. Thus one can set up a real null tetrad (l, n, x, y) at m with l above and which is adapted, in turn, to each of the allowable Segre types above for R_{ab}. In this tetrad l is a null eigenvector of F_{ab}. For the Segre type $\{(211)\}$ with eigenvalue zero one has $R_{ab} = \mu l_a l_b$ at m $(0 \neq \mu \in \mathbb{R})$ and substituting this and the above relation $F_{ab}l^b = \nu l_a$ $(\nu \in \mathbb{R})$ into (10.37) yields $\nu = c$ $(\neq 0)$. From this it follows that $F(m)$ is either timelike or non-simple. For the Segre type $\{(31)\}$ with zero eigenvalue one has $R_{ab} = 2\alpha l_{(a}x_{b)}$ and $F_{ab}l^b = \beta l_a$ $(\alpha, \beta \in \mathbb{R},\ \alpha \neq 0)$ at m. Equation (10.37) then reveals

$$\alpha(2c - \beta)2l_{(a}x_{b)} + 2\alpha l_{(a}p_{b)} = 0 \tag{10.39}$$

where $p_a = F_{ba}x^b$. Now since $\alpha \neq 0$, (10.39) shows that $l_{(a}x_{b)}$ and $l_{(a}p_{b)}$ and hence p_a and x_a are proportional, that is, $F_{ba}x^b = \lambda x_a$ $(\lambda \in \mathbb{R})$. Then a contraction with x^a shows that $\lambda = 0$. It follows that $p_a = 0$ and from (10.39) that $\beta = 2c$ $(\neq 0)$. Hence $F(m)$ is a simple bivector and since $\beta \neq 0$, it is timelike.

A similar argument can be applied to the Weyl tensor at m using (10.38) assuming that it is non-zero (and hence Petrov type **N** or **III**) at m. First it is noted that ϕ_{t*} preserves the principal null directions of C at m. So for the type **N** case one can write out (10.38) in a null tetrad (l, n, x, y) suited to the canonical forms (7.73) and (7.74) with l spanning the repeated principal null direction of C and with $F_{ab}l^b = \rho l_a$ $(\rho \in \mathbb{R})$. If $F(m)$ is null, so that it takes the form $2l_{[a}q_{b]}$ at m with q spacelike and orthogonal to l at m, a direct substitution of this and (7.73) into (10.38) quickly reveals that $cC_{abcd} = 0$ and hence the contradiction that the Weyl tensor vanishes at m. Thus $F(m)$ is non-null and one can use the freedom in the tetrad for the canonical type **N** form to arrange that $F(m)$ takes the form $2al_{[a}n_{b]} + 2bx_{[a}y_{b]}$ $(a, b \in \mathbb{R})$. On substituting into (10.38) and contracting with $x^b x^d$ one finds that $a = c$ whereas a contraction with $x^b y^d$ shows that $b = 0$. So $F(m)$ takes the form $2cl_{[a}n_{b]}$ and is timelike. In the type **III** case ϕ_{t*} preserves the repeated and the non-repeated principal null directions of C at m and so if these are spanned by l and n, respectively, one can choose a null tetrad (l, n, x, y) at m realising the type **III** canonical form (7.75) for C at m and for which l and n are null eigenvectors of $F(m)$. So $F(m)$ takes the form $2a'l_{[a}n_{b]} + 2b'x_{[a}y_{b]}$. On substituting these expressions into (10.38) and contracting with $x^b y^d$

one finds that $b' = 0$ and $a' = 2c$. Hence $F(m)$ takes the form $4cl_{[a}n_{b]}$ and is timelike. One thus has the following theorem ([160] - see also [164]).

Theorem 10.5 *Let M be a space-time and X a proper homothetic vector field on M satisfying (10.21). Suppose X has a zero at $m \in M$, $X(m) = 0$.*

i) Either m is an isolated zero or else the set of zeros of X in some normal neighbourhood U of m consists of (part of) a null geodesic through m and this latter possibility will occur if and only if there exists a (necessarily) null $v \in T_mM$ satisfying $F_{ab}v^b = -cv_a$ at m.

ii) If $F(m)$ is zero, null or spacelike, m is an isolated zero (by part (i)) and the Weyl tensor and Ricci tensor (and hence the curvature tensor) vanish at m. If $F(m)$ is non-simple the Weyl tensor vanishes at m.

iii) All Ricci and Weyl eigenvalues vanish at m, so that, at m, the Ricci tensor is either zero or has Segre type $\{(211)\}$ or $\{(31)\}$ each with zero eigenvalue and the Petrov type is either **III**, **N** *or* **O**.

iv) If at m the Ricci tensor is of Segre type $\{(211)\}$ (with zero eigenvalue) a null tetrad (l, n, x, y) exists at m with l a null Ricci eigenvctor and $F(m)$ takes the form $2cl_{[a}n_{b]} + 2dx_{[a}y_{b]}$ ($d \in \mathbb{R}$). Hence, at m, $F_{ab}n^b = -cn_a$, and the zero m is not isolated (by part (i)). If at m the Ricci tensor has Segre type $\{(31)\}$ a null tetrad (l, n, x, y) exists at m with l a null Ricci eigenvector and $F(m)$ takes the form $4cl_{[a}n_{b]}$. Hence the zero m is isolated.

v) If at m the Petrov type is **N** *a null tetrad (l, n, x, y) exists at m where l spans the repeated principal null direction of C and $F(m)$ takes the form $2cl_{[a}n_{b]}$. The zero m of X is not isolated. If at m the Petrov type is* **III** *a null tetrad (l, n, x, y) exists at m with l spanning the repeated, and n the non-repeated, principal null direction of C and $F(m)$ takes the form $4cl_{[a}n_{b]}$. Hence the zero m is isolated.*

As a consequence of this theorem it follows that if the Weyl and Ricci tensors are non-zero at m then $F(m)$ is timelike and the repeated principal null direction of C and the null Ricci eigendirection coincide with each other and with one of the principal null directions of $F(m)$ (spanned by l). If C is of type **III** at m then the two principal null directions of C coincide with those of $F(m)$. Thus the possibilities at m are either (i) the

Ricci tensor is zero *or* has Segre type $\{(211)\}$ with zero eigenvalue *and* the Weyl tensor is either zero *or* of Petrov type **N** (but not both zero) in which case $F(m)$ takes the form $2cl_{[a}n_{b]} + 2dx_{[a}y_{b]}$ (with d vanishing if the Petrov type is **N**) and the zeros of X form part of a null geodesic through m with tangent proportional to n at m, or (ii) the Ricci tensor is either zero *or* has Segre type $\{(31)\}$ with zero eigenvalue *and* the Weyl tensor is either zero or of Petrov type **III** (but not both zero) in which case $F(m)$ takes the form $4cl_{[a}n_{b]}$ and the zero m isolated or (iii) the Ricci and Weyl tensors each vanish at m. In the first of these cases the obvious null tetrad at m is mapped according to $\phi_{t*}l = e^{2ct}l$, $\phi_{t*}n = n$, $\phi_{t*}x = e^{ct}(\cos(dt)x - \sin(dt)y)$, $\phi_{t*}y = e^{ct}(\sin(dt)x + \cos(dt)y)$ and one can construct the matrices A and $\exp tA$ as in the Killing case discussed earlier and show that one can choose normal coordinates v, u, x, y based on this tetrad (the use of the symbols x and y for both a coordinate function and a tangent vector should cause no confusion) so that the components of X and its integral curves beginning at (a_0, a_1, a_2, a_3) are given in this coordinate domain by

$$t \to (a_0 e^{2ct}, a_1, e^{ct}(a_2 \cos dt - a_3 \sin dt), e^{ct}(a_2 \sin dt + a_3 \cos dt))$$
$$X^a = (2cv, 0, cx - dy, cy + dx,) \tag{10.40}$$

The zeros of X in this domain are then given by $v = x = y = 0$. In the second of these cases a similar argument shows that $\phi_{t*}l = e^{3ct}l$, $\phi_{t*}n = e^{-ct}n$, $\phi_{t*}x = e^{ct}x$, $\phi_{t*}y = e^{ct}y$ and that in the associated normal coordinates and domain, the components of X and its integral curve beginning at (a_0, a_1, a_2, a_3) are

$$t \to (a_0 e^{3ct}, a_1 e^{-ct}, a_2 e^{ct}, a_3 e^{ct}), \quad X^a = c(3v, -u, x, y,) \tag{10.41}$$

with the isolated zero of X at the origin. The above discussion allows of the possibility that $F(m)$ is non-simple at the (non-isolated) zero m of X with the Ricci tensor of type $\{(211)\}$ with zero eigenvalue and the Weyl tensor vanishing at m. Each of the possibilities described here can occur as will be seen at the end of this section. It is noted from (10.40) and (10.41) that, as $t \to -\infty$, the only integral curves of X that get arbitrarily close to a zero m of X are those lying in a $u =$ constant region through m in the normal coordinate domain.

Now let X be a proper homothetic vector field with a zero at m (so that (10.21) holds *with* $c > 0$). One can say more than theorem 10.5(ii) in the case when $F(m)$ is zero, spacelike, null or non-simple and of the form $F(m) = 2al_{[a}n_{b]} + 2bx_{[a}y_{b]}$ with $c \pm a$ each positive. In fact, in each of these cases, the linearisation matrix A has eigenvalues each of which has positive real part. It follows from the theory of differential equations [38] that m is a *source* for X and that there exists an open neighbourhood U of m such that for each $p \in U$, $\phi_t(p) \to m$ as $t \to -\infty$. It now follows by either a slight adaptation of [165] or by direct use of the maps ϕ_t [160] that the curvature tensor vanishes on some open neighbourhood of m. Thus theorem 10.5 can be completed by the following result [164], [160]

Theorem 10.6 *Let M be a space-time, let X be a proper member of $H(M)$ satisfying (10.18) and let $m \in M$ be a zero of X. If $F(m)$ is such that the linearisation matrix A has eigenvalues each of which has positive real part (and this includes the cases when $F(m)$ is zero, spacelike or null), m is isolated and M is flat in some open neighbourhood of m.*

Proof. A brief sketch proceeds along the following lines (and that m is isolated in each case is clear since the linearisation of X at m is non-singular). Suppose X is a proper homothetic vector field on M vanishing at m and whose bivector vanishes at m, $F(m) = 0$. Since now $X^a{}_{,b}(m) = c\delta^a_b$ one has for any $k \in T_mM$, $\phi_{t*}k = e^{ct}k$. So set up normal coordinates x^a in some neighbourhood V of m with respect to some basis of T_mM. The relation (10.26) shows that in this coordinate system ϕ_t is represented by $x^a \to e^{ct}x^a$ and so if $p \in V$ is regarded as fixed in V and if $q \equiv \phi_t(p) \in U$, then the linear map $\phi_{t*} : T_pM \to T_qM$ has matrix $e^{ct}I_4$. Now the relation $\phi_t{}^*g = e^{2ct}g$ shows that for $u, v \in T_pM$, $g_q(\phi_{t*}u, \phi_{t*}v)$ $(=(\phi_t{}^*g)_p(u,v))$ $= e^{2ct}g_p(u,v)$ and so replacing u and v by the appropriate tangent vectors $\frac{\partial}{\partial x^a}$ at p one sees that $g_{ab}(q) = g_{ab}(p)$. Since V may be chosen such that q $(=\phi_t(p)) \to m$ as $t \to -\infty$ for each $q \in V$, it follows that the metric components are constant on V (and equal to their values at m). Thus the metric is flat on V. The argument when $F(m)$ is spacelike or null is the same except for the extra complications in the expression for the action of ϕ_t this time using a basis for T_mM adapted in a natural way to $F(m)$. Again one shows that in some normal coordinate neighbourhood V of m the metric tensor components are constant and so V is flat. A variant of this method can be employed in the case when $F(m)$ is non-simple and takes the form $F(m) = 2al_{[a}n_{b]} + 2bx_{[a}y_{b]}$ as described above with $c \pm a > 0$. One can choose normal coordinates from the basis l, n, x, y of T_mM in some open neighbourhood V of m with V chosen so that $p \in V \Rightarrow q(=\phi_t(p)) \in V$ and $\phi_t(p) \to m$ as $t \to -\infty$. The coordinate representation of ϕ_t is obtained from (10.26) after calculating the effect of ϕ_{t*} on the above basis at m, which is $\phi_{t*}l = e^{(c+a)t}l$, $\phi_{t*}n = e^{(c-a)t}n$,

$\phi_{t*}x = e^{ct}(\cos(bt)x - \sin(bt)y)$, $\phi_{t*}y = e^{ct}(\sin(bt)x + \cos(bt)y)$. Now let R' represent the curvature tensor with components R_{abcd}. Then since $\mathcal{L}_X R' = 2cR'$ ($\Leftrightarrow \phi_t{}^* R' = e^{2ct} R'$) one has $R_q{}'(\phi_{t*}e_0, \phi_{t*}e_1, \phi_{t*}e_2, \phi_{t*}e_3)$ $(= (\phi_t{}^* R')_p(e_0, e_1, e_2, e_3)) = e^{2ct} R_p{}'(e_0, e_1, e_2, e_3)$ for any members $e_a \in T_m M$. Thus by choosing for the e_a various combinations and orderings of l, n, x, y one can obtain information about the components of R' knowing that the components of $R_q{}'$ converge to those of $R'(m)$ as $t \to -\infty$ and those of R'_p are fixed constants. Thus choosing for $e_0, \ldots, e_3$ the members l, n, x, y in that order gives $R_{1234}(q) = e^{-2ct} R_{1234}(p)$ and so to avoid this component becoming infinite at m one must have $R_{1234}(p) = 0$. Similarly choosing the ordered basis members x, l, x, l one finds $\cos^2(bt) R_{3131}(q) - \sin(2bt) R_{3141}(q) + \sin^2(bt) R_{4141}(q) = e^{-2(c+a)t} R_{3131}(p)$. Since the left hand side of this expression is bounded as $t \to -\infty$ and $c + a > 0$ one finds that $R_{3131}(p) = 0$ The other calculations are similar and it follows that $R'(p) = 0$ and hence R' vanishes on an open neighbourhood of m. □

The above permits a general discussion of the behaviour of integral curves of Killing and homothetic vector fields in the neighbourhood of a zero m of that vector field in terms of $F(m)$ and can be used to show which of these curves get ‘arbitrarily close’ to m (see, e.g. [166]).

Now let $X \in H(M)$ satisfying (10.21). Then as shown in section 10.3, $\sigma \equiv X^a X_a$ satisfies (10.24) and so along an integral curve of X with parameter t, $\frac{d\sigma}{dt} = 2c\sigma$. Hence, along such an integral curve of X, $\sigma = Ae^{2ct}$ where $A \in \mathbb{R}$ (and $A = 0$ corresponds to the case where X is null along this curve). Next, the equations $\mathcal{L}_X Ricc = 0$ and $\mathcal{L}_X g = 2cg$ are equivalent to $\phi_t{}^* Ricc = Ricc$ and $\phi_t^* g = e^{2ct} g$ for the maps ϕ_t associated with X and show that the associated maps ϕ_{t*} map eigenvectors of $Ricc$ at m to eigenvectors of $Ricc$ at $\phi_t(m)$ on the curve. In fact, if ϕ_t is an integral curve of X beginning at m_0 and if k_0 is an eigenvector of $Ricc$ at m_0 with eigenvalue γ_0 let $k(t)$ be a vector field defined on some neighbourhood of m_0 on this curve by $\mathcal{L}_X k = 0$, $k(0) = k_0$. Then $\mathcal{L}_X(g_{ab}k^a k^b) = 2cg_{ab}k^a k^b$ and so the nature (timelike, spacelike or null) of $k(t)$ is preserved along the curve and the vector field $R_{ab}k^b(t) - \gamma_0 e^{-2ct} g_{ab}k^b(t)$ defined on the same piece of this curve has zero Lie derivative with repect to X there and also vanishes at m. Hence it vanishes on the curve, showing that $k(t)$ is an eigenvector of $Ricc$ for each t with eigenvalue $\gamma(t) \equiv \gamma_0 e^{-2ct}$. Thus $\frac{d\gamma}{dt} = -2c\gamma$ for appropriate t and, in particular, γ is constant if $c = 0$ (*i.e.* if X is Killing). It follows that *the Segre type of Ricc (including degeneracies) is the same at each point of an integral curve of X with eigenvalues behaving as $\gamma(t)$ along the* *integral curves of* X. It also follows that $\sigma\gamma$ is constant along an integral curve of X [160], [166].

For $X \in H(M)$ the equations (10.29) still hold and the first of these gives $\mathcal{L}_X C_{abcd} = 2cC_{abcd}$. Also, from (7.11) one finds $\mathcal{L}_X G_{abcd} = 4cG_{abcd}$. The work in section 7.3 and a calculation similar to that above shows that the Petrov type of C is the same at each point of an integral curve of X and that, along such a curve, any Weyl eigenvalue δ behaves as $\delta(t) = \delta_0 e^{-2ct}$ so that, again, $\delta\sigma =$ constant along the curve. Once more, δ is constant if $c = 0$ (*i.e.* if X is Killing). Similar remarks apply to the eigenvalues of the curvature tensor and thus, for any homothetic vector field X, the *ratio* (where defined) of two (Ricci, Weyl or curvature) eigenvalues is constant along an integral curve of X.

The existence of a proper homothetic vector field on M with a non-isolated zero m imposes rather strict conditions on M. In fact, an adaptation of the argument used in the proof of theorem 10.6 this time applied to the Ricci and Weyl tensors, using (10.29) (i) and (iii) gives the following result [164] [160].

Theorem 10.7 *If a non-flat space-time M admits a proper homothetic vector field with a non-isolated zero at m then some neighbourhood of m is isometric to a plane wave.*

The metric of such a space-time in a global coordinate system on $\mathbb{R}^4$ can be put into the form (8.11) where [55]

$$H(u, x, y) = a(u)x^2 + b(u)y^2 + c(u)xy \tag{10.42}$$

for smooth functions a, b and c. From (8.12) the Ricci tensor, if non-zero, has Segre type $\{(211)\}$ with zero eigenvalue and the vacuum condition on (10.42) is $a \equiv -b$ [41], [55]. The Weyl tensor, if non-zero, is of Petrov type **N**. The conformally flat condition on (10.42) is $a \equiv b$ and $c \equiv 0$ [55]. In theorem 10.5(ii) it was shown that the homothetic bivector of X at m is non-simple the Weyl tensor vanishes at m. For a plane wave with $F(m)$ non-simple at the (non-isolated) zero m of X the techniques used in the proof of theorem 10.6 can be used to show that the Weyl tensor vanishes on the $u =$ constant submanifold through m of an appropriate normal coordinate neighbourhood of m. But, by continuity, $F(m)$ will also be non-simple on some open subset W, containing m, of the subset of zeros of X. It follows that the Weyl tensor will vanish on the open neighbourhood of m formed by the union of all the $u =$ constant submanifolds through points of W. Hence *one has conformal flatness in some open neighbourhood of m.*

As a slight variant of theorem 10.7 it is noted that if there exist $X \in H(M)$ and $Y \in K(M)$ with X proper and Y non-trivial and $X(m) = Y(m) = 0$ then from theorems 10.3, 10.5 and 10.7 either the Weyl tensor vanishes at m or some neighbourhood of m is a plane wave.

The metric (8.11) (and (10.42)) always admits (at least) a 5-dimensional Lie algebra of Killing vector fields containing the covariantly constant null vector field with components $l^a = g^{ab}u_{,b}$ [41]. This vector field spans the repeated principal null direction of the Weyl tensor and the null Ricci eigendirection where the latter two tensors do not vanish. In the type **N** case, this metric admits the proper homothetic vector field (10.40) (with $d = 0$) which vanishes along the null geodesic $v = x = y = 0$. The homothetic bivector at any of these zeros is timelike and is described in theorem 10.5. In the conformally flat case, the homothetic vector field (10.40) is admitted and $\dim K(M)$ is at least six because of the existence of an extra Killing vector field (which admits zeros) of the type (10.35) with $\alpha = 0$. By taking linear combinations of such a vector field with the above homothetic vector field one can find a homothetic vector field which vanishes (and has non-simple homothetic bivector) at $v = x = y = 0$ as promised earlier.

An example of a metric admitting a homothetic vector field with an isolated zero was given in [165] (cf. [166]). The metric in a global coordinate system on $\mathbb{R}^4$ is

$$ds^2 = e^{ux}dudv + dx^2 + dy^2 \tag{10.43}$$

This space-time is everywhere of Petrov type **III** and has Ricci (and hence energy-momentum) tensor of Segre type $\{31\}$ or its degeneracy. The homothetic vector field (10.41) is admitted and with unique zero $v = u = x = y = 0$. The vector fields $\frac{\partial}{\partial v}$ and $\frac{\partial}{\partial y}$ are Killing, the former being null and the latter spacelike and covariantly constant. There are no other independent homothetic vector fields. [It is remarked that the metric (10.43) has holonomy type R_{10} [161]. This follows since the covariant constancy of $\frac{\partial}{\partial y}$ means that the holonomy type is either R_2, R_3, R_4, R_6 or R_{10} and since it is everywhere of Petrov type **III** the first three of these types are excluded by theorem 8.12. If the holonomy type is R_6 then a global recurrent nowhere zero null vector field k is admitted and this is, from the work preceeding (8.21), a Ricci eigenvector field. But $\frac{\partial}{\partial v}$ is hypersurface orthogonal, having covariant components $e^{ux}u_{,a}$, and is also a null Ricci eigenvector field (see the end of section 10.7). Thus by the uniqueness of the null Ricci eigendirection for (10.43), $\frac{\partial}{\partial v} = fk$ for some global smooth nowhere zero function $f : M \to \mathbb{R}$. Thus $\frac{\partial}{\partial v}$ is simultaneously Killing and recurrent and this easily leads to the contradiction that it is covariantly constant.]

It is interesting to compare the above results with the situation for positive definite metrics. Suppose (M, g) is an n-dimensional manifold ($n \geqslant 2$) admitting a positive definite metric and let X be proper homothetic vector field on M with a zero at m. Then for the linearisation A of X at m (see (10.25)) there is no $k \in T_m M$, $k \neq 0$, such that $A^a{}_b k^b = 0$ because, from (10.21), this would imply $F_{ab} k^b = c k_a$ and hence the contradiction $c k_a k^a = 0$ ($\Rightarrow c = 0$). Hence *m is necessarily isolated* and it is easily checked that there is an open neighbourhood of m in which all integral curves of X get arbitrarily close to m. Further, similar techniques to those used in theorem 10.6 (or see [20]) show that *some neighbourhood of m is flat*. It also turns out that *if X is proper homothetic and M is geodesically complete then X necessarily admits a zero* [20]. For space-times, however, the plane waves and the metric (10.43) show that a zero m of a proper homothetic vector field need not be isolated and, whether isolated or not, the metric need not be flat in some neighbourhood of it (and even if m is isolated, there need not exist a neighbourhood of m in which all integral curves of X get arbitrarily close to m, as is the case for (10.43)). In fact, these last two examples are easily generalised to an n-dimensional manifold ($n \geqslant 4$) admitting a Lorentz metric by taking a metric product of the space-time with flat Euclidean space $\mathbb{R}^p$ for appropriate p. Examples of metrics exhibiting similar features also exist for Lorentz manifolds of dimension two (where appropriate) or three [161].

10.5 Orbit Structure

In keeping with the description of symmetries of a space-time M in terms of vector fields on M, the following general results are useful. Let S be a non-trivial vector space of global smooth vector fields on M. Continuing the discussion of sections 4.13 and 5.12 recall that, for $m \in M$, $S_m = \{X(m) : X \in S\} \subseteq T_m M$ is the subspace of $T_m M$ consisting of all members of S evaluated at m and, that for $0 \leqslant p \leqslant 4$, $V_p = \{m \in M : \dim S_m = p\}$. Then (4.8) reveals a disjoint decomposition of M in the form

$$M = V_4 \cup \operatorname{int} V_3 \cup \operatorname{int} V_2 \cup \operatorname{int} V_1 \cup \operatorname{int} V_0 \cup V \tag{10.44}$$

where V is closed and $\operatorname{int} V = \emptyset$. Since M is connected, S will give rise to a *distribution* $m \to S_m$ on M (in the sense of Fröbenius, *i.e.* $\dim S_m$ is constant on M) if and only if V is empty. This decomposition can now be refined [91] (but see [92] for corrections to several errors in [91] and where in the latter the last sentence of theorem 2 is not proven).

For each $m \in M$ either S_m is trivial or of dimension 4 or, if of dimension 1, 2 or 3, is either spacelike, timelike or null. For $p = 1, 2$ or 3 define subsets $\tilde{S}_p$, $\tilde{T}_p$ and $\tilde{N}_p$ by

$$\begin{aligned}
\tilde{S}_p &= \{m \in M : \ S_m \text{ is spacelike and } \dim S_m = p\} \\
\tilde{T}_p &= \{m \in M : \ S_m \text{ is timelike and } \dim S_m = p\} \\
\tilde{N}_p &= \{m \in M : \ S_m \text{ is null and } \dim S_m = p\} \\
&\left(\Rightarrow V_p = \tilde{S}_p \cup \tilde{T}_p \cup \tilde{N}_p\right)
\end{aligned} \tag{10.45}$$

Theorem 10.8 *Let M be a space-time and S a non-trivial vector space of global, smooth vector fields on M. Then M may be disjointly decomposed as*

$$M = V_4 \cup \bigcup_{p=1}^{3} \operatorname{int} \tilde{S}_p \cup \bigcup_{p=1}^{3} \operatorname{int} \tilde{T}_p \cup \bigcup_{p=1}^{3} \operatorname{int} \tilde{N}_p \ \cup \operatorname{int} V_0 \cup Z' \tag{10.46}$$

where the disjointness of the decomposition defines the (necessarily closed) subset Z' of M and where $\operatorname{int} Z' = \emptyset$.

Proof. As before V_4 is necessarily open and it remains only to establish that $\operatorname{int} Z' = \emptyset$. So let U be a non-empty open subset of M such that $U \subseteq Z'$. By disjointness, $U \cap V_4 = \emptyset$ and the rank theorem shows that V_4, $V_4 \cup V_3$, $V_4 \cup V_3 \cup V_2$, etc. are open in M. Hence the subset W where

$$W \equiv U \cap (V_4 \cup V_3) = U \cap V_3 = U \cap (\tilde{S}_3 \cup \tilde{T}_3 \cup \tilde{N}_3) \tag{10.47}$$

is open in M. Suppose $W \neq \emptyset$. Then U cannot be disjoint from $\tilde{S}_3$ *and* $\tilde{T}_3$ because this would then imply that $U \cap \tilde{N}_3$ (and hence $U \cap \operatorname{int} \tilde{N}_3$) is open and non-empty (since W is), contradicting the disjointness assumption. So suppose $U \cap \tilde{S}_3 \neq \emptyset$ and let $q \in U \cap \tilde{S}_3$. Then $q \in U \cap \tilde{S}_3 \subseteq W$ and W is open with $\dim S_m = 3$ for each $m \in W$. Hence there exist $X, Y, Z \in S$ such that $X(q), Y(q)$ and $Z(q)$ span the spacelike subspace S_q of $T_q M$ and thus an open neighbourhood W' of q exists such that for each $q' \in W'$, $X(q'), Y(q')$ and $Z(q')$ span $S_{q'}$ and $S_{q'}$ is spacelike. Thus $W' \subseteq \operatorname{int} \tilde{S}_3$ and the contradiction $U \cap \operatorname{int} \tilde{S}_3 \neq \emptyset$ follows, and similarly if $U \cap \tilde{T}_3 \neq \emptyset$. Hence $W = \emptyset$. One then repeats the argument using the open subset $U \cap (V_4 \cup V_3 \cup V_2) = U \cap V_2$, and so on, finally achieving $U \cap V_2 = U \cap V_1 = \emptyset$.

Thus $U \subseteq V_0$ and , since U must be disjoint from int V_0, one finds $U = \emptyset$. It follows that int $Z' = \emptyset$ and the proof is complete. □

One can refine this decomposition by combining it with the decompositions of M in terms of Petrov type and energy-momentum tensor Segre type (theorems 7.15 and 7.16 to get the following general decomposition theorem [91], [92].

Theorem 10.9 *Let M be a space-time and let S be a non-trivial vector space of global smooth vector fields on M. Then M may be disjointly decomposed into a finite number of open subsets on each of which the dimension of S_m and (for $\dim S_m = 1$, 2 or 3) the nature (spacelike, timelike or null) of S_m, the Petrov type of the Weyl tensor and the Segre type (including degeneracies) of the energy-momentum tensor are constant, together with a closed subset which has empty interior. Thus the union of these open subsets is an open dense subset of M.*

Proof. From theorems 10.8, 7.15 and 7.16 one may write decompositions

$$\begin{aligned} M &= B_1 \cup B_2 \cup \cdots \cup B_m \cup B' \\ M &= C_1 \cup C_2 \cup \cdots \cup C_n \cup C' \\ M &= D_1 \cup D_2 \cup \cdots \cup D_p \cup D' \end{aligned} \tag{10.48}$$

for positive integers m, n, p, open subsets B_i, C_i, D_i of M on which, respectively, the Petrov type, energy-momentum tensor Segre type (including degeneracies) and the dimension and nature (where applicable) of $S_m (m \in M)$ are constant and closed subsets B', C', D' of M each of which has empty interior. Now let

$$B = \bigcup_{i=1}^{3} B_i, \quad C = \bigcup_{i=1}^{3} C_i, \quad D = \bigcup_{i=1}^{3} D_i \tag{10.49}$$

$$\left(\Rightarrow B \cap C \cap D = \bigcup_{i,j,k} B_i \cap C_j \cap D_k \right)$$

Then $M = (B \cap C \cap D) \cup (B' \cup C' \cup D')$ is a disjoint union and since B', C' and D') are closed and nowhere dense, so also is $(B' \cup C' \cup D')$. Thus $B \cap C \cap D$ is open and dense in M and this disjoint union, bracketed in (10.49), is the one required by the theorem. □

Theorems 10.8 and 10.9 clearly apply when S is any of $A(M), H(M)$ or $K(M)$ and since they are finite-dimensional so also do theorems 4.13 and 5.14 and their consequences can now be stated.

Theorem 10.10 *Let M be a space-time and S be one of the algebras $A(M), H(M)$ or $K(M)$. Then each orbit of S is a leaf of M and a maximal integral manifold of S.*

A consideration of the linear map $X \to X(m)$ from $A(M)$ to T_mM where m lies on an orbit O of $A(M)$ shows that $\dim A(M) = \dim S_m + \overset{*}{A}_m = \dim O + \dim \overset{*}{A}_m$ and similarly for $H(M)$ and $K(M)$. One consequence of these last three theorems applied to $A(M)$, $H(M)$, $K(M)$ (and certain other finite-dimensional "symmetry" algebras to be discussed later) is the following remark. In the quest for exact solutions of Einstein's equations one usually imposes some or all of the following: (i) M has the same Petrov type everywhere, (ii) M has the same energy-momentum Segre type (including degeneracies) everywhere, (iii) some non-trivial Lie algebra of "symmetries" of one of the above types is admitted whose orbits are of the same dimension and nature everywhere. The above theorems show that *the conditions (i)-(iii) can be achieved in some open neighbourhood of "almost any" point of M* (*i.e.* of any point in the open dense subset $B \cap C \cap D$ of M). It is also remarked here that, since any $X \in A(M)$ is identically zero on M if it vanishes over a non-empty open subset of M (theorem 10.1), then in the decompositions (10.44) and (10.46) applied to $A(M)$, $\text{int}\, V_0 = \emptyset$.

It is also remarked at this point that the behaviour of Ricci, Weyl and curvature eigenvalues on orbits of $K(M)$ and $H(M)$ is controlled by the work at the end of section 10.4. In particular, such eigenvalues are constant over orbits of $K(M)$ and their ratios (where sensible) are constant over orbits of $H(M)$. The Segre type (including degeneracies) of the Ricci tensor and the Petrov type of the Weyl tensor are the same at each point of an orbit of $K(M)$ or $H(M)$.

Before the next theorem is stated some definitions are required. If N is a submanifold of the space-time M with $\dim N = 1$, 2 or 3 then the *nature* (spacelike, timelike or null) of the subspace of T_mM tangent to N for appropriate $m \in M$ may change with m. If, however, each of these subspaces are spacelike (respectively timelike, null) then N is called *spacelike* (respectively *timelike*, *null*). It follows from theorem 4.16 that if N is spacelike (respectively, timelike) then the space-time metric g on M induces a smooth metric h on N which is positive definite (respectively, Lorentz). In fact, if $i : N \to M$ is the usual inclusion map then $h = i^*g$. If X is any smooth vector field in M tangent to N (section 4.11) there is a natural smooth vector field $\tilde{X}$ on N induced by X and given by $i_*\tilde{X} = X$. Retaining this notation but now with N being an orbit O of S, if $\tilde{c}$ is an integral curve of $\tilde{X}$ then $c \equiv i \circ \tilde{c}$ is an integral curve of X (section 4.11). Conversely, if c is an integral curve of X and if $\tilde{c}$ is defined by $c = i \circ \tilde{c}$ then $\tilde{c}$ is an integral curve of $\tilde{X}$. To see this, briefly, one has that $c : I \to M$ is smooth for some open interval I of $\mathbb{R}$ and its range lies in O and is represented by the map $\tilde{c} : I \to O$ and is smooth since O is a leaf of M (theorem 10.10). The condition $X \circ c = \dot{c}$ (section 4.10) then shows that $c_* \circ \frac{\partial}{\partial t} = (i \circ \tilde{c})_* \circ \frac{\partial}{\partial t} = i_* \circ \tilde{c}_* \circ \frac{\partial}{\partial t} = X \circ c = X \circ i \circ \tilde{c} = i_* \circ \tilde{X} \circ \tilde{c}$. Thus, since i is an immersion (and so i_* is injective) $\tilde{c}_* \circ \frac{\partial}{\partial t}\ (= \dot{\tilde{c}}) = \tilde{X} \circ \tilde{c}$ and $\tilde{c}$ is an integral curve of $\tilde{X}$ Then, if ϕ_t and $\tilde{\phi}_t$ represent the local flows of X and $\tilde{X}$, respectively, $\phi_t \circ i = i \circ \tilde{\phi}_t$.

Theorem 10.11 *Let M be a space-time and suppose O is an orbit of $H(M)$ or $K(M)$ with $\dim O =$ 1, 2 or 3.*

i) O is either spacelike, timelike or null and the Segre type of the Ricci tensor (including degeneracies) and the Petrov type are constant on O.

ii) Let O be spacelike or timelike with induced metric h and let X be homothetic on M (thus X is tangent to O). Then if $i : N \to M$ is the natural inclusion map (so that $h = i^* g$), the associated vector field $\tilde{X}$ on O ($i_* \tilde{X} = X$) is homothetic with respect to h and has the same homothetic constant as X. Thus if X is Killing on M then $\tilde{X}$ is Killing on O. If, in addition, X is non-trivial and $\dim O = 3$, then $\tilde{X}$ is non-trivial.

iii) For O non-null the map $\theta : H(M) \to H(O)$ (or $K(M) \to K(O)$) given by $\theta(X) = \tilde{X}$ is a Lie algebra homomorphism under the Lie bracket operation. If $\dim O = 3$, θ is one-to-one and in this case $H(M)$ (respectively $K(M)$) is Lie isomorphic to a subalgebra of $H(O)$ (respectively $K(O)$). The map θ is not necessarily onto.

Proof. (i) Let $p, q \in O$. Then there exist local diffeomorphisms ϕ_t of the appropriate type (homothetic or Killing) such that a combination of them of the form (5.12), say f, which is also of that type, satisfies $f(p) = q$. Now using the invariance property in theorem 5.14 it follows that if $u, v \in T_p M$ are tangent to O at p then $f_* u$ and $f_* v$ are tangent to O at q and

$$\left(f^{-1*} g\right)_q (f_* u, f_* v) = g_p(u, v) \tag{10.50}$$

where $g_p = g(p)$, etc. Since for the cases considered $(f^{-1*} g)_q$ either equals g_q or some positive multiple of it, the inner products of u and v at p and $f_* u$ and $f_* v$ at q are either both zero or of the same sign. It follows that the nature (spacelike, timelike or null) of O is the same at p and q (see section 6.1) and the result follows.

(ii) Let $X \in H(M)$ with associated vector field $\tilde{X}$ on O and let ϕ_t and $\tilde{\phi_t}$ be the local diffeomorphisms associated with X and $\tilde{X}$, respectively, so that $i \circ \tilde{\phi_t} = \phi_t \circ i$. Then

$$\tilde{\phi_t}^* h = \tilde{\phi_t}^* (i^* g) = (i \circ \tilde{\phi_t})^* g = (\phi_t \circ i)^* g = i^* (\phi_t^* g) \tag{10.51}$$

Thus if X is homothetic on M, then $\tilde{X}$ is homothetic on O and if X is Killing on M, $\tilde{X}$ is Killing on O (and possibly $\tilde{X} \equiv 0$ on O). However, if $\dim O = 3$, then $\tilde{X}$ cannot be identically zero on O because theorem 10.3 would then imply that $X \equiv 0$ on M. To see this, suppose $X \in K(M)$ vanishes on the orbit O through (the non-isolated zero) m with $\dim O = 3$ and let V be an open neighbourhood of m such that the zeros of X in V constitute a 2-dimensional regular submanifold N of V. Let $O' \equiv O \cap V$ so that O' is an open subset and hence an open submanifold of O and hence a submanifold of M. Since $O' \subseteq V$ and V is regular, O' is a submanifold of V (section 4.11 example (ix)). But $O' \subseteq N \subseteq V$ with O' a submanifold of V and N a regular submanifold. Thus O' is a submanifold of N and one has the contradiction $\dim O$ $(= \dim O') \leqslant 2$. A similar argument applies for X a proper member of $H(M)$ using theorem 10.5. For then if X vanishes on any non-null orbit O through m let V be an open neighbourhood of m such that the zeros of X constitute a null 1-dimensional regular submanifold N of V. The above argument shows that O' is a submanifold of N and hence $\dim O$ $(= \dim O') = \dim N = 1$. But O is non-null and N null and a contradiction easily follows.

(iii) Let X, Y be members of $H(M)$ or $K(M)$ with associated vector fields $\tilde{X}$ and $\tilde{Y}$ on O. Then the relation $i_*[\tilde{X}, \tilde{Y}] = [X, Y]$ (section 4.9) shows that θ, which is clearly a linear map, is a Lie algebra homomorphism. If $\dim O = 3$, the previous argument shows that $\tilde{X} \equiv 0 \Rightarrow X \equiv 0$ and so θ is one-to-one. That θ may not be onto will be discussed at the end of this section (see *e.g.*, [167]). □

For the remainder of this section, following [167], only orbits of $K(M)$ (Killing orbits) will be considered. Some relations between the dimension of $K(M)$ and the possible orbit dimension will be derived. However, in order to do this a distinction must be made between those orbits which are in a sense 'stable' with respect to their dimension and nature and those which are not. It is clear from the previous theorem that the dimension and *nature* (spacelike, timelike or null) of an orbit O is the same at each point of O (and from now on the dimension and nature of O will be collectively referred to as its *type*). Thus if $1 \leqslant \dim O \leqslant 3$ then O is entirely contained in one of the subsets $\tilde{S}_p, \tilde{T}_p$ or $\tilde{N}_p$ ($1 \leqslant p \leqslant 3$) of M. So let O be an orbit of $K(M)$ with $1 \leqslant \dim O \leqslant 3$. Then O will be called *stable* if it is entirely contained in one of the subsets $\operatorname{int}\tilde{S}_p, \operatorname{int}\tilde{T}_p$ or $\operatorname{int}\tilde{N}_p$. Thus if O is stable then nearby orbits are of the same type as O. Stable orbits are well-behaved (with some expected anomalies for null orbits) but non-stable orbits need not be. The orbit O will be called *dimensionally stable* if it is entirely contained in $\operatorname{int} V_p$ ($1 \leqslant p \leqslant 3$). In this case, nearby orbits have the same dimension as O. Clearly if O is stable it is dimensionally stable (but not conversely). The remarks on rank in section 3.11 show that if O is an orbit of $K(M)$ of maximum dimension then it is dimensionally stable and if O is an orbit of $K(M)$ which is *not* dimensionally stable then for each $p \in O$ there exists a non-trivial member $X \in K(M)$ such that $X(p) = 0$. It is noted here that each component of the associated open subset V_4 is necessarily a 4-dimensional orbit and, conversely, each 4-dimensional orbit is a component of V_4 ([168] - see also [30], [27]). If, in the decomposition (10.44), S is one of the algebras $A(M)$, $H(M)$ or $K(M)$ and $M = V_4$ (*i.e.* $\dim S_m = 4$ for each $m \in M$) then, since M is connected, there is a single 4-dimensional orbit M and S is called *transitive* or *homogeneous*.

The previous theorem shows that Killing vector fields induce similar vector fields in the associated non-null orbit geometry but that *independent such vector fields on M may not induce independent vector fields in the orbit unless the orbit is 3-dimensional*. Examples will be given later.

Theorem 10.12 *Let M be a space-time.*

i) If O is an orbit of $K(M)$ such that $O \cap \operatorname{int}\tilde{S}_p \neq \emptyset$ (respectively, $O \cap \operatorname{int}\tilde{T}_p \neq \emptyset$ or $O \cap \operatorname{int}\tilde{N}_p \neq \emptyset$) for some p, $1 \leqslant p \leqslant 3$, then $O \subseteq \operatorname{int}\tilde{S}_p$ (respectively, $O \subseteq \operatorname{int}\tilde{T}_p$ or $O \subseteq \operatorname{int}\tilde{N}_p$) and so O is stable. If $O \cap \operatorname{int} V_p \neq \emptyset$ then $O \subseteq \operatorname{int} V_p$ and so O is dimensionally stable.

ii) If M is decomposed as in (10.46) with respect to the Lie algebra of Killing vector fields $K(M)$ then $\tilde{S}_3$ and $\tilde{T}_3$ are open subsets of M. In other words, each 3-dimensional spacelike or timelike orbit for $K(M)$ is stable.

Proof. (i) Suppose O is such an orbit and $O \cap \operatorname{int} S_p \neq \emptyset$ and let $m \in O \cap \operatorname{int} S_p$. Let m' be any other point of O. Then there exists an open neighbourhood U of m in M and an associated local diffeomorphism f of the form (5.12) such that f is defined on U, $f(m) = m'$ and with $U \subseteq \operatorname{int}\tilde{S}_p$.

Thus the orbits through the various points of U are p-dimensional and spacelike and this is clearly true for the orbits through the various points of the open neighbourhood $f(U)$ of m'. Thus $m' \in f(U) \subseteq \text{int}\, \tilde{S}_p$ and so $O \subseteq \text{int}\, \tilde{S}_p$. A similar argument applies in the timelike and null cases. The final sentence of (i) is now clear.

(ii) Let O be a 3-dimensional spacelike orbit and let $m \in O$. Then one constructs a Gaussian coordinate neighbourhood U of m ([35]) such that each point of U lies on a timelike geodesic with tangent vector $k^a(t)$ and affine parameter t and which intersects O at m' at which point it is orthogonal to O. If $X \in K(M)$ then Killing's equation and the geodesic equation reveal that [35]

$$\frac{d}{dt}(X_a k^a) = (X_a k^a)_{;b} k^b = X_{a;b} k^a k^b = 0 \tag{10.52}$$

and so, since $(X^a k_a)(m') = 0$, $X^a k_a = 0$ along the geodesic. Since this is true for any $X \in K(M)$ it follows that the orbits of $K(M)$ through the various points of U are orthogonal to a timelike vector at that point and hence have dimension at most three. However, the rank theorem reveals a neighbourhood V of m through each point of which the orbit dimension is at least three. It follows that $U \cap V \subseteq \tilde{S}_3$ and hence, from part (i), that $O \subseteq \text{int}\, \tilde{S}_3$ and so O is stable. A similar argument applies to the case of timelike orbits. □

It follows from part (i) of this theorem that in the decomposition (10.46) *the open dense subset $M \setminus Z'$ of M is a union of stable orbits and Z' a union of unstable orbits* (and with similar obvious comments regarding the decomposition (10.44) and dimensionally stable and unstable orbits).

The next theorem gives some relations between the dimension of $K(M)$ and the possible associated orbits.

Theorem 10.13 *Let M be a space-time. Then the following hold.*

i) If there exists a null 3-dimensional orbit of $K(M)$ then

$$3 \leqslant \dim K(M) \leqslant 7.$$

If, however, there exists a null, dimensionally stable 3-dimensional orbit of $K(M)$ or a non-null 3-dimensional orbit of $K(M)$, then

$$3 \leqslant \dim K(M) \leqslant 6.$$

ii) *If there exists a 2-dimensional null orbit of $K(M)$ then*

$$2 \leqslant \dim K(M) \leqslant 5$$

whilst the existence of a non-null, 2-dimensional orbit of $K(M)$ implies that

$$2 \leqslant \dim K(M) \leqslant 4.$$

If there exists a dimensionally stable, 2-dimensional orbit of any nature then

$$2 \leqslant \dim K(M) \leqslant 3.$$

iii) *For a 1-dimensional orbit O of $K(M)$ the respective situations for O null, O non-null and O dimensionally stable (of any nature) are*

$$1 \leqslant \dim K(M) \leqslant 5 \ , \ 1 \leqslant \dim K(M) \leqslant 4 \ \textit{and} \ \dim K(M) = 1.$$

Proof. (i) If any 3-dimensional orbit O exists, then $\dim K(M) \geqslant 3$. Suppose, in addition, that O is null (and not necessarily stable). Let $m \in O$ and let X', Y' and Z' be independent members of $K(M)$ in some coordinate neighbourhood U of m such that $X'(m')$, $Y'(m')$ and $Z'(m')$ are independent tangent vectors at each $m' \in U$ and which span O on $O' \equiv O \cap U$. Then the smooth vector field k in U with components $k^a = \epsilon^a{}_{bcd} X'^b Y'^c Z'^d$ (with ϵ as in section 7.2) is orthogonal to X', Y' and Z' (and nowhere zero) in U and hence is null on O' and tangent to O there, since O is null. Now let $X \in K(M)$, $X \not\equiv 0$ on M. Then X is tangent to O and hence orthogonal to k on O'. Thus $(X^a k_a)_{;b} p^b = 0$ on O' where p^a are the components of the tangent vector to any curve in O'. Suppose $X(m) = 0$. Then evaluating the previous equation at m using (10.20) shows that the Killing bivector F of X satisfies $F_{ab} k^a p^b = 0$ at m for any $p \in T_m M$ tangent to O at m. It follows that $F_{ab} k^b = \lambda k_a$ at m ($\lambda \in \mathbb{R}$) and that at most four independent bivectors at m can have this property (since k is null). Thus $\dim \overset{*}{K}_m \leqslant 4$ and hence (from the relation immediately after theorem 10.10) $\dim K(M) \leqslant 7$.

If, however, the orbit O is 3-dimensional and either null *and* dimensionally stable or non-null (and hence stable by theorem 10.12 (ii)) let $m \in O$ and U an open neighbourhood of m such that each orbit intersecting U is the same dimension as O (as above) and so that there exists a smooth vector field k on U which is everywhere orthogonal to the orbits in U. Then if $0 \not\equiv X \in K(M)$, $X^a k_a = 0$ on U and so if $X(m) = 0$, one evaluates $(X^a k_a)_{;b} = 0$ at m to find that the Killing bivector F of X satisfies

$F_{ab}k^b = 0$ at m. Thus by arguments similar to those above, at most three independent bivectors have this property and hence $\dim \overset{*}{K}_m \leqslant 3$ and so $\dim K(M) \leqslant 6$. It is remarked here that, from theorem 10.11 (iii), if O is non-null the map $K(M) \to K(O)$ given by $X \to \tilde{X}$ is injective and so, since $\dim K(O) \leqslant 6$, one again achieves the result without appealing to the dimensional stability of O.

The proofs of (ii) and (iii) are similar and are omitted. The final part of (iii) is just the result that if $X, Y \in K(M)$ and $X = \lambda Y$ for some function λ on some non-empty open subset of M then λ is constant and $X = \lambda Y$ on M [35]. □

It follows from this theorem that if $\dim K(M) > 7$ the only possible orbits are 4-dimensional and so there is a single orbit O equal to M. If $\dim K(M) = 9$ one has $\dim \overset{*}{K}_m = 9 - 4 = 5$ which is impossible since the Lorentz algebra has no 5-dimensional subalgebras. *Thus* $\dim K(M) = 9$ *is impossible* (see end of chapter). If $\dim K(M) = 8$ then $\dim \overset{*}{K}_m = 4$ and theorem 10.4 shows that the tensor E and the Weyl tensor vanish in (7.10) and so M is of constant curvature and admits, *locally*, a 10-dimensional Killing algebra. (It is not clear to the present author that this case is impossible as is sometimes claimed unless, for example, M is simply connected, in which case these local Killing vector fields can be globally extended to give the contradiction $\dim K(M) = 10$ [169], [170].)

It also follows from theorems 10.2 and 10.4 that if $H(M)$ admits proper homothetic vector fields then if $\dim H(M) > 8$ then M is of constant curvature and that $\dim H(M) = 10$ is impossible. In fact, if $\dim H(M) > 8$ then $\dim K(M) > 7$ and M is the only (Killing) orbit and has constant curvature. But this means that for each $m \in M$ there is a proper member of $H(M)$ vanishing at m. To see this, let $X^1, \cdots, X^4 \in K(M)$ such that $X^1(m), \cdots, X^4(m)$ span the Killing orbit at m (i.e. T_mM). If X is proper homothetic there exist $\lambda_1, \cdots, \lambda_4 \in \mathbb{R}$ such that $X(m) = \sum \lambda_i X^i(m)$ and then $X - \sum \lambda_i X^i$ is a non-trivial member of $H(M)$ vanishing at m. Theorem 10.5 then shows that the Ricci scalar is zero and so M is flat.

It also follows from the proof of theorem 10.13 that if O is a *dimensionally stable* orbit through $m \in M$ with $\dim O = 2$ or 3 and if X is a non-trivial member of $K(M)$ with $X(M) = 0$ then the Killing bivector F of X is *simple* at m. Thus *if* $\dim K^*_m = 1$, *the isotropy at* m *is not a screw motion* (section 6.2). This last result (which is inapplicable if $\dim O = 1$) is also true if $\dim O = 4$ but the proof is more involved ([171] - see also [55]).

The proof of theorem 10.13 shows this result to be also true if $\dim O = 2$ with O null and not necessarily dimensionally stable.

Theorem 10.13 suggests that differences will occur in the maximum value for $\dim K(M)$ for a given orbit type depending on whether the orbit is stable or not. The following examples show that, at least in some cases, these differences can actually occur.

Let $M_1 = \mathbb{R}^2$ with positive definite metric g_1 given by $e^{x^2+y^2}(dx^2+dy^2)$ so that $K(M_1)$ is 1-dimensional and spanned by the Killing vector field $x\frac{\partial}{\partial y} - y\frac{\partial}{\partial x}$ which has a single zero at the origin O. Then let $M_2 = (-1,\infty)\times\mathbb{R}$ with Lorentz metric g_2 given by $-dt^2 + 2dtdz + tdz^2$ so that $K(M_2)$ is 1-dimensional being spanned by the nowhere vanishing Killing vector field $\frac{\partial}{\partial z}$ with 1-dimensional orbits which are timelike for $-1 < t < 0$, null for $t = 0$ and spacelike for $t > 0$. Finally, let $M_3 = \mathbb{R}^2$ with the usual 2-dimensional Minkowski metric g_3 so that $K(M_3)$ is 3-dimensional. Now let $M' = M_1 \times M_2$ with Lorentz metric product $g_1 \otimes g_2$. Then (as will be explained in the next section) $K(M')$ is the vector space sum $K(M_1) + K(M_2)$ and is thus 2-dimensional. However, in M', the submanifolds of the form $N' = \{0\} \times N$ where N is any Killing orbit in M_2, are each 1-dimensional Killing orbits in M' which are spacelike (respectively, timelike or null) if N is spacelike (respectively, timelike or null) and dimensionally unstable. All other orbits in M' are 2-dimensional, dimensionally stable and can be spacelike (and stable), timelike (and stable) or null (and unstable). Next let $\tilde{M} = M_1 \times M_3$ with Lorentz metric product $g_1 \otimes g_3$. In this case, $K(\tilde{M})$ is the vector space sum $K(M_1)+K(M_3)$ and is hence 4-dimensional. However, in $\tilde{M}$ the submanifold $\tilde{N} = \{0\} \times M_3$ is a 2-dimensional timelike Killing orbit and all other orbits are 3-dimensional and timelike.

These examples indicate what can happen if orbits (*e.g.* the orbits N' and the orbit $\tilde{N}$) are not stable or dimensionally stable. The map θ in theorem 10.11 (iii), here represented by the map $K(M') \to K(N')$ for N' non-null or $K(\tilde{M}) \to K(\tilde{N})$, is not injective. Essentially, this highlights the fact that *independent Killing vector fields in M need not give rise to independent Killing vector fields on O*. However, it can be shown [167] that a non-trivial Killing vector field on M cannot, in this way, give rise to the zero vector field on any 3-dimensional orbit or on any dimensionally *stable* orbit of whatever dimension. Hence, if O is a non-null stable orbit, then independent Killing vector fields on M give rise to independent Killing vector fields on O. [Thus combining this with theorem 10.11 for any *non-null stable* orbit O it is seen that the map $\theta : H(M) \to H(O)$ (or $K(M) \to K(O)$) given in that theorem is one-to-one.] Further examples can be obtained by considering the following space-time metrics each defined on the manifold $\mathbb{R}^4$ with coordinates (t, x, y, z) (restricted by $t > 0$ in (10.55))

$$ds^2 = -dt^2 + e^{x^2+y^2+z^2}(dx^2 + dy^2 + dz^2) \quad (10.53)$$

$$ds^2 = dz^2 + e^{-t^2+x^2+y^2}(-dt^2 + dx^2 + dy^2) \quad (10.54)$$

$$ds^2 = -dt^2 + t dx^2 + e^{2t}dy^2 + e^{3t}dz^2 \quad (10.55)$$

For each of the space-times (10.53) and (10.54), $\dim K(M) = 4$, with $K(M)$ being spanned by the sets of global Killing vector fields $\{(1,0,0,0), (0,0,z,-y), (0,z,0,-x), (0,y,-x,0)\}$ and $\{(0,0,0,1), (x,t,0,0), (y,0,t,0), (0,-y,x,0)\}$, respectively. Then (10.53) admits the 1-dimensional timelike orbit $x = y = z = 0$ and (10.54) admits the 1-dimensional spacelike orbit $t = x = y = 0$. Clearly these orbits are neither stable nor dimensionally stable since all other orbits are 3-dimensional and stable, being timelike for (10.53) and timelike, spacelike or null for (10.54). The metric (10.55) has $\dim K(M) = 3$ with $K(M)$ spanned by $\frac{\partial}{\partial x}$, $\frac{\partial}{\partial y}$ and $\frac{\partial}{\partial z}$ and with (stable) orbits given by the 3-dimensional spacelike submanifolds of constant t. Each orbit O is flat in its induced geometry and has $\dim K(O) = 6$. This is an example where the map θ in theorem 10.11 (iii) is not onto.

The examples (together with some obvious variants) of the last two paragraphs show that many of the possibilities left open for unstable orbits in theorem 10.13 can occur. Each of the possibilities for *stable* or *dimensionally stable* orbits in this theorem can be achieved [55] with the following exceptions. If O is 3-dimensional and spacelike (and hence stable) then $\dim K(M) \neq 5$. This follows because if $\dim K(M) = 5$ then for $m \in O$, $\dim \overset{*}{K}_m = 5 - 3 = 2$. But $\overset{*}{K}_m$ would then be a 2-dimensional subalgebra of the Lie algebra of $SO(3)$ which is impossible. The type **N** plane waves discussed earlier (see also section 10.7) reveal examples with $\dim K(M) = 5$ and with O 3-dimensional, null and stable. The situation with $\dim K(M) = 5$ and O 3-dimensional and timelike is not clear since the existence of such an example would lead for $m \in O$ to $\overset{*}{K}_m$ being a 2-dimensional subalgebra of the 3-dimensional Lorentz group and such subalgebras exist. However, it would force O to be of constant curvature in its induced geometry and since O is stable (theorem 10.12(ii)) the non-empty open subset $T_3 \subseteq M$ of all orbits of this type can be shown [167] to have the property that for m in some open dense subset of T_3 there is an open subset U in M such that $m \in U \subset T_3$ and $\dim K(U) \geq 6$. In fact U is conformally flat, being a "spacelike version" of the F.R.W. metric (10.84).

10.6 Space-Times admitting Proper Affine Vector Fields

If a space-time M admits a proper affine vector field then, from section 10.2, M admits a global second order symmetric tensor field which is covariantly constant but which is not *trivial* (*i.e.* is not a constant multiple of the metric g). The set S of all second order symmetric covariantly constant tensors on M was discussed in section 9.2. Now suppose for the remainder of this section that M *is simply connected*. (If one is working locally one may always *assume* a local coordinate domain is chosen simply connected). Then from theorem 9.2 it follows that, since one requires $\dim S > 1$, the holonomy type of M is either $R_2, R_3, R_4, R_6, R_7, R_8, R_{10}, R_{11}$ or R_{13} (the flat case R_1 is evident from (10.19)). Space-times of these types can now be discussed. First, the relation between the holonomy type and the dimension of S given in theorem 9.2 gives rise to the following result. Let $X_1, \cdots, X_k$ be affine vector fields on M and suppose the members of S with which they are associated are $h_1, \cdots, h_k$. Suppose the set $\{g, h_1, \cdots, h_k\}$ is a dependent set in the finite-dimensional vector space S. Then there exist $\alpha, \alpha_1, \cdots, \alpha_k \in \mathbb{R}$ not all zero such that $\alpha g + \sum \alpha_i h_i$ is the zero member of S. Thus $\sum \alpha_i X_i$ is a member of $A(M)$ satisfying (10.15) with $h = -\alpha g$ and is hence homothetic or, if $\alpha = 0$, Killing. The next theorem now follows [101], [172]

Theorem 10.14 *Let M be a simply connected space-time. If the holonomy type of M is R_9, R_{12}, R_{14} or R_{15} (and these are collectively generic since the type R_{15} is - see theorem 8.9) no proper affine vector fields exist on M (i.e. $A(M) = H(M)$).If the holonomy type of M is $R_6, R_7, R_8, R_{10}, R_{11}$ or R_{13} and if two (or more) affine vector fields exist on M, some linear combination of them is homothetic. If the holonomy type of M is R_2, R_3 or R_4 and if four (or more) affine vector fields exist on M, some linear combination of them is homothetic.*

To say this another way, let S' denote the subset (clearly a subspace) of members of S which arise from an affine vector field through (10.8). Let $f : A(M) \to S$ be the linear map associating $X \in A(M)$ with $h \in S$. Then S' is the range of f and $K(M)$ the kernel of f. It follows that S' is isomorphic to the quotient space $A(M)/K(M)$ and so if $\dim S' = m_1$ and $\dim A(M) = r(\geqslant m_1)$ then $\dim K(M) = r - m_1$. Thus if $H(M) = K(M)$ then in any basis of $A(M)$ one may, by taking appropriate linear combinations, choose m_1 proper affine and $r - m_1$ Killing vector fields. In addition, if $g \in S'$ (equivalently, if M admits a proper homothetic vector field) one may (theorem 10.2 (iii)) arrange from such a basis for $A(M)$ that $m_1 - 1$ are proper affine and $r - m_1 + 1$ homothetic (of which $r - m_1$ may be chosen Killing). It is remarked that S and S' may, but need not, be equal. It is also clear from the discussion following theorem 10.13 that if $\dim A(M) = 20$, M is flat.

Now let (M, g) be a simply connected space-time of holonomy type R_{13} and let u be the global timelike covariantly constant vector field on M uniquely defined by it being unit and future pointing (see section 7.1) so that in any coordinate system, $u^a u_a = -1$ and $u^a{}_{;b} = 0$. Suppose X is an affine vector field on M. Then use of table 9.1 for the possibilities for the covariantly constant tensor field h in (10.8) gives in any coordinate system

$$X_{a;b} = \alpha g_{ab} + \beta u_a u_b + F_{ab} \tag{10.56}$$

where $\alpha, \beta \in \mathbb{R}$ and F is the affine bivector. Define a global smooth function κ on M by $\kappa = g(X, u)$. Then in coordinate form, $\kappa = X^a u_a$ and the Ricci identity and (10.7) give

$$u^a{}_{;b} = 0 \Rightarrow R_{abcd} u^a = 0 \Rightarrow \kappa_{;ab} = 0 \tag{10.57}$$

Hence $d\kappa$ (with components $\kappa_a = \kappa_{,a}$) is a global covariantly constant 1-form on M and so by uniqueness (*i.e.* $\dim V = 1$ in theorem 9.2) in any coordinate system $\kappa_a = \gamma u_a (\gamma \in \mathbb{R})$. But $\kappa_a u^a = X^b{}_{;a} u_b u^a = \beta - \alpha$ and so $\gamma = \alpha - \beta$ and $\kappa_a = (\alpha - \beta) u_a$. Then comparing this result with the expression for κ_a calculated from definition using (10.56) gives $F_{ab} u^b = 0$ and so F, if not zero, is a simple spacelike bivector on M. Next construct global vector fields k and k' on M by

$$k' = \kappa u, \quad k = X + k' \tag{10.58}$$

so that k is orthogonal to u, $g(u, k) = 0$. Then in any coordinate system

$$k'^a{}_{;b} = (\alpha - \beta) u^a u^b \quad (\Rightarrow k'^a{}_{;bc} = 0 = R^a{}_{bcd} k'^d) \tag{10.59}$$

$$k^a{}_{;b} = \alpha(\delta^a{}_b + u^a u^b) + F^a{}_b \quad (\Rightarrow k^a{}_{;bc} = F^a{}_{b;c} = R^a{}_{bcd} X^d = R^a{}_{bcd} k^d) \tag{10.60}$$

(where (10.15) has been used) and so k and k' are global affine vector fields on M by (10.7). From the work on holonomy invariant distributions in chapter 8, it is clear that there exists through each $m \in M$ a 3-dimensional submanifold arising from the holonomy invariant distribution defined by the subspace of $T_m M$ orthogonal to $u(m)$. The vector field u is everywhere orthogonal, and the vector field k is everywhere tangent, to such a submanifold. These submanifolds are everywhere spacelike and have positive definite metric $q = i^*g$ from g where i is the associated inclusion map. The decomposition of X into k and k' is now seen to be the decomposition of X "tangent" and "orthogonal" to these hypersurfaces (section 4.11).

Now let $m \in M$ and let (t, x^α) be one of the special coordinate systems of M described in section 8.3 with domain U and in which the metric takes the form (8.3). Then the vector field u can be taken to restrict, on U, to the vector field $\frac{\partial}{\partial t}$ with components $(1, 0, 0, 0)$ and the above holonomy invariant submanifolds' intersection with U are described by $t =$ constant and their induced metric q has components $q_{\alpha\beta}$ $(\alpha, \beta = 1, 2, 3)$ in the chart with coordinates x^α which is naturally induced on them by the original chart (t, x^α) on U. It is easily checked that the components $q_{\alpha\beta}$ equal the 'space' components of the tensor $g_{ab} + u_a u_b$. Since $F_{ab}u^b = 0$, one has in these coordinates $F_{a0} = 0$ and the components $F_{\alpha\beta}$ are the components of the skew-symmetric tensor $G = i^*F$ in the hypersurfaces $t =$ constant. The hypersurfaces $t = t_1$ and $t = t_2$ $(t_1, t_2 \in \mathbb{R})$ in U are isometric with respect to their induced metrics under the isometry $\theta : (t_1, x^\alpha) \to (t_2, x^\alpha)$. The Christoffel symbols and curvature tensor components have the properties described in the discussion following (8.3).

Now k is everywhere tangent to these holonomy invariant hypersurfaces and so (section 4.11) a global smooth vector field $\tilde{k}$ exists in any of them such that $i_*\tilde{k} = k$ and whose components in the above chart satisfy $\tilde{k}^\alpha = k^\alpha$ (more precisely, the hypersurface vector field $\tilde{k}$ is "independent" of the hypersurface). This is because if θ is the above isometry linking two such hypersurfaces and $\tilde{k}_1$ and $\tilde{k}_2$ are the associated vector fields in these hypersurfaces, then $\theta_*\tilde{k}_1 = \tilde{k}_2$). But then, if one denotes a covariant derivative in the hypersurfaces by a stroke, (10.60) becomes

$$\tilde{k}_{\alpha|\beta} = \alpha q_{\alpha\beta} + G_{\alpha\beta} \tag{10.61}$$

and so $\tilde{k}$ is a *homothetic* (possibly Killing) vector field in the hypersurface geometry. Further, if $\tilde{k}$ is identically zero on any of these hypersurfaces then, from (10.61), G and hence F would vanish on this hypersurface and $\alpha \equiv 0$ on M. Thus k would vanish on this hypersurface and, from (10.60), so also would the covariant derivative of k. Thus k is a *Killing* vector field on M which vanishes together with its covariant derivative at some point of the hypersurface and then theorem 10.2 shows that $k \equiv 0$ on M. It follows that either $k \equiv 0$ on M (and hence $X = -k' = -\kappa u$) or each such hypersurface admits a non-trivial homothetic vector field. Since these hypersurfaces will, 'in general', not admit such a vector field, the general situation here is that the only affine vector fields are parallel to u. One can say more here. Since M is simply connected, there exists a global function $f : M \to \mathbb{R}$ such that the covariant vector field $\tilde{u}$ associated with u (with local components u_a) satisfies $\tilde{u} = df$ (section 4.16). It follows (section 4.11, example (v)) that the surfaces of constant f are submanifolds (hypersurfaces) of M orthogonal to u and coincide with the above hypersurfaces. It then follows that the vector field $X = fu$ on M satisfies (10.56) with $\alpha = 0, F = 0, \beta = 1$ in any coordinate system, and so is *proper affine* on M. In fact, for this vector field X, $\kappa = -f$ (and in the above special coordinate systems, the surfaces of constant f and t coincide). From this it follows that the Lie algebra $A(M)$ of affine vector fields on M consists of vector fields of the form $(\mu + \nu f)u + k$, where $\mu, \nu \in \mathbb{R}$ and k is tangent to the surfaces of constant f and gives rise to a homothetic vector field in their induced geometries. Also, from theorem 10.13, any affine vector field on M is a linear combination of fu and a homothetic vector field on M. Further, since u and fu are independent, $\dim A(M) \geqslant 2$.

Now let U be a special coordinate domain $U = I \times V$ where I is an open interval of $\mathbb{R}$, V is a connected 3-dimensional manifold with positive definite metric $q = i^*g$ and i is any of the inclusions $V_t \to M$ with $V_t = t \times V, t \in I$. If $\tilde{X}$ is a homothetic vector field on (any of the isometric submanifolds) V_t, let X be the associated vector field on U which is everywhere tangent to V (i.e. to the appropriate V_t) and satisfies $i^*\tilde{X} = X$ and also $X^a u_a = 0$. Then it follows that X is affine on U. To see this, let ϕ_r and ψ_s be the local flows of X and u, respectively. Then clearly, $\phi_r \circ \psi_s = \psi_s \circ \phi_r$ and hence from (5.14) $\mathcal{L}_u X = [u, X] = 0$. Now $\mathcal{L}_{\tilde{X}} q = 2cq$ ($c \in \mathbb{R}$) and so applying (5.15) one finds

$$i^*(\mathcal{L}_X g) = \mathcal{L}_{\tilde{X}}(i^*g) = \mathcal{L}_{\tilde{X}} q = 2cq = i^*(2cg) \tag{10.62}$$

Thus $i^*(\mathcal{L}_X g - 2cg) = 0$ and it is easily checked from the definition of i^* that the only solution to this equation is

$$\mathcal{L}_X g - 2cg = u_a v_b + v_a u_b \tag{10.63}$$

where v is a covector field on U. Thus

$$X_{a;b} = cg_{ab} + u_{(a}v_{b)} + F_{ab} \tag{10.64}$$

for some bivector field F on U. Since $u^a{}_{;b} = 0$, $X^a u_a = 0$ and $[X, u] = 0$ one has $u_a X^a{}_{;b} = X^a{}_{;b} u^b = 0$. Then contractions of (10.64) with u^a and

u^b, respectively, show that $F_{ab}u^b = 0$ whilst a contraction with $u^a u^b$ shows that $v = (2c + \delta)u$ with $\delta \equiv u^a v_a = -c$. Thus $v = cu$ and then (10.64) and (10.15) show that X is affine on U (and, in fact, the vector field $X - cfu$ on M is homothetic on M with homothetic constant c). It follows that the vector field $(a + bt)u + X$ is affine on U for $a, b \in \mathbb{R}$. In fact, if u is the only (up to a constant scaling) covariantly constant vector field on U (and it may not be since although M is of holonomy type R_{13}, U may admit a further holonomy reduction) the above argument and the preceding one show that the Lie algebra $A(U)$ of affine vector fields on U consists of precisely the vector fields of the form $(a + bt)u + X$, where X arises from a homothetic vector field on V as described above. Thus in this case $A(U) = A' \oplus H(V)$ where A' is the Lie algebra of vector fields spanned by $u(\equiv \frac{\partial}{\partial t})$ and $t\frac{\partial}{\partial t}$ and $H(V)$ is the homothety algebra of V, and $\dim A(U) = 2 + \dim H(V)$.

Now consider the linear map $f : A(M) \to A(U)$ obtained by restricting affine vector fields on M to U. By theorem 10.1 (ii), f is injective and so $\dim A(M) \leqslant \dim A(U)$ for each such U. Hence, $2 \leqslant \dim A(M) \leqslant 2 + \dim H(V)$ for all possible hypersurfaces $t = t_0$ in all possible special neighbourhoods U. The inequality essentially results from the possibility of a member of $A(U)$ not being globally extendible to M (see section 10.7). The fact that the holonomy type of M is R_{13} requires that M is not flat. Hence, there exists a domain such as U above in which the curvature tensor is nowhere the zero tensor and thus the induced curvature tensor on the hypersurfaces of constant t is nowhere the zero tensor. It follows that for this domain U, $\dim H(V) \leqslant 6$ and so $2 \leqslant \dim A(M) \leqslant 8$. Each extreme in this inequality can be achieved by $\dim A(M)$ by choosing N in the following paragraph appropriately .

If, in addition to be being simply connected M is also geodesically complete, one has a global metric product $M = \mathbb{R} \times N$ (theorem 8.8) with N a 3-dimensional, connected, simply connected, geodesically complete manifold admitting a positive definite metric (and noting that $\mathbb{R}$ is the only 1-dimensional, connected, paracompact, simply connected, geodesically complete manifold [21]).

A similar argument to that above then shows that $A(M) = A_1 \oplus H(N)$, where A_1 is the Lie algebra spanned by u and fu. Thus $\dim A(M) = 2 + \dim H(N)$. Since N has a positive definite metric and is geodesically complete, it follows that N admits no *proper* homothetic vector fields [20] (see the end of section 10.4). Thus $H(N) = K(N)$ and $A(M) = A_1 \oplus K(N)$. Also M cannot admit a proper homothetic vector field otherwise, from

(10.56) and (10.61), N would. Another consequence of M being geodesically complete is that each member of $A(M)$ is a complete vector field [20]. It then follows from Palais' theorem 5.10 that *the Lie algebras $K(M), H(M)$ and $A(M)$ arise from connected Lie group actions on M.*

Returning to the general case with M being simply connected but not necessarily geodesically complete, one can easily modify the above arguments to show that if $X \in K(M)$ then, in the special coordinate domain U (again assumed to admit only one independent covariantly constant vector field u) $\alpha = \beta = 0$ in (10.56) and $X = k - k'$ as in (10.58) with $k' = \kappa u$ ($\kappa \in \mathbb{R}$) and $\tilde{k} \in K(V)$ from (10.61). Also, every member of $K(V)$ gives rise to a member of $K(U)$ in this way. Thus $K(U) = K(V) \oplus \mathbb{R}$ and $\dim K(U) = \dim K(V) + 1$. One can also see that $1 \leqslant \dim K(M) \leqslant \dim K(U) = \dim K(V) + 1$. Hence $1 \leqslant \dim K(M) \leqslant 7$ and by appropriate choice of N in an above paragraph each extreme is possible. If M is also geodesically complete so that $M = \mathbb{R} \times N$ as above, then $K(M) = K(N) \oplus \mathbb{R}$ and $\dim K(M) = 1 + \dim K(N)$. It is remarked that if $X \in K(M)$ then $g(X, u)$ is constant on M and $\{X \in K(M) : g(X, u) = 0\}$ is a subalgebra of $K(M)$ Lie isomorphic to $K(N)$.

Now suppose that M is a simply connected space-time of holonomy type R_3, R_6 or R_{10}; then a (not necessarily unique) global covariantly constant spacelike vector field x results. The analysis of $A(M)$ in this case is similar to the R_{13} case above but the following two points should be borne in mind. Firstly, in the R_3 or R_6 cases, the hypersurfaces orthogonal to x exhibit further holonomy decomposition in that M admits a recurrent or constant null vector field orthogonal to x. Secondly, these hypersurfaces are now timelike with induced Lorentz metric and may admit proper homothetic vector fields. More details of this and other aspects of this section can be found in [101], [172], [161].

Now let M be a simply connected space-time of holonomy type R_7. Then from section 8.3 one has the existence of two global recurrent vector fields l and n on M scaled so that $g(l, n) = 1$ on M. About any $m \in M$, one can introduce a special coordinate neighbourhood U for this holonomy type and a global null tetrad l, n, x, y on U. In this type, the vector space of covariantly constant second order symmetric tensors on M has dimension two and (restricted to U) is spanned by any two of the tensors P, Q and the metric g, where on each such U

$$P_{ab} = 2l_{(a}n_{b)}, \quad Q_{ab} = x_a x_b + y_a y_b, \quad g_{ab} = P_{ab} + Q_{ab} \tag{10.65}$$

It should be noted that since g, l and n are globally defined, so also are P and Q. The curvature tensor on U takes the form (sections 8.3 and 9.3)

$$R_{abcd} = aM_{ab}M_{cd} - b^*M_{ab}{}^*M_{cd} \quad \left(M_{ab} = 2l_{[a}n_{b]}, \quad {}^*M_{ab} = 2x_{[a}y_{b]}\right) \tag{10.66}$$

where M and $^*M$ are covariantly constant and where a depends only on the coordinates in the timelike holonomy-invariant submanifold of decomposition in U and b on the coordinates of the corresponding spacelike submanifold.

Now let X be an affine vector field on M. Then from (10.8) and the previous paragraph, one has in U

$$X_{a;b} = \alpha g_{ab} + 2\beta l_{(a}n_{b)} + F_{ab} = (\alpha + \beta)P_{ab} + \alpha Q_{ab} + F_{ab} \tag{10.67}$$

where $\alpha, \beta \in \mathbb{R}$ and F is the affine bivector. Now since P and Q are globally defined on M, one can define global vector fields k and k' by the (everywhere) local relations $k^a = P^a{}_bX^b$ and $k' = Q^a{}_bX^b$ so that $X = k + k'$. Then one can compute the derivatives of k using (10.67) and then (10.10) and (10.66)

$$k_{a;b} = (\alpha + \beta)P_{ab} + P_{ac}F^c{}_b \tag{10.68}$$

$$k_{a;bc} = P_{ad}R^d{}_{bce}X^e = aP_{ad}M^d{}_bM_{ce}k^e = R_{abcd}k^d \tag{10.69}$$

Equation (10.69) shows that k *is also an affine vector field on* M and that $k_{(a;b)}$ is covariantly constant and symmetric on M. Hence, there exist $\mu, \nu \in \mathbb{R}$ such that

$$k_{(a;b)} = \mu g_{ab} + \nu P_{ab} \tag{10.70}$$

Substituting this into (10.68) and contracting with x^b (respectively y^b) shows that $\mu = 0$ on M and that $F^a{}_bx^b$ (respectively $F^a{}_by^b$) lies in the span of x and y at each point of U. Thus $F_{ab}l^ax^b = F_{ab}l^ay^b = F_{ab}n^ax^b = F_{ab}n^ay^b = 0$ and so, on U

$$F_{ab}x^b = -\rho y_a, \quad F_{ab}y^b = \rho x_a, \quad F_{ab}l^b = \sigma l_a, \quad F_{ab}n^b = -\sigma n_a \tag{10.71}$$

for functions ρ and σ on U. A consistency check using (10.68) and (10.70) reveals that $\nu = \alpha + \beta$ on M. From (10.71) F, if not zero, is a non-null bivector whose canonical pair of blades are spanned by the pairs (l, n) and (x, y) at each point of U and so on M

$$F_{ab} = \sigma P_a{}^cF_{cb} + \rho Q_a{}^cF_{cb} \tag{10.72}$$

On substituting this into (10.68) one finds

$$k_{a;b} = (\alpha + \beta)P_{ab} + \sigma P_a{}^c F_{cb} \tag{10.73}$$

A similar analysis for the vector field k' reveals

$$k'_{a;bc} = R_{abcd}k'^d, \quad k'_{a;b} = \alpha Q_{ab} + \rho Q_a{}^c F_{cb} \tag{10.74}$$

and hence k' is also affine on M. Thus the original affine vector field X decomposes uniquely (in the above sense) into affine vector fields k and k' on M which are, respectively, tangent to the timelike and spacelike submanifolds of decomposition and $[k, k'] = 0$. It follows from (10.73) and (10.74) and remarks after (8.5) that, in the respective induced submanifold geometries, k and k' give rise in a similar way to the previous case to homothetic vector fields. Since in general these submanifolds will not admit non-trivial homothetic vector fields, one sees that *M will not, in general, admit a non-trivial affine vector field*. Now consider the special coordinate domain U which itself is the metric product of 2-dimensional submanifolds U_1 and U_2 with U_1 timelike and U_2 spacelike (and $U = U_1 \times U_2$) and where it is again assumed that U has no further holonomy reduction, that is, U has holonomy type R_7. For any affine vector field X on M, its restriction to U can be decomposed, as above, into the sum of two vector fields which may be regarded as homothetic vector fields in U_1 and U_2. Conversely, if $\tilde{X}_1$ and $\tilde{X}_2$ are homothetic vector fields in U_1 and U_2, respectively, one can use the method employed in the holonomy type R_{13} case, with the inclusion maps $U_1 \to U$, $U_2 \to U$, to show that the associated vector fields X_1 and X_2 in U tangent to U_1 and U_2, respectively, are each affine vector fields on U. Hence $X \equiv X_1 + X_2$ is affine and so $A(U)$ is isomorphic to $H(U_1) \oplus H(U_2)$ and for any such U, $\dim A(U) = \dim H(U_1) + \dim H(U_2)$. Further, the restriction (linear) map $A(M) \to A(U)$ discussed in the previous case is again injective. Also, if $X \in K(M)$ then $\alpha = \beta = 0$ in (10.67) and (10.73) and (10.74) reveal induced members of $K(U_1)$ and $K(U_2)$ and a repeat of the above argument shows that if $\tilde{X} \in K(U_1)$ or $K(U_2)$, the associated $X \in K(U)$. Thus, $K(U)$ is isomorphic to $K(U_1) \oplus K(U_2)$. Now if W is a connected 2-dimensional manifold with metric h of any signature, $\dim K(W) \leqslant 3$ and if this maximum is achieved W is of constant curvature. Also if $\dim K(W) = 2$ then a third independent local Killing vector field is admitted by W and again W is of constant curvature [35]. Use of theorem 10.2 and an obvious modification of theorem 10.5 then shows that if W admits a proper homothetic vector field and if $\dim H(W) \geqslant 3$ then each point of W is a zero of a proper member of $H(W)$ and W is flat. Since neither U_1 nor U_2 is flat, these remarks together with the preceding results combine to show that $A(U)$, $A(M)$, $K(U)$ and $K(M)$ all have dimension at most six and that if a *proper* member of $A(U)$ exists, one of $H(U_1)$ or $H(U_2)$ admits a proper member and so $\dim A(U) \leqslant 5$. It is noted also that if a proper member of $H(U)$ exists the respective projections onto U_1 and U_2 are proper members of $H(U_1)$ and $H(U_2)$ with the same homothetic constant, and conversely, and so in this case $\dim H(U) \leqslant 4$. [In fact, one can easily check, following the ideas of chapter 9, that for a 2-dimensional *non-flat* manifold W, $A(W) = H(W)$].

The situation when M is geodesically complete can be discussed in a similar way to that of the previous example. One finds that M is the metric product $N_1 \times N_2$ where N_1 is a timelike and N_2 a spacelike 2-dimensional geodesically complete manifold and $A(M)$ is isomorphic to $H(N_1) \oplus H(N_2)$. Also, $A(M)$ consists of complete vector fields and (theorem 5.10) arises from a connected Lie group action on M. Similar comments apply to the Killing structure and $K(M)$ is isomorphic to $K(N_1) \oplus K(N_2)$. The dimension results above can now be applied but it is remarked that a geodesically complete 2-dimensional manifold with positive definite metric and which is non-flat cannot admit a proper homothetic vector field [20].

Finally, consider the case when M is a simply connected space-time with holonomy type R_4 (see 8.7). The construction is similar to the previous cases and need only be briefly sketched. In this case, one has independent global covariantly constant vector fields u and x with u timelike and x spacelike. So the associated covector fields $\tilde{u}$ and $\tilde{x}$ satisfy $\tilde{u} = df$ and $\tilde{x} = dh$ where f and h are functions on M. Thus u, x and $hu - fx$ are independent members of $K(M)$. The vector space of covariantly constant symmetric tensor fields on M is 4-dimensional, being spanned by the tensors $g, \tilde{u} \otimes \tilde{u}, \tilde{x} \otimes \tilde{x}$ and $\tilde{u} \otimes \tilde{x} + \tilde{x} \otimes \tilde{u}$. Three independent *proper* affine vector fields are always admitted (no linear combination of which is homothetic) and can be chosen as fu, hx and $hu + fx$. The associated covariantly constant tensor fields are the last three listed here. Any other proper affine vector field is a linear combination of these and a homothetic vector field on M (theorem 10.14).

Let X be an affine vector field on M. Then in any coordinate system of M from the definition of an affine vector field and table 9.1

$$X_{a;b} = \alpha g_{ab} + \beta u_a u_b + \gamma x_a x_b + 2\delta x_{(a} u_{b)} + F_{ab} \tag{10.75}$$

where F is the affine bivector and $\alpha, \beta, \gamma, \delta \in \mathbb{R}$. Define global functions κ and κ' on M by $\kappa = g(X, x)$ and $\kappa' = -g(X, u)$ and then define global vector fields k and k' on M by

$$X = k + k', \quad k' = \kappa x + \kappa' u \tag{10.76}$$

so that k is orthogonal to u and x, $g(k, u) = g(k, x) = 0$. Then $\kappa_{;ab} = \kappa'_{a;b} = 0$ and if F is not zero at $m \in M$, it is non-null there with one canonical blade spanned by u and x and the other being its orthogonal complement. Then (10.75) and (10.76) give

$$k'_{a;b} = (\alpha + \gamma)x_a x_b + (\beta - \alpha)u_a u_b + 2\delta x_{(a}u_{b)} + F'_{ab} \tag{10.77}$$

$$k_{a;b} = \alpha(g_{ab} + u_a u_b - x_a x_b) + H_{ab} \tag{10.78}$$

where F' and H are simple bivectors whose blades are orthogonal, that of F' being spanned by u and x where F' is not zero. Then $[k, k'] = 0$ and

$$k'^a{}_{;bc} = 0 (= R^a{}_{bcd}k'^d), \quad k^a{}_{;bc} = X^a{}_{;bc} = R^a{}_{bcd}X^d = R^a{}_{bcd}k^d \tag{10.79}$$

and so k and k' are affine vector fields on M. Also, k is tangent to the holonomy invariant submanifolds orthogonal to u and x (*i.e.* of constant f and h) and gives rise to a homothetic vector field in the induced geometry on them. Also, since $\kappa_{a;b} = \kappa'_{;ab} = 0$, one sees that $\kappa = \mu f + \nu h + \rho$ and $\kappa' = \mu' f + \nu' h + \rho\prime$ with $\mu, \nu, \rho, \mu', \nu', \rho\prime \in \mathbb{R}$ and a general expression for any $X \in A(M)$ now emerges from (10.76). In fact, from the material in chapter 8 one may choose an open neighbourhood $U = U_1 \times U_2$ of any $m \in M$ in an obvious notation with U_1 flat and u and x tangent to U_1. Then $A(U) = A(U_1) \oplus H(U_2)$ and $K(U) = K(U_1) \oplus K(U_2)$ and one has $6 \leqslant \dim A(U) \leqslant 9$, $3 \leqslant \dim K(U) \leqslant 6$, $6 \leqslant \dim A(M) \leqslant 9$ and $3 \leqslant \dim K(M) \leqslant 6$. If M is geodesically complete, then $A(M)$ consists of complete vector fields and arises from a connected Lie group action as before.

The situation for the holonomy type R_2 is very similar.

If the space-time M is simply connected and has holonomy type R_3, R_4, R_8 or R_{11}, a covariantly constant null vector field l exists on M and the associated covector field $\tilde{l}$ satisfies $\tilde{l} = df$ for some smooth function $f : M \to \mathbb{R}$. In this case, $X \equiv fl$ is a proper affine vector field on M. If no other independent covariantly constant vector fields are admitted by M (*i.e.* if the holonomy type is R_8 or R_{11}) then any other affine on M is a linear combination of X and a homothetic vector field on M. If another (independent) covariantly constant vector field is admitted (holonomy types

R_3 or R_4) then the situation is one of the cases discussed earlier.

10.7 Examples and Summary

The relationship between the proper affine symmetry a space-time may or may not possess, its holonomy type and its curvature rank and class (chapter 9) shows immediately that many well known space-times can not admit a proper affine vector field. In fact, it follows from the definition of an affine vector field and theorem 9.2 (ii) that space-times of holonomy type R_9, R_{12}, R_{14} and R_{15} cannot admit any proper affine vector field. It also follows that a necessary condition for M to admit a proper affine vector field is that its curvature tensor has rank at most three at each $m \in M$ otherwise its holonomy type is R_{14} or R_{15}. Theorems 8.10 and 8.11 are helpful in this respect if the energy-momentum tensor is known. For example, if M represents a non-flat vacuum space-time then theorem 8.10 shows that the holonomy type must be R_8 in order that a proper affine vector field be admitted. This forces M to admit a global covariantly constant null vector field and so M is locally a (vacuum) pp-wave (c.f. Trumper's theorem in [41]). That one actually exists follows from the last paragraph of the last section.

It is remarked here in passing that there is a close relationship between those space-times which admit a proper affine vector field X and those whose Levi-Civita conections do not determine their metric up to a constant conformal factor. This can be seen from the discussion in section 10.1 using a local flow ϕ_t of X. If $\phi_t \to \phi_t(U)$ for some open subset U of M then, since X is proper, the metrics g and $\phi_t^* g$ on U are not related by a constant conformal factor but give rise to the same Levi-Civita connection on U [101].

Returning to some examples of the material of this chapter (and recalling from chapter 1 that some elementary knowledge of exact solutions of Einstein's equations is assumed, and is well documented, for example, in [55], [173], [48] , [63]), consider the exterior Schwarzschild (vacuum) metric in the conventional (t, r, θ, ϕ) coordinates

$$ds^2 = -(1 - \frac{2m}{r})dt^2 + (1 - \frac{2m}{r})^{-1}dr^2 + r^2(d\theta^2 + \sin^2\theta d\phi^2) \qquad (10.80)$$

where m is a positive constant representing the mass of the central particle and $r > 2m$. Here the Petrov type is **D** everywhere and the Killing algebra is 4-dimensional and spanned by

$$\frac{\partial}{\partial t}, \quad \frac{\partial}{\partial \phi}, \quad \sin\phi\frac{\partial}{\partial\theta} + \cos\phi\cot\theta\frac{\partial}{\partial\phi}, \cos\phi\frac{\partial}{\partial\theta} - \sin\phi\cot\theta\frac{\partial}{\partial\phi} \tag{10.81}$$

The first of these, k $(= \frac{\partial}{\partial t})$, is timelike and the others, representing the spherical symmetry of (10.80), are spacelike. The first two of (10.81) indicate that the metric (10.80) is independent of t and ϕ. Also $k_a = (\frac{2m}{r} - 1)t_{,a}$ and so k is hypersurface orthogonal (section 4.16). The Killing orbits are 3-dimensional submanifolds of constant r and are timelike and stable. If one considers the extension of (10.80) to $r > 0$ given by the advanced Eddington-Finkelstein metric (see, e.g. [48]) in the usual (v, r, θ, ϕ) coordinates

$$ds^2 = -(1 - \frac{2m}{r})dv^2 + 2dvdr + r^2(d\theta^2 + \sin^2\theta d\phi^2) \tag{10.82}$$

the above remarks still apply with obvious trivial changes (and $\frac{\partial}{\partial v}$ and $\frac{\partial}{\partial t}$ agree on the region $r > 2m$ after the coordinate change $v = r + t + 2m\log(r - 2m)$) except that the (3-dimensional) Killing orbits are timelike and stable for $r > 2m$, null and unstable for $r = 2m$ and spacelike and stable for $r < 2m$. Further, the holonomy type is R_{15} and no proper affine vector field can be admitted. It also follows [174] (see chapter 11) that no proper homothetic vector field is admitted. Hence for this metric, $A(M) = K(M)$ and is spanned by (10.81).

A similar analysis may be carried out for the Reissner-Nordstrom metric representing the space-time of a spherically symmetric particle with mass m and charge e (a non-null Einstein-Maxwell field) and given by

$$ds^2 = -(1 - \frac{2m}{r} + \frac{e^2}{r^2})dt^2 + (1 - \frac{2m}{r} + \frac{e^2}{r^2})^{-1}dr^2 + r^2(d\theta^2 + \sin^2\theta d\phi^2) \tag{10.83}$$

for (say) either $e^2 > m^2$ or $e^2 \leqslant m^2$ *and* $r > m + \sqrt{m^2 - e^2}$. That no proper homothetic vector fields exist was shown in [175] (see chapter 11) and the fact that the curvature tensor, generically, has rank six (holonomy type R_{15}), together with the remarks above, show that no proper affine vector fields exist. Thus $A(M) = K(M)$ and $K(M)$ is as in (10.81).

The Kerr metric (see *e.g.* [48]), for which $\dim K(M) = 2$, can admit no proper homothetic vector field [176] and being a vacuum metric, it follows again from remarks above that no proper affine vector field is admitted. For the de Sitter and anti-de Sitter metrics, the constant curvature condition means that the curvature rank is six everywhere and so there are no proper affine vector fields. That there are no proper homothetic vector fields follows from the fact that since, in this case, M has constant curvature, for each $m \in M$ there exists an open neighbourhood U of m such that $\dim K(U) = 10$. Also there are $X_1, \cdots, X_4 \in K(U)$ such that $X_1(m), \cdots, X_4(m)$ span $T_m M$. Thus, if X is proper homothetic, then either $X(m) = 0$ or there exist $\alpha_i \in \mathbb{R}$ $(1 \leqslant i \leqslant 4)$ such that $X(m) = \sum \alpha_i X_i(m)$. Then $Y \equiv X - \sum \alpha_i X_i$ is proper homothetic and $Y(m) = 0$. Thus for each $m \in M$, there is a proper homothetic vector field which vanishes at m. Theorem 10.5 (iii) then gives the contradiction that the (constant) curvature of M is zero. The Gödel metric (8.15) has holonomy type R_{10} and the Einstein static metric (8.14) has holonomy type R_{13} and hence may admit proper affine vector fields in accordance with section 10.6. For the Gödel metric, $\dim K(M) = 5$ with $\frac{\partial}{\partial z}$ covariantly constant in (8.15) and $K(M)$ is transitive with a single orbit M. For each $m \in M$, $\dim \overset{*}{K}_m = 1$ and the Petrov type is **D** everywhere. The Einstein static metric also has a transitive algebra $K(M)$ with $\dim K(M) = 7$, $\dim \overset{*}{K}_m = 3$ for each $m \in M$ and $\frac{\partial}{\partial t}$ covariantly constant. It is conformally flat from theorem 10.4. Since in each of these cases the Killing algebra is transitive, there can be no proper homothetic vector field. This follows because the existence of such a vector field would, upon taking appropriate linear combinations with members of $K(M)$ as before, mean that each point would be the zero of some proper homothetic vector field. Theorem 10.5 (iii) then contradicts the fact that each is a (non-trivial) perfect fluid space-time. Thus for the Eintein static metric, $\dim A(M) = 8$ and for the Gödel metric, $\dim A(M) = 6$ (c.f. section 10.6). For the Bertotti-Robinson type metrics [55], the holonomy type is R_7 and the description of $A(M)$ is as in section 10.6 and proper affine vector fields may exist. Next consider the Friedmann-Robertson-Walker (F.R.W.) metrics given in the usual coordinates and domain by

$$ds^2 = -dt^2 + R^2(t)(d\chi^2 + f^2(\chi)(d\theta^2 + \sin^2\theta d\phi^2)) \tag{10.84}$$

where the standard distinctions $k = \pm 1$ or 0 are defined by $f(\chi) = \sin\chi$ $(k = 1)$, $f(\chi) = \chi$ $(k = 0)$ and $f(\chi) = \sinh\chi$ $(k = -1)$. A special case of (10.84) is the Einstein static type. Excluding this special case one has $\dim K(M) = 6$ and the orbits are the 3-dimensional spacelike stable submanifolds of constant t (for full details, see [177]). For all F.R.W. metrics $\dim \overset{*}{K}_m = 3$ for each $m \in M$ and so, from theorem 10.4, M is conformally flat. For the non-Einstein static F.R.W. metrics, a proper homothetic vector field may (but, in general, does not) exist for each of $k = \pm 1$ or 0 [177], [178] (see chapter 12). The plane wave metrics (8.11) with H as in (10.42) are at each point of Petrov type **N** or **O** and always admit five independent Killing vector fields tangent to the null hypersurfaces of constant u (and including the covariantly constant vector field $\frac{\partial}{\partial u}$ [41], [55]). They also always admit the proper homothetic vector field (10.40) (with $d = 0$ in the type N case) which is also tangent to these hypersurfaces. If M is an everywhere type **N** plane wave then, in the general case, these are the only members of $H(M)$ and so $\dim K(M) = 5$, $\dim H(M) = 6$ and the orbits of $K(M)$ and $H(M)$ coincide and are 3-dimensional, null and stable. For each $m \in M$, $\overset{*}{K}_m$ is a type R_8 subalgebra of the Lorentz algebra. In the type **N** case, and for special choices of H, there can be (at most) one further independent Killing vector field [41], [55] and a single Killing orbit M (the so-called *type* **N** *homogeneous plane waves*). In the conformally flat case, there is *always* a sixth Killing vector field admitted which is also tangent to the hypersurfaces of constant u, the orbits of $K(M)$ (and $H(M)$) being as in the type **N** general case but now with each $\overset{*}{K}_m$ a (3-dimensional) type R_{11} subalgebra of the Lorentz algebra. For special choices of H, a seventh independent Killing vector field (and no more) may be admitted [55] and then there is a single orbit M (the so-called *conformally flat homogeneous plane waves*).

For completeness, it is remarked that a space-time M is called *stationary* if it admits a global nowhere zero timelike Killing vector field and *static* if it admits a global nowhere zero timelike hypersurface orthogonal Killing vector field. Thus if M is static it is stationary. For a stationary space-time, one can choose a coordinate system so that the Killing vector field in question is $\frac{\partial}{\partial t}$ and the metric tensor components are independent of t. If M is static one can, in addition to the above, arrange in the metric that if t is the coordinate x^0, then $g_{a0} = 0$ $(a = 1, 2, 3)$. It can be seen from section 4.16 that if X is Killing and hypersurface orthogonal, its associated bivector satisfies $F_{[ab}X_{c]} = 0$ and so at each $p \in M$, $F(p)$, if not zero, is simple and its blade contains $X(p)$. If a Killing vector field is a gradient, then it is covariantly constant. From the above, it is seen that the Schwarzschild and Reissner-Nordstrom metrics are static. The Kerr metric is stationary but not static.

A very extensive list of metrics possessing symmetries such as those discussed above can be found in [55]. Some additional work of interest in the context of this chapter can be found in [179]-[193].

A brief remark should be included about the relation between global and local symmetries. Symmetries have so far been regarded as *global* vector fields on M. One may also wish to consider, for physical reasons, "local" symmetries which are vector fields satisfying some symmetry condition but which are defined on some open subset U of M and may not be the restriction to U of some global vector field (of the same symmetry type) on M, that is, it may not be globally extendible to M as a global vector field of the same symmetry type. This has been discussed using two different approaches [169], [170] and, roughly speaking, says that if the local symmetries are of the type discussed in this chapter and are "the same" everywhere and M is simply connected, then global extension of each symmetry is possible.

Another topic which has attracted some interest and which concerns symmetries is that of symmetry inheritance. This arises in the following way. Symmetries in general relativity are vector fields which preserve (in the sense of section 10.1) some geometrical object on space-time. Einstein's field equations are relations between the space-time geometry and certain basic physical descriptors which collectively make up the energy-momentum tensor. To what extent do these vector fields preserve these physical descriptors (*i.e.* to what extent do the latter inherit the geometrical symmetries)? For example, if M represents an Einstein-Maxwell space-time and X is a Killing vector field on M, does X preserve the Maxwell tensor F (*i.e.* is $\mathcal{L}_X F = 0$)? Of course, in asking this question, one must take care to specify the *form* of the tensor F (*i.e.* the positions of the indices) although when $X \in K(M)$ this is clearly not necessary. Another example arises when M represents a perfect fluid space-time and $X \in K(M)$. Does X preserve the fluid flow vector and the pressure and energy functions? In these examples, it is clear that the energy-momentum tensor T is preserved, $\mathcal{L}_X T = 0$, but more is required to answer the above questions.

This problem will not be pursued in any depth here. A few remarks and references must suffice, a selection of the latter being [194]-[203], [85] (with [197]-[199] being particularly instructive). Regarding the first (Einstein-Maxwell) example above, it does *not* follow that $X \in K(M) \Rightarrow \mathcal{L}_X F = 0$ and counterexamples for both the null and non-null Einstein-Maxwell cases are known ([197]-[199] and [201]-[203]) [It is remarked at this point that *if X is a hypersurface orthogonal Killing vector field on M then X is an eigenvector field of the Ricci tensor on M.* This result was proved when X is everywhere timelike in [41]. To prove it for any Killing vector field X one notes (briefly) that the condition $X_{[a;b}X_{c]} = 0$ implies $F_{[ab}X_{c]} = 0$ and so $F_{ab} = 2X_{[a}P_{b]}$ for some local covector field P. Then on substituting this into (10.7) and contracting with X^c one finds that, if $Q_a \equiv P_{a;b}X^b$, then $X_{[a}Q_{b]} = -(X^cP_c)X_{[a}P_{b]}$. Hence, $Q_a = \alpha X_a + \beta P_a$ for functions α and β and where $\beta = -X^cP_c$. A contraction of (10.7) with g^{ac} then completes the proof. It follows that if M is static, it can not be a null Einstein-Maxwell field since the associated Ricci tensor has Segre type $\{(211)\}$ with eigenvalue zero and admits no timelike eigenvectors. It is also true that a static space-time M is of Petrov type **O**, **I** or **D** at each $m \in M$] ([41]-see also section 11.6). For the second (perfect fluid) case with fluid flow vector field u, isotropic pressure p and energy density ρ as in (7.7), and with $p + \rho$ nowhere zero on M, $X \in K(M) \Rightarrow \mathcal{L}_X u = 0$ and $\mathcal{L}_X p = \mathcal{L}_X \rho = 0$ [198]. In fact, for $X \in K(M)$, the resulting condition $\mathcal{L}_X T = 0$ implies the more general result [85] (see section 10.4) that the Segre type of T (including degeneracies) is preserved along the integral curves of X and that any eigenvalue α of T is constant along these curves (*i.e.* $\mathcal{L}_X \alpha = 0$). From this, certain results are available for eigenvector fields of T which include the above quoted results for perfect fluids.

It is also remarked that the material on zeros of homothetic vector fields can be used to study the possibilities for the dimension of the Lie algebra $H(M)$ [204] and the associated orbits [205]. This last reference also includes some results on the (non-trivial) orbits associated with $A(M)$, a brief summary of which will now be given. The orbits associated with $H(M)$ and $A(M)$ can be studied using similar techniques to those used for $K(M)$ and using similar criteria for stability and dimensional stability. It turns out that, given $K(M) \neq H(M)$, the orbits associated with $K(M)$ and $H(M)$ through some $p \in M$ are, in general, distinct with the homothetic orbit exceeding the Killing orbit in dimension by one. There could be an "accidental" equality of a Killing and a homothetic orbit such as the metric (10.43) exhibits. This example satisfies $\dim K(M) = 2$, $\dim H(M) = 3$ with Killing orbits everywhere 2-dimensional null and stable whilst the homothetic orbits are either 2-dimensional, null and dimensionally unstable and given by $u = x = 0$ (and this coincides with a Killing orbit), 3-dimensional, null, dimensionally stable but not stable and given by $u = 0$, $x > 0$, or $u = 0$, $x < 0$, with the remaining orbits 3-dimensional, timelike and stable. However, if O is simultaneously a Killing and a homothetic orbit *and is dimensionally stable as a homothetic orbit then some neighbourhood of each $p \in O$ is isometric to a plane wave* (the common orbits then being either 4-dimensional or 3-dimensional, null and stable). For a simply connected space-time not of holonomy type R_7 the affine and homothetic orbits coincide whereas in the R_7 case these orbits either coincide or the affine orbit through $p \in M$ is one greater in dimension than the homothetic orbit through p. Again, in the R_7 case, if O is a coincident orbit of $H(M)$ and $A(M)$ and is dimensionally stable as an $A(M)$ orbit, then the orbits of each of $H(M)$ and $A(M)$ coincide and are dimensionally stable through each point of some neighbourhood of any point of O. It is remarked that the R_7 type of space-time possesses another anomalous behaviour in that, amongst simply connected space-times, it is the only type which admits a covariantly constant type $(0,2)$ tensor field which is not a constant multiple of g (*i.e.* a non-trivial member of S in the language of section 9.2) and yet need not admit any affine vector field (section 10.6). The affine and homothetic orbits, like the Killing orbits, have the same nature (spacelike, timelike or null) at each point and an orbit of either of these algebras which is of maximum dimension is dimensionally stable in that algebra. If O is an orbit of either of these algebras which is *not* dimensionally stable then for each $p \in O$ some non-trivial member of that algebra vanishes at p. Finally it is pointed out that a detailed study of homothetic isotropy can be achieved by a consideration of the conformal algebra $c(1,3)$. Further details of these points can be found in [205].

Finally some comments can be added regarding the discussion of the dimension of $K(M)$ after theorem 10.2 and again in section 10.5. Several results quoted in the relativistic literature are sometimes unreliable since, amongst other things, they rely on classical proofs involving positive definite metrics. Concerning theorem 10.2(iii), it is clear that an elementary generalisation of the space-time argument will reveal the bracketed result of this theorem when M admits a metric of arbitrary signature. Also if such an M satisfies $\dim K(M) = \frac{1}{2}n(n+1)$ then for any $m \in M$ and orbit O through m the result

$$\dim K(M) = \dim O + \dim \overset{*}{K}_m \tag{10.85}$$

following theorem 10.10 and the facts that $\dim O \leqslant n$ and $\dim \overset{*}{K}_m \leqslant \frac{1}{2}n(n-1)$ show that $\dim O = n$ and $\dim \overset{*}{K}_m = \frac{1}{2}n(n-1)$. From the latter it is easy to check that M has constant curvature. Conversely, if M has constant curvature the general form of the metric (see [18]) leads to a *local* Killing algebra of dimension $\frac{1}{2}n(n+1)$ [206]. Another result often quoted is that $\dim K(M)$ can not have dimension $\frac{1}{2}n(n+1) - 1$ and it is attributed to Fubini [207]. However, so far as the present author can tell, Fubini's proof applies only in the positive definite case. To extend to the general situation for M above one notes, briefly [167], that if $\dim K(M) = \frac{1}{2}n(n+1)-1$ and if $m \in M$ with O an orbit through m, then (10.85) shows that O has dimension n or $n-1$. In the former case one necessarily has $\dim \overset{*}{K}_m = \frac{1}{2}n(n-1)-1$. Now for an orthogonal group $O(r,s)$ (section 2.5) with at least one of r and s equal to three or more, there can be no proper Lie subgroup of dimension $\frac{1}{2}n(n-1)-1$. In the latter case $\overset{*}{K}_m$ is forced to have its maximum dimension $\frac{1}{2}n(n-1)$. But the existence of an $(n-1)$-dimensional orbit would force restrictions on the dimension of $\overset{*}{K}_m$ similar to those encountered in the proof of theorem 10.13 and a contradiction follows. This establishes Fubini's theorem for all signatures except $(0,2), (1,1), (1,2)$ and $(2,2)$.

Chapter 11

Conformal Symmetry in Space-times

11.1 Conformal Vector Fields

Let M be a space-time with metric g and let X be a smooth global vector field on M with associated local diffeomorphisms ϕ_t. The condition that each such ϕ_t is a local *conformal* diffeomorphism is that, on the appropriate domains, $\phi_t{}^* g = fg$ for some positive function f. This is equivalent to the condition $\mathcal{L}_X g = 2\phi g$ for some smooth function $\phi : M \to \mathbb{R}$ called the *conformal function* of X (section 5.13). A vector field X, with any, and hence each, of these properties is called *conformal*. An equivalent condition for X to be conformal is that in any coordinate domain

$$X_{a;b} = \phi g_{ab} + F_{ab} \qquad (F_{ab} = -F_{ba}) \tag{11.1}$$

where F is the *conformal bivector* of X.

If one computes $X_{a;bc}$ from (11.1) and uses the Ricci identity (4.33) on X (and the abbreviation $\phi_a \equiv \phi_{,a}$) one finds

$$X_d R^d{}_{abc} = 2g_{a[b}\phi_{c]} + 2F_{a[b;c]}. \tag{11.2}$$

If one adds (11.2) to the equation obtained from (11.2) by replacing the indices abc by bca, respectively, and uses the symmetries of the curvature tensor, one gets

$$-X_d R^d{}_{cab} = 2g_{c[b}\phi_{a]} + 2F_{ab;c} + F_{bc;a} - F_{ac;b}. \tag{11.3}$$

Then by anti-symmetrising over the indices abc in (11.3) one finds $F_{[ab;c]} = 0$ which, on substitution back in (11.3) followed by some rearrangement, gives the following equations

$$F_{ab;c} = R_{abcd}X^d - 2\phi_{[a}g_{b]c} \tag{11.4a}$$
$$X_{a;bc} = R_{abcd}X^d + g_{ab}\phi_c + g_{ac}\phi_b - g_{bc}\phi_a. \tag{11.4b}$$

Next, take the covariant derivative of (11.4a) and use the Ricci identity (4.35) on F to get

$$\begin{aligned} F_{ad}R^d{}_{bce} + F_{db}R^d{}_{ace} = X^d{}_{;e}R_{abcd} - X^d{}_{;c}R_{abed} + X^d(R_{abcd;e} + R_{abde;c}) \\ + 2g_{c[a}\phi_{b];e} - g_{e[a}\phi_{b];c} \end{aligned} \tag{11.5}$$

If one substitutes from (11.1) into (11.5) and contracts over the indices b and e it follows that

$$\begin{aligned} F_{ad}R^d{}_c + F_{de}R^d{}_{ac}{}^e = 2\phi R_{ac} + 2\phi_{a;c} + X^d R_{ac;d} + (\phi^b{}_{;b})g_{ac} \\ + F_{de}R_a{}^e{}_c{}^d + F^d{}_c R_{ad}. \end{aligned} \tag{11.6}$$

A further contraction over the indices a and c then shows that

$$6\phi^a{}_{;a} = -2\phi R - R_{,a}X^a. \tag{11.7}$$

Finally, a substitution of (11.7) into (11.6) gives, after some calculation and use of the fact that $R_{dace} + R_{dcae}$ is symmetric in the indices d and e,

$$\phi_{a;b} = -\frac{1}{2}L_{ab;c}X^c - \phi L_{ab} + R_{c(a}F_{b)}{}^c \qquad (L_{ab} = R_{ab} - \frac{1}{6}Rg_{ab}). \tag{11.8}$$

It is a straightforward calculation to unravel (11.8) and show it to be equivalent to the following expression for the Lie derivative of the Ricci tensor with respect to X

$$\mathcal{L}_X R_{ab} = -2\phi_{a;b} - (\phi^c{}_{;c})g_{ab}. \tag{11.9}$$

By skew-symmetrising over the indices a and b in (11.4b) one easily obtains (11.4a). Also for *any* vector field X on M let $h_{ab} \equiv 2X_{(a;b)}$ ($= \mathcal{L}_X g_{ab}$) and assume that X satisfies (11.4b) for some function $\phi : M \to \mathbb{R}$. Then symmetrising this last equation over the indices a and b reveals that $h_{ab;c} = 2\phi_{,c}g_{ab} = (2\phi g_{ab})_{;c}$ and so $(h_{ab} - 2\phi g_{ab})_{;c} = 0$. Thus if M admits no covariantly constant second order symmetric tensors apart from (constant) multiples of g (*i.e.* $\dim S = 1$ in the notation of section 9.2), this gives $\mathcal{L}_X g_{ab} = \psi g_{ab}$ ($\psi : M \to \mathbb{R}$) and so X is conformal. Thus *if* $\dim S = 1$, *the condition that X is conformal is equivalent to X satisfying (11.4b).* From theorem 9.2 the condition that $\dim S = 1$ is generic and hence so is this equivalence. In this context it is noted that, from chapter 10, any *proper* affine vector field on M would satisfy (11.4b) for $\phi \equiv 0$ on M but fail to be conformal (cf. theorem 10.14).

If X is assumed only C^3, it follows from (11.1) that ϕ and F are C^2 and then (11.8) makes sense and shows that ϕ is C^3. Then (11.4a) shows that F is C^3 and (11.1) reveals that X is C^4. By repeating this argument one sees that X is smooth and that *the assumptions X is C^3 and X is C^∞ are equivalent.* Also, a contraction of (11.1) with $X^a X^b$ shows that if $f = X^a X_a$ and $x^a(t)$ is an integral curve of X then $\frac{df}{dt} = 2\phi f$. Hence, from Picard's theorem, *the nature (timelike, spacelike or null) of X is constant along an integral curve of X.*

The set of all conformal vector fields on M is denoted by $C(M)$ and clearly forms a real vector space. Use of result (vi) in section 4.17 then reveals that $C(M)$ is a Lie algebra (the *conformal algebra*) under the bracket operation. Also, the equations (11.1), (11.4a) and (11.8) lead to a system of first order differential equations for the quantities X, ϕ, $d\phi$ and F (cf. the affine case in section 10.2) and so a conformal vector field on M is uniquely determined by the values of these quantities at any $m \in M$ (or, equivalently, by inspection of (11.1) and (11.46), by the quantities X, ∇X and $\nabla(\nabla X)$ at any $m \in M$). It follows that $\dim C(M) \leq 4+1+4+6 = 15$ and the following theorem is established.

Theorem 11.1 *Let M be a space-time and $C(M)$ the set of conformal vector fields on M. Then*

i) the members of $C(M)$ are characterised by their satisfying (11.1),
ii) each member of $C(M)$ is uniquely determined by specifying at some (any) $m \in M$ the quantitites X, ϕ, $d\phi$ and F or, alternatively, X, ∇X and $\nabla(\nabla X)$. Hence if X vanishes over a non-empty open subset of M or if X, ϕ, F and $d\phi$ vanish at some point of M or if X, ∇X and $\nabla(\nabla X)$ vanish at some point of M, X vanishes on M,
iii) $C(M)$ is a finite-dimensional Lie algebra of smooth vector fields on M and $\dim C(M) \leq 15$. *(For a connected n-dimensional manifold M with metric of any signature,* $\dim C(M) \leq \frac{1}{2}(n+1)(n+2)$.*)*

It is remarked that if X and Y are non-trivial conformal vector fields on M satisfying $X = fY$ for a smooth function $f : M \to \mathbb{R}$ then f is constant (cf. [35]). To see this, briefly, one has $\mathcal{L}_Y g = 2\psi g$ and $\mathcal{L}_{fY} g = 2\phi g$ for smooth functions ψ and ϕ. When the second of these is expanded and the first substituted in, a rank condition immediately gives $f\psi = \phi$ and then $Y_a f_{,b)} = 0$. Since Y is nowhere zero on an open dense subset U of M (theorem 11.1(ii)) it follows that $f_{,a} = 0$ on U and hence on M and so, since M is connected, f is constant on M.

If $g' \equiv e^\sigma g$ for σ a smooth function $\sigma : M \to \mathbb{R}$, then g' is a smooth Lorentz metric on M conformally related to g. It then follows (section 5.13) that for any smooth vector field on M

$$\mathcal{L}_X g' = X(\sigma) e^\sigma g + e^\sigma \mathcal{L}_X g \tag{11.10}$$

and so X *is conformal with respect to* g' *if and only if it is conformal with respect to* g. Now for Minkowski space-time, with the usual notation and coordinates x, y, z, t, the Lie algebra $C(M)$ can be represented by its basis (see, *e.g.* [208])

$$\begin{aligned} P_a &= \frac{\partial}{\partial x^a} & M_{ab} &= x_a \frac{\partial}{\partial x^b} - x_b \frac{\partial}{\partial x^a} \\ H &= x^a \frac{\partial}{\partial x^a} & K_a &= 2x_a H - (x_b x^b) P_a \end{aligned} \tag{11.11}$$

where $x^a = (t, x, y, z)$ and $x_a = (-t, x, y, z)$. Thus $\dim C(M) = 15$ in this case. It follows that a conformally flat space-time admits a 15-dimensional Lie algebra of conformal vector fields locally. Conversely, it will be shown later that if $\dim C(M) = 15$ (in fact, if $\dim C(M) > 7$) then M is conformally flat.

A conformal vector field X is called *proper* if it is not homothetic (*i.e.*, in (11.1), ϕ is not constant on M). It is called *special conformal* if the conformal function satisfies $\phi_{a;b} = 0$. The subset $SC(M)$ of $C(M)$ of special conformal vector fields on M (the *special conformal algebra* of M) is easily checked to be a subalgebra of $C(M)$ because if $X, Y \in SC(M)$ with conformal functions ϕ and ψ, respectively, then $\mathcal{L}_{[X,Y]} g = 2\chi g$ with $\chi = X(\psi) - Y(\phi) = \psi_a X^a - \phi_a Y^a$ and then the Ricci identity and (11.4b) give

$$\phi_{a;b} = \psi_{a;b} = 0 \Rightarrow \phi_a R^a{}_{bcd} = \psi_a R^a{}_{bcd} = 0 \Rightarrow \chi_{a;b} = 0. \tag{11.12}$$

If X and fX are conformal ($f : M \to \mathbb{R}$) then (11.1) shows that $f_{,a} X_b + X_a f_{,b} = \chi' g_{ab}$ for some $\chi' : M \to \mathbb{R}$ and hence by comparing the rank, as matrices, on each side of this equation it follows that $\chi' \equiv 0$ and f is constant.

Thus conformal vector fields on M are those vector fields whose associated local diffeomorphisms preserve the "conformal structure" of space-time. It is noted here from (11.8) that every conformal vector field in

Minkowski space is special conformal. Also *if X is conformal on M (satisfying (11.1)) with respect to g then, from (11.10), it is Killing on M with respect to the conformally related metric $g' = e^{\sigma}g$ if and only if $2\phi + \sigma_{,a}X^a = 0$ in any coordinate domain.* Thus if X is conformal on M and $X(m) \neq 0$ then by employing theorem 4.12(i) it follows that one can always find a solution σ of this equation on some open neighbourhood U of m and hence a metric $e^{\sigma}g$ on U for which X is Killing. This is an important point since it (locally at least) reduces the difficult problem of a conformal symmetry to the less difficult one of a Killing symmetry. The difficulties with conformal symmetry arise in the neighbourhoods of *zeros* of conformal vector fields where they lack the "linearity" exhibited by affine vector fields. This problem will be discussed in more detail later.

If $X \in C(M)$ and $X(m) \neq 0$ then, again using theorem 4.12(i), one may choose an open coordinate domain U of m in which $X^a = \delta_1^a$ and then the condition $\mathcal{L}_X g = 2\phi g$ leads to $g_{ab,1} = 2\phi g_{ab}$ and so $g_{ab} = \chi q_{ab}$ on U where $\chi = e^{2\int \phi dx^1}$ and q is a Lorentz metric on U whose components are independent of x^1 (so that X is Killing on U with respect to q). Conversely, if g can be written in this form on U with q independent of x^1 then $X^a = \delta_1^a$ is a conformal vector field with respect to g.

11.2 Orbit Structure

In the study of the structure of the orbits associated with the (finite-dimensional) Lie algebra $C(M)$ the general discussion in section 10.5 can be taken over, where appropriate, to give the following result.

Theorem 11.2 *Let M be a space-time with conformal algebra $C(M)$. Then M may be decomposed as in (10.44) where A is now $C(M)$, and also as in theorem 10.8. Each orbit O of $C(M)$ is a leaf of M and a maximal integral manifold of $C(M)$ and if $\dim O = 1, 2$ or 3, its nature (timelike, spacelike or null) is the same at each of its points. Again, if $\dim O = 1, 2$ or 3 and O is not null, and if $X \in C(M)$ with X not identically zero (so that X is tangent to O), the associated vector field $\tilde{X}$ on O (section 4.11) is a (possibly trivial) conformal vector field in the geometry of O with induced metric from the metric g on M. The associated map $\theta : C(M) \to C(O)$ given by $\theta : X \to \tilde{X}$ is a Lie algebra homomorphism. Finally, if O is an orbit of $C(M)$ satisfying $O \cap \text{int}\, \tilde{S}_p \neq \phi$ (respectively, $O \cap \text{int}\, \tilde{T}_p \neq \phi$ or $O \cap \text{int}\, \tilde{N}_p \neq \phi$) for some p, $1 \leq p \leq 3$, then $O \subseteq \text{int}\, \tilde{S}_p$ (respectively, $O \subseteq \text{int}\, \tilde{T}_p$ or $O \subseteq \text{int}\, \tilde{N}_p$).*

If C is the Weyl tensor with local components $C^a{}_{bcd}$ and if X is a conformal vector field on M with local diffeomorphisms ϕ_t, the argument in section 10.1 shows, since C is unchanged after a conformal change of metric (and so $\phi_t^* C = C$), that $\mathcal{L}_X C = 0$ or, in the usual component form (with $\mathcal{L}_X g_{ab} = 2\phi g_{ab}$),

$$\mathcal{L}_X C^a{}_{bcd} = 0 \qquad (\Leftrightarrow \mathcal{L}_X C_{abcd} = 2\phi C_{abcd}). \tag{11.13}$$

Now recalling the tensor G in (7.11) it follows that

$$\mathcal{L}_X G_{abcd} = 4\phi G_{abcd}. \tag{11.14}$$

Now let C' represent the Weyl tensor with indices C_{abcd} and let ϕ_t be a local diffeomorphism of the conformal vector field X whose domain includes $m \in M$. Then [20]

$$[\phi_s^*(\mathcal{L}_X C')]\,(m) = \left[-\frac{d}{dt}(\phi_t^* C')(m)\right]_{t=s} \tag{11.15}$$

and a similar equation holds for G. From the second equation in (11.13) and (11.15) it can be shown that (see, *e.g.* [40])

$$(\phi_s^* C')\,(m) = C'(m) \exp\left[-2\int_0^{-s} (\phi \circ \phi_t)(m) dt\right] \tag{11.16}$$

and also a similar equation for G with the "2" inside the square bracket in (11.16) replaced by "4". Now let Q be any eigenbivector of the Weyl tensor at m with eigenvalue $\lambda_0 \in \mathbb{C}$ (section 7.3). Then at m

$$C_{abcd}Q^{cd} - \lambda_0 G_{abcd} Q^{cd} = 0. \tag{11.17}$$

Now define a smooth bivector Q_t at $\phi_t(m)$ by $Q_t = \phi_t^{-1*}Q$ and consider the tensor at $\phi_s(m)$ given by

$$C'(s)Q_s - \lambda_0 \exp\left[2\int_0^{-s} (\phi \circ \phi_t)(m) dt\right] G(s)Q_s \tag{11.18}$$

with components

$$C_{abcd}Q_s^{cd} - \lambda_0 \exp\left[2\int_0^{-s} (\phi \circ \phi_t)(m) dt\right] G_{abcd}Q_s^{cd}. \tag{11.19}$$

If one operates on the left hand side of (11.17) with ϕ_s^* and uses (11.16), the corresponding equation for G and λ_0 extended as a smooth function along the integral curve of X as in (11.20) one obtains zero for the result.

Thus (11.18) and (11.19) are zero for each s and so Q_s is an eigenbivector of the Weyl tensor at $\phi_s(m)$ with eigenvalue λ_s, where

$$\lambda_s = \lambda_0 \exp\left[2\int_0^{-s}(\phi\circ\phi_t)(m)dt\right] \quad (= \lambda_0 \exp\left[2\int_0^{-s}\phi(t)dt\right]). \quad (11.20)$$

This result describes the algebraic behaviour of the Weyl tensor along the integral curves of a conformal vector field [175],[166]. In particular, it displays the exponential behaviour of the Weyl eigenvalues along such curves and from it one can recover the analogous results for Killing and homothetic vector fields given in section 10.4 by setting ϕ = constant. It also shows that the Weyl eigenbivector-eigenvalue structure is unchanged along such curves and hence that *the Petrov type is constant along such curves.* In fact, a short extension of the argument in section 10.1 regarding the local diffeomorphisms ϕ_t of a conformal vector field X on M can be used to show that if $\ell \in T_mM$ is a repeated (respectively, non-repeated) principal null direction of the Weyl tensor at m then $\phi_{t*m}\ell$ is a repeated (respectively non-repeated) principal null direction of the Weyl tensor at $\phi_t(m)$. Since there are only finitely many of each type of these null directions at m if $C(m) \neq 0$, then, for an open subset U on which C never vanishes and if ℓ is now a smooth null vector field on U which is a principal null direction of C at each point of U, it follows that $\phi_{t*m}\ell(m) \propto \ell(\phi_t(m))$ for appropriate t and hence that

$$\mathcal{L}_X\ell = \psi\ell \quad (11.21)$$

for some smooth function $\psi : U \to \mathbb{R}$.

The above result on the behaviour of the Weyl eigenvalues along integral curves of conformal vector fields has several applications. As mentioned earlier it confirms that any Weyl eigenvalue is constant over any *Killing* orbit and hence constant on M if $K(M)$ is transitive. Also (11.20) shows that if O is a Killing orbit of Petrov type **I**, **II** or **D** at some (and hence any) point of O (so that at least one Weyl eigenvalue is a non-zero constant on O) then any conformal vector field X tangent to O must have its conformal function ϕ identically zero on O and if, in addition, X is homothetic, it must be Killing. The above remarks together with similar ones on Ricci eigenvalues for homothetic vector fields given in section 10.4 (and a few technical details [175]) leads to the following result.

Theorem 11.3 *Let M be a space-time. Then the following hold.*

i) The Petrov type of M is constant on each orbit of $C(M)$ and for each $X \in C(M)$, (11.21) holds in any open subset U of M upon which C never vanishes and on which ℓ is a smooth principal null direction of C.

ii) If O is an orbit of $K(M)$ and the Petrov type is **I**, **II** *or* **D** *at some (and hence every) point of O then if $X \in C(M)$ is tangent to O, the associated function ϕ vanishes on O. In particular, if $K(M)$ is transitive, any $X \in C(M)$ is Killing (*i.e. *$C(M) = K(M)$).*

iii) If O is an orbit of $K(M)$ and if at some (and hence every) point of O either the Petrov type is **I**, **II** *or* **D** or *the Ricci tensor is any non-zero Segre type except $\{(31)\}$ with zero eigenvalue, or $\{(211)\}$ with zero eigenvalue then if $X \in H(M)$ is tangent to O it is Killing and, in particular, if $K(M)$ is transitive any $X \in H(M)$ is Killing (i.e. $H(M) = K(M)$).*

For example, if (M, g) is the Gödel space-time (8.15), $K(M)$ is transitive and the Petrov type is **D**. Thus $C(M) = H(M) = K(M)$ and no proper homothetic or conformal vector fields exist. Similarly, the Einstein static metric (8.14) has $K(M)$ transitive and, although conformally flat, possesses a non-zero Ricci eigenvalue. Thus for this space-time, $H(M) = K(M)$.

To see the importance of the exclusion clauses in theorem 11.3 consider first the plane wave spacetimes (8.11) restricted by (10.42). Here, either $K(M)$ is transitive or else the orbits of $K(M)$ are everywhere null and 3-dimensional but in either case a proper homothetic vector field is admitted tangent to the hypersurfaces of constant u. For this metric, however, the Petrov type is, at each point, **N** or **O** and the Ricci tensor at each point is either zero or has Segre type $\{(211)\}$ with zero eigenvalue. Next consider the metric (10.43). Here, the Killing orbits are everywhere 2-dimensional and null and the orbits of $H(M)$ consist of two 3-dimensional null orbits $u = 0$, $x > 0$ and $u = 0$, $x < 0$, a 2-dimensional null orbit $u = x = 0$ and all the other orbits are 3-dimensional and timelike. The null 2-dimensional orbit coincides with a Killing orbit and the proper homothetic vector field admitted by this metric is thus tangent to this Killing orbit. However, on this orbit, the Petrov type is **III** and the Ricci tensor type is $\{(31)\}$ with zero eigenvalue. A little more effort will show that the Petrov type is **III** everywhere and that no proper conformal vector fields are admitted. As a final example in the "opposite" direction one notes that, for a "general" F.R.W. metric (10.84) where $\dim K(M) = 6$, if a proper homothetic vector field is admitted, it can not lie in the 3-dimensional spacelike Killing orbits.

Here, although the Weyl tensor vanishes everywhere, some Ricci eigenvalue is non-zero at each point.

11.3 Fixed Point Structure

Let X be a non-trivial conformal vector field on M such that $X(m) = 0$ so that $m \in M$ is a zero of X. If ϕ_t represent the local diffeomorphisms associated with X then m is a *fixed point* of ϕ_t, $\phi_t(m) = m$, and the map $\phi_{t*m} : T_mM \to T_mM$ is represented with respect to the basis $(\frac{\partial}{\partial x^a})_m$ of T_mM for some coordinate domain containing m with coordinates x^a, by the transpose of the matrix $\exp tA$ (section 10.4) where

$$A^a{}_b = \left(\frac{\partial X^a}{\partial x^b}\right)_m. \tag{11.22}$$

It follows from (11.1) that

$$A^a{}_b = (\phi\delta^a{}_b + F^a{}_b)_m. \tag{11.23}$$

The problem now is that, unlike in the affine case (section 10.4), one cannot guarantee that X is linearisable in some coordinate domain containing m. However, a study of the zeros of conformal vector fields is quite useful [166]. The *isotropy subalgebra* $\overset{*}{C}_m$ of $C(M)$ at m is the subset $\{X \in C(M) : X(m) = 0\}$ and is easily checked to be a subalgebra of $C(M)$. As before (cf. section 10.4) one finds $\dim C(M) = \dim \overset{*}{C}_m + \dim(\text{orbit of } C(M) \text{ through } m)$.

Remarks in section 11.1 reveal that any Killing or homothetic vector field with a zero at m in a space-time becomes a proper conformal vector field with respect to some appropriate conformally related metric. Thus all features of zeros encountered for members of $H(M)$ exist for zeros of $C(M)$ provided they are, in an obvious sense, "conformally invariant". In particular the results regarding the *algebraic* behaviour of the Weyl tensor at a zero of a non-trivial Killing (respectively a proper homothetic) vector field carry over without change, from (11.13), at a zero m of a non-trivial conformal vector field satisfying $\phi(m) = 0$ (respectively $\phi(m) \neq 0$). The results regarding the Ricci tensor do not, however, since this tensor is not "conformally invariant" (see (11.9)). These and other results can now be collected together. For this purpose it is convenient to make the following definitions. A zero m of a conformal vector field X is called *isometric* if the conformal function ϕ in (11.1) satisfies $\phi(m) = 0$ and *homothetic* if

$\phi(m) \neq 0$. Further, if $\sigma : U \to \mathbb{R}$ is some smooth function on some open neighbourhood U of m then X is conformal with respect to the metric $g' = e^{\sigma} g$ on U and, since m is a zero of X, the conformal functions ϕ and ϕ' and conformal bivectors F_{ab} and F'_{ab} of X with respect to g and g', respectively, are related at m by (see (11.34))

$$\phi'(m) = \phi(m), \quad F'^{a}{}_{b}(m) = F^{a}{}_{b}(m), \quad F'_{ab}(m) = e^{\sigma(m)} F_{ab}(m). \quad (11.24)$$

So the type of the zero m and the algebraic nature of the conformal bivector at m is, in this sense, conformally invariant. It is also remarked that if $X, Y \in \overset{*}{C}_m$ then $[X, Y]$ always has m as an isometric zero. This follows from the relation immediately before (11.12) (cf. theorem 10.2(iv)).

Theorem 11.4 *Let M be a space-time and X ($\not\equiv 0$) a conformal vector field on M. Let $m \in M$ be a zero of X.*

i) If A in (11.23) is non-singular, the zero m is isolated.
ii) If m is an isometric zero and $F(m) \neq 0$ then if $F(m)$ is non-simple, timelike or spacelike (i.e. $F(m)$ is non-null) the Weyl tensor C is either zero or of Petrov type **D** *at m and, in the latter case, the timelike blade of $F(m)$ or $\overset{*}{F}(m)$ (whichever is timelike) or the timelike member of the canonical pair of blades of $F(m)$ if $F(m)$ is non-simple, contains the two principal null directions of C at m. If $F(m)$ is null, C is either zero or type* **N** *at m and in the latter case the principal null direction of $F(m)$ and of C coincide at m.*
iii) If m is a homothetic zero and $F(m) \neq 0$ then all Weyl eigenvalues vanish at m and the Petrov type at m is **O**, **N** *or* **III**. *If $F(m)$ is null, spacelike or non-simple, C vanishes at m. If C is of type* **N** *at m then a null tetrad (ℓ, n, x, y) exists at m for which ℓ spans the repeated principle null direction of C and $F(m)$ takes the form $2\phi(m)\ell_{[a}n_{b]}$. If C is of type* **III** *at m then a null tetrad (ℓ, n, x, y) exists at m with ℓ spanning the repeated, and n the non-repeated, principal null direction of C, $F(m)$ takes the form $4\phi(m)\ell_{[a}n_{b]}$ and the zero m of X is isolated.*
iv) If $F(m) = 0$ the Weyl tensor vanishes at m.

Proof. Part (i) follows from theorem 4.1 after establishing the existence of a coordinate neighbourhood U of m on which the map $U \to \mathbb{R}^4$ given by $m' \to X(m')$ is bijective and parts (ii) and (iii) follow from results given in theorems 10.3 and 10.5. For part (iv) suppose $C(m) \neq 0$. Then if m is a homothetic zero one expands (11.13) using (11.1) (see (10.38)) and the result follows immediately. If m is an isometric zero then [209] the conditions assumed immediately give $X^a{}_{;b}(m) = 0$ and $\phi_{a;b}(m) = 0$ from (11.1) and (11.8). Thus (11.4a) gives

$$F_{ab;c}(m) = -(\phi_a g_{bc} - \phi_b g_{ac})(m) \tag{11.25}$$

and so from (11.4b)

$$X_{a;bc}(m) = (g_{ab}\phi_c + g_{ac}\phi_b - g_{bc}\phi_a)(m). \tag{11.26}$$

Now take the covariant derivative of (11.13) and evaluate at m to get

$$C^a{}_{bce}X^e{}_{;df} + C^a{}_{bed}X^e{}_{;cf} + C^a{}_{ecd}X^e{}_{;bf} - C^e{}_{bcd}X^a{}_{;ef} = 0. \tag{11.27}$$

Then substitute (11.26) into (11.27) and use the algebraic symmetries of C and an obvious contraction to get, at m,

$$\phi_a C^a{}_{bcd} = 0. \tag{11.28}$$

This equation means that, at m, either ϕ_a is zero or null (section 7.3). But $\phi_a(m) = 0$ would, when allied with the other conditions at m, force X to be trivial (theorem 11.1(ii)) and so ϕ_a is null. But then (11.28) implies that (section 7.3)

$$C_{ab[cd}\phi_{e]} = 0. \tag{11.29}$$

A back substitution of (11.28) and (11.29) and again of (11.26) into (11.27) finally yields $C_{abcd}\phi_e = 0$ at m and so the Weyl tensor vanishes at m. □

The following theorem reveals the maximum dimension for $C(M)$ in non-conformally flat space-times [166], [209], [210].

Theorem 11.5 *Let M be a space-time which is not conformally flat. Then* $\dim C(M) \leq 7$.

In more detail, if at any point $m \in M$ the Petrov type is **I** *or* **II**, *then* $\dim C(M) \leq 4$, *if at any point the Petrov type is* **III** *then* $\dim C(M) \leq 5$ *and if at any point the Petrov type is* **D** *then* $\dim C(M) \leq 6$. *If the Petrov type is* **N** *at any point of M then* $\dim C(M) \leq 7$. *It can be added here that if M has Petrov type* **III** *over a non-empty open subset of M then* $\dim C(M) \leq 4$. *The proof of this cannot be given until after theorem 11.6.*

[One may add to this, for completeness, the situation for $K(M)$ and $H(M)$ when M is not conformally flat and which can be gathered together from section 10.4. One has $\dim K(M) \leq 6$ *and* $\dim H(M) \leq 7$ *and, in more detail, if at any $m \in M$ the Petrov type is* **I** *or* **II**, $\dim K(M) \leq 4 \geq \dim H(M)$, *if at any $m \in M$ the Petrov type is* **III**, $\dim K(M) \leq 4$,

$\dim H(M) \leq 5$, *if at any* $m \in M$ *the Petrov type is* **D**, $\dim K(M) \leq 6 \geq \dim H(M)$ *and if at any* $m \in M$ *the Petrov type is* **N**, $\dim K(M) \leq 6$ *and* $\dim H(M) \leq 7$.*]*

Proof. Since M is not conformally flat choose $m \in M$ where the Weyl tensor is not zero and consider the isotropy algebra $\overset{*}{C}_m$ at m. By taking linear combinations of the members of a basis for $\overset{*}{C}_m$, if necessary, one can arrange that all except at most one member of this basis has m as an isometric zero. Now from theorem 11.4(ii) one can see that only two independent bivectors are available as potential conformal bivectors for isometric zeros at m (two null bivectors if the Petrov type at m is **N** and a timelike-spacelike dual pair of bivectors if it is type **D**). Thus, by taking appropriate linear combinations of members of $\overset{*}{C}_m$ one can arrange that all but at most three members of $\overset{*}{C}_m$ satisfy (in the notation of 11.1) $X(m) = 0$, $\phi(m) = 0$, $F(m) = 0$ and from theorem 11.4(iv) this is possible only for the trivial member of $\overset{*}{C}_m$. It follows that $\dim \overset{*}{C}_m \leq 3$ and hence from a formula shortly after (11.23), $\dim C(M) \leq 3 + 4 = 7$.

If the Petrov type at some $m \in M$ is **I** or **II** then theorem 11.4 shows that $\dim \overset{*}{C}_m = 0$ and so $\dim C(M) \leq 4$. The same theorem shows that if at some $m \in M$ the Petrov type is **III** (respectively **D**) then $\dim \overset{*}{C}_m \leq 1$ (respectively $\dim \overset{*}{C}_m \leq 2$) and so $\dim C(M) \leq 5$ (respectively $\dim C(M) \leq 6$). If the Petrov type at m is **N** then $\dim \overset{*}{C}_m \leq 3$ and the proof is complete. It is worth stressing here that, as a consequence of this argument, if $m \in M$ and the Weyl tensor is not zero at m then $\dim \overset{*}{C}_m \leq 3$. The rest of the proof follows in a similar fashion from theorems 10.3 and 10.5. For example, if the Petrov type at m is **I** or **II** then $\dim \overset{*}{K}_m = \dim \overset{*}{H}_m = 0$ whereas for type **III**, $\dim \overset{*}{K}_m = 0$, $\dim \overset{*}{H}_m \leq 1$, for type **D**, $\dim \overset{*}{K}_m \leq 2 \geq \dim \overset{*}{H}_m$ and for type **N**, $\dim \overset{*}{K}_m \leq 2$, $\dim \overset{*}{H}_m \leq 3$, and the result follows. □

11.4 Conformal Reduction of the Conformal Algebra

For any vector field X on a space-time M there is, about any $m \in M$ for which $X(m) \neq 0$, a coordinate neighbourhood U in which X has everywhere the simple component representation $X^a = \delta^a_1$ (theorem 4.12(i)). If, however, $X(m) = 0$, no such U exists but one might, at least, hope for such a neighbourhood U in which the components X^a of X were *linear* functions of the coordinates (so that X is *linearisable* - see section 10.4). Unfortunately, even this might fail as will be seen later.

It is immediately seen from section 10.4 that *any affine vector field is linearisable* (about any of its zeros). Another result (the *Sternberg linearisation theorem* - see, *e.g.* [211]) which is occasionally useful says that if X is a vector field on M with a zero at m and is such that the eigenvalues $\lambda_1, \ldots, \lambda_4 \in \mathbb{C}$ of the matrix $X^a{}_{,b}(m)$ $(= X^a{}_{;b}(m))$ counted properly (*i.e.* with their algebraic multiplicities as roots of the associated characteristic equation) have the property that

$$\lambda_i \neq m_1\lambda_1 + \cdots + m_4\lambda_4 \qquad (1 \leq i \leq 4) \tag{11.30}$$

whenever the m_i are non-negative integers satisfying $2 \leq m_1 + \cdots + m_4$, then X is linearisable about m. The advantages which arise when a vector field X is linearisable about one of its zeros m are clear. For example, the linear expressions for X^a enable the other zeros of X (at least in some neighbourhood of m) to be studied conveniently and also the integral curves of X are, locally, the solutions of the linear equations $\frac{dx^a}{dt} = A^a{}_b x^b$ where $A^a{}_b = X^a{}_{,b}(m)$ is a *constant* matrix. The first of these shows that if U is a coordinate neighbourhood of m on which X is linearisable then either m is isolated or the totality of zeros of X in U constitute a regular submanifold of (the open submanifold) U found by solving the linear equations $X^a = 0$ in U. Unfortunately, proper conformal vector fields may not be linearisable about one of their zeros. To see this let M be Minkowski space-time with the usual coordinates $x^a = (t, x, y, z)$ and metric η and let X be the global, smooth vector field with components [166] (cf. (11.11))

$$X^a = \frac{1}{2}(\eta_{cd}x^c x^d)k^a - (\eta_{cd}x^c k^d)x^a \tag{11.31}$$

where $k^a = (0, 0, 0, 1)$. Then X is conformal, satisfying (11.1) with $\phi = -\eta_{ab}x^a k^b$ and $F_{ab} = 2k_{[a}x_{b]}$, and has an isometric zero at the coordinate origin m. The totality of zeros of X on M is that set of points whose coordinates satisfy $\eta_{ab}x^a x^b = \eta_{ab}x^a k^b = 0$ (the "null cone" at the origin of the 3-dimensional Minkowski space $z = 0$). However, since the conformal function and bivector of X also vanish at m, one now has $X^a{}_{,b}(m)$ $(= X^a{}_{;b}(m)) = 0$ and so if X were linearisable in some coordinate neighbourhood of the origin it would be identically zero on this neighbourhood and hence (theorem 11.1(ii)) on M. Thus X is not linearisable in any neighbourhood of the origin. It is remarked here that, in this example,

$X(m) = 0$, $\phi(m) = 0$ and $F(m) = 0$ and also that M is conformally flat (cf. theorem 11.4(iv)).

It was shown in section 11.1 that if $X \in C(M)$ and if $m \in M$ with $X(m) \neq 0$ then there exists an open neighbourhood U of m and a function $\sigma : U \to \mathbb{R}$ such that X is Killing on U with respect to the metric $e^{\sigma} g$ on U. Extending this one may ask if one might still achieve this result if $X(m) = 0$ and then, if one can achieve such a conformal reduction of $C(M)$ to Killing vector fields in some neighbourhood of m *simultaneously* (that is, can one choose U and σ as above so that *each* member of $C(M)$ is Killing with respect to $e^{\sigma} g$ on U). If this were the case then $C(M)$ could be reduced to a Killing algebra, at least locally, with respect to some metric conformally related to the original space-time metric. Then each $X \in C(M)$ would be linearisable about any of its zeros since any Killing vector field is linearisable. It is clear, however, from the example of the previous paragraph that such a strong result is too much to expect. Also, it is clear even without this example that such a result must fail since when it is applied to a conformally flat space-time one would achieve local Killing algebras of dimension fifteen and which exceeds the maximum allowed (theorem 10.2).

One is thus tempted to change the question by insisting that the original space-time is not conformally flat. However, one still does not have a theorem because of the possibility of the existence of a proper homothetic vector field with a zero. Such a zero would, necessarily, be a homothetic zero whilst after a reduction, should this be possible, it would become a zero of a Killing vector field and (11.24) is contradicted (*i.e.* $\phi(m) \neq 0$ and $\phi'(m) = 0$). Thus, for the metric (10.43), no neighbourhood U and function σ exist about any of the zeros of the proper homothetic vector fields admitted. However, in this metric, the orbits of $H(M)$ vary in nature (null or timelike) and dimension (two or three) as described earlier. This suggests a restriction on the variation of the orbit nature and dimension over M may be required. But such a variation may involve a variation in the nature of $\overset{*}{C}_m$ over M and this could give rise to a variation in the Petrov type over M. Hence a restriction on the Petrov type may also be needed.

Apart from a technical clause (the lack of local conformal vector fields which cannot be globally extended to M) one has now effectively achieved a statement of the reduction theorem. Various parts of this theorem (often ignoring some of the difficulties mentioned above, particularly those concerning zeros of members of $C(M)$ and not stressing that the function σ may only exist locally) were given in [212]-[214]. The statement here is taken from [209] where the proof can be found.

Theorem 11.6 *Let M be a space-time with metric g and conformal algebra $C(M)$ and suppose that either there is a single orbit (equal to M) associated with $C(M)$ or that these orbits are everywhere of the same nature (timelike, spacelike or null) and of the same dimension r $(1 \leq r \leq 3)$. Suppose also that the Petrov type is the same at each point of M and that there does not exist any conformal vector field on some proper open subset V of M $(\emptyset \neq V \neq M)$ which cannot be globally extended to a member of $C(M)$. Then*

i) for each $m \in M$ there exists an open neighbourhood U of p and a function $\sigma : U \to \mathbb{R}$ such that the restrictions of the members of $C(M)$ to U constitute a Lie algebra of special conformal vector fields on U with respect to the metric $e^{\sigma}g$ on U,

*ii) if the (constant) Petrov type of M is not **O** then the open set U and function σ in (i) can be chosen so that the restrictions of the members of $C(M)$ to U constitute a Lie algebra of homothetic vector fields on U with respect to the metric $e^{\sigma}g$ on U,*

*iii) if the Petrov type of M is not **O** and if the original space-time is not locally conformally related to a plane wave about any point of m the above objects U and σ may be chosen so that the restrictions of the members of $C(M)$ to U constitute a Lie algebra of Killing vector fields on U with respect to the metric $e^{\sigma}g$ on U.*

It is remarked here that the construction of the neighbourhood U and function σ when the Petrov type (at each point) of M is **I**, **II** or **D** is immediate [214] since then there exists in some neighbourhood U of any $m \in M$ a smooth nowhere-zero function $\lambda : U \to \mathbb{C}$ which is, at each point of U, an eigenvalue of the Weyl tensor at that point. Then (11.20) shows that $\frac{d\lambda}{ds} + 2\phi\lambda = 0$ holds along an integral curve of any member of $C(M)$ satisfying (11.1) and the same equation is also satisfied by the real and imaginary parts of λ. Since $\lambda(m) \neq 0$ one may thus assume the existence of a real valued positive function $a : U' \to \mathbb{R}$ where U' is open and $m \in U' \subseteq U$ satisfying this equation. Then if $g' = ag$ on U' one has (see (11.10)) for any $X \in C(M)$

$$\mathcal{L}_X g' = (a_{,a}X^a + 2\phi a)g = \left(\frac{da}{ds} + 2\phi a\right) g = 0. \tag{11.32}$$

Thus the restriction to U' of each member of $C(M)$ is Killing on U' with respect to g'.

Theorem 11.6 covers the (local) simultaneous conversion of each member of $C(M)$ to a homothetic vector field with respect to the metric $g' = e^\sigma g$ on some neighbourhood U of m for any $m \in M$. It should be noted that, from the last restricting condition of this theorem, the resulting vector fields on U constitute the *total* Lie algebra of conformal vector fields on U with respect to g'. Also, theorems 11.6 and 10.9 can be combined to show that the restricting conditions required in theorem 11.6, with the exception of that regarding non-extendible local conformal vector fields, can be achieved in some neighbourhood of any point in an open dense subset W of M (*i.e.* "almost everywhere" in M). Thus if this condition regarding the absence of non-extendible local conformal vector fields is retained and if, in addition, the Weyl tensor does not vanish over any non-empty open subset of M, *the conclusions of theorem 11.6(ii) and (iii) hold in some open neighbourhood of any m belonging to an open dense subset of M (i.e. in some open neighbourhood of "almost any" point of M).*

Returning to the unproved part of theorem 11.5 suppose U ($\emptyset \neq U \subseteq M$) is open and the Petrov type is **III** at each point of U. Then it is clear from the previous paragraph that some open subset $W \subseteq U$ satisfies the conditions of theorem 11.6 except, possibly, the absence of local (inextendible) conformal vector fields. Now suppose $\dim C(M) = 5$ (and so $\dim \overset{*}{K}_m = 0$ and $\dim \overset{*}{C}_m = 1$). If a local conformal vector field X exists say defined on some connected open subset W' of W then X, together with the members of a basis for $C(M)$ restricted to W' are independent conformal vector fields in $C(W')$. Otherwise X could be written as a linear combination of these (restricted) members of $C(M)$ on W' and this would show that X was extendible to M. Thus $\dim C(W') \geq 6$ contradicting theorem 11.5 applied to W' which would give $\dim C(W') \leq 5$. Thus the conditions of 11.6 are satisfied on W and so local conformally related metrics exist in W whose Petrov type is still **III** but whose Killing algebra is 5-dimensional. This contradicts the second part of theorem 11.5.

For individual vector fields in $C(M)$ the following should be noted. Let $X \in C(M)$, let $m \in M$ with $X(m) = 0$ and suppose an open neighbourhood U of m and a function $\sigma : U \to \mathbb{R}$ exist such that X, *restricted to* U, *is homothetic with respect to* $g' = e^\sigma g$ *on* U. (U and σ always exist if $X(m) \neq 0$.) Then, from section 10.4 X *is linearisable about* m. Now, using a semi-colon and a stroke for the covariant derivatives arising in the usual way from g and g', respectively, on U one has, in an obvious notation, on

U (and with c constant)

$$X^a{}_{;b} = \phi\delta^a_b + F^a{}_b \qquad X^a{}_{|b} = c\delta^a_b + F'^a{}_b. \tag{11.33}$$

A subtraction of the equations in (11.33) and using $\{\ \}$ and $\{\ \}'$ for the Christoffel symbols constructed from g and g', respectively, one has

$$X^c P^a_{bc} = (\phi - c)\delta^a_b + (F^a{}_b - F'^a{}_b) \qquad \left(P^a_{bc} = \left\{ {a \atop bc} \right\} - \left\{ {a \atop bc} \right\}'\right) \tag{11.34}$$

(which, incidentally, when evaluated at m reveals (11.24)). Now take a ($g-$) covariant derivative of (11.34) to get

$$(F^c{}_d + \phi\delta^c_d)P^a_{bc} + X^c P^a_{bc;d} = \phi_d\delta^a_b + F^a{}_{b;d} - F'^a_{b;d}. \tag{11.35}$$

Then evaluate (11.35) at m and contract over the indices a and b to get (since $P^a_{ac} = 2\sigma_{,c}$) at m

$$4\phi_d = (F^c{}_d + \phi\delta^c_d)P^a_{ac} = 2X^c{}_{;d}\sigma_{,c}. \tag{11.36}$$

Thus for such a conformal vector field, $\phi^a(m)$ $(\equiv g^{ab}\phi_{,b}(m))$ *is in the range of the linear map* $T_mM \to T_mM$ *with matrix* $X^a{}_{;b}(m)$ $(= X^a{}_{,b}(m))$. Now consider the following statements for $X \in C(M)$, $m \in M$, $X(m) = 0$. (i) There exists an open neighbourhood U of m and a function $\sigma : U \to \mathbb{R}$ such that X, restricted to U, is *homothetic* (and *Killing* if, in addition, m is an isometric zero of X) with respect to the metric $g' \equiv e^\sigma g$. (ii) X is linearisable about m and (iii) $\phi^a(m)$ is in the range of the linear map $T_mM \to T_mM$ with matrix $X^a{}_{,b}(m)$ $(= X^a{}_{;b}(m))$. The above remarks show that (i) $\Rightarrow$ (ii) and (i) $\Rightarrow$ (iii). In fact it can be shown that *these three statements are equivalent* [215],[216] - see also [210].

The zeros of a conformal vector field X can be classified according to the nature and eigenvalues of $F(m)$ and the value of $\phi(m)$. This classification is tabulated (with some changes of notation) in [166] where further details can be found. The results of the previous paragraph may now be utilised to obtain the following result about individual conformal vector fields [166],[210].

Theorem 11.7 *Let M be a space-time with metric g and X a non-trivial conformal vector field on M satisfying (11.1). Let $m \in M$ be a zero of X and suppose the Weyl tensor is not zero at m. Then with the possible exception of the special cases when the Petrov type at m is N* and *where the conformal function ϕ and bivector F satisfy* either $\phi(m) = 0$ *and* $F(m)$ *is null,* or $\phi(m) \neq 0$ *and* $F(m)$ *is timelike, there exists an open neighbourhood*

U of m and a function $\sigma : U \to \mathbb{R}$ such that the restriction of X to U is homothetic with respect to the metric $e^\sigma g$ on U (and so X is linearisable at m).

11.5 Conformal Vector Fields in Vacuum Space-Times

Let M be a non-flat *vacuum* space-time and let X be a *proper* conformal vector field on M. Let V be the open dense subset of M on which the curvature tensor does not vanish. Then (11.8) shows that $d\phi$ is covariantly constant on M (and so X is *special conformal*) and hence, since M is connected and X *proper* conformal, it follows that $d\phi$ is nowhere zero on M. Thus in any coordinate domain in V the Ricci identity and vacuum condition give $0 = \phi_a R^a{}_{bcd} = \phi_a C^a{}_{bcd}$ and so at each point of V the Petrov type is **N** (section 7.3) and ϕ^a is proportional to the unique repeated principal null direction of the Weyl tensor at that point. *It follows, since $\phi_{a;b} = 0$, that each point of V admits a neighbourhood which, with the original metric g on M restricted to it, is a (vacuum) pp-wave.*

One can say more here. If Y is another proper conformal vector field on M with conformal function ψ then $d\psi$ is covariantly constant and nowhere zero on M. Thus $d\phi$ and $d\psi$ are null and proportional on V and hence on M since V is dense in M. Hence $d\phi = \chi d\psi$ for some smooth function $\chi : M \to \mathbb{R}$. It then follows since $d\phi$ and $d\psi$ are covariantly constant and M is connected that χ is constant on M. So $\phi = \chi\psi + c$ for some constant c and it follows that *$X - \chi Y$ is homothetic on M*. Thus $\dim C(M) \leq \dim H(M) + 1$ and the following result is established [121] (see also [217]).

Theorem 11.8 *Let M be a non-flat vacuum space-time whose curvature tensor is nowhere zero on the open dense subset V of M. Suppose M admits a proper (necessarily special) conformal vector field X. Then each point of V admits an open neighbourhood in which the metric g on M restricts to a pp-wave. Also if $X, Y \in C(M)$ then some linear combination of them is homothetic and so* $\dim C(M) \leq \dim H(M) + 1$. *For such space-times, $C(M) = SC(M)$.*

This theorem is somewhat loosely paraphrased as "the only vacuum space-times admitting proper conformal vector fields are the *pp*-waves". An alternative proof of this theorem can be obtained by restricting X to the subset V and letting ϕ_t represent a local diffeomorphism of X with connected, open domain $U \subseteq V$ and range $\phi_t(U) \subseteq V$. Let g_1 and g_2 be the restrictions of the metric g on M to U and $\phi_t(U)$, respectively. Then following the ideas of section 10.1, g_1 and $\phi_t^* g_2$ are *vacuum* metrics on U which are conformally related on U and a short argument using theorem 9.5 completes the proof.

If M is an everywhere type **N** vacuum plane wave (see (8.11) and (10.42)) then (section 10.6), *in general*, $\dim K(M) = 5$ and $\dim H(M) = 6$ with the orbits for $K(M)$ and $H(M)$ being the 3-dimensional null hypersurfaces of constant u [41], [55], [218]-[220]. There are special cases (the type **N** *homogeneous* (vacuum) plane waves) where $\dim K(M) = 6$ and $\dim H(M) = 7$ and where $K(M)$ and $H(M)$ admit the single orbit M. For these homogeneous plane waves no proper conformal vector fields are possible (theorem 11.5). For the non-homogeneous (vacuum) plane waves a proper conformal vector field may be admitted in special cases [218],[219] but it cannot be tangent to the homothetic orbits of constant u [166]. (This gives a very slight generalisation of theorem 11.3). An extensive treatment of Killing symmetry for such metrics is given in [220].

It is remarked here that for a plane wave space-time the Ricci tensor takes the form $R_{ab} = \rho(u)\ell_a\ell_b$ for some real valued function ρ [41],[55]. If one constructs the metric $g' = e^{\sigma(u)}g$ then g' is conformally related to the original metric g on the open subset U of M where σ is defined and is in fact a *vacuum* metric on U provided σ is chosen to satisfy $\ddot{\sigma} - \frac{1}{2}\dot{\sigma}^2 = \rho(u)$ as is easily checked from a formula in [55]. Further, it follows from (11.10) that any vector field X tangent to the hypersurfaces of constant u (and so $u_{,a}X^a = 0$) and which is proper homothetic (respectively, Killing) with respect to g is proper homothetic (respectively, Killing) with respect to g'. Thus it follows that g' is a plane wave. Hence *any plane wave is locally conformal to a vacuum plane wave* and hence, in this local sense, has the same Lie algebra of conformal vector fields, from (11.10).

As a final remark it is known that geodesically complete vacuum plane waves exist [41] and so all members of the affine algebra $A(M)$ for such space-times are complete vector fields [20]. Hence, from theorem 5.10, $K(M)$ and $H(M)$ arise from Lie group actions on M.

11.6 Other Examples

The difficulties associated with the study of proper conformal vector fields have inevitably resulted in the lack of interesting metrics in general relativity possessing them. Of course, one can always start with a known metric which possesses Killing vector fields and, by means of a conformal change of metric, convert some or all of these vector fields to proper conformal vector fields using (11.10). But the energy-momentum tensor of the new metric will differ from that of the old one, usually in a complicated way, and the physical interpretation will, in general, be lost. Thus one usually has to resort to a direct integration of (11.1).

There is, however, a technique which, for certain space-times, reduces the problem of finding space-time conformal vector fields to a much simpler problem. First, consider those space times which have a reducible holonomy group (*i.e.* those of holonomy type $R_2 - R_4$, $R_6 - R_{14}$ in chapter 8). A detailed study has been made of the algebra $C(M)$ for such space-times in [221] following earlier work in [222],[223]. The following theorem, whose proof can be found in these references, summarises these results and uses the notation of section 8.3.

Theorem 11.9

i) Let M be a holonomy type R_{13} space-time which admits a special coordinate domain and metric (8.3) satisfying the condition that its (isometric) 3-dimensional submanifolds of constant t, with their induced geometry, admit no non-trivial conformal vector field. Then if X is a conformal vector field on M, $X = ku$, where k is constant and $u = \frac{\partial}{\partial t}$, and so X is a (covariantly constant) Killing vector field. A similar, obviously modified, result holds in the corresponding type R_{10} case.

ii) If M is any space-time of holonomy type R_2, R_4 or R_7 and M' is the (open) subset of M on which the Weyl tensor is non-zero then any conformal vector field on M is homothetic on each component of M' (and hence on M if M' is dense in M).

iii) If M is any space-time of holonomy type R_3 or R_6 and if M' is the open subset of M on which the Weyl tensor is non-zero then there are open subsets U and V of M such that any conformal vector field on M is Killing when restricted to U, special conformal when restricted to V and $U \cup V$ is open and dense in M' (and hence in M if M' is dense in M).

It is further remarked here that for the other reducible holonomy types R_6, R_8, R_9, R_{11}, R_{12} and R_{14} (and assuming M is not conformally flat), any point in the open subset of M on which the Weyl tensor does not vanish admits a coordinate neighbourhood W and a null recurrent vector field ℓ on W (section 8.3) such that $\ell_a = u_{,a}$ for some $u : W \to \mathbb{R}$. Then the conformal function ϕ of any conformal vector field on M, when restricted to W, satisfies $\phi = \phi(u)$.

Thus if one can show that a given metric g is conformally related to one of the reduced holonomy metrics discussed above, the algebra $C(M)$ of g is identical to that of its "conformal holonomy type". For example, consider the class of spherically symmetric metrics defined on a chart with coordinates t, r, θ, ϕ and given by

$$ds^2 = -e^{\nu(r,t)}dt^2 + e^{\mu(r,t)}dr^2 + r^2(d\theta^2 + \sin^2\theta d\phi^2). \tag{11.37}$$

This is clearly conformally related (by the conformal factor r^2) to a metric of the R_7 holonomy type (8.5) which is the product of the usual 2-sphere in Euclidean space $\mathbb{R}^3$ and a 2-dimensional manifold M' with a Lorentz metric. Now, assuming that the region under consideration is such that the Weyl tensor does not vanish at any point of it, theorem 11.9(ii) shows that this latter metric admits no proper conformal vector field. In fact, (see section 10.6), the algebra $C(M)$ for this latter metric is just $H(M)$ and is easily calculated. The 2-sphere has a 3-dimensional Killing algebra consisting of the usual Killing vector fields on the 2-sphere. That there are no proper homothetic vector fields follows from its non-zero constant curvature and an obvious 2-dimensional reworking of theorem 10.5. This lack of a homothetic vector field means that the related type R_7 metric (8.5) admits no proper homothetic vector field and so its Killing algebra is the conformal algebra $C(M)$ for (11.37). If X is a member of this algebra, (11.10) with $\sigma = \log r^2$ will reveal whether it is Killing, homothetic or proper conformal for (11.37). Any Killing vector field X in the 2-sphere clearly goes to a Killing vector field in (11.37) (as expected from spherical symmetry) since, then, $X(\sigma) = \sigma_{,a}X^a = 0$. The only other conformal vector fields for (11.37) would arise from Killing vector fields in M' (for example, the static case, when ν and μ are independent of t, so that $\frac{\partial}{\partial t}$ is Killing. This general procedure can be used to show, for example, that the Schwarzschild metric admits no proper conformal or homothetic vector fields and can be applied to find the conformal algebra of several other space-times [174]. (That no proper conformals exist in the Schwarzschild metric also follows from theorem 11.8.)

Returning to those holonomy reduced space-times which admit covariantly constant vector fields and noting that such vector fields are hypersurface-orthogonal conformal vector fields one can now show that the existence of this latter type of vector field in a space-time with metric g is conformally invariant and that g is (locally or globally) conformally related to one of these above holonomy types [221].

Theorem 11.10 *Let M be a space-time which admits an everywhere timelike (respectively, spacelike) hypersurface-orthogonal conformal vector field X. Then M is globally conformally related to a space-time for which X is a covariantly constant timelike vector field, i.e. to one of holonomy type R_1, R_4 or R_{13} (respectively, for which X is a covariantly constant spacelike vector field, i.e. one of holonomy type R_1, R_2, R_3, R_4, R_6 or R_{10}). If M admits an everywhere null hypersurface-orthogonal conformal vector field X then M is locally conformally related to a space-time for which X is a covariantly constant null vector field, i.e. to one of holonomy type R_1, R_3, R_4, R_8 or R_{11}.*

Proof. The hypersurface-orthogonality of X gives, from (11.1), the equivalent conditions in some coordinate domain of any $m \in M$ (see the end of section 4.16)

$$X_{[a;b}X_{c]} = 0, \quad F_{[ab}X_{c]} = 0, \quad X_a = \chi\psi_{,a}. \tag{11.38}$$

Of these, the second says that at each $m \in M$ either $F = 0$ or F is simple and its blade contains $X(m)$. The third is a local condition for functions χ and ψ on some neighbourhood of each $m \in M$. For the cases when X is everywhere spacelike or everywhere timelike, define the positive function $\rho = |g(X, X)|$ on M and then define the metric g' on M by $g' = \rho^{-1}g$ where g is the original metric on M. Then from (11.1)

$$\rho_{,a}X^a = \pm(g_{ab}X^aX^b)_{;c}X^c = 2\phi\rho \tag{11.39}$$

and

$$\mathcal{L}_X g' = (2\rho^{-1}\phi - \rho^{-2}(\rho_{,a}X^a))g = 0 \tag{11.40}$$

and so X is Killing with respect to g' and $|g'(X, X)| = 1$. Using a stroke for a covariant derivative with respect to g' and defining $X'_a = g'_{ab}X^b$ one then has after a short calculation

$$X'_{a|b} = G_{ab}(= -G_{ba}) \quad \text{and} \quad G_{ab}X^b = 0 \tag{11.41}$$

for some bivector G. The second equation in (11.41) follows since $g'_{ab}X^aX^b = \pm 1$ and so $X^aX'_{a|b} = X^aG_{ab} = 0$. Then $X'_a = \rho^{-1}X_a$ also satisfies (11.38) after replacing the g-covariant derivative with a g'-covariant derivative and so $G_{[ab}X'_{c]} = 0$. This last condition and the second equation in (11.41) shows, since X is nowhere null or zero on M, that $G \equiv 0$ on M

and the result follows. If X is null, one uses the local (third) condition in (11.38) where, since X is nowhere zero on M, one may assume $\chi > 0$ on some coordinate domain of any point of M. On this domain U define the metric $\tilde{g}_{ab} = \chi^{-1} g_{ab}$ and the covector field $\tilde{X}_a \equiv \tilde{g}_{ab} X^b = \psi_{,a}$. Using a double stroke for a $\tilde{g}$-covariant derivative one immediately has $\tilde{X}_{[a\|b]} = 0$ and hence, since X is still conformal with respect to $\tilde{g}$, $\tilde{X}_{a\|b} = \sigma \tilde{g}_{ab}$ for some $\sigma : U \to \mathbb{R}$. A contraction of this last equation with X^a and use of the fact that X is null with respect to $\tilde{g}_{ab}$ reveals $\sigma = 0$ on U. Hence X is $\tilde{g}$-covariantly constant on U. □

This theorem gives a conformally invariant way of describing certain local holonomy decompositions for which theorem 11.9 then gives information about $C(M)$. It also gives some information on the Petrov type of space-times with hypersurface-orthogonal conformal vector fields since the Petrov types are conformally invariant (theorem 8.12). One immediate consequence of this is the well-known result [41] that *any static space-time* (section 10.7) *is everywhere of Petrov type* **I**, **D** *or* **O**. It also follows that if M admits a null hypersurface-orthogonal conformal vector field X then M is algebraically special at each point with X as a repeated principal null direction.

Theorem 11.10 can be generalised by removing the restriction that $g(X, X)$ be everywhere of the same sign (or zero). But then one must make a topological decomposition of M and global results are replaced by local ones. It is perhaps of interest to point out that in the proof of theorem 11.10, for X non-null, the steps leading to X being a Killing vector field of the *global* metric g' on M did not require the assumption that X be hypersurface-orthogonal. This result is a slight extension of theorem 11.6.

In section 11.5, vacuum space-times were discussed. If one considers space-times whose energy-momentum tensor consists purely of a non-zero cosmological constant term then every conformal vector field is Killing unless the space-time has non-zero constant curvature [224]. For a vacuum space-time which possesses an asymptotically flat restriction, again every conformal vector field is Killing [176]. A study of the situation for various other energy-momentum tensors was given in [225]-[228].

11.7 Special Conformal Vector Fields

The special conformal algebra of M, $SC(M)$, consisting of those members of $C(M)$ satisfying (11.1) with $\phi_{a;b} \equiv 0$ on M, was introduced in section

11.1. In Minkowski space (11.8) confirms that $SC(M) = C(M)$ in this case and so proper (*i.e.* non-homothetic) members of $SC(M)$ exist there. (As pointed out earlier the same equation confirms that $SC(M) = C(M)$ for any vacuum space-time.) However, it will be shown in this section, following [229] (see also [222]) that the existence of proper members of $SC(M)$ is very restrictive. The following result is noted in passing.

Theorem 11.11 *Let M be a space-time and $X \in C(M)$. Then the following are equivalent:*
(i) $X \in SC(M)$ (ii) $\mathcal{L}_X R_{ab} = 0$ (iii) $\mathcal{L}_X G_{ab} = 0$ (iv) $\mathcal{L}_X R^a{}_{bcd} = 0$.

Proof. The equivalence of (i) and (ii) follows quickly from (11.9). Also the condition $\mathcal{L}_X g_{ab} = 2\phi g_{ab}$ implies that $\mathcal{L}_X g^{ab} = -2\phi g^{ab}$ and also

$$\mathcal{L}_X R = -6\phi^{c}_{;c} - 2\phi R. \tag{11.42}$$

Then (i) and (ii) show that $\mathcal{L}_X R = -2\phi R$ and hence that $\mathcal{L}_X (R g_{ab}) = 0$. The condition (iii) then follows. Conversely, if (iii) is assumed to hold and is written out in full using the definition of the Einstein tensor and (11.9) and then contracted with g^{ab} one finds $\phi^{a}_{;a} = 0$. This together with (11.42) then gives $\mathcal{L}_X (R g_{ab}) = 0$ and hence (ii). So (i), (ii) and (iii) are equivalent. Next, it is clear that (iv) implies (ii) so suppose (i), (ii) and (iii) hold. Then (11.13) and (11.14) hold and so $\mathcal{L}_X (R G^a{}_{bcd}) = 0$. Also, from (7.10) and (7.11), $\mathcal{L}_X \tilde{R}_{ab} = 0$ and $\mathcal{L}_X E_{abcd} = 2\phi E_{abcd}$ hold and hence $\mathcal{L}_X E^a{}_{bcd} = 0$. The condition (iv) now follows. □

One consequence of this result is a more immediate proof than the one given earlier of the fact that $SC(M)$ is a Lie algebra. Another is that, from conditions (ii) and (iii), the Segre type including degeneracies, of the Ricci and energy-momentum tensors is preserved along any integral curve of $X \in SC(M)$. One can also use arguments similar to those at the beginning of section 11.2 to show that the behaviour of Weyl *and* energy-momentum tensor eigenvalues along integral curves of any $X \in SC(M)$ is given by $\mathcal{L}_X \sigma = -2\phi\sigma$ for any such eigenvalue σ. A further consequence of (ii) and (iii) is that the behaviour of the eigenvalues of the Weyl *and* energy-momentum tensors at any zero m of a non-trivial member of $X \in SC(M)$ is as described in the Killing case if $\phi(m) = 0$ and in the homothetic case if $\phi(m) \neq 0$.

Now suppose M admits a *proper* member X of $SC(M)$. Then the associated conformal function leads to a nowhere-zero covariantly constant vector field given (in any coordinate system of M) by $\phi^a \equiv g^{ab}\phi_b$. Then define a global vector field on M with components Z^a where $Z_a \equiv F_{ba}\phi^b$. Then Z_a is a global gradient since, from (11.1)

$$Z_a = (X_{b;a} - \phi g_{ba})\phi^b = Z_{,a} \qquad (Z = X_a\phi^a - \frac{1}{2}\phi^2) \tag{11.43}$$

and also $Z_a\phi^a \equiv 0$ on M. Next, using (11.4a) and the equation $R_{abcd}\phi^d = 0$ (since $\phi^a_{;b} = 0$), one has

$$Z_{a;b} = F_{ca;b}\phi^c = -(\phi_c\phi^c)g_{ab} + \phi_a\phi_b. \tag{11.44}$$

Since ϕ^a is covariantly constant it will be of the same nature (spacelike, timelike or null) at each point of M.

Now suppose ϕ^a is *non-null*. Then $\phi_c\phi^c$ is a non-zero constant on M and (11.44) and theorem 10.1 show that Z^a is *a global proper affine vector field on* M which cannot vanish over any non-empty open subset of M since $Z_{a;b}$ cannot. Let $U \subseteq M$ be the open dense subset of M on which Z^a is not zero. Since M is, as usual, taken to be non-flat, the curvature tensor will, without loss of generality (since the intersection of two open dense subsets of M is open and dense), also be taken to be non-zero over U. Then since $\phi_a Z^a \equiv 0$ on M and ϕ^a is nowhere null or zero on M it follows that $Z^a(m)$ and $\phi^a(m)$ are independent members of T_mM for each $m \in U$. Now (11.44) and the Ricci identity show that $R_{abcd}Z^d = 0$ and also $R_{abcd}\phi^d = 0$. Thus the curvature tensor has rank one on U (see section 9.3) and hence has rank at most one on M (by the rank theorem in section 3.11). Thus one may write $R_{abcd} = \chi G_{ab}G_{cd}$ for some nowhere-zero function χ and (necessarily) simple bivector G on U, the bivector G having a blade orthogonal to that of the simple bivector $\phi_{[a}Z_{b]}$ on U. Now (11.44) shows that $Z^a Z_{a;b}$ is nowhere zero on U and so Z^a cannot become null over any non-empty open subset of U. Thus, without loss of generality, one may assume Z^a doesn't become null over U. So, since ϕ^a is nowhere null and $Z^a\phi_a = 0$ on U it follows (section 7.2) that the bivector $\phi_{[a}Z_{b]}$ (and hence G) is nowhere null on U. These bivectors thus constitute an orthogonal spacelike-timelike pair on U. The timelike member of this pair then determines two independent null vectors ℓ and n at each $m \in U$ each of which is a Ricci eigenvector at m with eigenvalues equal. This follows from the above expression for R_{abcd} and the possibilities for G. Thus, using the theory of chapter 7, one sees that ℓ and n are each repeated principal null directions of the Weyl tensor at m and hence the Petrov type is **D** or **O** at each point of U (and can be checked to be type **D** from the proof of theorem 8.12). The Ricci tensor is also non-zero on U (hence M cannot be a vacuum space-time) and in fact has Segre type $\{(1,1)(11)\}$ at each point of U. It is remarked here that because of the property $\phi^a_{;b} = 0$ of the vector field ϕ^a, the holonomy group of M is necessarily reducible.

The vector fields ϕ^a and ψ^a determine a 2-dimensional distribution of the Fröbenius type on U which is integrable since, from (11.44), $[\phi^a, Z^a] = 0$. The resulting integral manifolds are non-null (and will be referred to as the *submanifolds associated with* X) and the orthogonal distribution to it is also non-null and integrable and its integral manifolds are characterised by the constancy of the functions ϕ and Z. Thus this geometrical discussion shows that any *proper* member of $SC(M)$ whose associated vector field ϕ^a is not null uniquely determines the vector fields ϕ^a and Z^a, the open dense subset U and the 2-dimensional distribution that they span on U. Now construct from X the vector field Y by

$$Y^a = \lambda(-\phi Z^a + (\frac{1}{2}\phi^2 + Z)\phi^a) \tag{11.45}$$

where $\lambda = (\phi_c \phi^c)^{-1}$ is constant. Then from (11.43) and (11.44)

$$Y_{a;b} = \phi g_{ab} + 2\lambda\phi_{[a}Z_{b]}. \tag{11.46}$$

Thus Y is also a proper member of $SC(M)$ with the same conformal function ϕ as X. It is also clear from (11.40) that the vector field Z'^a constructed from Y (just as Z^a was constructed from X) satisfies $Z'^a = Z^a$ and so the subset U of M and the submanifolds associated with Y are identical to those associated with X. But now Y is actually tangent to these submanifolds. In fact, Y satisfies $Y_{[a;b}Y_{c]} = 0$ and so is hypersurface orthogonal (being proportional to the gradient of the function $Z\phi^{-1} - \frac{1}{2}\phi$ where this function is defined). So, starting from the existence of $X \in SC(M)$ one is guaranteed a proper member of $SC(M)$ tangent to the submanifolds associated with X. Now return to the original proper member $X \in SC(M)$ and construct from it the vector field W given by

$$W^a = [X^a, \phi^a] = \phi\phi^a - Z^a \qquad (\Rightarrow W_{a;b} = \lambda^{-1}g_{ab}). \tag{11.47}$$

The implication in brackets, which follows from (11.44) shows that W *is a proper homothetic vector field* tangent to the submanifolds associated with X and is, in fact, a gradient, being the gradient of $\frac{1}{2}\phi^2 - Z$.

To understand the geometry of the situation it is useful to prove a type of uniqueness result for $SC(M)$ and which is quite general, that is, it is not restricted by any non-null or null assumptions on ϕ^a. So let X and X' be any two proper members of $SC(M)$ with respective conformal functions ϕ and ϕ' and bivectors F and F'. Then $\phi^a_{;b} = \phi'^a_{;b} = 0$ and so either $\phi'^a = a\phi^a$ $(0 \neq a \in \mathbb{R})$ or ϕ'^a and ϕ^a are nowhere proportional on M. In the former case it follows that $X' - aX$ is homothetic on M. In the latter case suppose one of them, say ϕ^a, is not null and construct U and Z^a as was done earlier. Then ϕ^a, ϕ'^a and Z^a are each non-trivial solutions for k of the equation $R_{abcd}k^d = 0$ at each point of M. If these vector fields give rise to independent tangent vectors at any point $m \in M$ they will do so over some open neighbourhood V of m and the curvature tensor will vanish on V contradicting the non-flatness of M. Thus there are functions α and β on M such that, on M, $Z_a = \alpha\phi_a + \beta\phi'_a$ and, since ϕ_a and ϕ'_a are nowhere proportional, α and β are smooth (Cramer's rule). But $Z_{a;bc} = 0$ from (11.44) and so $\alpha_{;ab} = \beta_{;ab} = 0$ on M. Hence $\alpha_{,a}$ and $\beta_{,a}$ are each either zero or, by the same argument, a similar combination of ϕ_a and ϕ'_a. These possibilities lead to expressions for $Z_{a;b}$ which have rank zero and two, respectively, contradicting the expression (11.44) for $Z_{a;b}$ (which is clearly a matrix of rank three everywhere on M). (It is noted in passing here that this contradiction arose from the facts that ϕ^a is non-null and $\phi'^a_{;b} = 0$ and did not depend on ϕ'^a arising from a (proper) member of $SC(M)$. Hence if $X \in SC(M)$ with X proper and ϕ^a non-null *no other independent covariantly constant vector field can be admitted by* M.) Now suppose, still in the above latter case, that ϕ^a and ϕ'^a *are both null*. Then it is easily checked from the Ricci identity and theorem 7.12 that the Petrov type is **D** at each $m \in M$ with ϕ^a and ϕ'^a as repeated principal null directions at each point of M (the type **O** case being impossible since the curvature rank is still one). Then when (11.21) is applied to ϕ'^a one gets $F_{ab}\phi'^b = \nu\phi'_a$ for some function ν on M. When this last equation is covariantly differentiated and (11.4a) is used one gets

$$(\phi_c\phi'^c)g_{ab} = \phi_a\phi'_b + \phi'_a\nu_{,b}. \tag{11.48}$$

A consideration of the rank of each side of (11.48) gives the contradiction that $\phi_a\phi'^a = 0$ on M (since ϕ_a and ϕ'_a are null and not proportional). The conclusion is that if M admits two proper members $X, X' \in SC(M)$ with conformal functions ϕ and ϕ' then $\phi'_a = a\phi_a$ $(a \in \mathbb{R})$ and *some linear combination of* X *and* X' *is homothetic* (and $\phi' = a\phi + b$, $b \in \mathbb{R}$). This is the "uniqueness" result alluded to earlier since it says that given one proper member $X \in SC(M)$ any other member of $SC(M)$ is a linear combination of X and a member of $H(M)$. Said another way, $\dim SC(M) \leq \dim H(M) + 1$ and the covariantly constant vector field ϕ^a

determined by each member of $SC(M)$ is the same up to constant multiples.

Again consider the case when there exists a proper member $X \in SC(M)$ *with* ϕ^a *non-null.* If $X' \in SC(M)$ with associated function ϕ' and bivector F' then $\phi'_a = a\phi_a$ $(a \in \mathbb{R})$ and $X' = aX + Q$ where Q is homothetic on M (with homothetic bivector H) and

$$Q_{a;b} = cg_{ab} + H_{ab} \qquad (\Rightarrow F'_{ab} = aF_{ab} + H_{ab}). \tag{11.49}$$

Now define a function γ on M by $\gamma = Q^a\phi_a$ and apply (11.4b) to Q to get $\gamma_{;ab} = 0$. Since ϕ^a is non-null then, as remarked earlier, it is the only independent covariantly constant vector field on M. Thus $\gamma_{,a} = d\phi_a$ $(d \in \mathbb{R})$. Then the first equation in (11.49) when contracted with ϕ^a shows that $H_{ab}\phi^b \propto \phi_a$ and hence, since ϕ^a is non-null, $H_{ab}\phi^b = 0$ on M. The second equation in (11.49) then shows that the vector Z^a and Z'^a associated with X and X' as in (11.43) satisfy $Z'^a = aZ^a$. Thus the submanifolds associated with proper members X and X' are identical.

The geometrical interpretation is now clear. If M admits a proper member X of $SC(M)$ with associated ϕ^a non-null then as shown earlier X may not be tangent to its associated submanifolds but a proper member $Y \in SC(M)$ always exists which is, as is also the proper homothetic vector field W. If X is not tangent to these submanifolds, the above uniqueness result implies the existence of a homothetic vector field S (such that $Y = aX + S$, $a \in \mathbb{R}$) which is clearly not tangent to these submanifolds. Since a proper homothetic vector field W tangent to these submanifolds is already guaranteed, a Killing vector field independent of ϕ^a is also guaranteed. Such a vector field K can be directly constructed as

$$K^a = \lambda^{-1}X^a - (X^c\phi_c)\phi^a + \phi Z^a \qquad (\Rightarrow K_{a;b} = \lambda^{-1}F_{ab} - 2\phi_{[a}Z_{b]}) \tag{11.50}$$

and is clearly Killing and not tangent to the associated submanifolds if X is not.

If $X \in SC(M)$ is proper and ϕ^a not null then ϕ^a is the unique independent covariantly constant vector field on M and M is locally metrically decomposable (chapter 8). It was indicated in [229] how this decomposition could be used to construct such examples of M (and this was carried out in [230]). Such a decomposition was also used to place upper and lower bounds on the dimension of the various symmetry algebras for such space-times [229]. Briefly, $\dim K(M) \leq 4$. (For if $\dim K(M) \geqslant 5$ the only possibilities for the stable orbits are that they are of dimension 3 and 4 (theorem 10.13) and each contradicts the Segre type of the Ricci tensor -

the latter since it implies that each $m \in M$ is a zero of a proper homothetic vector field - see table 10.2 and theorem 10.5.) Hence from the above discussion $\dim H(M) = \dim K(M) + 1 \leq 5$. Hence, from the "uniqueness" argument above, $\dim SC(M) = \dim H(M) + 1 \leq 6$. Since an independent Killing, proper homothetic and proper member of $SC(M)$ exist, one achieves $3 \leq \dim SC(M) \leq 6$. It is also remarked (see theorem 10.14) that $\dim A(M) = \dim H(M) + 1$ and that $3 \leq \dim A(M) \leq 6$. It should be noted that these bounds only apply if a proper member of $SC(M)$ exists and ϕ^a is non-null.

If ϕ^a is null the situation is less clear. The "uniqueness" argument above still holds (and M could be a vacuum space-time - see theorem 11.8). One problem is that it is not clear whether another independent covariantly constant vector field exists. Another problem is that in this case M could be conformally flat (whereas if ϕ^a is non-null, conformal flatness is equivalent to flatness). Further details can be found in [229].

Chapter 12

Projective Symmetry in Space-times

12.1 Projective Vector Fields

In chapter 10 the concept of (local) geodesic preserving transformations which also preserved the affine parameter was studied. These were *affine symmetries*. In this chapter the more general concept of (local) geodesic preserving transformations which do not necessarily preserve affine parameters will be considered. These are the *projective symmetries* of space-time. The justification for their study is based on the importance of geodesics in Einstein's theory.

Let M be a space-time with Levi-Civita connection ∇ and X a global smooth vector field on M. Then, as before, decompose the covariant derivative of X in any coordinate system as

$$X_{a;b} = \frac{1}{2}h_{ab} + F_{ab} \qquad (h_{ab} = \mathcal{L}_X g_{ab} = h_{ba},\ F_{ab} = -F_{ba}). \tag{12.1}$$

Let ϕ_t be the local diffeomorphisms associated with X. Then X is called *projective* if and only if each ϕ_t maps a geodesic of M into a geodesic of M, *i.e.* if U is the domain of ϕ_t and c is a geodesic in U then $\phi_t \circ c$ is a geodesic in $\phi_t(U)$. Recalling the definition of $\mathcal{L}_X\nabla$ from section 10.2 it can be shown [159] (see also [231]) that X is projective if and only if there exists a 1-form ψ on M such that for arbitrary vector fields Y and Z on M

$$\mathcal{L}_X\nabla(Y, Z) = \psi(Y)Z + \psi(Z)Y. \tag{12.2}$$

Using (12.2), (10.5) and (10.6) this can be shown equivalent to the condition that in any coordinate system

$$X_{a;bc} = R_{abcd}X^d + 2g_{a(b}\psi_{c)}. \tag{12.3}$$

The fact that ∇ is a metric connection gives the extra information that ψ *is a closed* 1-*form* and hence locally a gradient [35]. The 1-form ψ is called the *projective* 1-*form* of X. If M is simply connected ψ is a global gradient on M [25].

Symmetrising (12.3) over the indices a and b and using (12.1) gives

$$h_{ab;c} = 2g_{ab}\psi_c + g_{ac}\psi_b + g_{bc}\psi_a. \tag{12.4}$$

Conversely, assuming (12.1) and (12.4) and using the Ricci identity and the strategy in section 11.1 one can get back to (12.3) and also to the consequence

$$F_{ab;c} = R_{abcd}X^d + g_{c[a}\psi_{b]}. \tag{12.5}$$

Thus, given the decomposition (12.1), the conditions (12.3) and (12.4) are equivalent to each other and to X being projective. The bivector F is called the *projective bivector* of X.

Now define a tensor P with components $P_{ab} = X_{a;b}$. If X is projective then (12.3) gives a convenient expression for $P_{ab;c}$ for substituting into the Ricci identity (4.35) for P. A short calculation then reveals

$$\begin{aligned} P_{eb}R^e{}_{acd} + P_{ae}R^e{}_{bcd} = P^e{}_dR_{abce} - P^e{}_cR_{abde} - R_{abcd;e}X^e \\ +g_{ac}\psi_{b;d} - g_{ad}\psi_{b;c}. \end{aligned} \tag{12.6}$$

This is nothing but an expression for the Lie derivative of the curvature tensor (and after an obvious contraction, for the Ricci tensor also). Thus

$$\mathcal{L}_X R^a{}_{bcd} = \delta^a_d\psi_{b;c} - \delta^a_c\psi_{b;d} \qquad \mathcal{L}_X R_{ab} = -3\psi_{a;b}. \tag{12.7}$$

Also, permuting the indices a, b and c in (12.4) gives the condition

$$3h_{ab;c} - h_{bc;a} - h_{ca;b} = 4g_{ab}\psi_c. \tag{12.8}$$

This is, in fact, equivalent to (12.4) as can be demonstrated by adding to twice (12.8) the two equations obtained from (12.8) by replacing the indices a, b, c by b, c, a and c, a, b, respectively.

Regarding the differentiability of X (and recalling the assumed smoothness of M and the metric g) X has, as in the affine case, been so far assumed smooth. However if one assumes X to be C^3 then (12.1) shows that h and F are C^2 and then (12.8) shows that ψ_a is C^1. The second equation in (12.7) then shows that ψ_a is C^2 and then (12.4) and (12.5) show that h and F are C^3 and hence, from (12.1), that X is C^4. Repetition of this process shows that X is, in fact, smooth. So the assumptions that X is C^3 and that X is C^∞ are equivalent. The smoothness of h, F and the projective 1-form ψ then follows from (12.1) and (12.8).

The equations (12.1), (12.4), (12.5) and the second of (12.7) lead to a system of first order differential equations for the quantities X^a, h_{ab}, F_{ab} and ψ_a, respectively, in a way very similar to that described in detail for affine vector fields in section 10.2. It follows that a projective vector field X on M is uniquely determined by the values of X^a, h_{ab}, F_{ab} and ψ_a at some (any) point of M. From (12.1) and (12.3) this is equivalent to specifying X, ∇X and $\nabla(\nabla X)$ at some (any) point of M. Also it is clear from (12.1) and (12.4) that the set of all projective vector fields constitutes a real vector space, denoted by $P(M)$, and from the defining properties of the maps ϕ_t, and (5.14), that $P(M)$ is closed under the Lie bracket operation and is hence a Lie algebra. Combining these last pieces of information shows that $\dim P(M) \leq 4 + 10 + 6 + 4 = 24$. The Lie algebra $P(M)$ will be referred to as the *projective algebra*. Thus one has

Theorem 12.1 *Let M be a space-time and $P(M)$ the set of projective vector fields on M. Then*

i) the members of $P(M)$ are characterised by their satisfying (12.1) and (12.3) or (12.1) and (12.4),
ii) each member X of $P(M)$ is uniquely determined by specifying at any point of M either X, h, F and the projective 1-form, ψ, or, X, ∇X and $\nabla(\nabla X)$. Hence if $X \in P(M)$ vanishes over a non-empty open subset of M it vanishes on M,
iii) $P(M)$ is a finite-dimensional Lie algebra of smooth vector fields on M and $\dim P(M) \leq 24$. $A(M)$ is a subalgebra of $P(M)$.

If the upper bound in theorem 12.1(iii) is achieved then M is of constant curvature. Conversely if M is of constant curvature then this maximum dimension is achieved locally (*i.e.* for each $m \in M$ there is some open neighbourhood U of m such that $\dim P(U) = 24$) [158]. For Minkowski space-time $\dim P(M) = 24$ and a basis for $P(M)$ may be chosen as the 20 affine vector fields in (10.19) together with the four projective vector fields whose components are given in the usual global coordinates t, x, y, z by (see, *e.g.* [232])

$$X^a = \sum_{b=1}^{4} A_{ab}\theta_b \qquad A_{ab} = \begin{pmatrix} t^2 & tx & ty & tz \\ xt & x^2 & xy & xz \\ yt & yx & y^2 & yz \\ zt & zx & zy & z^2 \end{pmatrix} \tag{12.9}$$

where the θ_a are four arbitrary constants which, once chosen to specify a projective vector field, are then easily checked to be the components of its projective 1-form. It can be seen from (12.9) that if X is a non-trivial projective vector field on M and $\lambda : M \to \mathbb{R}$ such that λX is projective then λ need not be constant. For example, let $X^a = tY^a$ with $Y^a = (t, x, y, z)$. Then X and Y are projective (and, in fact, Y is homothetic).

There is an interesting tensor which plays a similar role in studying connections as the Weyl tensor does for conformal changes of the metric. It is called the *projective tensor*, denoted by W, and defined in any coordinate system of a space-time by (see, *e.g.* [35])

$$W^a{}_{bcd} = R^a{}_{bcd} - \frac{1}{3}(\delta^a_c R_{bd} - \delta^a_d R_{bc}). \tag{12.10}$$

It has the property that if ∇ and ∇' are two symmetric connections on M with the same geodesics then the expressions (12.10) constructed from ∇ and ∇' are equal (see, *e.g.* [35]). It easily follows, by considering the local flow maps, that *if X is a projective vector field on M, $\mathcal{L}_X W^a{}_{bcd} = 0$.* Unfortunately, the converse of this result fails since (see chapter 13) in a vacuum space-time the condition $\mathcal{L}_X W^a{}_{bcd} = 0$ means that X is a *curvature collineation* and the collection of such vector fields need not be finite-dimensional. It also follows from (12.10) and (7.10) that, at any $m \in M$, (i) $W^a{}_{bcd} = C^a{}_{bcd} \Leftrightarrow E_{abcd} = 0$ and (ii) $W^a{}_{bcd} = R^a{}_{bcd} \Leftrightarrow R_{ab} = 0$. To prove (i) ($\Rightarrow$) the fact that $W^a{}_{bcd} = C^a{}_{bcd}$ implies that W_{abcd} is skew-symmetric in the indices a and b, is useful. Thus if the conditions in (i) hold at each $m \in M$, M is an Einstein space whilst if the conditions in (ii) hold at each $m \in M$, M is a vacuum space-time. Equations (12.10) and (7.10) also show that $W = 0$ at each $m \in M$ if and only if M has constant curvature.

A projective vector field X uniquely determines the symmetric tensor h_{ab} $(= \mathcal{L}_X g_{ab})$ in (12.1) and, from (12.8) uniquely determines ψ_a. Let X' be another projective vector field with associated tensors h'_{ab} and ψ'_a. If $h'_{ab} = ch_{ab}$ $(c \in \mathbb{R})$ then from (12.8) applied to X and X' one easily sees that $\psi'_a = c\psi_a$ and that $X' - cX$ is *Killing*. If, on the other hand, $\psi'_a = c\psi_a$ is assumed $(c \in \mathbb{R})$ then $h'_{ab} - ch_{ab}$ is covariantly constant and $X' - cX$ is *affine*. Thus for a projective vector field X, h_{ab} uniquely determines ψ_a whilst ψ_a determines h_{ab} up to a covariantly constant symmetric tensor field on M. If $h_{ab} \equiv 0$ on M, X is Killing and if $\psi_a = 0$ on M, X is affine. If X is projective but not affine it is called *proper projective* (and (12.9) gives examples of proper projective vector fields). If X is projective and $\psi_{a;b} \equiv 0$

on M, X is called *special projective*. (Again the examples in (12.9) are special projective and, from the basis for $P(M)$ with M Minkowski space given there, or from (12.7) it follows that in this case every projective vector field is special projective.) Also, from (12.7) every projective vector field in a vacuum space-time is special projective (in fact, affine - see theorem 12.3). Also, it follows from (12.7) that, for a projective vector field X, the conditions that it is (i) special projective, and (ii) satisfies $\mathcal{L}_X R_{ab} = 0$ (*i.e.* X is a Ricci collineation - see chapter 13) are equivalent. From this it immediately follows that the *set* $SP(M)$ *of special projective vector fields on* M *is a Lie algebra of smooth vector fields on* M *called the special projective algebra and is a subalgebra of* $P(M)$ *containing* $A(M)$.

The Lie algebra $P(M)$, since it is finite-dimensional, has a well behaved orbit structure. In fact, theorems 10.8 and 10.10 go through with S replaced by $P(M)$ and theorem 10.10 becomes

Theorem 12.2 *Let M be a space-time. Then each orbit of $P(M)$ is a leaf of M and a maximal integral manifold of $P(M)$.*

12.2 General Theorems on Projective Vector Fields

If X is a projective vector field on M the controlling equation (12.4) is not easy to handle. This section investigates certain situations where the Lie algebra can be determined and explores some techniques for doing this. These techniques revolve around finding convenient algebraic expressions for h_{ab} and $\psi_{a;b}$. A particularly simple example of this technique is utilised in the following theorem [233] (part of which was given in [234],[63] by different techniques).

Theorem 12.3 *Let M be a non-conformally flat Einstein space (so that the Weyl tensor of M does not vanish over any non-empty open subset of M). Then M does not admit a proper projective vector field. In particular a non-flat vacuum space-time admits no proper projective vector fields.*

Proof. Let M be decomposed as in theorem 7.15 into its Petrov types (but now with $\text{int}\,\mathbf{O} = \emptyset$) as

$$M = U \cup \text{int}\,\mathbf{N} \cup Z \tag{12.11}$$

where

$$U = \mathbf{I} \cup \text{int}\,\mathbf{II} \cup \text{int}\,\mathbf{D} \cup \text{int}\,\mathbf{III} \tag{12.12}$$

and Z has empty interior. Since M is an Einstein space two results from the previous section show that $W^a{}_{bcd} = C^a{}_{bcd}$ on M and that if X is a projective vector field on M (satisfying (12.1) and (12.4)) then $\mathcal{L}_X C^a{}_{bcd}$ $(= \mathcal{L}_X W^a{}_{bcd}) = 0$. Thus taking the Lie derivative with respect to X of the identity $g_{a(e}C^e{}_{b)cd} = 0$ and using the result $\mathcal{L}_X g_{ab} = h_{ab}$ one finds

$$h_{ae}C^e{}_{bcd} + h_{be}C^e{}_{acd} = 0. \tag{12.13}$$

Thus on contracting (12.13) with any complex bivector H^{cd} one sees that for each (complex) bivector G in the range of the Weyl tensor (*i.e.* each G of the form $G_{ab} = C_{abcd}H^{cd}$) $h_{ae}G^e{}_b + h_{be}G^e{}_a = 0$. One can now take the real and imaginary parts of this equation and employ theorem 9.1 to find eigenvectors of h with the same eigenvalue. One way to achieve this is to deal with each Petrov type in turn and use the results of sections 7.3 and 7.4. If, for example, the Petrov type is **III** at the point p in question then (7.75) shows that the real and imaginary parts of M and V are in the range of C. Then theorem 9.1 shows that in the obvious real null tetrad, ℓ, n, x and y are eigenvectors of h with the same eigenvalue. Thus $h_{ab} \propto g_{ab}$ at p. For Petrov types **I**, **D** and **II** at p, (7.77), (7.79) and (7.80) show in a similar way that the same result for h follows. Hence one must have $h_{ab} = \alpha g_{ab}$ on the open subset U for some smooth function α. Then using (12.4) on U one has

$$\alpha_{,c}g_{ab} = 2g_{ab}\psi_c + g_{ac}\psi_b + g_{bc}\psi_a \tag{12.14}$$

and a contraction over a and b yields $\alpha_{,a} = \frac{5}{2}\psi_a$. On substituting back into (12.14) one then finds

$$g_{ab}\psi_c = 2g_{ac}\psi_b + 2g_{bc}\psi_a. \tag{12.15}$$

If ψ_a is non-zero at some $m \in U$ then there exists $k \in T_mM$ such that $k^a\psi_a \neq 0$. If one contracts (12.15) with k^c and compares the rank (as a matrix) of each side of the resulting equality one arrives at a contradiction. Hence ψ_a is zero at m and hence everywhere on U. (In fact, quite generally, if $h_{ab} = \alpha g_{ab}$ with $\alpha : M \to \mathbb{R}$ and h satisfying (12.4) then α is constant.) Now consider the subset $\operatorname{int}\mathbf{N}$ of M. The expression for a type **N** Weyl tensor (section 7.3) and the above argument for the bivectors G now reveal that at each $m \in \operatorname{int}\mathbf{N}$ and any null tetrad ℓ, n, x, y at m, with ℓ spanning the repeated principle null direction of the Weyl tensor, ℓ, x and y are eigenvectors of h with the same eigenvalue. Thus (section 7.5) $h_{ab} = \beta g_{ab} + \gamma\ell_a\ell_b$ on $\operatorname{int}\mathbf{N}$ for smooth functions β and γ (ℓ is smooth from theorem

7.14, β is smooth since $h^a{}_a$ is smooth and the smoothness of γ is now clear). Again one substitutes into (12.4) and finds $\beta_{,a} = \frac{5}{2}\psi_a$ and a back substitution and contraction with ℓ^a, again using a rank argument, shows that $\psi_a = 0$ on $\operatorname{int} \mathbf{N}$. Thus $\psi_a = 0$ on the open dense subset $U \cup \operatorname{int} \mathbf{N}$ of M and hence on M. It follows from (12.4) that $h_{ab;c} = 0$ on M and hence that X is affine and not proper projective. This completes the proof. $\square$

At this point it is convenient to introduce some further consequences of the existence of projective vector fields. The following theorem is, in a sense, a generalised version of the technique used in the last theorem [235] (and for an earlier special case see [233]).

Theorem 12.4 *Let M be a space-time and for real bivectors G and H at $m \in M$ let $A = G + iH$ be a complex eigenbivector of the curvature tensor with eigenvalue $c = a + ib$ ($a, b \in \mathbb{R}$) so that $R_{abcd}A^{cd} = cA_{ab}$. Let X be a projective vector field on M satisfying (12.1) and (12.4).*

i) If A and hence c are real (H and b zero) and if A is simple (so that $A^{ab} = 2p^{[a}q^{b]}$ for $p, q \in T_mM$) then each vector in the blade of A (i.e. the span of p and q) is an eigenvector of the symmetric tensor P at m given by

$$P_{ab} = ah_{ab} + 2\psi_{a;b} \tag{12.16}$$

with the same eigenvalue.

ii) If A and hence c are real and A non-simple then from (7.39) A may be written in terms of a null tetrad ℓ, n, x, y at m in the form $A_{ab} = 2e\ell_{[a}n_{b]} + 2fx_{[a}y_{b]}$ with $e, f \in \mathbb{R}$, $e \neq 0$, $f \neq 0$. Then the 2-spaces of T_mM spanned by the vector pairs (ℓ, n) and (x, y) are eigenspaces of P.

iii) In the general case the real and imaginary parts G and H of A, the real numbers a and b and the symmetric tensors h_{ab} and $\psi_{a;b}$ at m satisfy the relations (12.19) below.

Proof. It should be noted first that the general expression for P in (12.16) is complex although as used in parts (i) and (ii) it is real because $c = a \in \mathbb{R}$.

The Ricci identity (4.35) for h_{ab} and (12.4) give

$$\begin{aligned} h_{ae}R^e{}_{bcd} + h_{be}R^e{}_{acd} \quad (= 2h_{ab;[cd]}) \\ = g_{ac}\psi_{b;d} - g_{ad}\psi_{b;c} + g_{bc}\psi_{a;d} - g_{bd}\psi_{a;c}. \end{aligned} \tag{12.17}$$

To establish (i) and (ii) contract (12.17) with A^{cd} (real) and rearrange to get

$$P_{ae}A^{e}{}_{b} + P_{be}A^{e}{}_{a} = 0. \tag{12.18}$$

Results (i) and (ii) now follow from theorem 9.1. To establish (iii) one contracts (12.17) with A^{cd} and performs similar rearrangements as in the previous case and then separates out the real and imaginary parts to get

$$\begin{aligned} P^{1}{}_{ae}G^{e}{}_{b} + P^{1}{}_{be}G^{e}{}_{a} &= P^{2}{}_{ae}H^{e}{}_{b} + P^{2}{}_{be}H^{e}{}_{a} \\ P^{1}{}_{ae}H^{e}{}_{b} + P^{1}{}_{be}H^{e}{}_{a} &= -P^{2}{}_{ae}G^{e}{}_{b} - P^{2}{}_{be}G^{e}{}_{a} \end{aligned} \tag{12.19}$$

where P^1 and P^2 are the real and imaginary parts of P,

$$P^{1}{}_{ab} = ah_{ab} + 2\psi_{a;b} \qquad P^{2}{}_{ab} = bh_{ab}. \tag{12.20}$$

□

This theorem provides algebraic information on the tensor P_{ab} (*i.e.* on h_{ab} and $\psi_{a;b}$) directly from the algebra of the curvature tensor. For those situations when the curvature tensor is highly degenerate in the algebraic sense the increase in dimension of the eigenspaces of P leads to especially simple expressions for P in terms of some tetrad closely associated with the curvature. A little more information is provided in the following corollary to theorem (12.4). Suppose $R_{abcd}A^{cd} = 0$ holds, for a (real) simple non-null smooth bivector A in some non-empty open region U of M. Then using (12.7)

$$\mathcal{L}_X(R^{a}{}_{bcd}A^{cd}) = 0 = (\delta^a_d\psi_{b;c} - \delta^a_c\psi_{b;d})A^{cd} + R^{a}{}_{bcd}\mathcal{L}_X A^{cd}. \tag{12.21}$$

Suppose A is spacelike so that in some (smooth) orthonormal tetrad (x, y, z, t) on U, and with no loss of generality, one has $A^{ab} = 2x^{[a}y^{b]}$ $(\Rightarrow A_{ab}A^{ab} = 2)$. Then (12.18) and theorem 12.4(i) with $c = 0$ show that x and y are eigenvectors of $\psi_{a;b}$ on U with eigenvalue α and so

$$\psi_{a;b} = \alpha g_{ab} + \beta z_a z_b + \gamma t_a t_b + 2\delta z_{(a}t_{b)} \tag{12.22}$$

for smooth functions α, β, γ and δ on U. A contraction of (12.21) with $A_a{}^b$ then reveals that $\alpha = 0$. A similar conclusion is reached if A is timelike. *Thus if the conditions of theorem 12.4(i) hold over a non-empty open subset U of M and if A is simple and non-null then, if $c = 0$ on U, $\alpha = 0$ on U.*

The link between the algebraic structure of the curvature tensor and that of h_{ab} and $\psi_{a;b}$ can now be used in conjunction with the classification scheme in section 9.3 to establish the following results [233].

Theorem 12.5 *Let M be a space-time such that for each $m \in M$ there exists an open neighbourhood U of m and a nowhere zero smooth vector field k on U such that $R_{abcd}k^d = 0$ everywhere on U. If X is a projective vector field on M then $\psi_{a;b} \equiv 0$ on M (i.e. X is necessarily special projective or affine on M).*

Proof. For each $k' \in T_mM$ there is an open neighbourhood V of m with $V \subseteq U$ and a vector field $\tilde{k}'$ on V such that $\tilde{k}'(m) = k'$. The (real) simple bivector $G^{ab} = 2k^{[a}\tilde{k}'^{b]}$ on V then satisfies $R_{abcd}G^{cd} = 0$ on V. Theorem 12.4(i) then shows that every member of T_mM is an eigenvector of $\psi_{a;b}$ at m with the same eigenvalue. Since at least one of these bivectors G is non-null at m the italicised remark following (12.22) shows that this eigenvalue is zero. Hence $\psi_{a;b} = 0$ at m and hence everywhere on M and the result follows. □

Theorem 12.6 *Let M be a space-time of (curvature) class C. Then any projective vector field on M is affine.*

Proof. Since space-times of class C satisfy the conditions of theorem (12.5) it follows that if X is a projective vector field on M then $\psi_{a;b} \equiv 0$ on M. Since M is connected it follows that either $\psi_a \equiv 0$ on M (and so X is affine from (12.4)) or ψ_a is nowhere zero on M. In the latter case the Ricci identity then shows that $R_{abcd}\psi^d \equiv 0$ on M and, by the class C condition, ψ^a is the unique tangent vector with this property at each $m \in M$ up to a scaling. Then (12.17) gives on M

$$h_{ae}R^e{}_{bcd} + h_{be}R^e{}_{acd} = 0 \tag{12.23}$$

and so by theorem 9.3(ii) there are smooth functions α and β such that

$$h_{ab} = \alpha g_{ab} + \beta\psi_a\psi_b. \tag{12.24}$$

On substituting this into (12.4) one finds

$$g_{ab}(\alpha_{,c} - 2\psi_c) = g_{ac}\psi_b + g_{bc}\psi_a - \beta_{,c}\psi_a\psi_b. \tag{12.25}$$

For any $m \in M$ choose $k \in T_mM$ such that $k^ak_a \neq 0$ and $k^a\psi_a = 0$ at m and contract (12.25) with k^ak^b to obtain $\alpha_{,a} = 2\psi_a$ on M. On substituting this back into (12.25) and contracting with k^a one finds $\psi_a \equiv 0$ on M and the result follows. □

Theorem 12.7 *Let M be a space-time of class B. Then any projective vector field on M is affine.*

Proof. From the work of section 9.3 there are, for this type, four simple bivectors G in some open neighbourhood of any $m \in M$ satisfying $R_{abcd}G^{cd} = 0$ and which, in the notation of section 9.3 may be taken as $\ell \wedge x$, $\ell \wedge y$, $n \wedge x$ and $n \wedge y$. It immediately follows from theorem 12.4(i) that ℓ, n, x and y are eigenvectors of $\psi_{a;b}$ with the same eigenvalue. Further the non-null simple bivector $G = (\ell + n) \wedge x$ also satisfies the above relation and so this common eigenvalue is zero (see remarks following (12.22)). Hence $\psi_{a;b} = 0$ on M. But this type of space-time admits no non-zero covariantly constant vector fields. Hence $\psi_a \equiv 0$ on M and X is affine. □

Theorem 12.8 *Let M be a space-time of class D. Then the number of independent global covariantly constant vector fields admitted by M is either none, one or two. If this number is either none or two any projective vector field on M is affine. If this number is one then a proper projective X may be admitted and, if so, X is special projective and unique in the sense that if X' is another proper (necessarily special) projective vector field then some linear combination of X and X' is affine.*

Proof. Let X be a projective vector field on M. Then, since for this class there exists a non-trivial solution (in fact, two independent such solutions) of (9.9), theorem (12.5) shows that $\psi_{a;b} = 0$ so that X is either affine or special projective. If M admits no (non-trivial) covariantly constant vector fields then $\psi_a \equiv 0$ on M and X is affine. So suppose X is special projective so that ψ_a is nowhere zero and covariantly constant on M. Again (12.17) leads to (12.23) and the Ricci identity gives $R_{abcd}\psi^d = 0$. So from theorem 9.3(i) it follows that for any $m \in M$ there is an open neighbourhood U of m such that, on U

$$h_{ab} = \alpha g_{ab} + \beta s_a s_b + \gamma \psi_a \psi_b + \nu(s_a \psi_b + \psi_a s_b) \qquad (12.26)$$

for a smooth covector field s and functions α, β, γ and ν on U (which can be shown smooth from the results of section 9.3) and where $R_{abcd}s^d = 0$ holds in U. Now if M admits a second independent covariantly constant vector field, the Ricci identity on this vector field shows that (its restriction to U) may, with no loss of generality, be taken as s in (12.26). With this assumed done, one substitutes (12.26) into (12.4) and contracts at any point $m \in U$ with $k^a k^b$ and with $k \in T_m M$ such that, at m, $k^a k_a \neq 0$ and $k^a \psi_a = k^a s_a = 0$. One finds that $\alpha_{,a} = 2\psi_a$ at m and a back substitution and contraction with k^a then shows that $\psi_a = 0$ at m. Thus $\psi_a \equiv 0$ on M and X is affine. This establishes the first part of the theorem. Now suppose M admits a unique covariantly constant vector field and let X and X' be proper (necessarily special) projective vector fields on M with associated tensors ψ_a, h_{ab} and ψ'_a, h'_{ab}, respectively. Then $\psi_{a;b} = \psi'_{a;b} = 0$ and so, by the uniqueness assumption, $\psi'_a = \lambda\psi_a$ $(0 \neq \lambda \in \mathbb{R})$. Thus the (projective) vector field $X' - \lambda X$ is easily seen to be affine from (12.4) and the proof is complete. □

In fact if M is a simply connected space-time of class D which admits a proper projective vector field then theorem 12.8 restricts the holonomy type of M to R_6, R_8, R_{10}, R_{11} and R_{13} and it can be shown that R_6 and R_8 are impossible [233].

12.3 Space-Times Admitting Projective Vector Fields

So far the theorems proved have been towards the non-existence of (proper) projective vector fields. Some cases where they do exist can now be discussed [233].

In the previous theorem the situation for a space-time of class D admitting exactly one independent global covariantly constant vector field was not completed except to say that any proper projective vector field admitted was necessarily special projective and was unique up to additive affine vector fields. So let M be such a space-time with (say) a covariantly constant *unit timelike* vector field u. If X is a proper projective vector field on M then $\psi_{a;b} = 0$ and since ψ_a is not zero one may assume after a constant rescaling of X, if necessary, that $\psi_a = u_a$. Then (12.26) holds on the open subset U of M with $\psi_a = u_a$ and s satisfying $R_{abcd}s^d = 0$ on U. One can always choose s so that $s_a s^a = 1$ and $\psi^a s_a = 0$ on U and it then follows that, on U

$$-\psi^a\psi_a = s^a s_a = 1, \quad \psi^a s_a = 0, \quad \psi^a{}_{;b} = 0, \quad s^a s_{a;b} = \psi^a s_{a;b} = 0 \quad (12.27)$$

A substitution of (12.26) into (12.4) and contractions firstly with $\psi^a\psi^b$ and secondly with $\psi^a s^b$ yield

$$\alpha_{,a} - \gamma_{,a} = 4\psi_a \qquad s_a = \nu_{,a}. \qquad (12.28)$$

Again on substituting (12.26) into (12.4), contracting with ψ^a and using (12.28) one finds

$$\nu s_{a;b} = \psi_a\psi_b - s_a s_b + g_{ab} \qquad (12.29)$$

which, from rank considerations, shows that ν is nowhere zero on U. Equation (12.28) shows that $\nu_{,a}$ is nowhere zero on U.

Now for any $m \in M$ one may construct a local coordinate system about m by first noting that a function t exists on some open neighbourhood of m such that $\psi_a = t_{,a}$. Also, from section 4.11 example (v), the subsets of this neighbourhood of constant t and ν are 2-dimensional submanifolds of M. Finally define coordinates x^2 and x^3 in these 2-dimensional submanifolds and let $x^0 = t$, $x^1 = \nu$. In this coordinate system the metric is

$$ds^2 = -dt^2 + d\nu^2 + g_{\alpha\beta}dx^\alpha dx^\beta \tag{12.30}$$

where $\alpha, \beta = 2, 3$. In this coordinate system one can compute $s_{a;b}$ in terms of the Christoffel symbols Γ^a_{bc} associated with (12.30) to find

$$s_{a;b} = s_{a,b} - \Gamma^c_{ab}s_c = -\Gamma^1_{ab} = \tfrac{1}{2}g_{ab,1} \tag{12.31}$$

and so, using (12.29)

$$t_{,a}t_{,b} - \nu_{,a}\nu_{,b} + g_{ab} = \frac{\nu}{2}g_{ab,1}. \tag{12.32}$$

From the usual metric completeness relation this equation is identically true unless a and b take values in $\{2,3\}$ and, together with the fact that the metric components are independent of t (since u is a Killing vector field on M) then show that $g_{\alpha\beta} = \nu^2 q_{\alpha\beta}$ where the $q_{\alpha\beta}$ are independent of t and ν (*i.e.* x^0 and x^1) and ν must now be restricted by $\nu > 0$ or $\nu < 0$. Thus the metric is

$$ds^2 = -dt^2 + d\nu^2 + \nu^2 q_{\alpha\beta}dx^\alpha dx^\beta \tag{12.33}$$

in some coordinate system about any point of M [233].

In one of the coordinate domains described above define three vector fields X, Y and Z by

$$X^a = (t^2, \nu t, 0, 0), \quad Y^a = (t, 0, 0, 0), \quad Z^a = (t, \nu, 0, 0). \tag{12.34}$$

Then $X = tZ$, $Y = -tu$ and, from (12.33) after a short calculation

$$\begin{gathered} Z_{a;b} = g_{ab}, \qquad Y_{a;b} = H_{ab} \quad (H_{ab} = -u_a u_b) \\ X_{a;b} = \tfrac{1}{2}h_{ab} + F_{ab} \quad (h_{ab} = 2tg_{ab} + 2Z_{(a}u_{b)}, \; F_{ab} = Z_{[a}u_{b]}. \end{gathered} \tag{12.35}$$

Thus Z is proper homothetic, Y is proper affine (since $H_{ab;c} = 0$) and X is easily checked to be proper (special) projective in this coordinate domain since h_{ab} satisfies (12.4) with $\psi_a = u_a$. Regarding this coordinate domain as the space-time M it follows that $\dim P(M) = \dim SP(M) \geq 4$ and from the projective, affine and homothetic "uniqueness" results (theorems 12.8, 10.2(iii) and (10.14) any further independent projective vector fields may be chosen to be Killing (since the holonomy type can not be R_2, R_3 or R_4).

The above analysis was carried out with the assumption that the unique independent covariantly constant vector field admitted by M was timelike. If it is spacelike (and chosen to be unit and denoted by y) then one may arrange that y and s are orthogonal independent vector fields satisfying (9.9) at each $m \in M$ and again it will be assumed that $\psi_a = y_a$. There are now three possibilities given by the nature of the 2-spaces at m spanned by y and s. Let A, B and C be the subsets of M at each point of which this 2-space is, respectively, null, timelike and spacelike. Then $M = A \cup B \cup C$ is a disjoint decomposition of M and locally on the closed subset A, s can be chosen null whereas locally on the open subsets B and C, s can be chosen, respectively, timelike and spacelike. Consider first the open subset $\operatorname{int} A$ of M where, if non-empty, one has in some coordinate neighbourhood of any $m \in \operatorname{int} A$

$$\psi^a \psi_a = 1, \quad \psi^a s_a = s^a s_a = 0, \quad \psi^a{}_{;b} = 0, \quad s^a s_{a;b} = \psi^a s_{a;b} = 0. \quad (12.36)$$

It follows that h_{ab} may be written in the form (12.26) above. A substitution into (12.4) and a contraction with $\psi^a s^b$ then gives the contradiction $\psi_a = 0$ and so $\operatorname{int} A = \emptyset$. So A has empty interior (and if $A = \emptyset$ then, since M is connected, either $M = B$ or $M = C$). On the subsets B and C the analysis and results are very similar to the previous case and a proper special projective vector field is admitted.

Now suppose the unique independent covariantly constant vector field is null (and taken to be ψ^a). Then at each $m \in M$ there is $s \in T_m M$ such that s^a and ψ^a are smooth independent solutions of (9.9). Two possibilities arise given by the nature of the 2-space spanned at m by these solutions. Let A and B be the subsets of M at each point of which this 2-space is null and timelike, respectively. Then $M = A \cup B$ is a disjoint decomposition of M with B open and A closed. On A, s^a may be chosen spacelike and orthogonal to ψ^a whilst on B, s^a may be chosen null and satisfying $\psi^a s_a = 1$. So consider $\operatorname{int} A$ and for each $m \in \operatorname{int} A$ choose an open neighbourhood U of m so that (12.26) holds on U. Then substitute into (12.4) and contract successively with ψ^a, $s^a s^b$ and g^{ab} to obtain $\psi_a = 0$ on U and so $\operatorname{int} A = \phi$, *i.e.* A has empty interior. Next for $m \in B$ choose an open neighbourhood of m on which (12.26) holds and substitute into (12.4). Successive contractions with $s^a s^b$, $\psi^a s^b$, g^{ab} and s^a give $\gamma_{,a} = 2s_a$, $\nu_{,a} = \psi_a$ and

$$\nu s_{a;b} = g_{ab} - \psi_a s_b - s_a \psi_b. \tag{12.37}$$

(and (12.37) together with the Ricci identity confirm the relation $R_{abcd}s^d = 0$). So choose local coordinates x^a with $x^0 = \nu$, $x^1 \equiv \omega = \frac{1}{2}\gamma$, and x^2 and x^3 coordinates in the 2-dimensional submanifolds of constant ν and γ to get a metric

$$ds^2 = 2d\nu d\omega + g_{\alpha\beta}dx^\alpha dx^\beta \tag{12.38}$$

where $\alpha, \beta = 2, 3$. An analysis similar to that yielding (12.33) now gives [233]

$$ds^2 = 2d\nu d\omega + \nu^2 q_{\alpha\beta}dx^\alpha dx^\beta \tag{12.39}$$

where $q_{\alpha\beta}$ is independent of ν and ω. For this metric define vector fields X, Y and Z by

$$X^a = (\nu^2, \nu\omega, 0, 0) \quad Y^a = (0, \nu, 0, 0) \quad Z^a = (\nu, \omega, 0, 0). \tag{12.40}$$

Then $Z_{a;b} = g_{ab}$ and so Z is proper homothetic, $Y_{a;b} = \psi_a\psi_b$ and so Y is proper affine and $X = \nu Z$ so X is (proper) special projective satisfying (12.1) and (12.4) with $h_{ab} = 2\nu g_{ab} + 2Z_{(a}\psi_{b)}$ and projective 1-form ψ_a. Thus regarding this coordinate domain as the space-time it follows that $\dim P(M) = \dim SP(M) \geq 4$ and from the uniqueness results used previously any further independent members of $P(M)$ may be chosen to be Killing.

It follows from theorem 12.3 that the Schwarzschild metric admits no proper projective vector fields and, since it admits no proper affine or homothetic vector fields (section 10.7) the only projective vector fields admitted are just the 4-dimensional Lie algebra of Killing vector fields (10.81). Exactly the same is true of the Reissner-Nordström metric which in the usual coordinates t, r, θ, ϕ is given by (10.83).

To establish this, note that the absence of (proper) homothetic and affine vector fields was established in section 10.7 and so it is sufficient to show that no proper projective vector fields are admitted [236]. To see this one notes that the curvature tensor components for (10.83) can be written in the usual 6-dimensional block form (for notation see chapter 7)

$$R^{ab}{}_{cd} = \mathrm{diag}(A, B, B, C, B, B). \tag{12.41}$$

The exact expressions for A, B and C, which are functions only of r, are not needed. All that is required is the knowledge that none vanishes nor are any two of them equal over any non-empty open subset of the space-time manifold M. Thus the curvature tensor has four independent real eigenbivectors $e^1 \wedge e^2$, $e^1 \wedge e^3$, $e^2 \wedge e^0$ and $e^3 \wedge e^0$ with real eigenvalue $B(p)$ at any $p \in M$ where e^0, e^1, e^2 and e^3 are, respectively, mutually orthogonal unit vectors at p in the directions of $\frac{\partial}{\partial t}$, $\frac{\partial}{\partial r}$, $\frac{\partial}{\partial \theta}$ and $\frac{\partial}{\partial \phi}$. Also $e^2 \wedge e^3$ and $e^1 \wedge e^0$ are real eigenbivectors with real eigenvalues $A(p)$ and $C(p)$, respectively, at p. Now let X be a projective vector field on M so that (12.1) and (12.4) hold for X^a and its corresponding tensors h_{ab} and ψ_a. Then theorem (12.4)(i) shows that e^0, e^1, e^2 and e^3 are everywhere eigenvectors of the symmetric tensor $Bh_{ab} + 2\psi_{a;b}$ with equal (real) eigenvalues and thus

$$Bh_{ab} + 2\psi_{a;b} = \beta g_{ab} \tag{12.42}$$

for some function β on M. Similarly e^2 and e^3 are eigenvectors of $Ah_{ab} + 2\psi_{a;b}$ with the same real eigenvalue, and similarly for e^0, e^1 and $Ch_{ab} + 2\psi_{a;b}$. Thus, on M (see section 7.5)

$$Ah_{ab} + 2\psi_{a;b} = Y_{ab} + \mu({e^2}_a {e^2}_b + {e^3}_a {e^3}_b) \tag{12.43}$$

$$Ch_{ab} + 2\psi_{a;b} = Z_{ab} + \nu({e^1}_a {e^1}_b - {e^0}_a {e^0}_b) \tag{12.44}$$

where Y_{ab} and Z_{ab} are, respectively, collections of symmetrised products of ${e^0}_a$ and ${e^1}_a$ and of ${e^2}_a$ and ${e^3}_a$ and μ and ν are functions on M and whose precise forms are not required. Since A, B and C are distinct over an open dense subset U of M, subtractions of (12.42) from each of (12.43) and (12.44) lead to the facts that over U e^0 and e^1 are eigenvectors of h_{ab} with equal eigenvalue and similarly for e^2 and e^3. Hence there exists functions ρ and σ on U satisfying

$$\begin{aligned} h_{ab} &= \rho({e^1}_a {e^1}_b - {e^0}_a {e^0}_b) + \sigma({e^2}_a {e^2}_b + {e^3}_a {e^3}_b) \\ &= \sigma g_{ab} + (\rho - \sigma)({e^1}_a {e^1}_b - {e^0}_a {e^0}_b) \end{aligned} \tag{12.45}$$

where the completeness relation $g_{ab} = \eta_{cd} {e^c}_a {e^d}_b$ has been used. Now substitute the second relation in (12.45) into (12.4) and contract successively with $\overset{0}{e}{}^a \overset{1}{e}{}^b$ and $\overset{2}{e}{}^a \overset{3}{e}{}^b$ to get (recalling that $\overset{0}{e}_a \overset{1}{e}{}^a = 0 \Rightarrow \overset{0}{e}_{a;b} \overset{1}{e}{}^a + \overset{0}{e}_a \overset{1}{e}{}^a{}_{;b} = 0$, etc.) $\psi_a \overset{0}{e}{}^a = \psi_a \overset{1}{e}{}^a = \psi_a \overset{2}{e}{}^a = \psi_a \overset{3}{e}{}^a = 0$ and hence that $\psi_a \equiv 0$ on U and hence on M. Then (12.4) reveals that X is affine (and so not proper projective) on M. This completes the proof. A similar argument to the above but with $A = C$ on M reveals the result (again) for the Schwarzschild metric. In fact for each of these metrics X must be Killing.

An argument using similar techniques can be used to show that null Einstein-Maxwell fields can never admit a proper projective vector field [237]. In this case the argument is more complicated because of the need to decompose the space-time into Petrov types.

For the FRW (conformally flat perfect fluid) models the above techniques enable the problem of the admission of projective vector fields to be approached in a similar manner [238]. Here, a very brief sketch of the argument will be given. The metric g in the usual coordinates is given by (10.84) where t ($= x^0$) is the cosmological time coordinate and $R(t)$ a function of t controlled by the Einstein field equations for a perfect fluid. The metric g induces a positive definite metric of constant curvature in the hypersurfaces of constant t. This constant curvature is represented in a conventional way by a parameter k which is zero when these hypersurfaces are flat and ± 1 otherwise (for details see *e.g.* [239]). The (timelike fluid) flow vector on this space-time M is $u = \frac{\partial}{\partial t}$ with components in the above coordinates given by $u^a = \delta^a{}_0$, $u_a = -\delta^0{}_a = -t_{,a}$ and which satisfies

$$u_{a;b} = \frac{\theta}{3}(g_{ab} + u_a u_b) \tag{12.46}$$

where the *expansion* $\theta \equiv u^a{}_{;a}$ of u is a function only of t and is related to R by $3\dot{R} = R\theta$ (and $\cdot \equiv \frac{d}{dt}$).

Two special cases of these models can be quickly considered. They are, firstly, the case when the space-time metric (10.84) is actually of constant curvature (the de-Sitter type metrics) and in these cases the situation is known [234]. The second case is the Einstein static type (8.14) which can, essentially, be realised by (10.84) with u covariantly constant, $u_{a;b} = 0$ (and hence, from (12.46), $\theta = 0$ and $R =$ constant). The curvature tensor for the general metric (10.84) takes the (6×6) form

$$R^{ab}{}_{cd} = \mathrm{diag}(f_1, f_1, f_1, f_2, f_2, f_2) \tag{12.47}$$

where f_1 and f_2 are functions only of t. Since M is assumed (as always) *non-flat* (so that the curvature tensor never vanishes over a non-empty open subset of M) it follows that the curvature rank is at least three over an open dense subset U of M. In the Einstein static case, $u_{a;b} = 0 \Rightarrow R^{ab}{}_{cd}u^d = 0$ and so the rank on U is three. Then theorem 12.6 confirms that no proper projective vector fields exist.

The case considered next is the "general" *local* situation with $M = I \times H$ with I some open interval of $\mathbb{R}$ (playing the role of an interval of time t) and H a 3-dimensional connected manifold of constant curvature (a connected open subset of the hypersurfaces of constant t). It will be assumed that no non-empty open subset of M is of the de-Sitter or Einstein static type and so the functions f_1 and f_2 in (12.47) are distinct on some open dense subset of M. Also M is assumed to admit a 6-dimensional Killing algebra $K(M)$ with 3-dimensional spacelike orbits of constant t each isometric to H. Then any $X \in K(M)$ induces a Killing vector field $\tilde{X}$ in an orbit $t = t_0$ with respect to the induced metric of that orbit (theorem 10.11(ii)) and the map $\chi \cdot X \to \tilde{X}$ is a linear map between the Lie algebras $K(M)$ and $K(H)$. It is also one-to-one, from theorem 10.11(iii), and onto (since $\dim H = 3 \Rightarrow \dim K(H) \leq 6$). Hence χ is a Lie algebra isomorphism (and so $\dim K(H) = 6$). Thus every Killing vector field in a hypersurface of constant t (with induced metric) can be regarded as arising naturally from a unique space-time Killing vector field.

Another feature of the hypersurface geometry of H which is important in this context is its homothetic symmetry. Suppose there exists a *proper* (local or global) homothetic vector field in the induced geometry in H. Then, since H has a transitive Killing algebra, it follows that for any $p \in H$ there is such a proper homothetic vector field Y defined in some neighbourhood U of p in H. Then either $Y(p) = 0$ or else $Y(p) \neq 0$ and there is a *Killing* vector field on H such that $X(p) = Y(p)$. Then $Z = X - Y$ ($\not\equiv 0$) is proper homothetic on U and $Z(p) = 0$. So for any $p \in H$ some (local or global) proper homothetic vector field on H vanishes at p. Then theorem 10.5(iii) modified so as to apply to the 3-dimensional positive definite case shows that all Ricci eigenvalues and hence the Ricci and Riemann tensors in the geometry of H vanish at each $p \in H$. Thus H is flat. It follows that *in the cases of FRW models with $k = \pm 1$ can there be no (hypersurface geometry) proper homothetic vector fields in H*. They exist, of course, if H is flat ($k = 0$).

The proof now proceeds (briefly) in the following steps. First let X be a projective vector field on M. Then (12.47) and theorem (12.4)(i) show that

$$h_{ab} = \alpha u_a u_b + \beta g_{ab} \tag{12.48}$$

for functions α and β. Then use of (12.4) and (12.46) and certain contractions shows that α and β are functions only of t, that $\alpha + \beta$ is constant, that $\psi_a = \lambda u_a$ with λ a function only of t and that $3\lambda = \alpha\theta$ and $\alpha_{,a} = -2\lambda u_a$.

Next write X as $X^a = \gamma u^a + q^a$ where the vector field q on M satisfies $q^a u_a = 0$ and γ is a function on M. The expression (7.7) solved for the Ricci tensor in terms of the fluid pressure and density (which depend only on t) and (12.7) can be used to show that γ is a function only of t. Now suppose γ vanishes over some non-empty open subset of M. Then it vanishes everywhere in the open subset $U = (t_0, t_1) \times H$ of M, for some $t_0, t_1 \in \mathbb{R}$ with $t_0 < t_1$, and so $X^a u_a = 0$ on U. Thus $X_{a;b} u^a u^b = 0$ and hence $h_{ab} u^a u^b = 0$ on U. This leads (recalling the exclusion clauses on M) to X being Killing on M. So if non-Killing members of $P(M)$ are being sought, one assumes X is not Killing and hence that γ does not vanish on any non-empty open subset of M. The expression $X^a = \gamma u^a + q^a$ is then substituted into (12.1) to show that q defines in a natural way an (induced geometry) homothetic vector field in the hypersurfaces of constant t (*i.e.* in H). In the cases $k = \pm 1$ no *proper* such homothetic vector fields (as shown earlier) and this leads to the fact that q is a Killing vector field on M and that $X^a - q^a = \gamma u^a$. Thus, up to Killing vector fields, one need only study projective vector fields of the form $X^a = \gamma(t) u^a$ in the cases $k = \pm 1$ (and for such vector fields the *projective bivector* F in (12.1) can then be checked to be identically zero. Equations (12.1) and (12.4) then lead to differential equations relating α, γ and θ from which it can be concluded that the FRW metric ($k = \pm 1$) does not, in general, admit proper projective vector fields but does so for certain special choices of the function $R(t)$. A similar conclusion can be achieved in the $k = 0$ case, but care is needed here since H now admits (local or global) proper homothetic vector fields. Full details are available in [238]. A by-product of this calculation is the finding of those FRW models which admit proper homothetic vector fields, a result which was already known (section 10.7). The FRW models are examples of conformally flat perfect fluid space-times. In fact theorem 12.4 may be used to show that *the only conformally flat perfect fluid space-times which admit proper projective vector fields are the special cases of the FRW models alluded to above* [240]. In this reference the fact that the assumptions made implied the existence of an equation of state between the pressure and density of the model was omitted. A proof of this fact can be found in [241].

12.4 Special Projective Vector Fields

Special projective vector fields were encountered in section 12.3 and also for Minkowski space in (12.9). In this section the general situation when a special projective vector field exists will be described. The ideas and results are very similar to those discussed for special conformal vector fields in section 11.7 and so only a summary will be given (for details see [233],[232]).

Suppose M is non-flat and admits a proper special projective vector field X satisfying (12.1) with projective 1-form ψ_a satisfying $\psi_{a;b} = 0$. Thus ψ_a is nowhere zero and $\psi_a \psi^a$ is constant on M. So suppose $\psi^a \psi_a \neq 0$ (*so that* ψ_a *is non-null on* M). Using the projective bivector F define a global vector field Y on M in components by $Y^a = F^{ba}\psi_b$ and use (12.5) to get

$$Y_{a;b} = -\tfrac{1}{2}\left[(\psi_c\psi^c)g_{ab} - \psi_a\psi_b\right] \qquad (\Rightarrow Y_{a;bc} = 0). \tag{12.49}$$

Thus from (10.8) and (10.10) Y is proper affine and is everywhere orthogonal to ψ^a. Just as in the conformal case Y^a and ψ^a have vanishing Lie bracket and give rise to a 2-dimensional integrable distribution over an open dense subset of M. It follows that $R_{abcd}\psi^d = R_{abcd}Y^d = 0$ and the geometry of the integral manifolds of this distribution is as in the special conformal case. In particular, the curvature rank is one over an open dense subset U of M where, it may be assumed without loss of generality, Y is non-zero and together with ψ^a gives rise to the above mentioned distribution. It now follows from theorem 12.8 that M admits a unique (up to a constant scaling) covariantly constant vector field and that if $X, X' \in SP(M)$ then some linear combination of them is affine and so $\dim SP(M)$ $(= \dim P(M)$ from theorem 12.8) $= \dim A(M)+1$ $(= \dim H(M)+2$ from theorem 10.14).

Now assume M is simply connected (or restrict to a simply connected region) so that one may write $\psi_a = \psi_{,a}$ for some function ψ. Then construct vector fields P and Q where

$$Q^a = -2(\psi_c\psi^c)^{-1}(Y^a - \tfrac{1}{2}\psi\psi^a) \qquad P^a = \psi Q^a. \tag{12.50}$$

Then from (12.49) one finds

$$Q_{a;b} = g_{ab}, \qquad P_{a;b} = \tfrac{1}{2}q_{ab} + Q_{[a}\psi_{b]} \tag{12.51}$$

where

$$q_{ab} = 2Q_{(a}\psi_{b)} + 2\psi g_{ab} \qquad (\Rightarrow q_{ab;c} = 2g_{ab}\psi_c + g_{ac}\psi_b + g_{bc}\psi_a) \tag{12.52}$$

Thus P is (proper) special projective, with the same projective 1-form as X and is tangent to the integral manifolds described above. Also Q is proper homothetic.

Now it is recalled that in Minkowski space-time a basis for $P(M)$ could be chosen for which only 4 of the 24 members were proper projective and, in fact, special projective (from (12.9)). A choice of these vector fields can be made by choosing, successively, $\theta_a = \delta_a^0, \ldots, \delta_a^3$ and then *each equals some function f multiplied by the homothetic vector field on Minkowski space with components (t, x, y, z) and where $f_{,a} = \theta_a$ is the associated projective 1-form.* For example, the choice $\theta_a = \delta_a^0$ gives, in (12.9), $X^a = t(t, x, y, z)$ and $f = t$ so that $f_{,a} = \theta_a$. Then $X_a = t(-t, x, y, z)$ and $h_{ab} = 2X_{(a,b)}$ is easily shown to satisfy (12.4) with $\psi_a = \theta_a$. *That such a choice is available generally follows from (12.50)-(12.52).*

The similarity between the studies of the Lie algebras $SC(M)$ and $SP(M)$ can be taken further. First assume M simply connected (or work on a simply connected coordinate domain of M). If M admits a *proper* member X of $SP(M)$ *with projective* 1-*form* ψ_a *non-null* (so that $0 \neq \psi_a\psi^a$ = constant), there is an associated homothetic vector field Q as in (12.50). Now Q_a and ψ_a are each local gradients and so (by simply connectedness) they are each global gradients $\psi_a = \psi_{,a}$, $Q_a = \alpha_{,a}$ for functions $\psi, \alpha : M \to \mathbb{R}$. The global vector field $E^a = \psi Q^a - \alpha\psi^a$ then satisfies $E_{a;b} = \psi g_{ab} + 2Q_{[a}\psi_{b]}$ and so is special (proper) conformal (with ψ playing the role of ϕ in (11.1)). Conversely, let M admit a (proper) *special conformal vector field X* satisfying (11.1) with $0 \neq \phi_a\phi^a$ = constant. Then define the global vector field $E'^a = (\phi_c\phi^c)^{-1}\phi W^a$ with W given as in (11.47). Then if $H_{ab} \equiv 2E'_{(a;b)}$ it follows that H_{ab} satisfies (12.4) with ψ_a replaced by ϕ_a. Hence E' is special (proper) projective. It is noted that this converse proof did not require the simply connected assumption. The following theorem is thus established.

Theorem 12.9 *The set of simply connected space-times which admit a special (proper) conformal vector field satisfying (11.1) with $\phi_a\phi^a \neq 0$ and those which admit a special (proper) projective vector field satisfying (12.1) and (12.4) with $\psi^a\psi_a \neq 0$ are, in the sense described above, identical.*

If (only) the simply connectedness assumption is dropped it is still true that those space-times which admit a proper member of $SC(M)$ also admit a proper member of $SP(M)$ but now those space-times which admit a proper member of $SP(M)$ have only been proved to admit a proper member of $SC(M)$ locally (*i.e.* in some open neighbourhood of any point of M). It is also noted that the theorem is false if the conditions $\phi_a\phi^a \neq 0 \neq \psi_a\psi^a$ are dropped. To see this note that the vacuum plane wave space-times can admit proper members of $SC(M)$ (section 11.5) with ϕ^a null but can not admit any proper projective vector field from theorem 12.3. Something, however, remains of these results in the case $\psi_a\psi^a = 0$. Suppose M admits a proper special projective vector field X with projective 1-form ψ_a satisfying $\psi_{a;b} = 0$ and $\psi^a\psi_a = 0$. Then (12.39) gives a local form for the metric and reveals an expression (12.40) for X in a coordinate domain U as $X = \nu Z$ where Z is homothetic and $Z_{a;b} = g_{ab}$. So again X can be chosen as ν times a proper homothetic vector field, where $\psi_a = \nu_{,a}$. Further, from (12.40) one can write $Z_a = Z_{,a}$ where $Z = \nu\omega$. Then the vector field $X'^a = \nu Z^a - Z\psi^a$ $(= \nu^2 s^a)$ satisfies $X'_{a;b} = \nu g_{ab} + 2Z_{[a}\psi_{b]}$ on U and is hence (proper and) special conformal on U with conformal function ν.

One can add a little more here to the uniqueness result established for $SP(M)$ in this section. Suppose M admits a proper member of $SP(M)$ whose projective 1-form satisfies $\psi_a\psi^a \neq 0$. Then the previous work shows that the curvature rank on M is at most one and so if X is *any* member of $P(M)$ it must be *special* by theorem 12.5. Hence in this case $P(M) = SP(M)$. Also, if M is simply connected and admits a proper member of $SP(M)$ with $\psi_a\psi^a \neq 0$, then since M admits covariantly constant, proper homothetic and proper affine vector fields, the uniqueness results so far given show that $\dim P(M) = \dim SP(M) \geq 4$ and $\dim SP(M) = \dim A(M) + 1 = \dim K(M) + 3$. The same argument as in section 11.7 shows that $\dim K(M) \leq 4$ and so $4 \leq \dim SP(M)$ $(= \dim P(M)) \leq 7$. If M is not simply connected this becomes a local result.

12.5 Projective Symmetry and Holonomy

Theorems 12.4-12.8 suggest a connection between the existence of projective vector fields and the curvature rank and hence the holonomy type of M. This will be explored further in this section. First it is convenient to return to the curvature tensor classification described in section 9.3 and to the (disjoint) decomposition of M into the subsets A, B, C, D and O. In that section it was stated that M could be disjointly decomposed as

$$M = A \cup \operatorname{int} B \cup \operatorname{int} C \cup \operatorname{int} D \cup \operatorname{int} O \cup Z \tag{12.53}$$

where A is open and Z is closed with $\text{int}\, Z = \emptyset$. This result will now be briefly established. To do this, some remarks on the range space B_m of the linear map f in section 9.3 will be useful.

(i) If $m \in A \cup B$ there does not exist $k \in T_mM$, $k \neq 0$, such that $F^a{}_b k^b = 0$ for all $F \in B_m$.
(ii) If $m \in A$, $\dim B_m \geq 2$ and if $\dim B_m \geq 4$ then necessarily $m \in A$.
(iii) If all members of B_m are simple, $\dim B_m \leq 3$ and if, in addition, $\dim B_m = 2$, there exists $k \in T_mM$, $k \neq 0$, such that $F^a{}_b k^b = 0$ for all $F \in B_m$.

Remarks (i) and (ii) follow from the definitions of the curvature classes. For (iii) if $\dim B_m = 2$ and all members of B_m are simple then let $\{F, G\}$ be a basis for B_m with F, G simple and $F + \lambda G$ simple for each $\lambda \in \mathbb{R}$. Applying condition (v) of theorem 7.1 to these bivectors gives $F^a{}_b {}^*G^b{}_c + G^a{}_b {}^*F^b{}_c = 0$. Then there exists $k' \in T_mM$, $k' \neq 0$, such that $F^a{}_b k'^b = 0$ and one contracts the previous equation with k'_a to obtain $(k'_a G^a{}_b)^* F^b_c = 0$. Thus either the vector $G^a{}_b k'^b$ is zero or it is in the blade of G and orthogonal to the blade of ${}^*F$ (and hence in the blade of F). In either case the blades of F and G intersect non-trivially and the vector k claimed in (iii) exists. Now suppose all members of B_m are simple and $\dim B_m \geq 4$. With F and G as above and with $H \in B_m$ such that F, G and H are independent, one may use the previous argument to show that there exist $p, q, r, s \in T_mM$ such that $F = p \wedge q$, $G = p \wedge r$ and either $H = q \wedge r$ or $H = p \wedge s$. Now repeat the argument with a fourth independent $K \in B_m$. It is easily checked now that one either contradicts the independence of F, G, H and K or the original assumption that all members of B_m are simple.

Now define the subspace $U_m \in T_mM$ associated with B_m to be the span of the union of the blades of members of B_m (including each of the canonical pair of blades for non-simple members of B_m - see section 7.2). Then if $m \in D$, $\dim B_m = 1$ and $\dim U_m = 2$, if $m \in C$, $\dim B_m = 2$ or 3 and $\dim U_m = 3$, if $m \in B$, $\dim B_m = 2$ and $\dim U_m = 4$ and if $m \in A$, $\dim B_m \geq 2$ and $\dim U_m = 4$. It is also useful to note that if $F \in B_m$, so that for some bivector F' at m, $f(F') = F$, there exists a smooth bivector $\tilde{F}$ on some open neighbourhood W of m such that $\tilde{F}(m) = F'$ and hence a smooth bivector $R^{ab}{}_{cd}\tilde{F}^{cd}$ on W which is in $B_{m'}$ for each $m' \in W$ and which equals F at m. From this it is easy to check that if $k \in U_m$ then (reducing W if necessary) there exists a smooth vector field $\tilde{k}$ on W such that $\tilde{k}(m') \in U_{m'}$ for each $m' \in W$ and $\tilde{k}(m) = k$. It follows (section 3.11) that there exists an open neighbourhood W' of m such that

$\dim U_{m'} \geq \dim U_m$ for each $m' \in W'$. Clearly, W' may be chosen so that one also has $\dim B_{m'} \geq \dim B_m$ for each $m' \in W'$.

To establish the decomposition (12.53) let $m \in A$ so that $\dim B_m \geq 2$ and $\dim U_m = 4$. Then the above argument reveals the existence of an open neighbourhood N_1 of m such that $\dim U_{m'} = 4$ for each $m' \in N_1$ and so $N_1 \subseteq A \cup B$. If $\dim B_m \geq 3$ one may take $N_1 \subseteq A$. If $\dim B_m = 2$, B_m must contain a non-simple member and hence two independent non-simple members whose canonical blade pairs do not coincide and so by continuity there is an open neighbourhood N_2 of m such that $N_2 \cap B = \emptyset$. Hence $N_1 \cap N_2 \subseteq A$. This argument shows that A is open. Now let $m \in B$ so that B_m contains a non-simple member. Continuity then shows that $B_{m'}$ contains a non-simple member for m' in some open neighbourhood N_3 of m and so $N_3 \subseteq A \cup B$. Thus $A \cup B$ is open. A simple curvature rank consideration then shows that $A \cup B \cup C$ and $A \cup B \cup C \cup D$ are open. Finally, to show $\operatorname{int} Z = \emptyset$ let $W \subseteq Z$ be open. Then by disjointness $W \cap A = \emptyset$. Also, since $A \cup B$ is open, $W \cap B = W \cap (A \cup B)$ is open and since Z is disjoint from $\operatorname{int} B$, $W \cap B = \emptyset$. Similarly, $W \cap C = W \cap D = W \cap O = \emptyset$ and so $W = \emptyset$. It follows that $\operatorname{int} Z = \emptyset$.

Before the final theorem is given it is instructive to recall equations (12.17) and (12.18) in the proof of theorem 12.4 for the case where A^{ab} is a real eigenbivector of the curvature tensor with zero eigenvalue ($c = 0$). The interplay between these equations is essentially that if "sufficient" solutions for A^{ab} of (12.18) exist then $\psi_{a;b} \propto g_{ab}$. But this last condition is equivalent to the vanishing of the right hand side of (12.17) and hence the left hand side. This latter condition then heavily restricts h according to theorem 9.3. This link was used in [233] to rule out proper projective vector fields in space-times of certain holonomy types but where $\dim B_m$ was assumed constant. With the help of the decomposition (12.53) this theorem can now be established without this restriction [242].

Theorem 12.10 *Let M be a non-flat simply connected space time.*

i) If M has holonomy type R_2, R_3, R_4, R_6, R_7, R_8 or R_{12} then M does not admit a proper member of $P(M)$.
ii) If M admits a proper member X of $P(M)$ then it has holonomy type R_{10}, R_{11} or R_{13} if and only if $X \in SP(M)$. If these equivalent conditions hold then some open dense subset of M is of class D.

Proof. For (i) let X be a projective vector field on M and let M be decomposed as in (12.53). The members of this decomposition will be considered separately in the case that they are not empty. It is then clear from theorems 12.6 and 12.7 that the projective 1-form ψ vanishes on $\operatorname{int} B$ and $\operatorname{int} C$. Next, the assumption regarding the possible holonomy types of M and the fact that, loosely speaking, B_m is at each $m \in M$ contained in the infinitesimal holonomy algebra and hence the holonomy algebra enables sufficiently many solutions G of $R_{abcd}G^{cd} = 0$ to be calculated for theorem 12.4 to be invoked at each $m \in A$ and to check that, on A, $\psi_{a;b} = \gamma g_{ab}$ for a smooth function $\gamma : A \to \mathbb{R}$. Then from (12.17) and theorem 9.3 it follows that $h_{ab} = \beta g_{ab}$ for a smooth function $\beta : A \to \mathbb{R}$. Thus (see the penultimate paragraph of this chapter) $\psi = 0$ on A. If $m \in \operatorname{int} D$ and $\psi(m) \neq 0$ then ψ is non-zero over some open connected and simply connected neighbourhood U of m and theorem 12.8 and the remark following its proof show that the holonomy type of U is either R_{10}, R_{11} or R_{13}. These subalgebras can not be contained in those of the stated holonomy types of M and so $\psi = 0$ on $\operatorname{int} D$. Since M is non-flat, $\operatorname{int} O = \emptyset$ and so ψ vanishes on the open dense subset $M \setminus Z$ of M and hence on M and X is not proper.

For (ii) let $X \in P(M)$ be proper and let M have holonomy type R_{10}, R_{11} or R_{13}. Then the appropriate definitions show that A and B are empty sets in (12.53) and so $M = C \cup \operatorname{int} D \cup Z$ with C open by the rank theorem. It now follows from theorem 12.6 that $\psi = 0$ on C and from theorem 12.5 that $\psi_{a;b} = 0$ on $\operatorname{int} D$. Thus $\psi_{a;b} = 0$ on $M \setminus Z$ and hence on M and so $X \in SP(M)$. One can also conclude because of this that, since X is proper, ψ_a cannot vanish anywhere on M and so $C = \emptyset$ and $M = D \cup Z$ with D open and dense in M. Conversely, if X is a proper member of $SP(M)$, it follows from section 12.4 that $M = D \cup Z$ in (12.53) and from theorem 12.8 that ψ is the unique independent global covariantly constant (co)vector field on M. When this restriction on the holonomy type is combined with part (i) of the theorem only the holonomy types R_{10}, R_{11} or R_{13} remain□

If the conditions and conclusions of part (ii) of this theorem hold then the metric is of the form (12.33) in the R_{13} case (or its equivalent for spacelike ψ in the R_{10} case) or (12.39) in the R_{11} case in some coordinate domain of any point of the open dense subset D of M. It is also remarked that if a non-flat simply connected space-time M admits a *proper* projective and a *proper* affine vector field it must have one of the holonomy types R_{10}, R_{11} or R_{13} as described above. This follows from part (i) of the above theorem and theorem 10.14.

Theorem 12.10 is essentially a complete description of projective symmetry for all holonomy types except R_9, R_{14} and R_{15}. Part (i) of this theorem can also be established by a direct curvature rank decomposition of M but the curvature class decomposition (12.53) leads to a more elegant proof. Some limited information on the holonomy types R_9, R_{14} and R_{15} is available [233],[242] (including a "local uniqueness" result in the R_9 and R_{14} types) and the F.R.W. metric example in section 12.3 is of holonomy type R_{15}. For examples in space-times of holonomy types R_9 and R_{14} consider the metric on a global chart u, v, x, y, restricted only by $u, v > 0$, and given by [242]

$$ds^2 = 2dudv + (uv^{-3})^{1/2}dv^2 + v^2 e^{f(x,y)}\left(dx^2 + dy^2\right) \tag{12.54}$$

for some smooth function f on $\mathbb{R}^2$. It can then be checked that the vector field with components $(1,0,0,0)$ is null and recurrent on M and, from a calculation of the curvature tensor and use of theorems 8.7 and 8.9(iii), that the holonomy type of (12.54) is R_9 if $\frac{\partial^2 f}{\partial x^2} + \frac{\partial^2 f}{\partial y^2} \equiv 0$ on M and R_{14} otherwise. Then the vector field X with components $(uv, v^2, 0, 0)$ can be shown to be *proper projective* on M with projective 1-form $\psi_a = (0,1,0,0)$.

So far, one has for the various symmetry algebras of M the inclusions $K(M) \subseteq H(M) \subseteq A(M) \subseteq P(M)$ and $H(M) \subseteq C(M)$. It can, in fact, be established that the fork on these inclusions really does occur at $H(M)$ in the sense that *a vector field X on a simply connected space-time M is simultaneously in $P(M)$ and $C(M)$ if and only if $X \in H(M)$*. To see this, briefly, let $X \in P(M) \cap C(M)$. Then from (12.1) and (11.1), $h_{ab} = \chi g_{ab}$ satisfies (12.4) for some function χ and 1-form ψ_a. A substitution into (12.4) and the usual contractions then shows that $\psi_a = 0$ and $\chi_{,a} = 0$ and hence that $X \in H(M)$. The converse is clear.

An alternative approach to projective symmetry has been given using an algebraic study of the tensor h in (12.1). Further details of this can be found, for example, in [63],[243].

Chapter 13

Curvature Collineations

13.1 Introduction

The symmetries discussed so far have a number of general properties. They were represented by finite-dimensional Lie algebras and hence had a well-behaved orbit structure and each vector field in the algebra was uniquely determined by its value and those of finitely many derivatives at some point of the space-time M and possessed nice differentiability properties. Unfortunately not all symmetries of the important geometrical objects of space-time have such pleasant properties and this chapter will, by considering symmetries of the curvature tensor, attempt to draw attention to some of these problems [244], [245], [246].

13.2 Curvature Collineations

Let M be a (smooth) space-time with (smooth) Lorentz metric g and associated curvature tensor $\tilde{R}$. Let X be a global vector field on M assumed for the time being to be C^1. If the C^1 local diffeomorphisms (section 5.11) associated with X are denoted by ϕ_t then X is called a *curvature collineation* if either and hence both of the following two equivalent (section 5.13) conditions hold [244]

$$\begin{aligned}
&\text{(i)} \qquad \mathcal{L}_X \tilde{R} = 0 \qquad (\text{i.e. } \mathcal{L}_X R^a{}_{bcd} = 0) \\
&\text{(ii)} \qquad \phi_t^* \tilde{R} = \tilde{R} \qquad \text{for any } \phi_t \text{ associated with } X
\end{aligned} \tag{13.1}$$

One problem concerning such a vector field X lies in the degree of differentiability assumed for X. Clearly one requires at least C^1 for (13.1) to make sense. So let ϕ_t be a local diffeomorphism associated with X with (open coordinate) domain U and (open) range $W \equiv \phi_t(U)$ as in section 10.1 with coordinates x^a on U and the functions $y^a = x^a \circ \phi_t^{-1}$ on W. If ϕ_t is *smooth* then W and y^a constitute a coordinate neighbourhood of M (since M is smooth) on which the curvature $\tilde{R}$ on M and the tensor $\phi_t^{-1*}\tilde{R}$ coincide according to (13.1). These coincident tensors on W are then, respectively, the curvature tensors of the (smooth) metrics g and $\phi_t^{-1*}g$ on W. If, on the other hand, ϕ_t is C^k $(1 \leq k < \infty)$ then W and y^a do not constitute a chart of M (since M is smooth). However $\phi_t^{-1*}\tilde{R}$ is still smooth with respect to M (since it coincides with $\tilde{R}$) and could be regarded as the curvature tensor of the metric $\phi_t^{-1*}g$ on W if this metric were C^2. This will be achieved if X and ϕ_t are C^3. If, however, curvature collineations which are not smooth are admitted then the set of all curvature collineations, whilst clearly a vector space, is not a Lie algebra since if X and Y are curvature collineations, $[X, Y]$ may not be differentiable.

In spite of the fact that curvature collineations that are not smooth may (and, as will be seen later, do) exist, little is lost in the general discussion in assuming them smooth. In what is to follow, it will be indicated where and how non-smooth ones exist. So *a curvature collineation is taken as a smooth global vector field X on M satisfying (13.1)* and the set of all such curvature collineations is, from a result in section 5.13, a Lie algebra denoted by $CC(M)$.

Further problems arise with curvature collineations when compared with the symmetries so far studied. For example, the affine (including Killing and homothetic), conformal and projective algebras are all finite-dimensional and their individual members are uniquely determined by specifying their values and those of finitely many derivatives at some point of M (and are identically zero on M if they vanish over some non-empty open subset of M). It will be seen in later examples that none of these results need hold for (smooth) curvature collineations.

The following theorem is immediate from theorem 11.11 and from (12.7).

Theorem 13.1 *Let M be a space-time.*

i) Any affine vector field is a curvature collineation.

ii) A conformal (respectively projective) vector field is a curvature collineation if and only if it is special conformal (respectively special projective).

Thus in the notation so far established, $K(M)$, $H(M)$ and $A(M)$ are subalgebras of $CC(M)$ and

$$C(M) \cap CC(M) = SC(M), \quad P(M) \cap CC(M) = SP(M) \tag{13.2}$$

A curvature collineation is called *proper* if it is not affine.

There is a convenient way of obtaining quite general information on the existence of curvature collineations. Let $X \in CC(M)$ and let the local diffeomorphisms associated with X be represented by the maps ϕ_t. As in section 10.1 let ϕ_t be a particular such map with domain U and range $W \equiv \phi_t(U)$. Now, as earlier, consider the tensor pullback map ϕ_t^{-1*} which transfers tensors on U to tensors on W. If g and $\tilde{R}$ are the (restricted) metric and associated curvature tensor on U from M, then $\phi_t^{-1*}g$ is a metric on W with associated curvature tensor $\phi_t^{-1*}\tilde{R}$. Since X is a curvature collineation $\phi_t^{-1*}\tilde{R}$ and $\tilde{R}$ are equal and have g and $\phi_t^{-1*}g$ as associated metrics. It follows that, on W, g and $\phi_t^{-1*}g$ are related, pointwise, according to theorem 9.3. Thus using theorem 9.4 one sees that if the curvature tensor of M is of class A at each point of M (and this is automatic if the curvature rank is at least four at each point of M) then g and $\phi_t^{-1*}g$ are conformally related with a constant conformal factor. So one has the following general result regarding curvature collineations [119] [247].

Theorem 13.2 *Let M be a space-time of curvature class A. Then every curvature collineation on M is a homothetic vector field.*

Furthermore, for a fixed space-time manifold M and considering the set of all smooth Lorentz metrics on M, it is generic for the metric (i.e. the space-time) considered to satisfy the condition that the curvature rank is everywhere at least four (theorem 7.17 (iii)). Hence *generically* (on space-times) *every curvature collineation is homothetic*.

It is remarked that an alternative proof of (part of) this theorem can be obtained by taking the Lie derivative of (9.6) and using (13.1) and theorem 9.4 [247], [244].

For a vacuum space-time which has constant curvature class this class must be A or C (and the latter case corresponds to Petrov type N since now the curvature and Weyl tensors are equal - see the proof of theorem 9.6). Thus the following theorem now follows from the previous one.

Theorem 13.3 *If M is a vacuum space-time with constant curvature class then either M is of Petrov type N or every curvature collineation on M is a homothetic vector field.*

13.3 Some Techniques for Curvature Collineations

One way of studying curvature collineations is suggested by section 9.3. One could consider examples of space-times of class B, C or D (in the notation of that section). It then follows from that section, the earlier discussion of this chapter and the definition of the Lie derivative in section 5.13 that if $X \in CC(M)$ the local expressions for $\mathcal{L}_X g$ are, from (9.10)-(9.12) for the above spacetime classes B, C and D, respectively,

$$\begin{aligned} \mathcal{L}_X g_{ab} &= \phi g_{ab} + 2\lambda l_{(a}n_{b)} = (\phi+\lambda)g_{ab} - \lambda(x_a x_b + y_a y_b) \\ \mathcal{L}_X g_{ab} &= \phi g_{ab} + \lambda w_a w_b \\ \mathcal{L}_X g_{ab} &= \phi g_{ab} + \mu u_a u_b + 2\nu u_{(a}v_{b)} + \lambda v_a v_b \end{aligned} \tag{13.3}$$

for smooth functions ϕ, μ, ν, λ (the smoothness following from the theory in section 9.3). This will now be applied to special types of spacetimes of classes C and D and essentially in full generality for class B.

First let M be a simply connected space-time (or open region of space-time) of class C *and* of holonomy type R_{13}. Then M admits a global, smooth, covariantly constant unit timelike vector field u. The associated covector field $\tilde{u}$ (with local components u_a) is a global closed 1-form and hence is exact since M is simply connected (section 4.16). Thus there exists a global smooth function $f : M \to \mathbb{R}$ such that $\tilde{u} = df$. For such spacetimes one can choose about any point a natural product coordinate domain U with coordinates x^a and then the hypersurfaces of constant x^0 in U coincide with the intersection of the hypersurfaces of constant f with U. Also one can arrange that, in U, $u^a = \delta^a_0$ ($\Rightarrow u_a = -\delta^0_a$) and the metric is (section 8.3)

$$ds^2 = -(dx^0)^2 + g_{\alpha\beta}dx^\alpha dx^\beta \tag{13.4}$$

where the $g_{\alpha\beta}$ depends only on x^α ($\alpha, \beta, \gamma = 1, 2, 3$). The Ricci identity shows that in any coordinate system in M, $R_{abcd}u^d = 0$ and so $k = u$ is, at each point of M, the unique independent solution of (9.9). In the special coordinates on U the components R_{abcd} and $R^a{}_{bcd}$ (and also Γ^a_{bc}) are zero if any of the indices a, b, c or d equals zero and the components $g_{\alpha\beta}$ are the induced metric components in the hypersurfaces of constant x^0 in U (in the coordinates x^α). Any two of these hypersurfaces are isometric under the map given in an obvious notation by $(x^\alpha, x^0) \to (x^\alpha, x^0 + t^0)$ for appropriate $t^0 \in \mathbb{R}$.

Now consider the existence of a global curvature collineation X on M.

The discussion will be given globally by a component argument in any coordinate system in M (but sometimes the special coordinate system in U is useful). Thus (13.1) holds and so in any coordinate system on M, using the second of (13.3), dropping a factor $\frac{1}{2}$ and relabelling

$$X_{a;b} = \phi g_{ab} + \lambda u_a u_b + F_{ab} \tag{13.5}$$

for smooth functions ϕ and λ and smooth bivector F on M. Then (13.1) gives

$$R^a{}_{bcd;e}X^e + R^a{}_{ecd}X^e{}_{;b} + R^a{}_{bed}X^e{}_{;c} + R^a{}_{bce}X^e{}_{;d} - R^e{}_{bcd}X^a{}_{;e} = 0. \tag{13.6}$$

A contraction of this result with u_a reveals

$$R^e{}_{bcd}(X^a{}_{;e}u_a) = 0. \tag{13.7}$$

So by the uniqueness of solutions to (9.9) one sees that $X^a{}_{;e}u_a$ is proportional to u_e and then (13.5) shows that $F_{ab}u^b \propto u_a$ and hence that $F_{ab}u^b = 0$. One can now make a global decomposition of X on M according to

$$X = X' - \alpha u \qquad \alpha = X^a u_a \tag{13.8}$$

where α is a global smooth function and X' a global smooth vector field on M. Clearly X' is everywhere orthogonal to u, $X'^a u_a = 0$, and is hence tangent to the submanifolds of constant f (and to those of constant x^0 within U). Then a contraction of (13.5) with u^a shows, since $F_{ab}u^b = 0$, that $\alpha_{,a} = (\phi - \lambda)u_a$ (or $d\alpha = (\phi - \lambda)df$) and so in the coordinate system in U, α and $(\phi - \lambda)$ are functions of x^0 only. Now since u is covariantly constant it is necessarily a curvature collineation and so $X = u$ satisfies (13.6). Substituting into (13.6) gives $R^a{}_{bcd;e}u^e = 0$ and it is then a straightforward matter to see from (13.6) that αu and hence (from (13.8)) X' are curvature collineations on M. Furthermore, (13.5) and (13.8) easily reveal that $[X', u] = 0$. This is equivalent to $\mathcal{L}_u X' = 0$ and has the geometrical interpretation that if χ_t represent the local isometries between the x^0 =constant hypersurfaces in U arising from the vector field $\frac{\partial}{\partial x^0}$ on U then X' (on U) is "invariant" under χ_{t*} (i.e.$\chi_{t*}X' = X'$) and the vector fields in each hypersurface of constant x^0 in U naturally induced by X' are mapped to each other by the restrictions of χ_{t*} to these hypersurfaces. Put more simply, in these coordinates $X'^0 = 0$ and the X'^α are independent of x^0. Each of these (isometric) hypersurfaces in U has a metric $g_{\alpha\beta}$ and curvature tensor $R^\alpha{}_{\beta\gamma\delta}$ (and the Christoffel symbols $\Gamma^\alpha_{\beta\gamma}$ where Γ^a_{bc} are the

Christoffel symbols on U). It is then easily checked from (13.6) by setting a, b, c and d equal to Greek indices α, β, γ and δ and noting that one can change the semi-colon derivative to a covariant derivative with respect to the hypersurface geometry (because of the properties of Γ^a_{bc} mentioned earlier) that the vector field $\tilde{X}'$ with components $\tilde{X}'^{\alpha} = X'^{\alpha}$ induced in these hypersurfaces from X' is a curvature collineation in the hypersurface geometry. But (13.5) reveals more. It projects onto each hypersurface in the form (using a stroke for a hypersurface covariant derivative)

$$X'_{\alpha|\beta} = \phi g_{\alpha\beta} + F_{\alpha\beta} \tag{13.9}$$

which can be calculated directly from (13.5) (In fact, if one uses (5.15) with ϕ, X, Y and T replaced, respectively, by $i, \tilde{X}, X$ and g where i is the inclusion map from the hypersurface to M and $h = i^*g$ is the hypersurface metric one gets from (13.5) $\mathcal{L}_{\tilde{X}}h = 2\phi h$ which is the symmetric part of (13.9)). Since one can construct a coordinate domain like U about any point of M it follows that the vector field induced in the hypersurfaces of constant f by X' is a conformal vector field in the induced geometry. Now if one takes the covariant derivative of (13.5), uses the Ricci identity on X, contracts with u^c and takes the symmetric part on the indices a and b one finds using $F_{ab}u^b = 0$ and after a simple rank argument that $\phi_a u^a = 0$ and so in the coordinates on U, ϕ is independent of x^0. Further, from the class C condition, the hypersurface curvature components $R_{\alpha\beta\gamma\delta}$ in the coordinates x^α in U have rank two or three at each point of M. The obvious adaptation of the work in section 9.3 applied in three dimensions shows that the curvature tensor $R^{\alpha}{}_{\beta\gamma\delta}$ determines its metric up to a constant conformal factor and so the vector field $\tilde{X}'$ induced in the hypersurfaces of constant f which is already known to be a curvature collineation in the induced geometry is homothetic in the induced geometry. Thus ϕ is a constant on M. Since it was shown earlier that $(\phi - \lambda)$ was, in each coordinate domain U, a function of x^0 only, it follows that λ also depends only on x^0 in each such domain and so λ depends only on f. One final point to note is that it is easily checked that $u, fu, \ldots, f^n u$ are curvature collineations on M by direct substitution into (13.6). But since $\tilde{u} = df$, f is not constant on M and so they are independent members of $CC(M)$ for each n. *Hence the Lie algebra CC(M) is infinite-dimensional* and each member can be decomposed as in (13.8). Similarly, restricting attention to U, one gets the Lie algebra of curvature collineations $CC(U)$ on U and each member of $CC(U)$ can be decomposed uniquely as in (13.8). Thus $CC(U)$ is the vector space direct sum of the vector space of all vector fields of the form $g(x^0)u$ for some arbitrary smooth function g (each of which is easily checked to be in $CC(U)$ - *cf.* [244]) and the finite-dimensional vector space of homothetic vector fields of each (any) hypersurface in U of constant x^0 together with its induced geometry. It is remarked that it is easily checked that the sum of such a homothetic vector field and a vector field of the form $g(x^0)u$ is in $CC(U)$.

It should be clear from the arbitrary functions involved in describing $CC(M)$ how one might choose $X \in CC(M)$ which is not C^∞, or not uniquely determined by its value together with those of any number of its covariant derivatives at some (any) point of M or is such that X is not identically zero on M but vanishes over some non-empty open subset of M. The function $f : \mathbb{R} \to \mathbb{R}$ given by $h(x) = 0$ $(x \leqslant 0)$, $h(x) = e^{-\frac{1}{x}}$ $(x > 0)$ is useful for the last two of these properties since it is a global smooth function on $\mathbb{R}$ and vanishes together with each of its derivatives for $x \leq 0$.

If M is also both simply connected and geodesically complete then it is a global metric product $M = N \times \mathbb{R}$ with N a 3-dimensional manifold with a positive definite metric. In this case the above discussion clearly simplifies.

If the conditions required of M are retained except that the holonomy type is changed to R_{10} (section 8.3) then the above results go through. One should, however, bear in mind that in this case u is spacelike and its orthogonal hypersurfaces are Lorentzian and that homothetic symmetry theory in such hypersurfaces is quite different from the positive definite case, in particular, in the fixed point theorems (*cf.* section 10.4).

Now let M be a space-time of class B. If M is assumed simply connected then (chapter 9) it is of holonomy type R_7. Then in any coordinate system of M the Riemann tensor can be written as in (9.8)

$$R_{abcd} = \alpha G_{ab} G_{cd} - \beta {}^*G_{ab} {}^*G_{cd} \tag{13.10}$$

where α and β are nowhere zero real valued functions on M and at each point G is a timelike bivector satisfying $G_{ab}G^{ab} = -2$ and ${}^*G$ its spacelike dual. The objects α, β, G and ${}^*G$ and the distributions spanned by G and ${}^*G$ are taken as smooth and G and ${}^*G$ are covariantly constant (section 9.3). The notation of section 9.3 will be used here with U a product neighbourhood in M and (l, n, y, z) a smooth null tetrad on U with l and n (respectively y and z) spanning the blade of G (respectively ${}^*G$). The metric can be written as in (8.5)

$$ds^2 = g_{AB}dx^A dx^B + g_{A'B'}dx^{A'}dx^{B'} \tag{13.11}$$

and then, in (13.10), $\alpha = \alpha(x^0, x^1)$ and $\beta = \beta(x^2, x^3)$ (section 9.3). Now let X be a curvature collineation on M. Then $\mathcal{L}_X R_{ab} = 0$ and from (13.10), the Ricci tensor is non-degenerate on M and so may be considered a smooth metric on M. It follows (see section 10.3) that dim $CC(M) \leqslant 10$ (In this case if X had originally been assumed C^3 (section 13.2) it would have necessarily been smooth since the Ricci tensor is). One can decompose X (restricted to the product manifold $U = T \times S$ where $T(S)$ is a 2-dimensional timelike (spacelike) submanifold (see section 9.3)) along the parallelisation l, n, y, z of U as

$$X = al + bn + cy + dz \tag{13.12}$$

where a, b, c and d are smooth real valued functions on U. Now the vector fields l and n can, with an abuse of notation, be regarded as vector fields on T and similarly for y and z on S. In the product coordinates used here it then follows that the components l^a and n^a are independent of x^2 and x^3 and y^a and z^a are independent of x^0 and x^1. A simple calculation in this coordinate system then shows there exist 1-forms p and q such that

$$\begin{aligned} l_{a;b} &= l_a p_b \quad n_{a;b} = -n_a p_b \quad y^a p_a = z^a p_a = 0 \\ y_{a;b} &= z_a q_b \quad z_{a;b} = -y_a q_b \quad l^a q_a = n^a q_a = 0 \end{aligned} \tag{13.13}$$

(In fact $l_{a;b} = l_a p_b \Rightarrow l_{[a;b}l_{c]} = 0 \Rightarrow l_a$ is hypersurface orthogonal and can thus be scaled so that $l_{a;b} \propto l_a l_b$).

Now substitute (13.12) into (13.6) using (13.13) and (13.10) and contract with ${}^*G^{cd}l^b y_a$ to get $d_{,a}l^a = 0$. A similar procedure, this time contracting with ${}^*G^{cd}n^b y_a$, reveals $d_{,a}n^a = 0$. Thus, in U, $d = d(x^2, x^3)$. Similar contractions with ${}^*G^{cd}l^b z_a$ and ${}^*G^{cd}n^b z_a$ reveal that $c = c(x^2, x^3)$. Further contractions with $G^{cd}y^b l_a$, etc., show that $a = a(x^0, x^1)$ and $b = b(x^0, x^1)$. Thus X can be decomposed as

$$X = X_T + X_S, \quad X_T = al + bn, \quad X_S = cy + dz \tag{13.14}$$

where X_T and X_S are smooth. It is then easily shown that

$$\mathcal{L}_y X_T = \mathcal{L}_z X_T = \mathcal{L}_l X_S = \mathcal{L}_n X_S = 0 \tag{13.15}$$

So X_T and X_S are vector fields in U tangent to the appropriate submanifolds of decomposition in U and the naturally induced vector fields in these submanifolds, $\tilde{X}_T$ and $\tilde{X}_S$ (section 4.11), which arise from them are identically related by the differentials of the local isometries which link these submanifolds. Now equation (13.6) because of (13.10) and (13.14) naturally decomposes on U into two equations one for each of the submanifolds of decomposition. In fact if one substitutes (13.14) into (13.6)and uses the information about the local decomposition following (8.5) and (9.9) one sees that $\tilde{X}_T$ and $\tilde{X}_S$ *are each curvature collineations in the induced geometry of the submanifolds of decomposition.* It also follows from (13.6) and (13.14) that X_T *and* X_S *are curvature collineations on* U. Conversely, a similar argument shows that if $\tilde{X}_T$ and $\tilde{X}_S$ are curvature collineations in (the induced geometry of) T and S then they can be naturally extended to vector fields X_T and X_S on U tangent to the obvious submanifolds and then X_T, X_S and hence $X_T + X_S$ are curvature collineations on U (but which may not necessarily be extendible to be global on M). Now the relationship between the Ricci and Riemann tensor components on a 2-dimensional manifold ($R^a{}_{bcd} = \delta^a_c R_{bd} - \delta^a_d R_{bc}$) shows that the condition that $\tilde{X}_T$ is a curvature collineation in (the induced geometry of) T is equivalent to the condition that the Lie derivative of the Ricci tensor in T with respect to $\tilde{X}_T$ is zero. This latter tensor for this space-time and in this coordinate system has the components R_{AB} of the space-time Ricci tensor. But $R_{AB} = \sigma_T g_{AB}$ where σ_T is the nowhere zero scalar curvature of T and so the above condition is equivalent to $\mathcal{L}_{\tilde{X}_T}(\sigma_T g_{AB}) = 0$. Similar comments in an obvious notation for $\tilde{X}_S$ lead to $\mathcal{L}_{\tilde{X}_S}(\sigma_S g_{A'B'}) = 0$. Hence the solutions for $\tilde{X}_S$ and $\tilde{X}_T$ each belong to Lie algebras of vector fields on S and T, respectively, of dimension ≤ 3 since $\sigma_T g_{AB}$ could be regarded as a metric on T (and similarly for $\sigma_S g_{A'B'}$). This restricts $\tilde{X}_T$ to being a conformal vector field for g_{AB} (and similarly for $\tilde{X}_S$). Incidentally, if $\tilde{X}_S$ is a Killing vector field for the induced metric $g_{A'B'}$ on S then $\mathcal{L}_{\tilde{X}_S}\sigma_S = 0$ and the above condition is automatically satisfied (and similarly for $\tilde{X}_T$). Thus $CC(U)$ is a direct sum of two vector spaces each of dimension ≤ 3 and hence dim $CC(U) \leq 6$. But $CC(M)$ has already been shown to be finite-dimensional since $CC(M)$ is contained in the (finite-dimensional) Lie algebra of vector fields which are Killing with respect to the Ricci tensor, the latter being regarded as a metric on M. From the general theory in section 10.2 it then follows that non-trivial members of $CC(M)$ cannot vanish on non-empty open subsets of M. Now consider the linear map $CC(M) \longrightarrow CC(U)$ given by restricting a member of $CC(M)$ to U. The previous section shows this map to be one to one (otherwise a non-trivial member of $CC(M)$ would vanish on U)

and so one can improve the above result to $\dim CC(M) \leq 6$.

Again if M is geodesically complete as well as simply connected then M is the global metric product of two 2-dimensional manifolds and the above argument clearly simplifies.

Now let M be a space-time of class D and consider the special case when M is simply connected and of holonomy group type R_2. So (section 8.3) M admits a global null tetrad of smooth vector fields (l, n, y, z) with l and n recurrent and y and z covariantly constant and in any coordinate system the curvature tensor takes the form

$$R_{abcd} = \mu G_{ab} G_{cd} \tag{13.16}$$

for a smooth nowhere zero function μ and bivector $G_{ab} = 2l_{[a}n_{b]}$ on M. The recurrence of l and n (13.13) implies that $G_{ab;c} = 0$ and then the Bianchi identity $R_{ab[cd;e]} = 0$ contracted first with y^d and then with z^d shows that $\mu_{,a}y^a = \mu_{,a}z^a = 0$. Let X be a curvature collineation on M. Then from (13.3)

$$X_{a;b} = \phi g_{ab} + \alpha y_a y_b + \beta z_a z_b + \gamma(y_a z_b + z_a y_b) + F_{ab} \tag{13.17}$$

for smooth functions α, β, γ and ϕ and smooth bivector F. A substitution of (13.17) into (13.6) using the results $R_{abcd}y^d = R_{abcd}z^d = 0$ gives

$$0 = R^a{}_{bcd;e}X^e - R^e{}_{bcd}F^a{}_e + R^a{}_{ecd}F^e{}_b + R^a{}_{bed}F^e{}_c + R^a{}_{bce}F^e{}_d + 2\phi R^a{}_{bcd} \tag{13.18}$$

When this is contracted with y_a one finds

$$R^e{}_{bcd}(F^a{}_e y_a) = 0 \qquad (\Rightarrow F_{ab}y^b = \rho y_a + \sigma z_a)$$

the implication (for functions ρ and σ) following from the fact that the only independent solutions to (9.9) at any point of M are y and z. But then $F_{ab}y^a y^b \equiv 0 \Rightarrow \rho \equiv 0$ on M and so $F_{ab}y^b = \sigma z_a$. Similarly one finds $F_{ab}z^b = -\sigma y_a$. Next define functions a and b on M by $a = X^a y_a$, $b = X^a z_a$. Then contractions of (13.17) first with y^a and then z^a give

$$a_{,b} = (\phi + \alpha)y_b + (\gamma - \sigma)z_b \qquad b_{,b} = (\phi + \beta)z_b + (\gamma + \sigma)y_b \tag{13.19}$$

Then define a global vector field X' on M as

$$X = X' + ay + bz \qquad (\Rightarrow X'^a y_a = X'^a z_a = 0) \tag{13.20}$$

so that X' is everywhere orthogonal to y and z. Now a straightforward calculation using (13.6), (13.19), (13.20) and the fact that since $y, z \in CC(M)$, (13.6) gives $R^a{}_{bcd;e}y^e = R^a{}_{bcd;e}z^e = 0$, shows that $ay + bz$ is in

$CC(M)$ and hence so is X'. One also finds from (13.17), (13.19) and (13.20) that $X'^a{}_{;b}y^b \equiv 0 \equiv X'^a{}_{;b}z^b$ and so, on M

$$[X', y] \equiv 0 \equiv [X', z] \qquad (\Rightarrow \mathcal{L}_y X' \equiv 0 \equiv \mathcal{L}_z X') \tag{13.21}$$

Similar arguments now reveal that

$$X'_{a;b} = \phi(g_{ab} - y_a y_b - z_a z_b) + F'_{ab} \tag{13.22}$$

$$F'_{ab} = F_{ab} + 2\sigma y_{[a} z_{b]} \qquad (F'_{ab}y^b = F'_{ab}z^b = 0)$$

Now since the global covector fields $\tilde{y}$ and $\tilde{z}$ with local components y_a and z_a are closed 1-forms and M is simply connected it follows that they are each exact. So there exist global smooth functions $f_1, f_2 \to \mathbb{R}$ such that $\tilde{y} = df_1$, $\tilde{z} = df_2$ and (13.20) shows that X' is tangent to the 2-dimensional submanifolds of M defined by the constancy of f_1 and f_2. One can now choose the natural product coordinates on a domain U of any $m \in M$ (as in section 8.3) and write the metric (with capital Latin letters taking values 0 and 1) in U as

$$ds^2 = g_{AB}dx^A dx^B + {dx^2}^2 + {dx^3}^2 \tag{13.23}$$

where g_{AB} depends only on x^0 and x^1 and where x^2 and x^3 can be chosen to conside with the functions with f_1 and f_2, respectively, on U. Thus the components g_{AB} are those of a general Lorentz metric on the submanifolds of constant f_1 and f_2. In U the only nonvanishing components of the connection and curvature tensor are Γ^A_{BC} and R_{ABCD} and these are the connection and curvature components corresponding to g_{AB}. Next one can construct in U from (13.17) an expression for $X_{a;bc}$ and insert into the Ricci identity for X. A contraction of the resulting identity with $y^b w^a w^c$ (where $w^a = l^a + n^a$) and use of the previous information on F yields $\phi_{,a}y^a = 0$ (and a similar obvious contraction yields $\phi_{,a}z^a = 0$. Thus ϕ depends only on x^0 and x^1 in U as also does μ in (13.16). Now if $\tilde{X}'$ represents the vector field in the submanifolds of constant f_1 and f_2 naturally induced by X' and with similar definitions for $\tilde{F}$, $\tilde{g}$ and $\tilde{\phi}$ one finds from (13.22) (or (5.15))

$$\tilde{X}'_{A|B} = \tilde{\phi}\tilde{g}_{AB} + \tilde{F}_{AB} \tag{13.24}$$

where, in U, $\tilde{g}_{AB} = g_{AB}$ and $\tilde{F}_{AB} = F_{AB}$ and a stroke denotes a covariant derivative in the induced geometry. Thus $\tilde{X}'$ *is a conformal vector field in the Lorentz submanifolds of constant* f_1 *and* f_2 and its restriction to submanifolds of constant x^2 and x^3 in the product domain is invariant under the natural isometry linking such submanifolds. Similarly, by replacing the indices a, b, c, d by A, B, C, D, respectively, in (13.6) one sees that $\tilde{X}'$ *is also a curvature collineation in the induced geometry of these submanifolds.* Then, just as in the last example, this condition is equivalent to the condition $\mathcal{L}_{\tilde{X}'} R_{AB} \equiv \mathcal{L}_{\tilde{X}'}(\nu g_{AB}) = 0$ where ν is the (nowhere zero) submanifold scalar curvature. Thus the solutions for $\tilde{X}'$ constitute a vector space of dimension ≤ 3 and include the Killing vector fields of the induced metric on these submanifolds. Conversely, if $\tilde{X}'$ is a curvature collineation in these submanifolds the associated vector field X' is a curvature collineation in M and hence so is X in (13.20). Thus the members of $CC(M)$ are those vector fields X of the form (13.20) where a and b may be taken as smooth functions on M arbitrary except from the conditions $a_{,c}l^c = a_{,c}n^c = b_{,c}l^c = b_{,c}n^c = 0$ (so that each is a function only of the coordinates x^2 and x^3 in a product chart domain U) and where $X' \in CC(M)$ and is tangent to the above submanifolds. It follows that $CC(M)$ and $CC(U)$ are infinite-dimensional.

If M is simply connected and geodesically complete then M is a metric product and the usual simplifications occur. If M is of holonomy type R_4 the discussion is similar.

13.4 Further Examples

The last section listed examples of space-times which admitted proper curvature collineations and which were, in some sense, "attached" to the holonomy decomposition. One may ask how crucial the holonomy decomposition was for the existence of proper members of $CC(M)$ and whether or not the curvature class of the metric (section 9.3) is more important. To be precise one could ask the following questions.

(i) Can a space-time of curvature class C, but which admits no non-trivial covariantly constant vector fields, admit a proper member of $CC(M)$? If so, could $CC(M)$ be infinite-dimensional and if $CC(M)$ could be finite-dimensional, will it then contain no proper members?

(ii) Can a space-time of curvature class D which does not admit two independent covariantly constant vector fields admit a proper member of $CC(M)$ and, if so, is $CC(M)$ necessarily infinite-dimensional? If arbitrary functions are involved in the general solution for members of $CC(M)$, how many such functions are there and how "arbitrary" are they? If $CC(M)$ is finite-dimensional does it contain proper members?

The answers to some of these questions will be given by means of examples, together with a result which is helpful in constructing and understanding them. Suppose M is a space-time and let k be a nowhere-zero smooth vector field in M such that (i) $k \in CC(M)$, (ii) k is hypersurface orthogonal and (iii) in any coordinate system on M, $R_{abcd}k^d = 0$. Then for any $m \in M$ there is a coordinate domain U containing m with coordinates t, x, y, z such that k is orthogonal to the hypersurfaces of constant t and $f(t)k$ is a curvature collineation on U for an arbitrary function f. To see this note that, since k is in $CC(M)$, then k satisfies (13.6). But since $R_{abcd}k^d = 0$ on U it follows that $f_{,d}R^d{}_{abc} = 0$ on U and so the vector field fk is then easily shown to satisfy (13.6).

This result is useful in exploring metrics of curvature class C with no non-trivial covariantly constant vector fields or metrics of curvature class D with less than two independent covariantly constant vector fields. The following examples supply some of the answers to the questions posed above. These (and other similar) examples can be found in [248]-[258] together with some earlier work on the existence of curvature collineations.

Example 1 Consider the class of F.R.W. models (see e.g. [239] [55]). In general the curvature rank for such models is six and so theorem 13.2 confirms that $CC(M)$ is just the 6- or 7-dimensional homothetic algebra $H(M)$ of M, or, for the de-Sitter type metrics, the 10-dimensional Killing algebra of M. There are also Einstein static type metrics, where the curvature rank is three and which satisfy (13.4) with $CC(M)$ as given in section 13.3. There is, however, another class of F.R.W. models which are of the general holonomy type R_{15} but whose curvature rank is three (class C). For such a space-time the line element is

$$ds^2 = -dt^2 + (at + b)^2 d\sigma^2 \tag{13.25}$$

where $a, b \in \mathbb{R}$, $a \neq 0$, $d\sigma^2$ represent a positive definite (constant curvature) metric in the hypersurfaces of constant t and if u represents the usual unit timelike fluid flow vector field, $u \equiv \frac{\partial}{\partial t}$, $R_{abcd}u^d = 0$ holds everywhere. No non-trivial covariantly constant vector fields are admitted but the components $R^a{}_{bcd}$ are independent of t. Hence $u \in CC(M)$ and the general solution for members of $CC(M)$ still contains an arbitrary function. In fact the members of $CC(M)$ are those vector fields of the form [257], [258]

$$X = f(t)\frac{\partial}{\partial t} + Z \tag{13.26}$$

where f is an arbitrary C^∞ function and Z is any member of the 6-dimensional Killing algebra $K(M)$. There is, of course, an obvious 'space-like' version of this example. The solution (13.26) was obtained by expressing the curvature tensor in terms of the tensors E and G (since the Weyl tensor is zero) using (7.11) and then using the classification of the tensor E (section 7.6) to examine the curvature rank. The metric (13.25) is the only F.R.W. model, except for the Einstein static types, to admit proper members of $CC(M)$.

Example 2 Consider the metric (discussed in section 12.3)

$$ds^2 = -dt^2 + dx^2 + x^2 g_{AB} dx^A dx^B \tag{13.27}$$

on a single chart t, x, x^2, x^3 with x^2, x^3 coordinatising some open subset of $\mathbb{R}^2$, $t \in \mathbb{R}$, $0 < x \in \mathbb{R}$ and g_{AB} $(A, B = 2, 3)$ independent of t and x (the latter representing a positive definite metric in each submanifold of constant t and x). This space-time has a curvature tensor of rank one everywhere (curvature class D) and admits exactly one independent covariantly constant vector field $\frac{\partial}{\partial t}$ but two independent solutions $\frac{\partial}{\partial t}$ and $\frac{\partial}{\partial x}$ to (9.9). In this case $CC(M)$ can be shown to consist of those vector fields of the form [256]-[258]

$$X = f_1(x,t)\frac{\partial}{\partial t} + f_2(x,t)\frac{\partial}{\partial x} + Z \tag{13.28}$$

for arbitrary functions f_1 and f_2 and vector field Z where Z is any space-time curvature collineation lying in the submanifolds of constant t and x. Since Z is tangent to these submanifolds it naturally induces a vector field $\tilde{Z}$ in them (and $\mathcal{L}_{\frac{\partial}{\partial t}} Z = \mathcal{L}_{\frac{\partial}{\partial x}} Z = 0$). The possible solutions for $\tilde{Z}$ lie in a vector space of dimension at most three and include the Killing algebra of the metric g_{AB}. Thus the general solution (13.28) shows that $CC(M)$ is infinite-dimensional.

Example 3 Consider the plane symmetric static space-time consisting of a single chart (some suitably restricted connected open subset of $\mathbb{R}^4$) and with metric

$$ds^2 = -(ax+b)^2 dt^2 + dx^2 + (cx+d)^2(dy^2 + dz^2) \tag{13.29}$$

where $a, b, c, d \in \mathbb{R}$, $a, c \neq 0 \neq ad - bc$. Here $x_{,d} R^d{}_{abc} = 0$ holds everywhere and the curvature rank and class are 3 and C, respectively. There are no non-trivial covariantly constant vector fields. The set $CC(M)$ consists of just the Killing vector fields $\frac{\partial}{\partial t}$, $\frac{\partial}{\partial y}$, $\frac{\partial}{\partial z}$ and $z\frac{\partial}{\partial y} - y\frac{\partial}{\partial z}$)

Example 4 Several early examples of computing $CC(M)$ for vacuum metrics were given. Of these the case of the vacuum *pp*-wave metric (8.11) is interesting. Here the curvature rank and class are 2 and C, respectively, the Petrov type is N (see theorem 13.3) and the general solution X^a for $CC(M)$ is [248]

$$\begin{aligned} X^0 &= \int [2c(u) - b(u)]du + g, \quad X^1 = c(u)x - d(u) + ay \\ X^2 &= c(u)y + e(u) - ax, \\ X^3 &= -[\frac{1}{2}(x^2 + y^2)c'(u) + d'(u)x + e'(u)y] + b(u)v + f(u) \end{aligned} \tag{13.30}$$

where b, c, d, e and f are arbitrary functions of u, $a, g \in \mathbb{R}$ and a prime means $\frac{d}{du}$.

In summary, it is seen that for curvature class C non-trivial covariantly constant vector fields need not exist and the general solution for members of $CC(M)$ may or may not involve arbitrary functions or be finite-dimensional. For the curvature class D only one independent covariantly constant vector field may be admitted. An example of this class where no such vector field is admitted or where $CC(M)$ is finite-dimensional is not known to the author.

Similar problems to those arising in the study of $CC(M)$ arise in the study of symmetries of the Ricci tensor (Ricci collineations) and of the energy-momentum tensor (matter collineations) for which the condition on the appropriate vector field X is $\mathcal{L}_X R_{ab} = 0$ and $\mathcal{L}_X T_{ab} = 0$, respectively. Such symmetries will not be discussed here. Some information on them can be found in [259]-[267].

Bibliography

1. G. Berkeley "The Principles of Human Knowledge", Collins, London, 1967.
2. E. Mach "The Science of Mechanics", Open Court Publishing Company, La Salle, Illinois, 1960.
3. E. Kretschmann, *Ann. Physik.*, **55**, 1917, 575.
4. J.L. Anderson, *Gen. Rel. Grav.*, **2**, 1971, 161.
5. A. Trautman in "Cosmology and Gravitation", ed. P.G. Bergmann and V. de Sabatta, Plenum Press, New York and London, 1980.
6. N. Rosen, *Phys. Rev.* **57**, 1940, 147.
7. C. Møller, *Mat. Fys. Skr. Dan. Vid. Selsk.*, **1**, No.10, 1961.
8. D. T. Finkbeiner II "Introduction to Matrices and Linear Transformations", Freeman, San Francisco and London, 1960.
9. G. Birkhoff and S. Maclane "A Survey of Modern Algebra", MacMillan, New York, 1961.
10. P. R. Halmos "Finite-Dimensional Vector Spaces", Van Nostrand, 1958.
11. I. D. Macdonald "The Theory of Groups", Oxford, 1968.
12. I. N. Herstein "Topics in Algebra", Blaisdell, 1964.
13. R. P. Geroch "Mathematical Physics", University of Chicago Press, 1985.
14. J. L. Kelley "General Topology", Van Nostrand Reinhold, 1955.
15. J. Dugundji "Topology", Allyn and Bacon, 1974.
16. M. Eisenberg "Topology", Holt, Reinhart and Winston, 1974.
17. I. M. Singer and J.A. Thorpe "Lecture Notes on Elementary Topology and Geometry", Scott, Foresman and Co., 1967.
18. J. A. Wolf "Spaces of Constant Curvature", Publish or Perish, 1974.
19. F. Brickell and R. S. Clark "Differentiable Manifolds", Van Nostrand, 1970.
20. S. Kobayashi and K. Nomizu "Foundations of Differential Geometry" Vol.I, Interscience, New York, 1963.
21. R. Abraham, J. E. Marsden and T. Ratiu "Manifolds, Tensor Analysis and Applications", Springer, 2nd ed., 1988.
22. M. Spivak "Differential Geometry" Vol.1, Publish or Perish, 1970.
23. M. Spivak "Calculus on Manifolds", Benjamin, 1965.
24. D. Martin "Manifold Theory", Ellis Horwood, 1991.
25. W. M. Boothby "An Introduction to Differentiable Manifolds and Rieman-

nian Geometry", Acadamic Press, 1975.
26. H. Whitney, *Ann. Maths.*, **37**, 1936, 645.
27. P. Stefan, *Proc. London Math. Soc.*, **29**, 1974, 699.
28. R. Hermann, International Symposium on Nonlinear Differential Equations and Nonlinear Mechanics, Academic Press, New York, 1963, 325.
29. P. Stefan, *J. London Math. Soc.*, **21**, 1980, 544.
30. H. J. Sussmann, *Trans. Am. Math. Soc.*, **180**, 1973, 171.
31. G. S. Hall and A. D. Rendall, *Int. Jn. Theor. Phys.*, **28**, 1989, 365.
32. R. P. Geroch, *J. Math. Phys.* **9**, 1968, 1739.
33. N. J. Hicks "Notes on Differential Geometry", Van Nostrand, 1971.
34. C. von Westenholz "Differential Forms in Mathematical Physics", North Holland, 1981.
35. L. P. Eisenhart "Riemannian Geometry", Princeton, 1966.
36. P. J. Higgins "An Introduction to Topological Groups", Cambridge University Press, Cambridge, 1974.
37. S. Sternberg "Lectures on Differential Geometry", Prentice-Hall, Englewood Cliffs, 1964.
38. M. W. Hirsch and S. Smale "Differential Equations, Dynamical Systems, and Linear Algebra", Academic Press, 1974.
39. R. S. Palais, Mem. Am. Math. Soc. No. 22, 1957.
40. M. S. Capocci, MSc. Thesis, University of Aberdeen, 1992.
41. J. Ehlers and W. Kundt in "Gravitation; an introduction to current research", ed. L. Witten, Wiley, New York, 1962, 49.
42. R. Shaw *Quart. J. Math.*, Oxford, **20**, 1969, 333.
43. R. Shaw and G. Bowtell *Quart. J. Math.*, Oxford, **20**, 1969, 497.
44. R. Shaw *Quart. J. Math.*, Oxford, **21**, 1970, 101.
45. J. F. Schell *J. Math. Phys.*, **2**, 1961, 202.
46. A. Trautman and F. A. E. Pirani in "Lectures on General Relativity", Prentice Hall, Englewood Cliffs, 1965.
47. R.K. Sachs *Proc. Roy. Soc. London*, **A264**, 1961, 309.
48. S. W. Hawking and G. F. R. Ellis "The Large Scale Structure of Space-Time", Cambridge University Press, 1973.
49. C. W. Misner, K. S. Thorne and J. A. Wheeler "Gravitation", Freeman, San Francisco, 1973.
50. R. M. Wald "General Relativity", University of Chicago Press, Chicago, 1984.
51. H. Stephani "General Relativity", Cambridge University Press, 1982.
52. L. Markus, *Ann. Math.*, **62**, 1955, 411.
53. R. Penrose "Techniques of Differential Topology in Relativity", SIAM, 1972.
54. M. W. Hirsch, *Ann. Math.*, **73**, 1961, 566.
55. H. Stephani, D. Kramer, M. A. H. MacCallum, C. Hoenselears and E. Herlt "Exact Solutions of Einstein's Field Equations", Cambridge University Press, 2003.
56. J. Géhéniau and R. Debever, *Bull. Acad. Roy. Belg. Cl. des Sc.*, **42**, 1956, 114.

57. G. S. Hall, Banach Centre Publications, **12**, P.W.N. Warsaw, 1984, 53.
58. J. L. Synge, U. of Toronto Studies in *Appl. Math.*, **1**, 1935, 1.
59. J. L. Synge "Relativity: The Special Theory", North Holland, Amsterdam, 1956.
60. H. S. Ruse, *Proc. Lond. Math. Soc.*, **41**, 1936, 302.
61. F. de Felice and C. J. S. Clarke "Relativity on Curved Manifolds", Cambridge University Press, 1990.
62. A. Z. Petrov, *Sci. Not. Kazan State Univ.*, **114**, 1954, 55. (For an English translation see *Gen. Rel. Grav.* **32**, 2000, 1665.)
63. A. Z. Petrov "Einstein Spaces", Pergamon, 1969.
64. J. Géhéniau, *C.R. Acad. Sci. (Paris)*, **244**, 1957, 723.
65. L. Bel, *Cah. de Phys.*, **16**, 1962, 59.
66. R. Debever, *C.R. Acad. Sci. (Paris)*, **249**, 1959, 1744.
67. F. A. E. Pirani, *Phys. Rev.*, **105**, 1957, 1089.
68. R. Penrose, *Anns. Phys.*, **10**, 1960, 171.
69. R. Penrose and W. Rindler "Spinors and Space-Time", Cambridge University Press, 1986.
70. S. Bononas, *Gen. Rel. Grav.*, **21**, 1989, 953.
71. R. V. Churchill, *Trans. Amer. Math. Soc.*, **34**, 1932, 784.
72. J. F. Plebanski, *Acta. Phys. Polon.*, **26**, 1964, 963.
73. A. Lichnerowicz "Theories Relativistes de la Gravitation et de l'Electromagnétisme", Masson, Paris, 1955.
74. A. Zajtz, Zeszyty, Naukowe Uniwersytetu, Jagiellonskiego LXXIV, 1964.
75. G. Ludwig and G. Scanlon, *Comm. Math. Phys.*, **20**, 1971, 291.
76. R. Penrose, Gravitatsya, A. Z. Petrov Festschrift volume, Naukdumka, Kiev, 1972, 203.
77. C. D. Collinson and R. Shaw, *Int. J. Theor. Phys.*, **6**, 1972, 347.
78. G. S. Hall, *J. Phys.* A, **9**, 1976, 541.
79. W. J. Cormack and G. S. Hall, *J. Phys.* A, **12**, 1979, 55.
80. W. J. Cormack and G. S. Hall, *Int. J. Theor. Phys.*, **20**, 1981, 105.
81. R. F. Crade and G. S. Hall, *Acta Phys. Polon.* B, **13**, 1982, 405.
82. G. Sobczyk, *Acta. Phys. Polon.* B, **11**, 1980, 579.
83. C. B. G. McIntosh, J. M. Foyster and A. W-C. Lun, *J. Math. Phys.*, **22**, 1981, 2620.
84. G. C. Joly and M. A. H. MacCallum, *Class. Quant. Grav.*, **7**, 1990, 541.
85. G. S. Hall, Proceedings of the Relativity Workshop, Balatonszéplak, Hungary, 1985, 49.
86. G. S. Hall, *Arch. Math.*, **4**, Scripta Fac. Sci. Nat. Ujep. Brunensis XVIII, 1982, 169.
87. G. S. Hall, *Arab. Jn. for Sci. and Eng.*, **9**, 1984, 87.
88. G. S. Hall and D. Negm, *Int. Jn. Theor. Phys.*, **25**, 1986, 405.
89. P. A. Goodinson and R. A. Newing, *J. Inst. Math. App.*, **6**, 1970, 212.
90. G. S. Hall, *J. Phys. A.*, **6**, 1973, 619.
91. G. S. Hall, *Class. Quant. Grav.*, **13**, 1996, 1479.
92. G. S. Hall "The Petrov Lectures", Proceedings of the School Volga, Kazan, Russia, 2000.

93. J. M. M. Senovilla and R. Vera, *Class. Quant. Grav.*, **16**, 1999, 1185.
94. G. S. Hall and S. Khan, unpublished work, 2000. See S. Khan, PhD. thesis, University of Aberdeen, 2001.
95. A. D. Rendall, *J. Math. Phys.*, **29**, 1988, 1569.
96. B. G. Schmidt, *Comm. Math. Phys.*, **29**, 1973, 55.
97. W. Ambrose and I. M. Singer, *Trans. Am. Math. Soc.*, **75**, 1953, 428.
98. H. Wu, *Illinois J. Math.*, **8**, 1964, 291.
99. A. Besse "Einstein Manifolds", Springer, 1987.
100. L. Berard Bergery and A. Ikemakhen, Proceedings of Symposia in Pure Mathematics, **54**, 1993, Part 2, 27.
101. G. S. Hall, *Gen. Rel. Grav.*, **20**, 1988, 399.
102. G. S. Hall, *J. Math. Phys.*, **32**, 1991, 181.
103. C. D. Collinson and P. N. Smith, *Gen. Rel. Grav.*, **19**, 1987, 95.
104. G. S. Hall, *Gen. Rel. Grav.*, **27**, 1995, 567.
105. R. Debever and M. Cahen, *Bull. Acad. Belg. Class. Sci.*, **47**, 1962, 491.
106. W. Beiglböck, *Zts. für Phys.*, **179**, 1964, 148.
107. G. S. Hall, *Lett. al. Nu. Cim.*, **9**, 1974, 667.
108. G. S. Hall and W. Kay, *J. Math. Phys.*, **29**, 1988, 428.
109. J. N. Goldberg and R. P. Kerr, *J. Math. Phys.*, **2**, 1961, 327.
110. R. P. Kerr and J. N. Goldberg, *J. Math. Phys.*, **2**, 1961, 332.
111. G. S. Hall and D. P. Lonie, *Class. Quant. Grav.*, **17**, 2000, 1369.
112. R. Ghanam and G. Thompson, *Class. Quant. Grav.*, **18**, 2001, 2007.
113. C. B. G. McIntosh and E. Van Leeuven, *J. Math. Phys.*, **23**, 1982, 1149.
114. A. G. Walker, *Quart. J. Maths. Oxford*, **20**, 1949, 135.
115. H. S. Ruse, *Quart. J. Maths. Oxford*, **20**, 1949, 218.
116. A. G. Walker, *Quart. J. Maths. Oxford* (2), **1**, 1950, 69.
117. E. M. Patterson, *Quart. J. Maths. Oxford* (2), **2**, 1951, 151.
118. G. S. Hall and C. B. G. McIntosh, *Int. J. Theor. Phys.*, **22**, 1983, 469.
119. G. S. Hall, *Gen. Rel. Grav.*, **15**, 1983, 581.
120. H. W. Brinkmann, *Math. Ann.*, **94**, 1925, 119.
121. G. S. Hall in "Classical General Relativity", ed. W.B. Bonnor, J. N. Islam and M. A. H. MacCallum, Cambridge University Press, 1984, 103.
122. N. Rosen, *Anns. Phys.*, **22**, 1963, 1.
123. G. S. Hall, unpublished manuscript of lecture at the Fourth Meeting on Mathematical Physics at the University of Coimbra, Portugal, 1984.
124. G. S. Hall and A. D. Rendall, *J. Math. Phys.*, **28**, 1987, 1837.
125. E. Ihrig, *Gen. Rel. Grav.*, **7**, 1976, 313.
126. G. S. Hall, *Gen. Rel. Grav.*, **16**, 1984, 79.
127. J. A. Thorpe, *J. Math. Phys.*, **10**, 1969, 1.
128. R. S. Kulkarni, *Ann. Maths.*, **91**, 1970, 311.
129. B. Ruh, PhD. Thesis, E.T.H. Zurich, 1982.
130. B. Ruh, *Math. Z.*, **184**, 1985, 371.
131. R.S. Kulkarni, *Int. J. Math. and Math. Sci.*, **1**, 1978, 137.
132. G. S. Hall and A. D. Rendall, *Gen. Rel. Grav.*, **19**, 1987, 771.
133. G. S. Hall, 1978, unpublished.
134. G. S. Hall, *Z. Naturforsch*, **33a**, 1978, 559.

135. G. S. Hall in Proceedings of the First Hungarian Relativity Workshop, ed. Z. Perjés, Balatonszéplak, 1985, 141.
136. G. S. Hall, A. D. Hossack and J. R. Pulham, *J. Math. Phys.*, **33**, 1992, 1408.
137. O. Kowalski, *Math. Zts.*, **125**, 1972, 129.
138. E. Ihrig, *J. Math. Phys.*, **16**, 1975, 54.
139. E. Ihrig, *Int. J. Theor. Phys.*, **14**, 1975, 23.
140. C. B. G. McIntosh and D. W. Halford, *J. Phys. A*, **14**, 1981, 2331.
141. A. D. Rendall, *Class. Quant. Grav.*, **5**, 1988, 695.
142. A. D. Rendall, *Jn. Geom. Phys.*, **6**, 1989, 159.
143. S. B. Edgar, *J. Math. Phys.*, **32**, 1991, 1011.
144. C. D. Collinson in "Galaxies, Axisymmetric Systems and Relativity", ed. M. A. H. MacCallum, Cambridge University Press, 1985.
145. S. B. Edgar, *J. Math. Phys.*, **33**, 1992, 3716.
146. G. Marmo, C. Rubano and G. Thompson, *Class. Quant. Grav.*, **7**, 1990, 2155.
147. G. Thompson, *Class. Quant. Grav.*, **10**, 1993, 2035.
148. G. K. Martin and G. Thompson, *Pac. J. Maths.*, **158**, 1993, 177.
149. J. K. Beem and P. E. Parker, *J. Math. Phys.*, **31**, 1990, 819.
150. G. S. Hall and A. D. Hossack, *J. Math. Phys.*, **34**, 1993, 5897.
151. E. Cartan "Leçons sur la Geometrie des Espaces de Riemann", Gauthier-Villars, Paris, 1946.
152. J. Ehlers in "E.B. Christoffel", ed. P.L. Butzer and F. Fehér, Birkhauser, Basel, 1981.
153. A. Karlhede, *Gen. Rel. Grav.*, **12**, 1980, 693.
154. A. Karlhede, *Gen. Rel. Grav.*, **12**, 1980, 963.
155. A. Karlhede and M.A.H. MacCallum, *Gen. Rel. Grav.*, **14**, 1982, 673.
156. M. P. Machado Ramos and J.A.G. Vickers, *Class. Quant. Grav.*, **13**, 1996, 1579.
157. M. P. Machado Ramos and J.A.G. Vickers, *Class. Quant. Grav.*, **13**, 1996, 1589.
158. K. Yano. 'The Theory of Lie Derivatives and its Applications', North Holland Amsterdam 1957.
159. W. Poor 'Differential Geometric Structures', McGraw Hill, 1981.
160. G. S. Hall. *Gen. Rel. Grav.*, **20** 1988, 671.
161. G. S. Hall, D. J. Low and J. R. Pulham. *J. Math. Phys.*, **35** 1994, 5930.
162. R. F. Crade and G. S. Hall. *Phys. Lett.* **75A** 1979, 17.
163. R. H. Boyer. *Proc. Roy. Soc.* **A** 311, 1969, 245.
164. D. Alexeevski. *Anns. Glob. Anal. and Geom.* **3** 1985, 59.
165. J. K. Beem *Letts. Math. Phys.* **2** 1978, 317.
166. G. S. Hall *J. Math. Phys.* **31** 1990, 1198.
167. G. S. Hall. *Class. Quant. Grav.* **20** 2003, 3745.
168. W. L. Chow. *Math. Ann.* **117** 1939, 98.
169. K. Nomizu. *Ann. Maths.* **72** 1960, 105.
170. G. S. Hall. *Class. Quant. Grav.* **6** 1989, 157.
171. B. G. Schmidt. PhD. thesis, University of Hamburg 1968.

172. G. S. Hall and J. da Costa. *J. Math. Phys.* **29** 1988, 2465.
173. M. P. Ryan and L. C. Shepley. 'Homogeneous Relativistic Cosmologies' Princeton University Press, 1975.
174. B. O. J. Tupper. *Class. Quant. Grav.* **13** 1996, 1679.
175. G. S. Hall. *Gen. Rel. Grav.* **32** 2000, 933.
176. D. Garfinkle. *J. Math. Phys.* **28** 1987, 28.
177. R. Maartens and S. D. Maharaj. *Class. Quant. Grav.* **3** 1986, 1005.
178. D. M. Eardley. *Comm. Math. Phys.* **37** 1974, 287.
179. M. L. Bedran and B. Lesche. *J. Math. Phys.* **27** 1986, 2360.
180. B. Lesche and M. L. Bedran. Revista. Brasileira de Física **18** 1988, 93.
181. R. Maartens. *J. Math. Phys.* **28** 1987, 2051.
182. C. D. Collinson. *J. Math. Phys.* **29** 1988, 1972.
183. S. A. Hojman. *J. Math. Phys.* **32** 1991, 234.
184. B. G. Schmidt. *Z. Naturforschg.* **220** 1967, 1351.
185. B. G. Schmidt. *Gen. Rel. Grav.* **2** 1971, 105.
186. R. Sigal. *Gen. Rel. Grav.* **5** 1974, 737.
187. K. Yano and T. Nagano. *Japan J. Math.* **29** 1959, 173.
188. C. B. G. McIntosh. *Gen. Rel. Grav.* **7** 1976, 199.
189. C. B. G. McIntosh. *Phys. Letts.* **69A** 1978, 1.
190. D. Kramer. *J. Phys.* A. **13** 1980, L43.
191. W. D. Halford and R. P. Kerr. *J. Math. Phys.* **21** 1980, 120.
192. W. D. Halford and R. P. Kerr. *J. Math. Phys.* **21** 1980, 129.
193. A. W-C. Lun., C. B. G. McIntosh and D. B. Singleton. *Gen. Rel. Grav.* **20** 1988, 745.
194. M. L. Wooley. *Comm. Math. Phys.* **31** 1973, 75.
195. J. R. Ray and E. L. Thompson. *J. Math. Phys.* **16** 1975, 345.
196. B. Coll. C. R. Acad. Sci. Paris A **280** 1975, 1773.
197. H. Michalski and J. Wainwright. *Gen. Rel. Grav.* **6** 1975, 289.
198. J. Wainwright and P. E. A. Yaremovich. *Gen. Rel. Grav.* **7** 1976, 345.
199. J. Wainwright and P. E. A. Yaremovich. *Gen. Rel. Grav.* **7** 1976, 595.
200. M. L. Wooley. *J. Phys.* A **10** 1977, 2107.
201. C. B. G. McIntosh. *Gen. Rel. Grav.* **9** 1978, 277.
202. C. B. G. McIntosh. *Gen. Rel. Grav.* **10** 1979, 61.
203. M. A. H. MacCallum and N. van den Burgh in 'Galaxies, Axisymmetric Systems and Relativity' ed. M. A. H. MacCallum, C. U. P. 1985, 138.
204. G. S. Hall and J. P. Steele. *Gen. Rel. Grav.* **22** 1990, 457.
205. G. S. Hall and M. T. Patel. Preprint, University of Aberdeen 2003.
206. H. P. Robertson and T. W. Noonan "Relativity and Cosmology", Saunders 1969.
207. G. Fubini. *Annali di Mathematica* **8** 1903, 39
208. Y. Choquet-Bruhat, C. De-Witt-Morette with M. Dillard-Bleick "Analysis, Manifolds and Physics", North-Holland, 1982.
209. G. S. Hall and J. D. Steele *J. Math. Phys.*, **32**, 1991, 1847.
210. G. S. Hall, M. Capocci and R. Beig *Class. Quant. Grav.* **14**, 1997, L49.
211. E. Nelson "Topics in Dynamics; I Flows", Princeton University and University of Tokyo Press, Princeton, 1969.

212. R. F. Bilyalov, *Sov. Phys.*, **8**, 1964, 878.
213. T. Siguri and S. Ueno *Tensor. N. S.*, **24**, 1972, 253.
214. L. Defrise-Carter *Comm. Math. Phys.*, **40**, 1975, 273.
215. R. Beig, preprint, University of Vienna, 1992.
216. M. S. Capocci *Class. Quant. Grav.*, **16**, 1999, 927.
217. C. D. Collinson and D. C. French *J. Math. Phys.*, **8**, 1967, 701.
218. H. Salazar, A. Garcia D and J.F. Plebanski *J. Math. Phys.*, **24**, 1983, 2191.
219. R. Maartens and S.D. Maharaj *Class. Quant. Grav.*, **8**, 1991, 503.
220. R. Sippel and H. Goenner *Gen. Rel. Grav.*, **18**, 1986, 1229.
221. M. S. Capocci and G. S. Hall *Gravitation and Cosmology*, **3**, 1997, 1.
222. A. A. Coley and B. O. J. Tupper *J. Math. Phys.*, **30**, 1989, 2616.
223. A. A. Coley and B. O. J. Tupper *J. Math. Phys.*, **33**, 1992, 1754.
224. D. Garfinkle and Q. Tian *Class. Quant. Grav.*, **4**, 1987, 137.
225. D. Eardley, J. Isenberg, J. Marsden and V. Moncrief *Comm. Math. Phys.*, **106**, 1986, 137.
226. R. Maartens and C.M. Mellin *Class. Quant. Grav.*, **13**, 1996, 1571.
227. J. Lewandowski *Class. Quant. Grav.*, **7**, 1990, L135.
228. R. Maartens and S.D. Maharaj *J. Math. Phys.*, **31**, 1990, 151.
229. G. S. Hall *Gen. Rel. Grav.*, **22**, 1990, 203.
230. J. Carot *Gen. Rel. Grav.*, **22**, 1990, 1135.
231. K. Yano "Integral formulas in Riemannian Geometry", Marcel Dekker, New York, 1970.
232. G. S. Hall and I. M. Roy *Gen. Rel. Grav.*, **29**, 1997, 827.
233. G. S. Hall and D. P. Lonie *Class. Quant. Grav.*, **12**, 1995, 1007.
234. A. Barnes *Class. Quant. Grav.*, **10**, 1993, 1139.
235. G. S. Hall in *Proc. 6th Int. Conf. on Differential Geometry and its Applications*, Masaryk University, Brno, Czech. Republic, 1996.
236. G. S. Hall *Class. Quant. Grav.*, **17**, 2000, 3073.
237. G. S. Hall and S. Khan *J. Math. Phys.*, **42**, 2001, 347.
238. G. S. Hall *Class. Quant. Grav.*, **17**, 2000, 4637.
239. G. F. R. Ellis "Relativistic Cosmology" in "General Relativity and Cosmology", Academic Press, New York, 1971.
240. G. S. Hall and M. T. Patel *Class. Quant. Grav.*, **19**, 2002, 2319.
241. A. A. Coley and D. J. McManus *Class. Quant. Grav.*, **11**, 1994, 1261.
242. G. S. Hall and D. P. Lonie. Preprint, University of Aberdeen, 2004.
243. A. V. Aminova *J. Soviet. Maths.*, **55**, 1991, 1995.
244. G. H. Katzin, J. Levine and W. R. Davies. *J. Math. Phys.* **10** 1969, 617.
245. G. S. Hall and J. da Costa. *J. Math. Phys.* **32** 1991, 2848.
246. G. S. Hall and J. da Costa. *J. Math. Phys.* **32** 1991, 2854.
247. C. B. G. McIntosh and D. W. Halford. *J. Math. Phys.* **23** 1982, 436.
248. P. C. Aichelburg. *J. Math. Phys.* **11** 1970, 2485.
249. C. D. Collinson. *J. Math. Phys.* **11** 1970, 818.
250. N. Tariq and B. O. J. Tupper. *Tensor.* **31** 1997, 42.
251. W. D. Halford, C. B. G. McIntosh and E. H. van Leeuwen. *J. Math. Phys.* **13** 1980, 2995.
252. C. D. Collinson and E. G. L. R. Vaz. *Gen. Rel. Grav.* **14** 1982, 5.

253. E. G. L. R. Vaz. and C. D. Collinson. *Gen. Rel. Grav.* **15** 1983, 661.
254. R. A. Tello-Llanos. *Gen. Rel. Grav.* **20** 1988, 765.
255. J. Carot and J. da Costa. *Gen. Rel. Grav.* **23** 1991, 1057.
256. A. H. Bokhari, A. R. Kashif and A. Qadir. *J. Math. Phys.* **41** 2000, 2167.
257. G. S. Hall and G. S. Shabbir. *Class. Quant. Grav.* **18** 2001, 907.
258. G. S. Hall and G. S. Shabbir. *Gravitation and Cosmology* **9** 2003, 134.
259. J. Carot, J. da Costa. and E. L. G. R. Vaz. *J. Math. Phys.* **35** 1994, 4832.
260. G. S. Hall, I. Roy and E. L. G. R. Vaz. *Gen. Rel. Grav.* **28** 1996, 299.
261. J. Carot, L. A. Nunez and U. Percoco. *Gen. Rel. Grav.* **29** 1997, 1223.
262. J. Carot and J. da Costa. *Fields Institute Communications* **15** 1997, 179.
263. A. Qadir and M. Ziad. *Nu. Cim.* **113B** 1998, 773.
264. G. Contreras, L. A. Nunez and U. Percoco. *Gen. Rel. Grav.* **32** 2000, 285.
265. M. Tsamparlis and P. S. Apostolopoulos. *Gen. Rel. Grav.* **32** 2000, 281.
266. U. Camci and A. Barnes. *Class. Quant. Grav.* **19** 2002, 393.
267. M. Sharif and S. Aziz. *Gen. Rel. Grav.* **35** 2003, 1093.

Index